T0183456

# Lecture Notes in Computer Science    12488

More information about this series at http://www.springer.com/series/7410

Xiaofeng Chen · Hongyang Yan ·
Qiben Yan · Xiangliang Zhang (Eds.)

# Machine Learning for Cyber Security

Third International Conference, ML4CS 2020
Guangzhou, China, October 8–10, 2020
Proceedings, Part III

Springer

*Editors*
Xiaofeng Chen
Xidian University
Xi'an, China

Hongyang Yan
Guangzhou University
Guangzhou, China

Qiben Yan
Michigan State University
East Lansing, MI, USA

Xiangliang Zhang
Division of Computer, Electrical
and Mathematical Sciences and Engineering
King Abdullah University of Science
Thuwal, Saudi Arabia

ISSN 0302-9743          ISSN 1611-3349  (electronic)
Lecture Notes in Computer Science
ISBN 978-3-030-62462-0          ISBN 978-3-030-62463-7  (eBook)
https://doi.org/10.1007/978-3-030-62463-7

LNCS Sublibrary: SL4 – Security and Cryptology

This Springer imprint is published by the registered company Springer Nature Switzerland AG
The registered company address is: Gewerbestrasse 11, 6330 Cham, Switzerland

# Preface

The Third International Conference on Machine Learning for Cyber Security (ML4CS 2020) was held in Guangzhou, China, during October 8–10, 2020. ML4CS is a well-recognized annual international forum for AI-driven security researchers to exchange ideas and present their works. This volume contains papers presented at ML4CS 2020.

The conference received 360 submissions. The committee accepted 118 regular papers and 40 short papers to be included in the conference program. The proceedings contain revised versions of the accepted papers. While revisions are expected to take the referees comments into account, this was not enforced and the authors bear full responsibility for the content of their papers.

ML4CS 2020 was organized by the Institute of Artificial Intelligence and Blockchain, Guangzhou University, China. Furthermore, ML4CS 2020 was supported by the Peng Cheng Laboratory, Shenzhen, China. The conference would not have been such a success without the support of these organizations, and we sincerely thank them for their continued assistance and support.

We would also like to thank the authors who submitted their papers to ML4CS 2020, and the conference attendees for their interest and support. We thank the Organizing Committee for their time and effort dedicated to arranging the conference. This allowed us to focus on the paper selection and deal with the scientific program. We thank the Program Committee members and the external reviewers for their hard work in reviewing the submissions; the conference would not have been possible without their expert reviews. Finally, we thank the EasyChair system and its operators, for making the entire process of managing the conference convenient.

September 2020

Fengwei Zhang
Jin Li

# Organization

## General Chairs

| | |
|---|---|
| Fengwei Zhang | Southern University of Science and Technology, China |
| Jin Li | Guangzhou University, China |

## Program Chairs

| | |
|---|---|
| Xiaofeng Chen | Xidian University, China |
| Qiben Yan | Michigan State University, USA |
| Xiangliang Zhang | King Abdullah University of Science and Technology, Saudi Arabia |

## Track Chairs

| | |
|---|---|
| Hao Wang | Shandong Normal University, China |
| Xianmin Wang | Guangzhou University, China |
| Yinghui Zhang | Xi'an University of Posts and Telecommunications, China |
| Wei Zhou | Yunnan University, China |

## Workshop Chairs

| | |
|---|---|
| Sheng Hong | Beihang University, China |
| Ting Hu | Queen's University, Canada |
| Nan Jiang | East China Jiaotong University, China |
| Liangqun Li | Shenzhen University, China |
| Jungang Lou | Hunan Normal University, China |
| Hui Tian | Huaqiao University, China |
| Feng Wang | Wuhan University, China |
| Zhi Wang | Nankai University, China |
| Tao Xiang | Chongqing University, China |

## Publication Chairs

| | |
|---|---|
| Hongyang Yan | Guangzhou University, China |
| Yu Wang | Guangzhou University, China |

## Publicity Chair

| | |
|---|---|
| Lianyong Qi | Qufu Normal University, China |

# Steering Committee

| | |
|---|---|
| Xiaofeng Chen | Xidian University, China |
| Iqbal Gondal | Federation University Australia, Australia |
| Ryan Ko | University of Waikato, New Zealand |
| Jonathan Oliver | Trend Micro, USA |
| Islam Rafiqul | Charles Sturt University, Australia |
| Vijay Varadharajan | The University of Newcastle, Australia |
| Ian Welch | Victoria University of Wellington, New Zealand |
| Yang Xiang | Swinburne University of Technology, Australia |
| Jun Zhang | Swinburne University of Technology, Australia |
| Wanlei Zhou | Deakin University, Australia |

# Program Committee

| | |
|---|---|
| Silvio Barra | University of Salerno, Italy |
| M. Z. Alam Bhuiyan | Guangzhou University, China |
| Carlo Blundo | University of Salerno, Italy |
| Yiqiao Cai | Huaqiao University, China |
| Luigi Catuogno | University of Salerno, Italy |
| Lorenzo Cavallaro | King's College London, UK |
| Liang Chang | Guilin University of Electronic Technology, China |
| Fei Chen | Shenzhen University, China |
| Xiaofeng Chen | Xidian University, China |
| Zhe Chen | Singapore Management University, Singapore |
| Xiaochun Cheng | Middlesex University, UK |
| Frédéric Cuppens | Télécom Bretagne, France |
| Changyu Dong | Newcastle University, UK |
| Guangjie Dong | East China Jiaotong University, China |
| Mohammed EI-Abd | American University of Kuwait, Kuwait |
| Wei Gao | Yunnan Normal University, China |
| Dieter Gollmann | Hamburg University of Technology, Germany |
| Zheng Gong | South China Normal University, China |
| Zhitao Guan | North China Electric Power University, China |
| Zhaolu Guo | Chinese Academy of Sciences, China |
| Jinguang Han | Queen's University Belfast, UK |
| Saeid Hosseini | Singapore University of Technology and Design Singapore |
| Chingfang Hsu | Huazhong University of Science and Technology, China |
| Haibo Hu | The Hong Kong Polytechnic University, Hong Kong |
| Teng Huang | Guangzhou University, China |
| Xinyi Huang | Fujian Normal University, China |
| Wenchao Jiang | Guangdong University of Technology, China |
| Lutful Karim | Seneca College of Applied Arts and Technology, Canada |

| | |
|---|---|
| Hadis Karimipour | University of Guelph, Canada |
| Sokratis Katsikas | Open University of Cyprus, Cyprus |
| Neeraj Kumar | Thapar Institute of Engineering and Technology, India |
| Ping Li | South China Normal University, China |
| Tong Li | Guangzhou University, China |
| Wei Li | Jiangxi University of Science and Technology, China |
| Xuejun Li | Anhui University, China |
| Kaitai Liang | University of Surrey, UK |
| Hui Liu | University of Calgary, Canada |
| Wei Lu | Sun Yat-sen University, China |
| Xiaobo Ma | Xi'an Jiaotong University, China |
| Fabio Martinelli | IIT-CNR, Italy |
| Ficco Massimo | Second University of Naples, Italy |
| Weizhi Meng | Technical University of Denmark, Denmark |
| Vincenzo Moscato | University of Naples, Italy |
| Francesco Palmieri | University of Salerno, Italy |
| Fei Peng | Hunan University, China |
| Hu Peng | Wuhan University, China |
| Lizhi Peng | Jinan University, China |
| Umberto Petrillo | Sapienza University of Rome, Italy |
| Lianyong Qi | Qufu Normal University, China |
| Shahryar Rahnamayan | Ontario Tech University, Canada |
| Khaled Riad | Guangzhou University, China |
| Yu Sun | Guangxi University, China |
| Yu-An Tan | Beijing Institute of Technology, China |
| Zhiyuan Tan | Edinburgh Napier University, UK |
| Ming Tao | Dongguan University of Technology, China |
| Donghai Tian | Beijing Institute of Technology, China |
| Chundong Wang | Tianjin University of Technology, China |
| Ding Wang | Peking University, China |
| Hui Wang | Nanchang Institute of Technology, China |
| Jianfeng Wang | Xidian University, China |
| Jin Wang | Soochow University, China |
| Licheng Wang | Beijing University of Posts and Telecommunications, China |
| Lingyu Wang | Concordia University, Canada |
| Tianyin Wang | Luoyang Normal University, China |
| Wei Wang | Beijing Jiaotong University, China |
| Sheng Wen | Swinburne University of Technology, Australia |
| Yang Xiang | Swinburne University of Technology, Australia |
| Run Xie | Yibin University, China |
| Xiaolong Xu | Nanjing University of Information Science and Technology, China |
| Li Yang | Xidian University, China |
| ShaoJun Yang | Fujian Normal University, China |
| Zhe Yang | Northwestern Polytechnical University, China |

| | |
|---|---|
| Yanqing Yao | Beihang University, China |
| Xu Yuan | University of Louisiana at Lafayette, USA |
| Qikun Zhang | Beijing Institute of Technology, China |
| Xiao Zhang | Beihang University, China |
| Xiaosong Zhang | Tangshan University, China |
| Xuyun Zhang | Macquarie University, Australia |
| Yuan Zhang | Nanjing University, China |
| Xianfeng Zhao | Chinese Academy of Sciences, China |
| Lei Zhu | Huazhong University of Science and Technology, China |
| Tianqing Zhu | China University of Geosciences, China |

## Track Committee

| | |
|---|---|
| Jin Cao | Xidian University, China |
| Hui Cui | Murdoch University, Australia |
| Rui Guo | Xi'an University of Posts and Telecommunications, China |
| Qi Li | Nanjing University of Posts and Telecommunications, China |
| Muhammad Baqer Mollah | Nanyang Technological University, Singapore |
| Ben Niu | Chinese Academy of Sciences, China |
| Fatemeh Rezaeibagha | Murdoch University, Australia |

# Contents – Part III

# Deep Convolutional Neural Network Based Image Segmentation for Salt Mine Recognition

Ming Tao[1]([envelope]) [iD], Xueqiang Li[1,2] [iD], and Kai Ding[1] [iD]

[1] School of Computer Science and Technology,
Dongguan University of Technology, Dongguan 523808, People's Republic of China
ming.tao@mail.scut.edu.cn, {lixq,dingkai}@dgut.edu.cn
[2] Institute of Artificial Intelligence and Blockchain, Guangzhou University, Guangzhou 510006,
People's Republic of China

**Abstract.** Scientifically, the identification of salt ore has definite practical significance for the exploitation of oil and gas. Traditionally, this is achieved by picking the salt boundaries with manual vision, which may introduce serious systematic bias. Nowadays, with the technological progress of machine vision used in image analysis, human effort has been replaced by machine capacity in salt mine recognition. Especially, with the in-depth application of deep learning technology in the field of machine vision, salt mine recognition using image analysis is revolutionizing with more acceptable efficiency and accuracy. To this end, with exploratory data analysis to mine the characteristics and data processing to increase the size of the image data for further enhancing the generalization capability of the designed model, a deep convolutional neural network based image segmentation model is investigated to achieve salt mine recognition in this paper. Concretely, a U-Net model integrated modified ResNet34 is first designed as a basic recognition model, and many attempts then are conducted to further optimizing the model according to the data characteristics, including adding auxiliary function, hypercolumn, scSE and depth supervision scheme. In addition, multiple loss functions are also attempted to be adapted to further improving the model generalization capacity. The numerical analysis and evaluation finally show the efficiency of the investigations on loss value and recognition accuracy.

**Keywords:** Deep learning · Convolutional neural network · Image segmentation · Salt mine recognition

## 1 Introduction

Science and practice show that large deposits of oil and gas also tend to form large deposits of salt beneath the earth's surface. Therefore, we can use the search for salt mines to explore the locations of oil and gas. In general, determining and segmenting the locations of large salt mines are to regionally segment the instance of salt mine and take the rest as foreground, which can be achieved by analyzing seismic images through very professional manual vision, which are subjective and inefficient. Fortunately, with the technological progress of machine vision used in image analysis, machine capacity

X. Chen et al. (Eds.): ML4CS 2020, LNCS 12488, pp. 1–10, 2020.
https://doi.org/10.1007/978-3-030-62463-7_1

has gradually replacing human effort in salt mine recognition, which could provide higher efficiency and accuracy. Nowadays, due to offering a convincing alternative by learning the problem specific features on their own, the in-depth development of deep learning technology in the field of image analysis has further promoting the application of machine vision in various applications, e.g., medical image segmentation, salt mine recognition, etc. In particular, as a frequently used deep learning framework, convolutional neural networks (CNN) is usually used for achieving image segmentation [1].

For accurate medical image segmentation used in diagnosis, surgical planning and many other applications, Wang et al. [2] proposed a deep learning-based interactive segmentation method to improve the results obtained by an automatic CNN and to reduce user interactions during refinement for higher accuracy. With the facilitated gradient flow and implicit deep supervision during training achieved by DenseNet, Dolz et al. [3] proposed HyperDenseNet as a 3-D fully CNN that extended the definition of dense connectivity to multi-modal segmentation problems. In [4], the great potential of applying CNN for salt-related interpretations was also demonstrated. Concretely, the benefits of CNN-based classification are demonstrated by using U-Net network structure, along with the residual learning framework ResNet, to delineate salt body with high precision.

To this end, a method of deep convolutional neural network based image segmentation is proposed for achieve salt mine recognition in this paper. Concretely, a U-Net model integrated modified ResNet34 is first designed as a basic recognition model, and many attempts then are conducted to further optimizing the model according to the data characteristics, and multiple loss functions are also attempted to be adapted to further improving the model generalization capacity.

The rest of this paper is organized as follows. In Sect. 2, exploratory data analysis and processing are conducted to mine the data characteristics and to increase the size of the image data for further enhancing the generalization capability of the designed model. In Sect. 3, a U-Net model integrated modified ResNet34 is designed and many attempts are conducted to further optimizing the model according to the data characteristics. In Sect. 4, multiple loss functions are attempted to be adapted in the salt mine image segmentation model to further improve the model generalization capability. In Sect. 5, numerical analyses are conducted to evaluate the efficiency of achieved proposals in this work. In Sect. 6, this paper is summarized and concluded.

## 2   Exploratory Data Analysis and Processing

The image data source for salt mine recognition is seismic images collected by the seismic reflection. In this work, the image data source is provided by GS Salt Identification Challenge conducted in Kaggle, where, the train.zip is the training set containing 4000 sample images and the corresponding masks, the test.zip contains 18000 testing sample images, the train.csv contains the sample images' IDs and the labels of salt in the training set, and the depth.csv contains the depth of the sample images. To better understand the data and mine the characteristics of the data, which may play a certain role in the optimization of the designed salt mine recognition model, salt mine coverage rate of samples in the training set (that is, the proportion of salt mine image pixels in the whole sample image) was analyzed.

As shown in Fig. 1(a), most of the coverage ratios of pixels in salt mine are between 0.0 and 0.1, however, there are 1,562 sample images without salt, which indicates that the data categories in the training set were unbalanced. The unbalance of data categories will cause the designed recognition model to prefer fitting the image data with more categories and ignore the image data with fewer categories, which adversely influences the prediction ability of the designed model. The depth distribution of the salt mine location, that is the imaged depth in the sample images are also analyzed in Fig. 1(b), which indicates that the depth distribution in the training set is basically the same as that in the testing set, and the generalization ability of the designed model could be guaranteed.

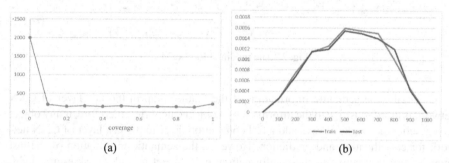

(a) (b)

**Fig. 1.** (a) Analysis of salt mine coverage rate of samples in the training set, and (b) the depth distribution of the salt mine location.

Since the original size of the sample image is 101*101, while the size of the commonly used input image in CNN training is 128*128 or 256*256, the bilinear interpolation mode of resize contained in OpenCV is used to adjust the image size. In addition, to make the characteristic values between different dimensions have some comparability in values, and make the network in the designed recognition model has a better learning capability for faster achieving the local optimal, normalization is executed on the pixel values of the sample images.

As the above exploratory data analysis, the unbalance of data categories in the training set will make the designed recognition model be easy to be overfitting on the training set, which adversely influences the prediction ability of the designed model. Therefore, to increase the size of the image data to further enhance the generalization capability of the designed model, the method of data augmentation, e.g., mirror flip, rotation, translation and other basic image operations, is employed to perform adjustments on sample images to obtain new image data [5].

## 3 Model Design and Optimization

In this work, since all the images in the employed data set are grey-scale images, U-Net [6], as one of the representative image segmentation model with a great advantage on segmenting gray-scale images, is used for designing a salt mine recognition model. A U-Net model consists of an Encoder and a Decoder. To further reduce the overfitting degree and ensure the generalization capability of the designed model, ResNet34-pretrained is used as the Encoder while removing the first MaxPooling down-sampling layer of ResNet34 to obtain a larger prediction feature map. The modified ResNet34 is shown in Fig. 2. Based on the designed U-Net enabled basic model, the following attempts are conducted to further optimizing the model according to the data characteristics.

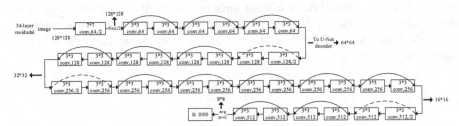

**Fig. 2.** The modified ResNet34.

Generally, the semantic information of the feature map in the last layer of CNN network is more sensitive to the information of object categories, therefore, many object classification networks always add a full connection layer to the last layer of CNN network for classification and prediction. However, the semantic information of the last layer is very insensitive to these information, e.g., object attitude, background light, object joints and position, etc. Additionally, the features of the last layer are relatively rough in space, which affects the accurate object location. Hence, for fine-grained analysis tasks, such as the image segmentation in this work, direct using the last layer of CNN is often not an optimal choice. Making full use of the feature maps in the shallow layers of CNN network could be an efficient alternative. To this end, as shown in Fig. 3, the hyper-column design is added to the U-Net model to optimize the designed recognition model [7]. Concretely, the feature maps in the upper-sampling layers of multiple stages in Decoder are bilinear interpolated to the same size and then combined in series as the final prediction features for fine-grained target location and object segmentation.

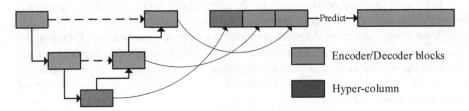

**Fig. 3.** U-Net model added the hyper-column.

Since the sample images in the employed data set contain images with salt and images without salt, if all the sample images can be classified into images with salt and images without salt in advance, and the segmentation operation could be only executed on the images with salt to improve the recognition efficiency. To this end, an extraction operation of a binary task branch is executed for the features at the last down-sampling stage in U-Net model [8]. The training principle of the auxiliary segmentation model is shown in Fig. 4, where the weight of the loss function value of the binary classification task is set to 0.05, which is relatively small because it will bring regularization and reduce the degree of overfitting.

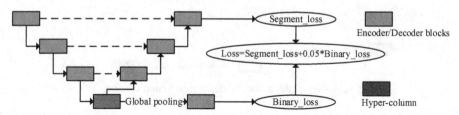

**Fig. 4.** U-Net model added the binary auxiliary function.

Subsequently, we consider the adjustment and optimization from the network learning to the feature map, and give more weight to the important features while reducing the weight for the unimportant features. In this work, comprehensive considering the advantages of both Spatial Squeeze and Channel Excitation (cSE) and Channel Squeeze and Spatial excitation (sSE) proven in the previous proposals, a combination named scSE is added to the U-Net model to optimize the designed recognition mode [9]. cSE is responsible for the feature reactivation in the channel dimension, while sSE is responsible for the feature reactivation in the spatial dimension, which can comprehensively control the importance of features. As shown in Fig. 5, the locations of the added scSE modules are behind the feature maps of the down-sampling or upper-sampling at each stage to monitor the importance of the features.

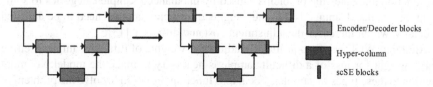

**Fig. 5.** U-Net model added scSE modules.

Finally, to further enhance the generalization capability of the recognition model and reduce the test error, a depth supervision scheme is added in the U-Net model shown in Fig. 6, where a partition loss function calculation is executed for the feature maps at each upper-layer sampling stage of Decoder and a weighted sum is obtained by multiple loss values [10]. This design considers not only the down-layer characteristics, but also the deep-layer characteristics, which could improve the performance of the model by learning more accurate edge position of the target object in terms of the features in multiple layers.

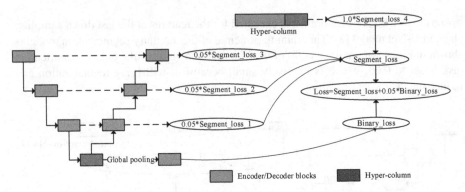

**Fig. 6.** U-Net model added the depth supervision scheme.

## 4  Loss Function Design and Optimization

To further improve the generalization capability of the designed recognition model, multiple loss functions are attempted to be adapted in the salt mine image segmentation model.

For ordinary target binary tasks, it is possible to use binary cross entropy (BCE) loss function [11, 12]. In every gradient return, BCE has the same attention to each image category without weight discrimination, so it is very susceptible to the influence of category imbalance. In the every gradient return, the logarithmic loss function has the same attention to each category without weight discrimination, so it is very susceptible to the influence of category imbalance. The exploratory data analysis indicates that the data categories in the training set were unbalanced. Therefore, to overcome this problem, the label generated by the designed image segmentation model is almost a mask that does not contain any location of salt mine. In addition, weighted cross-entropy (WCE) loss function alleviates the problems caused by unbalanced sample categories to some extent, but its disadvantage is that the weights of difficult samples need to be adjusted artificially, which increases the adjustment cost and difficulty [13].

Although WCE needs to artificially adjust the weights of difficult samples, setting a large weight for loss of a difficult sample is necessary to make the model pay more attention to these types of samples. Focal loss function proposed to solve the problem of the serious imbalance between positive and negative samples could make the network model to actively learn these difficult samples [14].

Dice loss function is derived from binary classification by introducing dice coefficient which is essentially an overlap measure of two samples overlap, or similarity of two contour areas [15]. If the value of dice coefficient is increasing while using Dice in the designed image segmentation model, the overlap between the predicted target location and the real target location is increasing as well.

For intersection over union (IOU) acting as a measure of the accuracy of an object in a given set of data, BCE cannot directly optimize the IOU, while Dice Loss can indirectly optimize the IOU. Therefore, under the premise of using BCE, combining Dice as multi loss could achieve multi-angle supervised model training. In addition, since BCE cannot directly optimize the IOU, some researches attempt to avoid using

BCE in the image segmentation model. As a representative method, Lovasz-Softmax Loss function (Lovasz for short) could directly optimize the IOU [16]. In this work, Lovasz is also used in the designed image segmentation model to directly optimize the IOU while effectively reducing loss.

## 5 Numerical Analysis and Evaluation

In this section, the following numerical analyses are conducted to evaluate the efficiency of achieved proposals in this work.

With the U-Net model and BCE + Dice loss function, the differences in the loss value and accuracy of model training under different data augmentation methods are shown in Fig. 7. From the comparisons, we can clearly see that these data augmentation operations, e.g., padding and horizontal mirror flip, could yield higher benefits, while rotate and direct resize would reduce returns, and a conclusion could be obtained that some data augmentation operations destroy the spatial structure information of the original image and increase the model learning difficulty.

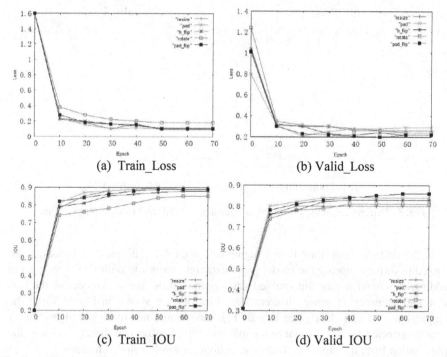

(a) Train_Loss  (b) Valid_Loss

(c) Train_IOU  (d) Valid_IOU

**Fig. 7.** Differences in the loss value and accuracy of model training under different data augmentation methods.

With the salt mint recognition model designed based on the U-Net, many attempts are conducted to further optimizing the model. The differences in the loss value and accuracy of different model optimization strategies are shown in Fig. 8. From the comparisons, we can clearly see that the segmentation of small objects and the positioning of object boundaries could be improved by adding hyper-column and scSE modules into the U-Net model, and the overfitting degree of the model on the training set can be effectively reduced by adding auxiliary functions and depth supervision scheme. In addition, the combination of hyper-column and scSE could achieve a greater improvement in model accuracy than that separately adding hyper-column or scSE module.

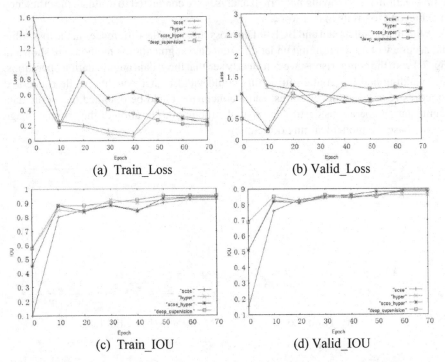

(a) Train_Loss        (b) Valid_Loss

(c) Train_IOU        (d) Valid_IOU

**Fig. 8.** Differences in the loss value and accuracy of different model optimizations.

In the designed salt mine image segmentation model, multiple loss functions are adapted to further improve the model generalization capability. With U-Net model and padding/horizontal mirror flip enabled data augmentation, the differences in the loss value and accuracy of using different loss functions are shown in Fig. 9. From the comparisons, we can clearly see that BCE is simple and fast to be calculated, but it is easily affected by the imbalance sample category. Lovasz can directly achieve the optimization from the model indicators with high returns but slow efficiency.

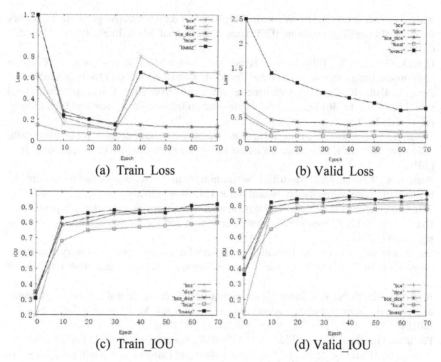

(a)  Train_Loss

(b)  Valid_Loss

(c)  Train_IOU

(d)  Valid_IOU

**Fig. 9.** Differences in the loss value and accuracy of using different loss functions.

## 6 Conclusion

In this paper, with the in-depth application of deep learning in image segmentation for salt mine recognition, a deep convolutional neural network based image segmentation model is investigated. With exploratory data analysis to mine the characteristics and data processing to increase the size of the image data, a basic recognition model is first designed based on U-Net model integrated modified ResNet34, and many model optimization methods then are used to further optimizing the model according to the data characteristics. Additionally, to further improving the model generalization capacity, multiple loss functions are investigated as well. Finally, numerical analyses are shown to evaluate the efficiencies on loss value and recognition accuracy.

**Acknowledgments.** This work was supported in part by the Natural Science Foundation of Guangdong Province (Grant No. 2018A030313014), the Guangdong University Key Project (2019KZDXM012), and the research team project of Dongguan University of Technology (Grant No. TDY-B2019009).

## References

1. Liu, F., Lin, G., Shen, C.: CRF learning with CNN features for image segmentation. Pattern Recogn. **48**(10), 2983–2992 (2015)

2. Wang, G., Zuluaga, M.A., Li, W., et al.: DeepIGeoS: a deep interactive geodesic framework for medical image segmentation. IEEE Trans. Pattern Anal. Mach. Intell. **41**(7), 1559–1572 (2018)
3. Dolz, J., Gopinath, K., Yuan, J., et al.: HyperDense-Net: a hyper-densely connected CNN for multi-modal image segmentation. IEEE Trans. Med. Imaging **38**(5), 1116–1126 (2018)
4. Zeng, Y., Jiang, K., Chen, J.: Automatic seismic salt interpretation with deep convolutional neural networks. In: 3rd International Conference on Information System and Data Mining, pp. 16–20 (2019)
5. Fawzi, A., Samulowitz, H., Turaga, D., et al.: Adaptive data augmentation for image classification. In: 2016 IEEE International Conference on Image Processing (ICIP), pp. 3688–3692 (2016)
6. Ibtehaz, N., Rahman, M.S.: MultiResUNet: rethinking the U-Net architecture for multimodal biomedical image segmentation. Neural Netw. **121**, 74–87 (2020)
7. Zhu, C., Zheng, Y., Luu, K., et al.: Weakly supervised facial analysis with dense hyper-column features. In: IEEE Conference on Computer Vision and Pattern Recognition Workshops, pp. 25–33 (2016)
8. Liu, Y., Jourabloo, A., Liu, X.: Learning deep models for face anti-spoofing: binary or auxiliary supervision. In: IEEE Conference on Computer Vision and Pattern Recognition, pp. 389–398 (2018)
9. Roy, A.G., Navab, N., Wachinger, C.: Recalibrating fully convolutional networks with spatial and channel "squeeze and excitation" blocks. IEEE Trans. Med. Imaging **38**(2), 540–549 (2018)
10. Pavlakos, G., Zhou, X., Daniilidis, K.: Ordinal depth supervision for 3D human pose estimation. In: IEEE Conference on Computer Vision and Pattern Recognition pp. 7307–7316 (2018)
11. Tao, M., Wei, W., Yuan, H., Huang, S.: Version-vector based video data online cloud backup in smart campus. Multimedia Tools Appl. **78**(3), 3435–3456 (2019)
12. Yi-de, M., Qing, L., Zhi-Bai, Q.: Automated image segmentation using improved PCNN model based on cross-entropy. In: International Symposium on Intelligent Multimedia, Video and Speech Processing, pp. 743–746 (2004)
13. Guerrero-Pena, F.A., Fernandez, P.D.M., Ren, T.I., et al.: Multiclass weighted loss for instance segmentation of cluttered cells. In: 25th IEEE International Conference on Image Processing (ICIP), pp. 2451–2455 (2018)
14. Lin, T.Y., Goyal, P., Girshick, R., et al.: Focal loss for dense object detection. In: IEEE International Conference on Computer Vision, pp. 2980–2988 (2017)
15. Sudre, C.H., Li, W., Vercauteren, T., Ourselin, S., Jorge Cardoso, M.: Generalised dice overlap as a deep learning loss function for highly unbalanced segmentations. In: Cardoso, M.J., et al. (eds.) DLMIA/ML-CDS -2017. LNCS, vol. 10553, pp. 240–248. Springer, Cham (2017). https://doi.org/10.1007/978-3-319-67558-9_28
16. Berman, M., Rannen Triki, A., Blaschko, M.B.: The lovász-softmax loss: a tractable surrogate for the optimization of the intersection-over-union measure in neural networks. In: IEEE Conference on Computer Vision and Pattern Recognition, pp. 4413–4421 (2018)

# Optimization of Collaborative Filtering Algorithm in Movie Recommendation System

Jiao Peng[✉] and Shu Gong

Guangdong University of Science and Technology, Dongguan, Guangdong, China
154711901@qq.com

**Abstract.** In the era of big data information explosion, people are faced with a large amount of information every day, and how to obtain the required content in a large amount of information. The appearance of intelligent recommendation system has brought great convenience to our life. The recommendation system can recommend corresponding functions, products and services according to users' past browsing information, enabling users to get their desired information data from massive data more efficiently. As an indispensable part of most people's entertainment life, movie recommendation has also become a very important part of Internet recommendation content. The collaborative filtering algorithm is used to realize the personalized recommendation of movies. However, in the process of movie recommendation, it is found that new users are only recommended movies based on their ratings without considering the attribute information between movies, which may lead to problems such as inaccurate recommendation accuracy. Therefore, this paper further optimizes the collaborative filtering algorithm and introduces the similarity calculation between movie attributes to improve the accuracy of movie recommendation.

**Keywords:** Movie recommendation system · Collaborative filtering algorithm · Optimization of algorithm

## 1 Introduction

In today's information explosion, there are various ways and means of obtaining information, as well as various kinds of information. Now what people spend the most time is no longer where to get information, but to find the content or information they are interested in among the numerous information, which is the so-called information overload problem. To solve this problem, recommendation systems are applied. Recommendation systems are ubiquitous in everyday web applications, such as online shopping, online bookstores, news apps, social networks, music websites, movie websites and so on. The system will make personalized content recommendations based on personal preferences, habits or needs. For example, open the news app. Because of the personalized recommendation function in the app, the front page of news is different for everyone. Collaborative filtering algorithm is the first personalization recommendation technology and the most widely used recommendation algorithm. The movie recommendation system makes use

© Springer Nature Switzerland AG 2020
X. Chen et al. (Eds.): ML4CS 2020, LNCS 12488, pp. 11–19, 2020.
https://doi.org/10.1007/978-3-030-62463-7_2

of this special information recommendation technology to recommend some movies that users may like or be interested in according to their previous information about watching movies and their rating information. However, in the process of using collaborative filtering algorithm, we found that if the movie recommendation is only based on the user's rating of the movie, the recommendation result is not accurate, and the recommendation accuracy and coverage may not be accurate enough. Therefore, collaborative filtering algorithm can be further optimized. In the process of movie recommendation, not only the rating information of the movie by users, but also the attribute information of the movie itself, such as the movie category and the similarity between the movie categories.

## 2   Collaborative Filtering Algorithm

### 2.1   Introduction of Collaborative Filtering Algorithm

Collaborative filtering algorithm is a recommended technique which is widely used and matures at present. Its working principle is to learn the user's interests and preferences according to the user's historical operation behavior, find out the user's neighbor users according to their interests and hobbies, determine the set of goods that the neighbor user likes, and then recommend to the user the information of goods in the set that the user has not bought.

The collaborative filtering algorithm is mainly divided into three steps:

(1) Create a "user-item" rating table

According to the user's purchase record and the score information of the purchased items, a "user-item" rating table is formed. Among them, The user set $U = \{u_1, u_2, \ldots, u_m\}$, where $u_i$ represents the user i, $T = \{t_1, t_2, \ldots, t_n\}$, where $t_i$ represents the item t. $R_{u_i,t_i}$ represents user $u_i$ buys item $t_i$ and gives a rating. The "user-item" rating table formed is shown in Table 1:

**Table 1.**   "user-item" rating table

|        | $t_1$          | $t_2$          | $\ldots$  | $t_n$          |
|--------|----------------|----------------|-----------|----------------|
| $u_1$  | $R_{u_1,t_1}$  | $R_{u_1,t_2}$  | $\ldots$  | $R_{u_1,t_n}$  |
| $u_2$  | $R_{u_2,t_2}$  | $R_{u_2,t_2}$  | $\ldots$  | $R_{u_2,t_n}$  |
| $\ldots$ | $\ldots$     | $\ldots$       | $\ldots$  | $\ldots$       |
| $u_m$  | $R_{u_m,t_1}$  | $R_{u_m,t_2}$  | $\ldots$  | $R_{u_m,t_n}$  |

(2) Form a neighborhood set

Neighbor set refers to the collection of users (items) with common characteristics with the target user. According to the data set in Table 1, the historical score records of each user (item) can be obtained from the "user-item" scoring matrix. Assuming that the target user is $u_v$, to calculate the set of neighbors of the user $u_v$, you should first calculate the similarity between the target user and other users. The higher the similarity between users, the closer the two users' interests and hobbies will be, and the greater the probability that they will share the same characteristics. Sort by similarity degree, take the first N users to form the neighbor set.

(3) Form a list of recommendations

Set the user's neighborhood set $U = \{u_1, u_2, \ldots, u_N\}$, and then add the collection of favorite items of each neighbor user to the list to form a recommended list set, which is recommended to the user.

Collaborative filtering recommendation algorithm can also be subdivided into two algorithms: user-based collaborative filtering and item-based collaborative filtering recommendation algorithm.

The main principle of user-based collaborative filtering algorithm: When the user needs personalized recommendation, according to the user's personal interests, hobbies or behavior habits and other information, calculate the similarity between users, and find the set of neighbors with the user. Then the information that each user likes in the neighbor set is added together to form a commodity set, and then the items in the commodity set are recommended to the user.

The main principle of item-based collaborative filtering algorithm: When a user needs personalized recommendation, he can usually analyze the historical behavior data of the goods he has purchased before, analyze the characteristics of the goods he has purchased, and then learn the characteristics of the items he likes. In the commodity set, according to the user's interest and the characteristics of the items to be recommended, a group of items with the greatest correlation is determined in the commodity set as the recommendation list and recommended to the user.

## 2.2  The Evaluation Index of Recommendation Algorithm

The evaluation index is used to measure the performance of collaborative filtering recommendation algorithm. The recommendation performance is mainly reflected in two aspects: prediction quality and recommendation quality. Recommendation algorithms usually use precision rate and recall rate to measure the quality of recommendation, and use coverage rate to measure the quality of recommendation prediction by recommendation system.

(1) Precision rate

The precision rate is the measure to evaluate the recommendation algorithm, which is used to evaluate the recommendation effect of the recommendation system. Precision rate is used to describe the ability of a recommendation system to predict user behavior. Generally, the coincidence rate between the recommendation list and user behavior given by the algorithm on the offline data set is calculated. The higher the coincidence rate is, the higher the accuracy will be.

$$Precision = \frac{\sum_{i=1}^{m} \left| T_{u_i} \cap recommender_{u_i} \right|}{\sum_{i=1}^{m} \left| T_{u_i} \right|} \tag{1}$$

Where, Precision refers to the precision of the recommendation algorithm, $T_{u_i}$ refers to the items purchased by user $u_i$, and $recommender_{u_i}$ refers to the items recommended to user by the system. $T_{u_i} \cap recommender_{u_i}$ refers to the items that user $u_i$ purchases which are recommended to the user by the recommendation system.

(2) Recall rate

Recall rate is a measure of how good a recommendation algorithm is. Recall rate measures the quality of the recommendation system by calculating the proportion of items purchased in the recommended items.

$$Recall = \frac{\sum_{i=1}^{m} |T_{u_i} \cap recommender_{u_i}|}{\sum_{i=1}^{m} |recommender_{u_i}|} \tag{2}$$

Where, Precision refers to the precision of the recommendation algorithm, $T_{u_i}$ refers to the items purchased by user $u_i$, and $recommender_{u_i}$ refers to the items recommended to user by the system. $T_{u_i} \cap recommender_{u_i}$ refers to the items that user $u_i$ purchases which are recommended to the user by the recommendation system.

(3) Coverage rate

Coverage rate is an important measure of a recommendation system. Coverage is calculated as the ratio of the number of items recommended to the user to the total number of items. Suppose the user set of the system is U, the list of items is represented by L, and the length of the list of items recommended to users is represented by R(u).

$$Coverage = \frac{|\sum_{u \in U} R(u)|}{|L|} \tag{3}$$

# 3   Optimization of User-Based Collaborative Filtering Algorithm

## 3.1   User-Based Collaborative Filtering Algorithm

The main content of user-based collaborative filtering recommendation algorithm is: when to recommend items to target users A, with target users will find A have similar preferences, user B, then the user B items like, the target users before A no goods, recommended to the user. For example, in Fig. 1, we can see that user $u_1$ watched the movie $m_1$ and $m_2$, user $u_2$ watched the movie $m_1$, $m_3$ and $m_4$, user $u_3$ watched the movie $m_1$, $m_5$ and $m_6$. Recommended new movies to user $u_1$ based on the above viewing records.

The relationship between movie names and movie types is shown in Table 2.

(1) According to the similarity between users, find the set of users with similar preferences to the target users.

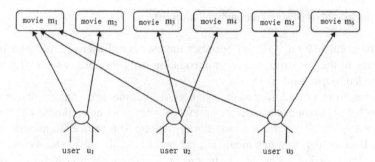

**Fig. 1.** User-based collaborative filtering recommendation

**Table 2.** Relationship between movie titles and movie types

| Name/type | Romance | Comedy | Drama | Action | Horror | Thriller | Fantasy | Crime |
|-----------|---------|--------|-------|--------|--------|----------|---------|-------|
| $m_1$ | 1 | 1 | 1 | 0 | 0 | 0 | 0 | 0 |
| $m_2$ | 1 | 0 | 0 | 1 | 0 | 0 | 0 | 0 |
| $m_3$ | 1 | 0 | 0 | 0 | 1 | 0 | 0 | 0 |
| $m_4$ | 1 | 0 | 0 | 0 | 0 | 1 | 0 | 0 |
| $m_5$ | 1 | 0 | 0 | 0 | 0 | 0 | 1 | 1 |
| $m_6$ | 1 | 0 | 0 | 0 | 0 | 0 | 1 | 1 |

For the first time, we need to calculate the similarity between users. We set user $u_i$ and $u_j$, $\mathbf{N}u_i$ represents the collection of items that user $u_i$ likes, and $\mathbf{N}u_j$ represents the collection of items that user $u_j$ likes. We can use the following formula to calculate the similarity between users.

$$\text{sim}\,(u_i, u_j) = \frac{\left|\mathbf{N}u_i \cap \mathbf{N}u_j\right|}{\sqrt{|\mathbf{N}u_i||\mathbf{N}u_j|}} \tag{4}$$

In Fig. 1, user $u_1$ likes to watch movie $m_1$, $m_2$, user $u_2$ likes to watch movie $m_1$, $m_3$ and $m_4$, user $u_3$ likes to watch movies $m_1$, $m_5$ and $m_6$. Now recommend movies to target user $u_1$. Calculate the user similarity between user $u_1$ and user $u_2$.

$$\text{sim}(u_1, u_2) = \frac{|\{m_1, m_2\} \cap \{m_1, m_3, m_4\}|}{\sqrt{|\{m_1, m_2\}||\{m_1, m_3, m_4\}|}} = \frac{|\{m_1\}|}{\sqrt{6}} = \frac{1}{\sqrt{6}}$$

The user similarity between user $u_1$ and user $u_3$ is calculated as follows.

$$\text{sim}(u_1, u_3) = \frac{|\{m_1, m_2\} \cap \{m_1, m_5, m_6\}|}{\sqrt{|\{m_1, m_2\}||\{m_1, m_5, m_6\}|}} = \frac{|\{m_1\}|}{\sqrt{6}} = \frac{1}{\sqrt{6}}$$

The similarity between user $u_1$ and user $u_2$ is calculated to be the same as that between user $u_1$ and user $u_3$.

(2) Forming neighbor sets and recommendation lists

By the above calculation, $\{u_2, u_3\}$ becomes the user set of user $u_1$. Add user $u_2$ and $u_3$'s movies to the list to form a recommendation list $\{m_3, m_4, m_5, m_6\}$,which is recommended to user $u_1$.

However, from Table 2, we can see that user $u_1$ likes movie $m_1$ and $m_2$, while movie $m_1$ and $m_2$ belong to romance, so user $u_1$ prefers Romance. User $u_2$ likes to watch movie $m_1$, $m_3$, and $m_4$, while movie $m_1$, $m_3$, and $m_4$ belong to romance, so user $u_2$ prefers romance. User $u_3$ likes to watch movie $m_1$, $m_5$, and $m_6$, while movie $m_5$, and $m_6$ belong to science fiction and crime movies, so user $u_3$ prefers science fiction and crime movies. To sum up, user $u_1$ and user $u_2$ prefer romance movies, while user $u_3$ prefers science fiction and crime movies. Therefore, user $u_1$ has a high similarity with user $u_2$, but a low similarity with user $u_3$.

Based on the above analysis, it can be seen that when using user-based collaborative filtering algorithm for movie recommendation, the attribute characteristics of items that users like need to be considered; otherwise, the similarity between users calculated is incorrect. Therefore, it is necessary to optimize the user-based collaborative filtering algorithm.

### 3.2  Optimization of User-Based Collaborative Filtering Algorithm

When the collaborative filtering algorithm based on users is optimized, the attribute information of items is added in the process of similarity calculation. The specific steps are as follows: firstly, the collaborative filtering algorithm is used to calculate the similarity between users, then the similarity of attribute feature vectors of the items that users like is calculated, and finally the two similarities are fused as the final similarity between users. Make recommendations based on the final similarity calculation.

(1) The first step is to calculate the similarity between users.

$$\text{sim}\,(u_i, u_j) = \frac{\left| \mathbf{N}u_i \cap \mathbf{N}u_j \right|}{\sqrt{\left| \mathbf{N}u_i \right| \left| \mathbf{N}u_j \right|}}$$

(2) Calculate the similarity of properties between films

Set the category set of the movie $A = \{a_1, a_2, \ldots, a_n\}$, and construct the movie's attribute feature vector according to the category to which the movie belongs. $\underset{a_{m_i}}{\rightarrow} = \{a_{i1}, a_{i2}, \ldots, a_{in}\}$, $\underset{a_{m_i}}{\rightarrow}$ represents film $m_i$ category information vector, as shown in Fig. 2, $a_{ij}$ value is 0 or 1, used to indicate whether movies $m_i$ belongs to the category $a_j$. Calculate the similarity between films as shown in formula (5).

$$\text{sim}\left(m_i, m_j\right) = \underset{a_{m_i}}{\rightarrow} * \underset{a_{m_j}}{\rightarrow} = \frac{\sum_{b=1}^{S}\left(a_{ib} * a_{jb}\right)}{\sqrt{\sum_{b=1}^{S} a_{ib}^2}\sqrt{\sum_{b=1}^{S} a_{jb}^2}} \tag{5}$$

**Fig. 2.** Comparison of precision before and after optimization

In the formula, b represents the category b attribute of the movie; S represents the total S attributes of the movie, and $sim(m_i, m_j)$ represents the similarity between movie $m_i$ and movie $m_j$.

(3) Calculate the similarity between users after optimization

The similarity between users after optimization mainly considers the similarity of item attributes while calculating the similarity of users. The specific calculation formula is shown in formula 6:

$$sim(u_i, u_j) = \alpha * sima + (1 - \alpha) * simb \qquad (6)$$

In the formula, $sim(u_i, u_j)$ is the similarity between the optimized users, *sima* is the similarity of item attribute characteristics, *simb* is the user similarity calculated by the traditional collaborative filtering algorithm, and $\alpha$ is the parameter to adjust the two similarities.

(4) Forming neighbor sets and recommendation lists

According to the optimized user similarity value, users with higher similarity value constitute neighbor sets, and the recommendation list is constructed to complete the user's movie recommendation.

## 4 Algorithm Evaluation Experiment

### 4.1 Data Acquisition

Using Movie Lens data set, the collection contains 209,171 movies rated by 526 users with a score of 0–5. The movie categories are mainly divided into 12 categories, including adventure, animation, children, drama, fantasy, romance, drama, action, crime, thriller, horror and so on. Category information for movie properties is indicated by 0 and 1 respectively to indicate whether the movie belongs to that category. The data in the set is

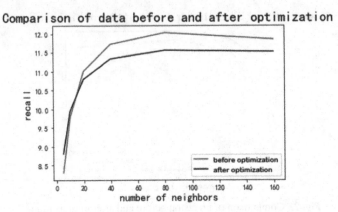

**Fig. 3.** Comparison of recall rate before and after optimization

divided into two parts, one as the training set and the other as the test set. Select different number of neighbors to test.

## 4.2 Algorithm Testing

Using the data collected above, the optimization algorithm based on user collaborative filtering was used to calculate the precision rate, recall rate and coverage rate before and after optimization. By comparing the three evaluation indexes before and after optimization, data simulation test was conducted according to formulas (1), (2) and (3). The specific test results are shown in Fig. 2, Fig. 3 and Fig. 4.

As can be seen from the following three figures, in the user-based collaborative filtering algorithm, the similarity of attributes and features of items is introduced to calculate the user similarity. After the algorithm is optimized, the precision and coverage of movie recommendation are significantly improved, but at the cost of the recall rate.

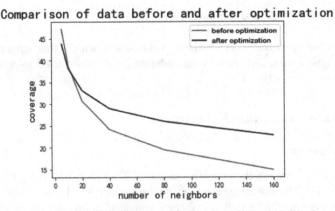

**Fig. 4.** Comparison of coverage before and after optimization

# 5   Conclusion

This article mainly adopted the user-based collaborative filtering algorithm to movie recommendation, but in the process of recommendation, find only according to the similarity between the user and the target collection of users, the user preference similarity recommendation precision is not accurate, so the film was introduced to the optimization algorithm in attribute characteristics of similarity between two neighbor set and form the suggestion list is calculated after synthetically, in order to enhance the precision in movie recommendation.

# References

1. Chen, J.F., Huang, L.S., Lin, G.T.: TV program recommendation strategy of collaborative filtering algorithm. J. Henan Inst. Eng. (Nat. Sci.) **32**(01), 61–65 (2020)
2. Jiao, J.F.: Research on collaborative filtering recommendation algorithm based on Spark. Comput. Program. Skills Maintenance **03**, 40–41 (2020)
3. Li, Z.F., Liu, Y.S., Li, C.T.: Research on news recommendation algorithm based on user behavior. Comput. Eng. Sci. **42**(03), 529–534 (2020)
4. Dong, Y.F., Zhu, C.S.: Collaborative filtering algorithm based on improved user attribute rating. Comput. Eng. Des. **41**(02), 425–431 (2020)
5. Ding, H.F., Huang, Q.S.: Research on personalized tourism recommendation algorithm based on attribute characteristics. Intell. Comput. Appl. **10**(01), 193–196 (2020)

# TagRec: Trust-Aware Generative Adversarial Network with Recurrent Neural Network for Recommender Systems

Honglong Chen[1(✉)], Shuai Wang[1], Nan Jiang[2], Zhe Li[1], Na Yan[1], and Leyi Shi[3]

[1] College of Control Science and Engineering, China University of Petroleum, Qingdao, China
chenhl@upc.edu.cn
[2] College of Information Engineering, East China Jiaotong University, Nanchang, China
[3] College of Computer Science and Technology, China University of Petroleum, Qingdao, China

**Abstract.** Recently recommender systems become more and more significant in the daily life. Although the recommender systems based on the generative adversarial network (GAN) are competent, the user trust information is seldom taken into consideration. In this paper, we propose a Trust-Aware Generative adversarial network with recurrent neural network for RECommender systems named *TagRec*, which makes use of the user trust information for top-N recommendation. In the framework, the discriminative model is a multi-layer perceptron to distinguish whether a sample is from the real data or fake data. The discriminator helps to guide the training of the generative model to make it fit the data distribution of the user trust information. The generative model is a recurrent neural network (RNN) with long short-term memory (LSTM) cells, aiming to confuse the discriminative model by generating samples as similar as possible to the real data. Through the adversarial training between the discriminative and generative models, the user trust information can be fully used to improve the recommendation performance. We conduct extensive experiments on real-word datasets to validate the effectiveness of the TagRec.

**Keywords:** Recommender systems · Generative adversarial network · Recurrent neural network · Trust-aware

## 1 Introduction

The current explosive growth of data makes the information overload more and more serious, which can be solved by recommender systems. Matrix factorization (MF) [7] is one of the most classical recommendation methods, in which the recommendation is based on the interactive information between users and items. Although MF has been widely used in recommender systems due to its simplicity and attractive accuracy, it still suffers from two main problems. The first one is the severe data sparsity, and the other one lies in its failing to well learn the deep representation of interactive information between the users and items. The performance of recommender systems will be severely constrained by the above two problems. The user trust information can be used to increase data volume based on the rating information for the data sparsity. Deep learning provides a potential solution for the deep representation of data, which can automatically learn the feature representation from heterogeneous data through multi-layer nonlinear network structure. In this paper, we intend to use the social information and deep learning in the recommender systems.

Recently, Goodfellow *et al.* [1] proposed Generative Adversarial Network (GAN), which learns to fit the distribution of given data by adversarial training. Inspired by GAN, IRGAN [3] is a unified framework that takes advantage of both generative model and discriminative model to apply the adversarial training in recommender systems. GraphGAN [4] also adopted the adversarial training and proposed a novel graph softmax to overcome the limitations of traditional softmax function. Both the IRGAN and GraphGAN use the policy gradient instead of gradient descent for the model optimization since the sampling of discriminative model is discrete. A novel GAN-based collaborative filtering framework called CFGAN [2] was proposed, which can solve the fundamental problem of the GAN-based methods, *i.e.*, the limitation of discrete item index generation. However, CFGAN ignores the potential of social information in recommender systems.

To address this problem, we propose a Trust-Aware Generative adversarial network with recurrent neural network for RECommender systems named *TagRec*, which combines the social information in the recommender systems. The proposed TagRec consists of two parts: a generative model $G$ and a discriminative model $D$. In the generative model, the dynamic recurrent neural network with long short-term memory (LSTM) cells is adopted. A user rating trust sequence is firstly constructed based on the user rating vector according to the user trust relationship. Then based on the historical user rating information and user trust information, the rating trust sequence is input to $G$ to generate a recommendation list for the user. In the discriminative model, the multi-layer perceptron is adopted. The real data and fake data (generated by the generator) are input to $D$, which aims to distinguish them. With the adversarial training between the discriminative and generative models, the discriminator helps to guide the training of the generative model to make it fit the data distribution of user trust information.

**The main contributions of this paper** are summarized as follows:

- By making use of the users' social relationship, we construct a user rating trust sequence based on the user rating vector, in which the similarity calculation is adopted to avoid the curse of dimensionality.
- We propose to integrate the generative adversarial network and recurrent neural network, which can extract both the rating information features and social information features, to handle the sparsity problem in the recommender systems.
- We implement the proposed TagRec model and conduct extensive experiments on two real-world datasets to validate its effectiveness.

The rest of this paper is organized as follows. In Sect. 2, we briefly review the related work. In Sect. 3, we introduce the proposed social recommendation method TagRec in details. In Sect. 4, we describe the experimental data, implementation details, evaluation metrics and benchmarks. In Sect. 5, we analyze the experimental results. In Sect. 6, we conclude this paper and put forward the future work.

## 2 Related Work

### 2.1 Recurrent Neural Network Based Recommendation

The Recurrent Neural Network (RNN) can be used to process the sequential data of the recommender systems. In [13], RNN is applied to the session-based recommender systems, which adopts Gated Recurrent Unit (GRU) as the RNN unit and takes the first clicked item as the initial input of the GRU unit. Each click of the user will produce a recommendation result that depends on all the previous clicks. A dynamic model is proposed in [6], which is combined with RNN to predict the future behavioral trajectories of users. This work adopts the long short-term memory (LSTM) as the basic RNN unit and uses LSTM to learn the states of the users and items respectively. In [15], Liu et al. propose a context-aware recurrent neural network to address the problem of context-aware sequential recommendation. In the proposed TagRec, we use RNN combined with GAN to process the sequential data in recommender systems.

### 2.2 Generative Adversarial Network Based Recommendation

Recently, GAN has been successfully applied to the recommendation tasks. IRGAN [3] is a unified framework, which takes advantage of both the generative model and discriminative model and can be used in the web search, item recommendation and question answering. In [16], RecGAN is proposed, which combines RNN and GAN to improve recommendation performance. Different from the proposed TagRec in this paper, RecGAN leverages RNN to extract time feature from the interactive information between the users and items, while TagRec is to process the social information. In [2], a new vector-wise mechanism is proposed to improve the training of recommendation methods based on policy gradient without considering the social information.

## 3 The Proposed Framework

In this paper, we focus on the top-N recommendation problem in recommender systems. Let $\mathcal{U} = \{u_1, u_2, \cdots, u_m\}$ represent the set of users and $\mathcal{I} = \{i_1, i_2, \cdots, i_n\}$ denote the set of items, where $m$ is the number of users and $n$ is the number of items respectively. Let $R = [R_{u,i}]_{m \times n}$ denote the ratings expressed by the users on items where $R_{u,i}$ is a real number that represents the preference of user $u$ on item $i$. The larger the value of $R_{u,i}$ is, the more user $u$ likes item $i$. The value of $R_{u,i}$ ranges from 1 to 5. We set $R_{u,i}$ to 1 if $R_{u,i} > 1$, and 0 otherwise. In addition to the rating matrix $R$, each user has a social trust matrix $T = [T_{u,v}]_{m \times m}$ where $T_{u,v}$ denotes the trust value that the trustor $u$ has on trustee $v$. Usually the trust value is either 0 or 1, where 0 means that user $u$ has no trust with user $v$ and 1 means that user $u$ completely trusts user $v$. The task of the recommender system is to use the social trust matrix $T$ and the existing values in the rating matrix $R$ to predict the missing values in $R$.

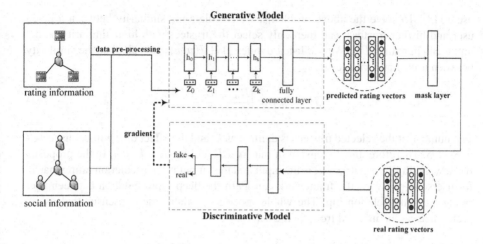

**Fig. 1.** The overall framework of the proposed TagRec.

## 3.1  An Overview of the Proposed Framework

The architecture of the proposed framework is shown in Fig. 1, which consists of three parts: data pre-processing, adversarial learning and negative sampling. Data pre-processing constructs a user rating trust sequence as the input of the proposed framework TagRec based on the user rating vector and social relationship. Adversarial learning consists of generative model and discriminative model. The generative model generates fake data as the negative sample of the discriminative model. The purpose of the discriminative model is to distinguish whether a sample is from the real data or generated data. Negative sampling is to solve the output polarizing problem of the generative model by adding a mask layer and a negative sampling function.

## 3.2  Data Pre-processing

In order to merge the user rating matrix $R$ and user trust matrix $T$ into one matrix without losing the deep representation between them, we exploit the user set $\mathcal{U}$. In fact, the user set $\mathcal{U}$ bridges the gap between the matrix $R$ and matrix $T$. Each user $u$ in the user set $\mathcal{U}$ is also in matrix $R$ and matrix $T$. This gave us the inspiration to merge the data. Each user $u$ has a rating vector of $n$ dimensions $[R_u]_n$ in matrix $R$ for all items. Each user $u$ has a trust vector of $m$ dimensions $[T_v]_m$ for all users. Therefore, for each element in the trust vector $[T_v]_m$ we can replace it with the $n$ dimensions rating vector corresponding to the user $u_m$. By this means, we can obtain a new matrix of $m \times m \times n$ dimensions for $m$ users in matrix $T$. We define this new matrix as the trustee rating matrix $TR = [TR_{u,v,i}]_{m,m,n}$. The trustee rating matrix successfully embeds trust relationship into user rating by deep representation. However, due to the following two factors, too much consideration of other trustees may even reduce the performance of recommender systems. First, it is computationally expensive. Second, even if a trustor has a lot of trustees, his/her preference will only be affected by some of the most trusted users [14]. To solve the above problems, we calculate the similarity between a given user and his/her trustees, and then only select the trustees with high similarity as the input data in matrix $TR$. We use Jaccard index [17] $J(u, u_m)$ to calculate the similarity between user $u$ and trustees $u_m$:

$$J(u, u_m) = \frac{|R_u \cap R_{u_m}|}{|R_u \cup R_{u_m}|}. \tag{1}$$

The number of the selected trustees is defined as $k$, and the effect of $k$ will be discussed in Sect. 5. Therefore, the dimension of matrix $TR$ will be $m \times k \times n$. In the proposed framework TagRec, matrix $TR$ is the input and the output is the prediction rating matrix for users. In this way, the framework can learn the deep representation between user rating and trust relationship. The whole process simulates the consensus that users' preference can be inferred from their trustees.

### 3.3 Adversarial Learning for Recommendation

The generative adversarial network consists of a generator $G$ and a discriminator $D$. Let $t_u$ and $r_u$ denote user $u$'s trust information and existing rating information respectively. Let $r_i$ and $\widetilde{r}_i$ denote the true and predicted ratings on item $i$ respectively. $\theta_G$ and $\phi_D$ are denoted as the model parameters of the generator $G$ and the discriminator $D$ respectively. The proposed framework is to learn the following two models:

Generator $G(r_i|t_u, r_u; \theta_G)$, which tries to approximate the real rating data distribution over items $p_{true}(r_i|t_u, r_u)$, and generate the most similar data distribution $p_\theta(r_i|t_u, r_u)$.

Discriminator $D(r_i, \widetilde{r}_i; \phi_D)$, which attempts to distinguish the real data distribution $p_{true}(r_i|t_u, r_u)$ from the fake data distribution $p_\theta(r_i|t_u, r_u)$ generated by $G$.

$$\min_{\theta_G} \max_{\phi_D} V(G, D) = \mathbb{E}_{r_i \sim p_{true}(r_i|t_u, r_u)} \left[ \log D\left(r_i|t_u, r_u; \phi_D\right) \right]$$
$$+ \mathbb{E}_{\widetilde{r}_i \sim p_\theta(r_i|t_u, r_u)} \left[ \log \left(1 - D\left(\widetilde{r}_i|t_u, r_u; \phi_D\right)\right) \right]. \tag{2}$$

Based on Eq. (2), the generator and the discriminator learn the optimal model parameters by iterative training.

**Generative Model.** The framework of TagRec adopts recurrent neural network (RNN) as the generative model. As mentioned in Sect. 3.2, a user's predicted ratings on items can be inferred from his or her trustees' ratings. Since the influence of trustees on users is cumulative, the function of recurrent neural network is similar with this situation. The RNN not only considers the input of the current moment, but also gives the network a 'memory' of the previous content. By taking advantage of this feature, RNN can capture the deep representation of the influence of each trustee on the user. As the user does not declare the trust relationship to all other users, the second dimension of the trustee rating matrix $TR$ is uncertain in practice. Therefore, we adopt the structure of dynamic RNN and use long short-term memory (LSTM) cells as the basic unit. In the data pre-processing stage, the trust information and rating information can be fused into a trustee rating matrix $TR = [TR_{u,v,i}]_{m,k,n}$. The TagRec takes the n-dimensional vectors $[TR_i]_n$ in the matrix $TR$ as the input of dynamic RNN at time $t$, where the maximum value of $t$ is equal to the number of the selected trustees $k$. LSTM unit consists of an input gate $i_t$, a forget gate $f_t$, an output gate $o_t$, and a state cell $c_t$. These gates can be calculated based on the previous hidden state $h_{t-1}$ and the current input $z_t$:

$$f_t, i_t, o_t = \sigma \left( W \left[ h_{t-1}, z_t \right] \right) \tag{3}$$

where $\sigma(\cdot)$ is the sigmoid activation function, $W$ is the weight parameter. A tanh layer is used to output the updated content $\widetilde{c_t}$ and the cell state $c_{t-1}$ at the previous moment $t-1$ is updated to the cell state $c_t$ at time $t$ :

$$\widetilde{c_t} = tanh\left( W_c \left[ h_{t-1}, z_t \right] \right) \tag{4}$$

$$c_t = f_t \times c_{t-1} + i_t \times \widetilde{c_t} \tag{5}$$

where $tanh(\cdot)$ is the hyperbolic tangent activation function, $W_c$ is the weight of the tanh layer. The hidden state at time $t$ can be given by:

$$h_t = o_t \times tanh\left( c_t \right). \tag{6}$$

In order to reconstruct the rating matrix from the deep representation, a full connected layer is defined as:

$$f_s = W_s h_t + b_s, \tag{7}$$

where $W_s$ and $b_s$ are the weight and bias of the full connected layer respectively and $f_s$ is the predicted rating matrix generated by the generator. For learning the optimal parameters $\theta^*$ of the generator, the framework needs to fix the discriminator's parameters and the generator can be optimized by minimizing the following function:

$$\theta^* = \arg \min_{\theta_G} \sum_{u \in \mathcal{U}} \mathbb{E}_{\widetilde{r_i} \sim p_\theta(r_i|t_u, r_u)} \left[ log \left( 1 - D \left( \widetilde{r_i} | r_u, t_u; \phi_D \right) \right) \right]. \tag{8}$$

Since the data generated by the generator in the TagRec is continuous vector in range $[0, 1]$, the gradient descent can be used for training directly.

**Discriminative Model.** The proposed framework adopts multi-layer perceptron as the discriminative model. The purpose of discriminative model is to distinguish the real data from the fake data generated by the generative model. This is actually a simple binary classification problem. Therefore, the discriminative model does not need too complex network, but only needs the non-linear learning ability. Multi-layer perceptron can learn the deep representation of the data well by changing the number of layers and neurons in the neural network [12], and it is easy to train with gradient descent. The discriminator takes the real user rating vector and the predicted rating vector generated by the generator as inputs. It should be noted that different from CFGAN, the real user rating vector defined in TagRec is not the original user rating vector, but the processed vector based on the original user rating vector plus the rating vector of the user's trustees. The function of discriminator can be summarized as follows:

$$a_l = \sigma\Big(W_l a_{l-1} + b_l\Big), \tag{9}$$

where $W_l$ denotes the $l$-th layer's weight and $b_l$ denotes the $l$-th layer's bias. $a_l$ denotes the output of the $l$-th layer and $a_{l-1}$ is the output of the previous layer. In particular, when $l = 1$, $a_{l-1} = a_0$ is the original input data of the discriminator. For learning the optimal parameters, the framework need to fix the generator's parameters and the discriminator can be optimized by maximizing the following function:

$$\phi^* = \arg\max_{\phi_D} \sum_{u \in \mathcal{U}} \Big( \mathbb{E}_{r_i \sim p_{true}(r_i | t_u, r_u)} \left[ \log D(r_i | t_u, r_u; \phi_D) \right]$$

$$+ \mathbb{E}_{\widetilde{r}_i \sim p_\theta(r_i | t_u, r_u)} \left[ \log(1 - D(\widetilde{r}_i | t_u, r_u; \phi_D)) \right] \Big), \tag{10}$$

where $\phi^*$ denotes the optimal parameters of the discriminator. The discriminator can be trained with gradient descent.

### 3.4 Negative Sampling

Through the data pre-process and adversarial learning illuminated above, the framework of TagRec can learn the deep representation between the rating and social information. But as described in CFGAN [2], the vector-wise method will face an output polarizing problem that the generator may simply predict all the outputs in the predicted rating matrix as 1 due to the lack of negative samples. Therefore, negative sampling is adopted and next we will introduce the solution in details.

As shown in the Fig. 1, similar to CFGAN, a mask layer is added on the top of the generator. The layer is defined as a matrix $e_u$, which has the same dimensions $m \times n$ as the output matrix of the generator. The mask layer matrix $e_u$ needs to be generated in two steps. Firstly, a matrix is defined as $e''_u$ which has $m \times n$ dimensions and all the entries in the matrix are 0. Each element $e''_{m,n}$ in the matrix corresponds to the rating that the user $u_m$ rated on the item $i_n$. For the elements corresponding to those items that the user and user's trustees have evaluated, the value is changed to 1 if it is greater than 1 and the new matrix is defined as $e'_u$. The output of the generator is multiplied by the corresponding elements of the matrix $e'_u$. By this step, the framework will only train the existing rating data and ignore the items that the users and corresponding trustees have not evaluated, which is similar to the principle of matrix factorization. Secondly, we select partial data from those items that the users and corresponding trustees have not evaluated as negative samples of the generator. The negative sampling ratio is defined as $s$. For the data sampled from unrated items, the corresponding element in the matrix $e'_u$ is changed from 0 to 1 and the final new mask matrix is defined as $e_u$. After the second step, the framework can not only train the existing data, but also propagate the gradient back to the negative samples during the training process. Since unrated items' rating is stored as 0 in the matrix $TR$, the sampled data is actually equivalent to negative samples of the generator during training. Finally, The output of the generator is multiplied by the corresponding elements of the matrix $e_u$. In this way, the generator can produce low values on the negative items and the reconstructed loss function of the generator is denoted as:

$$\theta^* = \arg\min_{\theta_G} \sum_{u \in \mathcal{U}} \Bigg( \mathbb{E}_{\widetilde{r}_i \sim p_\theta(r_i|t_u, r_u)} \Big[ \log\big(1 - D\left(\widetilde{r}_i \odot e_u|t_u, r_u; \phi_D\right)\big)$$

$$+ \alpha \cdot \sum_j \left(x_{uj} - \widetilde{x}_{uj}\right)^2 \Big] \Bigg), \tag{11}$$

where $\sum_j (x_{uj} - \widetilde{x}_{uj})^2$ is the regularization term, $\alpha$ is the regularization coefficient and $\odot$ presents the element-wise product. The reconstructed loss function of the discriminator is defined as follows:

$$\phi^* = \arg\max_{\phi_D} \sum_{u \in \mathcal{U}} \Bigg( \mathbb{E}_{r_i \sim p_{true}(r_i|t_u, r_u)} \left[ \log D(r_i|t_u, r_u; \phi_D) \right]$$

$$+ \mathbb{E}_{\widetilde{r}_i \sim p_\theta(r_i|t_u, r_u)} \left[ \log(1 - D(\widetilde{r}_i \odot e_u|t_u, r_u; \phi_D)) \right] \Bigg). \tag{12}$$

During the training stage, the discriminator and the generator are trained alternatively in an adversarial manner via Eq. (12) and Eq. (11), respectively. The pseudo code of the proposed TagRec is presented in Algorithm 1 as follows:

---

**Algorithm 1.** The procedure of TagRec
_____
**Input**: Trustee rating matrix $TR$, learning rate for generator $G$ and discriminator $D$, the real user rating vector, minibatch size for training.
**Output**: The predicted users' rating on items.
 1: Data pre-processing.
 2: Calculate similarity and select $k$ trustees.
 3: Initialize model parameters $G_\theta$ and $D_\phi$.
 4: **repeat**
 5:    **for** Each $u \in \mathcal{U}$ **do**
 6:        Negative sampling.
 7:    **end for**
 8:    **for** G-steps **do**
 9:        Sample minibatch of users.
10:        Generate predicted rating vector $\widetilde{r}_i$.
11:        Update $G_\theta$ by gradient descent.
12:    **end for**
13:    **for** D-steps **do**
14:        Use current $G_\theta$ to generate negative samples.
15:        Get positive samples sampled from true data.
16:        Update discriminator $D_\phi$ by gradient descent.
17:    **end for**
18: **until** Convergence.
_____

# 4 Experimental Setup

## 4.1 Datasets

**Table 1.** Characteristics of the datasets.

| Datasets | FilmTrust | Ciao |
|---|---|---|
| Users | 1,508 | 7,357 |
| Items | 2,071 | 99,746 |
| Ratings | 35,497 | 278,483 |
| Density of ratings | 1.14% | 0.0379% |
| Social relations | 1,853 | 111,781 |
| Sparsity of social relations | 0.0814% | 0.2065% |

We conduct the experiments based on two real-world datasets: FilmTrust [10] and Ciao [11]. Both datasets include user rating information of items and social information between users. The detailed statistics are summarized in Table 1. For each dataset, we randomly split the whole data into a training set (80%) and a testing set (20%).

## 4.2 Implementation Details and Evaluation Metrics

For the generator with dynamic RNN, we adopt a single-layer LSTM with 500 hidden neurons. We set the maximum number of iterations to 200. For both the generator and discriminator, we use stochastic gradient descent with learning rate $1 \times 10^{-4}$ to optimize the parameters. The number of the selected users $k$ is set to 5 and the negative sampling ratio $s$ is set to 0.001. In training process, we conduct mini-batch training with batch size 128. We employ two widely-used evaluation metrics to evaluate the top-N recommendation performance of the proposed TagRec, including Normalised Discounted Cumulative Gain ($NDCG@N$), Mean Average Precision ($MAP@N$). $N$ is set to 5, 10 and 20.

## 4.3 Comparison to Baselines

We compare the performance of the proposed TagRec with the following benchmarks.

BPR [8]: It provides personalized recommendations by optimizing the ordering between two items based on Bayesian analysis. It is a baseline for top-N recommendation.

SBPR [9]: It is an improved algorithm based on BPR, which uses social connections to improve accuracy of recommender systems.

TBPR [5]: It is also an improved BPR. The difference between TBPR and SBPR is that the former distinguishes the different effects of strong and weak ties on the recommender systems.

IRGAN [3]: It is the first work combining with generative adversarial networks for recommender systems.

CFGAN [2]: It is a vector-wise method, which points out the problem existing in the adversarial learning recommender system based on policy gradient.

# 5    Experimental Results and Analysis

## 5.1    Experimental Results

**Table 2.** Performance comparison of different recommender systems.

| Datasets | Metrics | Algorithms | | | | | |
|---|---|---|---|---|---|---|---|
| | | BPR | SBPR | TBPR | IRGAN | CFGAN | TagRec |
| Filmtrust | NDCG@5 | 0.3711 | 0.4135 | 0.4179 | 0.4571 | 0.4619 | **0.4718** |
| | NDCG@10 | 0.4232 | 0.4574 | 0.4673 | 0.5093 | 0.5120 | **0.5251** |
| | NDCG@20 | 0.4974 | 0.5039 | 0.5120 | 0.5307 | 0.5688 | **0.5824** |
| | MAP@5 | 0.3044 | 0.3397 | 0.3450 | 0.3622 | 0.3792 | **0.4179** |
| | MAP@10 | 0.3183 | 0.3586 | 0.3655 | 0.4038 | 0.4062 | **0.4282** |
| | MAP@20 | 0.3645 | 0.3871 | 0.4033 | 0.4403 | 0.4529 | **0.4755** |
| Ciaos | NDCG@5 | 0.0124 | 0.0131 | 0.0136 | 0.0145 | 0.0152 | **0.0158** |
| | NDCG@10 | 0.0139 | 0.0143 | 0.0152 | 0.0169 | 0.0189 | **0.0197** |
| | NDCG@20 | 0.0187 | 0.0211 | 0.0208 | 0.0223 | 0.0232 | **0.0241** |
| | MAP@5 | 0.0087 | 0.0096 | 0.0103 | 0.0115 | 0.0120 | **0.0127** |
| | MAP@10 | 0.0101 | 0.0110 | 0.0108 | 0.0121 | 0.0125 | **0.0132** |
| | MAP@20 | 0.0112 | 0.0123 | 0.0118 | 0.0127 | 0.0135 | **0.0142** |

The performance of all recommendation algorithms on two real-world datasets is shown in Table 2. From the experimental results, we can draw the following findings:

(1) TagRec outperforms BPR. This indicates that TagRec can learn data distribution better because the proposed TagRec based on deep learning can fit non-linear data distribution better.

(2) SBPR and TBPR outperform BPR. That is because BPR only uses the users' ratings on items, while SBPR and TBPR utilize both the users' social relations and ratings on items. At the same time, TagRec obtains better performance than SBPR and TBPR. This indicates that the TagRec with adversarial training is promising on the task of recommender systems.

(3) IRGAN and CFGAN can achieve better performance than SBPR and TBPR due to their excellent ability in learning representations. At the same time, the results show that CFGAN outperforms IRGAN. Both CFGAN and IRGAN are based on generative adversarial network. The difference lies in that IRGAN utilizes policy gradient

for optimization because it generates discrete data while CFGAN generates continuous values and is optimized by stochastic gradient descent. This indicates that discrete items sampling proposed in IRGAN can be further improved in generative adversarial network because the original GAN is designed for differentiable values and CFGAN proposes an appropriate method to solve this problem.

(4) The TagRec outperforms CFGAN in $MAP$ and $NDCG$. Compared with CFGAN, TagRec incorporates social relations. The results show that social relation is significant for recommender systems and the TagRec improves the performance of the recommender system.

## 5.2  Social Relations Analysis

In this section, we investigate the impact of the number of the selected trustees on our proposed approach. We take the FilmTrust dataset and $MAP@20$ as examples. Since the user has a maximum of 60 trustees in FilmTrust dataset, we set $k$ to 1, 3, 5, 10, 20, 30, 40, 50 and 60. As shown in Fig. 2, when we set $k$ to 5, the value of $MAP@20$ reaches the maximum value, and as the value of $k$ increases gradually, the value of $MAP@20$ decreases correspondingly at the same time. Although when $k$ is greater than 30, the value of $MAP@20$ is no longer reduced quickly due to data sparsity, we can find that the time consumption is still increasing. Therefore, selecting the most similar trustees by similarity calculation for a given user is necessary to keep a balance between recommendation performance and time consumption.

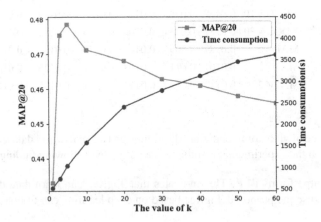

**Fig. 2.** Social relations analysis on FilmTrust dataset.

# 6  Conclusion and Future Work

In this paper, we propose a new recommendation method of combining social relationships based on generative adversarial networks. The deep representation between user-item interaction and social relations is learned by dynamic recurrent neural network with long short-term memory cells. The rating information and social relations are integrated into an input matrix and user similarity is calculated to further improve performance. The experiments on two real-world datasets demonstrate the effectiveness of our method.

In the future, we will focus on extracting reliable users' social relationships more effectively. Although users claim some trustees for himself, this data volume is small and does not necessarily have a positive effect on the recommendation systems. Therefore we plan to study the issue of trustees'confidence in social network in the future.

**Acknowledgments.** This work was supported in part by NSFC grant (No.61772551), the Fundamental Research Funds for the Central Universities (No.18CX07003A), Natural Science Foundation of Shandong Province, China (No.ZR2019MF034), Research Fund for the Creative Research Team of Young Scholars at Universities in Shandong Province, grant No.2019KJN019, the International Exchange Project of China University of Petroleum (No.UPCIEF202052).

# References

1. Goodfellow, I., et al.: Generative adversarial nets. In: Advances in Neural Information Processing Systems, Montreal, Canada, pp. 2672–2680 (2014)
2. Chae, D.K., Kang, J.S., Kim, S.W., Lee, J.T.: CFGAN: a generic collaborative filtering framework based on generative adversarial networks. In: 27th ACM International Conference on Information and Knowledge Management, pp. 137–146. ACM, Torino (2018)
3. Wang, J., et al.: Irgan: a minimax game for unifying generative and discriminative information retrieval models. In: 40th International ACM SIGIR Conference on Research and Development in Information Retrieval, pp. 515–524. ACM, Tokyo (2017)
4. Wang, H.W., et al.: Graphgan: graph representation learning with generative adversarial nets. In: 32nd AAAI Conference on Artificial Intelligence, New Orleans, USA (2018)
5. Wang, X., Lu, W., Ester, M., Wang, C., Chen, C.: Social recommendation with strong and weak ties. In: 25th ACM International on Conference on Information and Knowledge Management, pp. 5–14. ACM, Indianapolis (2016)
6. Wu, C.Y., Ahmed, A., Beutel, A., Smola, A.J., Jing, H.: Recurrent recommender networks. In: 10th ACM International Conference on Web Search and Data Mining, pp. 495–503. ACM, Cambridge (2017)
7. Koren, Y., Bell, R., Volinsky, C.: Matrix factorization techniques for recommender systems. Computer **42**(8), 30–37 (2009)
8. Rendle, S., Freudenthaler, C., Gantner, Z., Schmidt-Thieme L.: BPR: Bayesian personalized ranking from implicit feedback. In: 25th Conference on Uncertainty in Artificial Intelligence, Montreal, Canada, pp. 452–461 (2009)
9. Zhao, T., McAuley, J., King, I.: Leveraging social connections to improve personalized ranking for collaborative filtering. In: 23rd ACM International Conference on Information and Knowledge Management, pp. 261–270. ACM, Shanghai (2014)
10. Guo, G., Zhang, J., Yorke-Smith, N.: A novel bayesian similarity measure for recommender systems. In: 23rd International Joint Conference on Artificial Intelligence, Beijing, China, pp. 2619–2625 (2013)
11. Guo, G., Zhang, J., Thalmann, D., Yorke-Smith, N.: ETAF: an extended trust antecedents framework for trust prediction. In: International Conference on Advances in Social Networks Analysis and Mining, Beijing, China, pp. 540–547 (2014)
12. Li, P.J., Wang, Z.H., Ren, Z.C., Bing, L.D., Lam, W.: Neural rating regression with abstractive tips generation for recommendation. In: 40th International ACM SIGIR Conference on Research and Development in Information Retrieval, pp. 345–354. ACM, Tokyo (2017)

13. Hidasi, B., Karatzoglou, A., Baltrunas, L., Tikk, D.: Session-based recommendations with recurrent neural networks. In: 4th International Conference on Learning Representations, San Juan, Puerto Rico (2016)
14. Liu, C.Y., Zhou, C., Wu, J., Hu, Y., Guo, L.: Social recommendation with an essential preference space. In: 32nd AAAI Conference on Artificial Intelligence, New Orleans, USA (2018)
15. Liu, Q., Wu, S., Wang, D.Y., Li, Z.K., Wang, L.: Context-aware sequential recommendation. In: 16th International Conference on Data Mining, Barcelona, Spain, pp. 1053–1058 (2016)
16. Bharadhwaj, H., Park, H., Lim, B.Y.: Recgan: recurrent generative adversarial networks for recommendation systems. In: 12th ACM Conference on Recommender Systems, pp. 372–376. ACM, Vancouver (2018)
17. Jaccard, P.: The distribution of the flora in the alpine zone 1. New Phytologist **11**(2), 37–50 (1912)

# Software Entity Recognition Method Based on BERT Embedding

Chao Sun⑩, Mingjing Tang⑩, Li Liang⑩, and Wei Zou[✉]⑩

Yunnan Normal University, Kunming, China
zouwei@ynnu.edu.cn

**Abstract.** The global open source software ecosystem contains rich information in the field of software engineering. The existing analysis methods for the text content of the knowledge community in this field are mainly focus on the structural relationship and rule-based association and mining. This paper proposes a software entity recognition method based on BERT word embedding. Firstly, the BiLSTM-CRF model is constructed, and the entity recognition model is constructed by combining the word vector embedding in software engineering field. Then, the word vector in the input layer of the model is improved by introducing the BERT pre-training language model. In the process of pre-training of BERT, the pre-training data should be constructed based on the discussion content of Stack Overflow software Q & A community. Then, we use these data to pre-training the BERT model, so as to obtain the word vector representation suitable for software engineering field, improving the effect of entity recognition in software engineering field, and solving the problem that the traditional word vector embedding is mostly based on the general domain data training, which is not fully suitable for software engineering field, and can't well represent the context semantic information. At the same time, to solve the problem that there are few annotated data in the field of software, this paper tries to extends the data appropriately by the method of model prediction and dictionary matching, and carries out experimental test. Finally, this paper uses the method of deep learning to realize the entity recognition in the field of software engineering, so as to provide support for the extraction of software entities, the construction of software knowledge base, and the intelligent application of software engineering.

**Keywords:** Entity recognition · BERT model · Stack overflow

## 1 Introduction

In recent years, with the rapid development of social programming websites and software knowledge communities, software development has entered the era of networking and popularization [1]. For example, Stack Overflow, a well-known software knowledge community, provides more than 50 million software developers with software related Q & A platforms for development technology, configuration management, project organization, etc., and plays an important role in knowledge sharing and acquisition among software developers [2]. The basic function of these websites is to search for discussion

© Springer Nature Switzerland AG 2020
X. Chen et al. (Eds.): ML4CS 2020, LNCS 12488, pp. 33–47, 2020.
https://doi.org/10.1007/978-3-030-62463-7_4

about specific software entities (such as libraries, tools, APIs), so as to find the way to solve the errors and the use of tools. At present, most of the search methods in web sites use topic model, vector space model or neural network language model to index content, and process software entities in the same way as other conventional text content, which may lead to the content index is not very ideal, because the traditional information retrieval is not suitable for identifying specific entities in key areas. The most ideal form of information extraction and search in the software knowledge base is to realize the association between entities in the knowledge base in the form of knowledge map. The system can't only use the knowledge map to find the discussion about specific software entities, but also query and display some other information about the central entity [3]. In order to achieve this goal, it is necessary to identify specific software entities in the field of software engineering. However, most of the existing researches on named entity recognition (NER) are to identify and classify the proper names such as person name, place name and meaningful quantitative phrases such as time and date in the text, and there is not a good entity recognition model to be used in the field of software engineering.

Under such a realistic background, in order to build a knowledge map in the field of software engineering and support the development of relevant applications centered on software entities, this paper proposes a software entity recognition algorithm based on BERT [4] embedding. This method uses the social text of software engineering domain to pre-train the BERT model, so as to obtain the word vector embedding suitable for software engineering field in the process of entity recognition model construction. It solves the problem that the traditional word vector embedding is mostly based on the general domain data training, and can't well represent the context semantic information.

The organizational structure of this paper is as follows: Sect. 2 introduces the related work in this research field, and Sect. 3 describes the network model used to construct the entity identification method and the process of constructing extended data. Section 4 validates the classical BiLSTM-CRF model and the BERT-based entity recognition model and analyses the results. Finally, Sect. 5 summarizes the full text and proposes forward the next work direction.

## 2   Related Work

The main work of this paper is to implement entity recognition method in software engineering field based on the BERT word embedding tool, so as to provide support for software entity extraction, construction of software knowledge base and intelligent application of software engineering. There are many researches on social text in the field of software engineering. For example, the paper [5] studies the relevance of problems in stack overflow, and article [6] studies a framework to mine the usage scenarios of API from Stack Overflow. The paper [7] classifies the sentiment of text in Stack Overflow and JIRA by using multi tag method. The paper [8] constructs a technology called CAPS to classify API related posts in Stack Overflow. Although there are many researches on social text in software engineering, there are few researches on entity recognition in this field. We only find Ye et al.'s [3] article, which realizes entity recognition method in software engineering field by means of machine learning based on information such as dictionary and clustering characteristics. However, in the general domain or some other specific fields, there are many researches on entity recognition at home and abroad. In 2015, Santos et al. [9] adding character features to enhance the recognition effect on the basis of CNN-CRF model. In 2017, Strubell et al. [10] proposed to use Iterated Dilated Convolutional Neural Networks (IDCNN-CRF) for named entity recognition, which can speed up training and be more accurate. In 2018, Maimaiti et al. [11] proposed the BiLSTM-CNN-CRF model according to the characteristics of uygur language. Shen et al. [12] proposed a lightweight model based on CNN-CNN-LSTM structure, which uses deep active learning for named entity recognition task. Bharadwaj et al. [13] proposed an Attention neural network model. The model adds phonological features to the original BiLSTM-CRF model and uses attentional mechanism on character vectors to focus on and learn more effective characters. The model can be quickly applied to new language domains with little data or no data, so as to achieve cross-language transfer learning. In 2019, Zhang et al. [14] proposed to combine the generated antagonism network with the BiLSTM-Attention-CRF model for entity recognition. In 2020, Wang et al. [15] proposed a LSTM-CNN system for antagonism training to improve the generalization ability and robustness of the model, and improve the overfitting of the neural network. Entity recognition is also applied in a wide range of scenarios, Wang et al. [16] proposed a named entity recognition method based on bidirectional GRU and applied it to the task of meeting name recognition. Li et al. [17] applied the CNN plus BiLSTM-CRF model to biomedical field and obtained the highest F1 value in this field. Zhou et al. [18] proposed a SVM-BiLSTM-CRF model for entity extraction of judgment documents, which mainly extracts three types of entities: movable property, real property and intellectual property. The model uses SVM to judge sentences with keywords at first, and then input the sentences into BiLSTM-CRF model for extraction. Yang et al. [19] proposed two models, IndRNN-CRF and IDCNN-BiLSTM-CRF, for the text of online medical websites, and the experiments showed that these two models are superior to the classical BiLSTM-CRF model in this dataset.

The above research on entity recognition has made some progress in recognition effect and training time, and has been applied in the fields of meeting name, biomedicine, law and so on. However, due to the differences of language features and technical terms between different fields, it is better to have special word vector representation for different research fields, and the traditional word vector training models (such as: word2vec, glove, etc.) can't well represent the context information of words. Therefore, in the related research of entity recognition, we need to explore a new word vector training model.

As mentioned above, it is of great significance to study the expression of word vectors suitable for specific domain entity recognition model. Since the research on entity recognition in the field of software engineering only finds the related research in article [3], this paper compares the experimental results of article [3] to verify that the software entity recognition method based on BERT proposed in this paper can achieve better results. The main work of this paper is to firstly construct BiLSTM-CRF, a classic entity recognition model, to verify the entity recognition effect in the field of software engineering, and then introduce BERT word embedding tool to improve the word embedding representation. In the process of BERT introduction, text information in the field of software engineering should be acquired first, and pre-processing operations such as filtering and removing HTML tags. Then the text data were pre-trained by the BERT pretraining language model. Finally, the pre-trained model is applied to the entity recognition model to obtain the representation of the word vector, so as to improve the effect of the entity recognition model.

## 3   Main Models

We first tried the training effect of classical LSTM-CRF model in software engineering field, and improved the model by using bidirectional LSTM to encode semantic information, and According to the idea of transfer learning, the word embedded dictionary in the field of software engineering is introduced. Finally, the BERT word embedding is introduced into this basic model. Since our results are compared with those of article [3], the experiment uses the same data as article [3]. Finally, we use our method to expand the data.

### 3.1   Initial Model BiLSTM-CRF

Long Short-Term Memory (LSTM) neural network is a special kind of Recurrent Neural Network, which changes the cellular structure of RNN, can memorize the previous information and can apply the previous information in the current calculation. It has a good effect in processing the long distance sequence. Adding a layer of reverse LSTM on the basis of LSTM network can form bidirectional LSTM neural network (BiLSTM), in which the forward LSTM network can acquire and memorize the above information, and the reverse LSTM network can acquire the following information, so the bidirectional LSTM can more fully extract the context semantic information of the input statement [20]. The LSTM network structure is shown in Fig. 1:

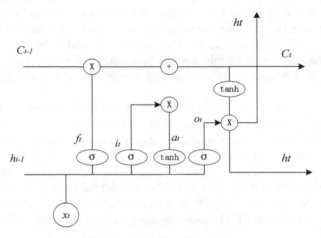

**Fig. 1.** LSTM network structure.

The specific calculation process is shown in the formula:

$$f_t = sigmoid\left(U_f x_t + W_f h_{t-1} + b_f\right) \tag{1}$$

$$i_t = sigmoid(U_i x_t + W_i h_{t-1} + b_i) \tag{2}$$

$$a_t = tanh(U_a x_t + W_a h_{t-1} + b_a) \tag{3}$$

$$C_t = C_{t-1} \cdot f_t + i_t \cdot a_t + b_i \tag{4}$$

$$o_t = sigmoid(U_o x_t + W_o h_{t-1} + b_o) \tag{5}$$

$$h_t = o_t \cdot tanh(C_t) \tag{6}$$

Where $i$ is the input gate, $f$ represent the forgetting gate, $o$ is the output gate, and $\bullet$ is the point multiplication operation. $x_t$ represents the input vector at time $t$. $h_t$ represents both the hidden state and the output vector, which Contains all valid information from the previous t-moment. $U$, $W$ represent the weight matrix connecting the two layers respectively. $b$ is the bias vector, $C$ is the state of memory unit, sigmoid and tanh are two different kinds of neuron activation functions [21].

BiLSTM network classifies individual words independently based on context features, but the results do not consider the constraints relationship between words in sentences. conditional random field (CRF) can consider the particularity between output tags, so it is necessary to access the CRF layer after the BiLSTM network to determine the final tagging sequence. The overall network structure of BILSTM-CRF is shown in the figure below (Fig. 2):

Where $wi$ is the vector converted from each word in the sentence, and after the bidirectional LSTM encoding, the output is the possible probability of each label. These probabilities are the input of CRF, and after being labeled at the sentence level of CRF layer, some restrictions are added to the output results of BiLSTM output layer to output the final label sequence. The original formula of CRF is as follows:

$$P(y|x) \propto \exp\left[\sum_{i,k} \lambda_k t_k(y_{i-1}, y_i, x, i) + \sum_{i,l} \mu_l s_l(y_{i-1}, x, i)\right] \tag{7}$$

The formula shows the probability that the random variable $Y$ is $y$ when the random variable is $x$, where $t_k$ is the transfer feature function, the conditional probability from the previous label to the current label is calculated, $s_l$ is the state feature function, $\ddot{e}_k$ and $ì_l$ represent the corresponding weights.

## 3.2 BERT Model

We improved on the basic model, and added the BERT pre-training language model before the BiLSTM layer. BERT is a new language representation model represented by the transformer's bidirectional encoder. By jointly adjusting the context in all layers,

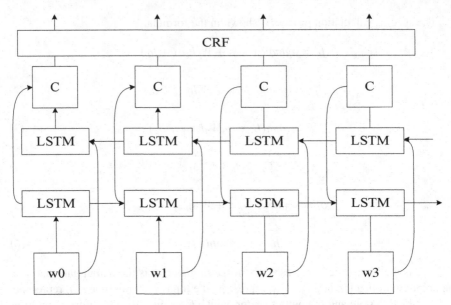

**Fig. 2.** BiLSTM-CRF network structure.

we can pre-train the deep bidirectional representation, which is suitable for natural language processing tasks (such as sequence tagging and text classification) at the word and sentence level. Compared with other word embedding models, it can obtain the semantic features at the sentence level and the context. The structure of BERT pre-training language model is shown in the Fig. 3:

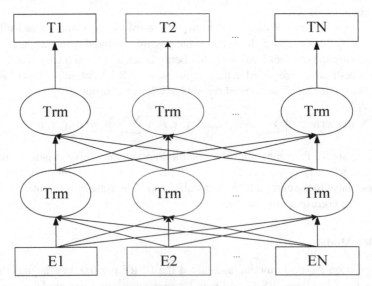

**Fig. 3.** structure of BERT model.

As shown in Fig. 4, the most important structure of BERT is the bidirectional transformer encoding structure. Its input is composed of three parts: the word information represented by token embeddings, the sentence information represented by segment embeddings and the location information represented by position embeddings. The input representation of BERT model as shown in Fig. 4:

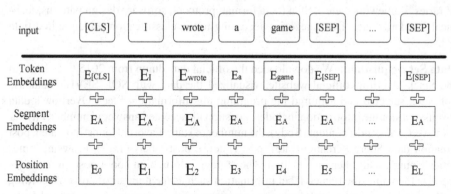

**Fig. 4.** Input representation of BERT model.

Two unique unsupervised tasks are used in the pre-training phase of BERT, the Masked Language Model task and the Next Sentence Prediction task..

"Masked language model" will randomly cover 15% of the words in each sentence, and use its context to make prediction. Due to the large number of covered words, in order to avoid some words never seen in the fine tuning stage, the following processing will be done to the covered words:

(1)  80% of them use [mask].
(2)  10% to find a random word to replace.
(3)  10% of the words remain unchanged.

"Next Sentence Prediction" select some sentence pairs A and B, Fifty percent of the data B is the next sentence in A, and the rest of the data is randomly combined. Adding such pre-training can learn the correlation between two sentences, making the model more suitable for natural language processing tasks that rely on the relationship between two sentences, such as QA, NLI, etc. [4].

### 3.3  Data Set Expansion

The paper [3] provides a small annotated data set for entity recognition in software engineering field, but the final learning effect of deep learning has a great relationship with the size of training data, and most deep learning models need large amount of training data. Social programming websites contain rich information about software engineering, we obtain the Q & A dump information from a specific social programming website, select the Q & A information according to the degree of attention of the Q & A information, the quality of the questions and answers, and pre prepare the text information. Then, according to our method, we annotate the corpus to supplement the small-scale annotated dataset. Entity classification we still classify entities into API (class, package, common methods, the built-in module, etc.), PL (object-oriented, scripts, web development and other programming languages), Plat (CPU instruction set, hardware architecture, operating system, etc.), Fram (development tools, development libraries, frameworks, etc.), Stan (data format, protocol, software design patterns, technical standards, etc.) five categories according to the classification principle in article [3].

**Data Set Acquisition.** We downloaded the official data dump from 2015 to 2018 from the Stack Overflow official website, visualized the data with SQL SEVER database software, selected and acquired the programming information. Stack Overflow's data discussion storage information includes: question id, accepted answer id number, answer (AnswerCount), post content (Body), the number of stars (FavoriteCount) indicated, post creation date Score (CreationDate), post (Score), post labels (Tags), browsing times (ViewCount), and other information. The storage information about post tag includes tag ID, Tag name, the related label information such as the number of posts. According to the number of posts on relevant tags, we determine the popularity of the tagged posts, and select the tags with the highest popularity to mark the posts. We determine the quality of the posts based on whether the questions are accepted, the score of the questions, the score of the answers and the number of times the posts are viewed, so as to select the posts with better quality. The specific process is as follows (Fig. 5):

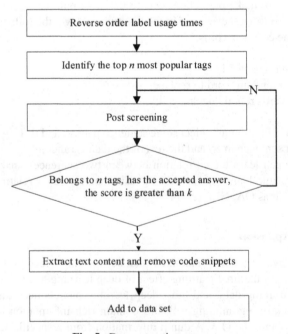

**Fig. 5.** Data extraction process

First, the tags are sorted according to the usage frequency and the top $n$ tags are determined. Then, the posts marked as the top $n$ tags with accepted answers and scores greater than $k$ are filtered from the post storage information. Finally, the post is parsed, removed tags and other related preprocessing operations and saved to the data set.

**Quick Labeling Training Set.** first of all to build the entity word dictionary corresponding to the entity category, extract the training set and preliminary annotate through the method of dictionary matching, and then on the basis of small-scale corpus application transfer learning methods constructing the basic entity recognition model, When the model recognition effect was good, the label of the unlabeled dataset was predicted by the model to expand the small dataset. Finally, the data marked by the dictionary and the data marked by the model prediction were integrated, it is applied to the entity recognition model which needs a larger corpus and has better recognition effect to further optimize the recognition effect of the basic model.

*Construct Entity Classification Dictionary.* Ye et al. [3] provide a similar data dictionary, but there are not only some words that are both entity words and common words in the dictionary, also some entity words appear in different entity categories due to the different context of use, and the dictionary is not very good for informal text discussion using different forms of entity words such as abbreviations, synonyms and spelling errors, so the dictionary can't meet our requirements. In order to build a dictionary for our matching and annotation methods, we improved it on the basis of the dictionary. Firstly, we use Wikipedia, stack overflow and other tools to supplement entity words. At the same time, we delete some common words and phrases by stopping thesaurus and common English dictionaries, and then manually browse and check to remove the undeleted common words. Then, in view of the use of entity abbreviations and synonyms in the discussion, we use the relevant knowledge base to expand the entity dictionary. For example, the SEthesaurus dictionary constructed in article [22], which excavates synonyms of common forms of software specific terms from informal software engineering discussions. The dictionary covers a large number of software specific terms, abbreviations, and synonyms just to meet our needs of expanding the dictionary. We use the thesaurus API provided by this article to query the abbreviations and synonyms of the words in the software dictionary we built, and then use the web crawler technology to crawl the words found and add them to our classification entity dictionary to expand the entity words. Finally, we compare the entities in several classification dictionaries, extract common entity words of multiple classifications and save them separately, so that we can check the accuracy of the entities manually and adjust them.

*Dataset Filtering.* Due to the large number of software entities, we can't guarantee to collect all entity words. In order to ensure that the data set Posts contain the entity words in our constructed dictionary, avoid the interference of a large number of non-entity words posts on our annotation effect, and increase the workload of later manual inspection, we will further screen the data set and select the ones containing the entity words in the software dictionary Post or sentence, delete useless information.

*Dictionary Matching Annotation.* Through the entity dictionary we built in the previous step, we preliminarily annotate the data set we obtained. We adopt the general text annotation strategy of BIO, where 'B-' plus the corresponding entity name indicates the beginning of the entity, 'I-' plus the corresponding and 'O' represents the non-entity. For example, the sentence "This only works on Windows 2000" should be marked "O O O O O O B-Plat I-Plat", where "Windows 2000" is an entity word of Plat class, so the first word of entity word begins with "B-" and adds Plat of corresponding entity category, while other parts of entity word begin with "I-" and adds Plat of corresponding entity category. The remaining nonentity words are labeled "O".

*Basic Model Forecast Annotation and Manual Integration.* In the early stage of the experiment, we firstly constructed the entity recognition model of BiLSTM-CRF, learned the features of small labeled data sets, and made label prediction on the dictionary labeled data. We compare the result of annotation by dictionary with the result of model prediction, mark the dictionary as "O" and predict the different annotation results marked as non-O at the same time, ignore the same result, and then check the different annotation results manually. Because the construction of data set is based on the entity dictionary, most entities have been marked, which greatly reduces the workload of manual inspection and improves the annotation efficiency. At the same time, it checks the words in the ambiguous dictionary manually to avoid the marking error of the ambiguous words.

## 4   Experiment and Analysis

### 4.1   Experimental Data

The experiment uses the small annotation data set in article [3] to compare the experimental results of BiLSTM-CRF model and model after the introduction of BERT word embedded tool. The data set adopts "B I O" annotation strategy, in which 'B-' plus the corresponding entity name indicates the beginning of the entity, 'I-' plus the corresponding entity name indicates the inside of the entity, and 'O' represents the non-entity. For example, the sentence "I developed a software with Android Studio" is labeled "O O O O O B-Fram I-Fram". The dataset consists five types entities, namely API, PL, Plat, Fram and Stan, with a total of 4,646 sentences and 70,570 words. The statistics of the number of various entities in the data set are shown in Table 1:

**Table 1.**  Entity count of small annotation dataset

|              | Total | API | PL  | Plat | Fram | Stan |
|--------------|-------|-----|-----|------|------|------|
| Entity count | 2578  | 983 | 604 | 91   | 590  | 291  |

In order to minimize the impact on the characteristics of the original manual annotation data, we expanded the label data to a scale of less than 1:1. The expanded dataset contains a total of 8600 sentences and 130640 words. The statistics of the number of entities in the expanded dataset are shown in Table 2:

**Table 2.** Entity count of expanded data set

|              | Total | API  | PL   | Plat | Fram | Stan |
|--------------|-------|------|------|------|------|------|
| Entity count | 7994  | 2144 | 1566 | 1134 | 2071 | 1079 |

## 4.2 Evaluation Index

We choose general entity recognition evaluation indexes, which include precision, recall and F1-Score. The calculation formula is as follows:

$$P = \frac{T_P}{T_P + F_P} \times 100\% \tag{7}$$

$$R = \frac{T_P}{T_P + F_n} \times 100\% \tag{8}$$

$$F1 = \frac{2PR}{P + R} \times 100\% \tag{9}$$

Where $T_P$ is the number of entities correctly identified by the model, $F_P$ is the number of unrelated entities identified by the model, and $F_n$ is the number of related entities not detected by the model.

## 4.3 Experimental Process and Results

First, we test the entity recognition effect of the basic model BiLSTM-CRF. We embed the universal glove word vector. We divide the datasets into 10 parts, one tenth of which is the test set each time, and the rest is the training set, loop executes ten times, and take the average value of the final ten evaluation indexes as the evaluation result of the model. The results are shown in Table 3:

**Table 3.** Experimental results of BiLSTM-CRF

| Label   | Precision | Recall | F1    |
|---------|-----------|--------|-------|
| API     | 55.23     | 68.86  | 61.27 |
| PL      | 91.69     | 88.01  | 89.64 |
| Plat    | 77.68     | 66.77  | 71.29 |
| Fram    | 54.42     | 64.89  | 59.14 |
| Stan    | 75.72     | 57.30  | 64.59 |
| **Overall** | **64.43** | **70.61** | **67.37** |

First, we improve the word vector. We introduce the word vector embedding Dictionary (SO_word2vec) of the software engineering field published in article [23].

SO_word2vec word embedding is obtained by training Stack Overflow more than 15 GB data posts with word2vec, which makes up for the shortcomings of the existing small annotation data set. The model training effect after the introduction of software engineering words is shown in Table 4:

**Table 4.** Experimental results of SO_word2vec-BiLSTM-CRF

| Label | Precision | Recall | F1 |
| --- | --- | --- | --- |
| API | 71.33 | 68.53 | 69.40 |
| PL | 92.88 | 95.68 | 93.72 |
| Plat | 79.75 | 78.15 | 77.86 |
| Fram | 73.20 | 72.24 | 72.56 |
| Stan | 88.30 | 73.10 | 75.67 |
| **Overall** | **78.40** | **76.86** | **77.54** |

In view of the traditional word vector representation method can't represent the polysemy of words and the excellent performance of the BERT pre-training language model in various natural language processing tasks, we add the BERT pre-training language model on the basis of the basic network model, use the two-way transformer structure of BERT to dynamically generate the context semantic representation of words in software engineering texts, so as to further improve the recognition effect of software engineering domain entity recognition model. There are two BERT pre-training language models in English, namely BERT-base and BERT-large. We adopted the BERT-base model, which has 12 layers in total, the hidden layer has 768 dimensions, and the 12-head model is adopted, with a total of 110 M parameters. First, we use the question and answer text in the field of software engineering to do further pre-training on the basis of this model, which makes the model more applicable to the field of software engineering. During the pre-training process, 2656719 questions and 5526559 answers from the top eight tags with the highest usage were extracted from the Stack Overflow website. After combining the questions and answers, a total of 7.93 GB file was obtained. After removing the code snippets and HTML tags, the remaining data size was 3.2 GB. These data are then pre-trained using the pre-training program provided by BERT, and after several trials, the pre-training parameters are finally set to: "max_Seq_Length":128, "learning_Rate":2e-4, "num_Train_Steps":600000. After the training is completed, word vectors corresponding to training data and test data were obtained from the pre-training model and then spliced with the word vectors obtained from SO_word2vec and then input into the BiLSTM-CRF model. In the process of model training, the optimizer uses "Adam". After many experiments, the final "learning_Rate" was set to 0.01 and the "droup_out" was set to 0.5. The final experimental results are shown in Table 5:

Finally, the expanded data is applied to the BiLSTM-CRF experimental model to make an attempt, and the final effect is shown in Table 6:

**Table 5.** Experimental results of BERT

| Label | Precision | Recall | F1 |
|---|---|---|---|
| API | 74.98 | 71.44 | 72.71 |
| PL | 93.45 | 94.13 | 93.72 |
| Plat | 76.70 | 82.33 | 78.26 |
| Fram | 76.10 | 76.29 | 75.96 |
| Stan | 84.43 | 72.45 | 77.55 |
| **Overall** | **80.71** | **78.40** | **79.51** |

**Table 6.** Experimental results of BERT

| Label | Precison | Recall | F1 |
|---|---|---|---|
| API | 70.24 | 70.53 | 70.44 |
| PL | 93.68 | 94.77 | 94.22 |
| Plat | 80.52 | 78.33 | 79.30 |
| Fram | 72.90 | 71.43 | 72.17 |
| Stan | 89.30 | 72.10 | 77.78 |
| **Overall** | **80.32** | **77.43** | **78.38** |

## 4.4 Experimental Analysis

By comparing the BiLSTM-CRF model based on the universal glove word vector with the BiLSTM-CRF model based on the word vector embedding trained in the field of software engineering, we can know that the word vectors in different fields have a great influence on the entity recognition model in this field, and the word embedding for specific fields can better express the characteristics of the words in this field than the universal word embedding vector, so as to better learn this Some characteristics. Compared with the experimental results of the entity recognition model embedded by BERT, it can be found that the experimental effect of the model is not only much better than the BiLSTM-CRF model based on the general word vector, but also about 2% higher than the F1 value of the experimental results embedded by the word vector in the field of software engineering, compared with article [3] final F1 value of 78.17%, the improvement is about 1.3%. which shows that the word vector generated by the BERT pre-training language model through context information can better represent the semantic information of words. At the same time, the attempt to expand the experiment is slightly higher than that of the original data, which indicates that the expansion of training data does have some influence on the experimental results.

## 5   Conclusion

In view of the lack of a good entity recognition model in the field of software engineering, a method of entity recognition in the field of software engineering based on deep learning is proposed. Firstly, the experimental effect of the classic entity recognition model, BiLSTM-CRF, is verified, and make improvement on the model, Finally, BERT pre-training language model is added to get word embedding suitable for software engineering, which greatly improves the final experimental effect of the model. Finally, the experiment of data expansion shows that appropriate expansion of small-scale data sets can improve the final model effect, and the method of data expansion needs further optimization and improvement.

Due to the limitation of hardware equipment, our BERT model is only pre-trained on a small data set. If the software engineering training data with the same training scale of SO_word2vec is applied to the existing model through the BERT training word embedding vector, the final experimental results will be greatly improved. At the same time, in the model, we can consider the introduction of graph convolution network (GCN), taking into account the syntax information of sentences, so as to more accurately identify the entity information in the software engineering text.

**Acknowledgment.** This work was supported by Yunnan Key Laboratory of Smart Education, Yunnan Innovation Team of Education Informatization for Nationalities, Scientific Technology Innovation Team of Educational Big Data Application Technology in University of Yunnan Province, and Kunming Key Laboratory of Education Informatization.

## References

1. Sen, R., Singh, S.S., Borle, S.: Open source software success: measure and analysis. Decis. Supp. Syst. **52**(2), 364–372 (2016)
2. Yin, G., et al.: A review of the research on software data mining technology for open source ecology. J. Softw. **29**(08), 2258–2271 (2018)
3. Ye, D.H., Xing, Z.C., Chee, F., Zi, A., Li, J., Nachiket, K.: Software-specific named entity recognition in software engineering social content. In: IEEE International Conference on Software Analysis, pp. 90–101 (2016)
4. Devlin, J., Chang, M.W., Lee, K., et al.: Bert: Pre-training of deep bidirectional transformers for language understanding. arXiv preprint arXiv:1810.04805 (2018)
5. Amirreza, S., Bowen, X., David, L., Solorio, T., Alipour, A.: Question relatedness on stack overflow: the task, dataset, and corpus-inspired models. In: Proceedings of the AAAI Reasoning for Complex Question Answering Workshop (2019)
6. Gias, U., Foutse, K., Chanchal, K.R.: Mining API usage scenarios from stack overflow. Inf. Softw. Technol. **122**, 106277 (2020)
7. Luis, A.C.D., Nik, B., Ioannis, K.: Classifying emotions in Stack Overflow and JIRA using a multi-label approach. Knowl.-Based Syst. **195**, 105633 (2020)
8. Ahasanuzzaman, M., Asaduzzaman, M., Roy, C.K., Schneider, K.A.: CAPS: a supervised technique for classifying Stack Overflow posts concerning API issues. Empirical Softw. Eng. **25**(2), 1493–1532 (2019). https://doi.org/10.1007/s10664-019-09743-4
9. Santos, C.N.D, Guimarães, V.: Boosting named entity recognition with neural character embeddings. Comput. Sci. (2015)

10. Strubell, E., Verga, P., Belanger D.: Fast and accurate entity recognition with iterated dilated convolutions. arXiv preprint arXiv:1702.02098 (2017)
11. Maimathiev, A.Y.F., Umur, S., Paridan, M.: Uighur named entity recognition based on bilstm-cnn-crf model. Comput. Eng. **44**(8), 230–236 (2008)
12. Shen, Y., Yun, H., Lipton, Z.C., et al.: Deep Active Learning for Named Entity Recognition, pp. 252–256 (2008)
13. Bharadwaj, A., Mortensen, D., Dyer, C.: Phonologically aware neural model for named entity recognition in low resource transfer settings. In: Conference on Empirical Methods in Natural Language Processing, pp. 1462–1472 (2016)
14. Zhang, H., Guo, Y.B., Li, T.: Domain named entity recognition combining GAN and BiLSTM-attention-CRF. Comput. Res. Dev. **56**(9), 1851 (2019)
15. Wang, J.N., Xu, W.J., Fu, X.Y., Xu, G.L., Wu, Y.R.: ASTRAL: adversarial trained LSTM-CNN for named entity recognition. Knowl.-Based Syst. **197**, 105842 (2020)
16. Wang, J., Zhang, R.D., Wu, C.S.: Named entity recognition method based on GRU. Comput. Syst. Appl. **27**(09), 18–24 (2008)
17. Li, L.H., Guo, Y.K.: Biomedical named entity recognition based on CNN-BLSTM-CRF model. Chin. J. Inf. (2018)
18. Zhou, X.L., Zhao, X.J., Liu, T.L., Zong, Z.X., Wang, Q.L., Li, J.Q.: Named entity identification method for property dispute based on SVM-BiLSTM-CRF model. Comput. Syst. Appl. **28**(01), 245–250 (2019)
19. Yang, W.M., Chu, W.J.: Named entity recognition of online medical Q & A texts. Comput. Syst. Appl. **28**(02), 8–14 (2019)
20. Li, G., Pan, R.Q., Mao, J., Cao, Y.J.: Entity identification of Chinese electronic medical records integrating BiLSTM-CRF network and dictionary resources. Modern Intell. **40**(04), 3–12 + 58 (2020)
21. Yang, X.M., et al.: Bidirectional LSTM-CRF for biomedical named entity recognition. In: 2018 14th International Conference on Natural Computation, Fuzzy Systems and Knowledge Discovery (ICNC-FSKD) (2018)
22. Chen, C.Y., Xing, Z.C., Wang, X.M.: Unsupervised software-specific morphological forms inference from informal discussions. In: 2017 IEEE/ACM 39th International Conference on Software Engineering (2017)
23. Vasiliki, E., Christos, C., Diomidis, S.: Word embeddings for the software engineering domain. In: 2018 ACM/IEEE 15th International Conference on Mining Software Repositories (2018)

# A Classification Method of Land Cover Based on Support Vector Machines

Kai Ding[1] ⓘ, Chisheng Wang[2](✉) ⓘ, Ming Tao[1] ⓘ, Huijuan Xiao[1] ⓘ, Chao Yang[2] ⓘ, and Peican Huang[1] ⓘ

[1] School of Computer Science and Technology, Dongguan University of Technology, Dongguan 523808, People's Republic of China
{dingkai,huangpc}@dgut.edu.cn, ming.tao@mail.scut.edu.cn, 768105108@qq.com
[2] Department of Urban Spatial Smart Sensing, Shenzhen University, Shenzhen 518060, People's Republic of China
wangchisheng@szu.edu.cn, yangchao161@sina.com

**Abstract.** In this paper, we develop a classification method of land cover based on support vector machines. As a case study, we choose five Landsat images to retrieve land cover maps in Shenzhen, China from 1979 to 2005. The classification method is based on support vector machines with assistance from visual interpretation. And then we take use of the complex network approach to analyze the character of land use-cover change from an overall perspective. The result shows that the main changes of land use-cover are different over time. The medium of bare land during the urban construction can hardly be witnessed, even though the time intervals are shorter than the two periods before. It reveals the transformation from vegetation to urban becomes faster. The transformation from vegetation to bare land is hard to be witnessed in the late stage. As bare land is the medium for transforming vegetation to urban land in Shenzhen during the past years from 1979 to 2005.

**Keywords:** Land cover change · Support vector machines · Remote sensing

## 1 Introduction

It plays an important role to study the rapid and effective land cover classification method for timely updating of land use information. The land cover is the result from the mutual effect between human activities and nature processes. Land cover change can directly bring climate change, affect biological species diversity and the ecological system [1–3]. The research on the land use-cover change (LUCC) can facilitate to understand the situation and mechanism of land environment. It is very important to the rational use of land resource and reconciliation of socio-economic development [4, 5]. Moreover, the rapid development of technologies on remote sensing provide big data and analysis tools for LUCC research. There have been plenty of studies in this field, which focus on LUCC monitor, mechanism, modeling, and so on [6–10].

Shenzhen belongs to Guangdong province, located in the east coast of Pearl River Delta. After the rapid development in recent decades, its economic aggregate has

© Springer Nature Switzerland AG 2020
X. Chen et al. (Eds.): ML4CS 2020, LNCS 12488, pp. 48–54, 2020.
https://doi.org/10.1007/978-3-030-62463-7_5

increased from 196 million in 1979 to 1151 billion in 2011. The development of remote sensing satellites in recent decades offers a good opportunity to capture this process. Studying the spatio-temporal character of Shenzhen LUCC has important implications to give support to government for making the land policy.

There are already some studies on Shenzhen LUCC research, including the LUCC character analysis, forcing driver analysis, dynamic modeling and prediction [11–14]. Some work focused on the modeling method, such as the model improvement and precision validation. Some only used two images while the time series of LUCC cannot be captured.

In this paper, we develop a classification method of land cover based on support vector machines. As a case study, we retrieve Shenzhen land cover maps in five timing points from 1979 to 2005, and then take use of the complex network approach to analyze the character of land use-cover change from an overall perspective. At last, the correlation between urban expansion and socio-economic factors are discussed.

We develop a classification method of land cover based on support vector machines. As a case study, we retrieve Shenzhen land cover maps in five timing points from 1979 to 2005, and then take use of the complex network approach to analyze the character of land use-cover change from an overall perspective. At last, the correlation between urban expansion and socio-economic factors are discussed.

## 2  Data and Method

### 2.1  Data Introduction

The research region covers most Shenzhen city and some surrounding waters. The data used in our study include the Shuttle Radar Topography Mission (SRTM) digital elevation models (DEMs) [15], and Landsat MSS/TM/ETM images (Table 1). The Landsat images are sourced from Global Land Cover Facility in University of Maryland (http://glcf.umd.edu/). The TM false color images in research region are showed in Fig. 1.

**Table 1.** Parameters of the used Landsat images.

| Date | Image type | Resolution | Bands | Wavelength | Orbit | Frame |
|------|-----------|-----------|-------|-----------|-------|-------|
| 1979-10-13 | MSS | 60 m | 4 | 0.5–1.1 μm | 131 | 044 |
| 1990-10-13 | TM | 30 m | 7 | 0.45–2.35 μm | 122 | 044 |
| 2000-09-14 | ETM+ | 30 m | 8 | 0.45–2.35 μm | 122 | 044 |
| 2003-01-10 | ETM+ | 30 m | 8 | 0.45–2.35 μm | 122 | 044 |
| 2005-11-23 | TM | 30 m | 7 | 0.45–2.35 μm | 122 | 044 |

### 2.2  Method

We group the land cover in Shenzhen into four classes, including urban land, vegetation, bare land and water. In some studies, the vegetation is further divided into subclasses

**Fig. 1.** The TM false color image in research region.

like forest land, wetland, orchard, and grassland [16]. With the development of remote sensing techniques, there are lots of classification methods, like decision tree, k-nearest neighbor, and artificial neural network [17–19]. Among these methods, support vector machines algorithm is a machine learning method based on statistic studying theory, which has good performance in the situation with limited training sample. It can solve some classification problems existing in other methods, such as small-sample, devilishly learning, and big-dimension [20]. Previous study has compared the support vector machines with maximum likelihood and artificial neural network in classification on Landsat image [21]. The results reveal support vector machines outperforms other methods in terms of better classification precision. We also tested other methods in our study. Therefore, we adopt support vector machines method to classify the Landsat images in this paper. Here Gaussian function is set as the kernel of support vector machines and the parameter $\gamma$ is set as the reciprocal of band number:

$$K(x_i, x_j) = e^{-\gamma \|x_i - x_j\|^2} \tag{1}$$

The previous classification methods cannot give a satisfactory solution to such problems. In order to ensure the classification accuracy, manual intervention was implemented on the classification. We corrected the affected areas by using of visual interpretation.

## 3   Results

First, we classify the Landsat images based on support vector machines with assistance from visual interpretation. We obtained five land cover maps during the period from 1979 to 2005 (Fig. 2).

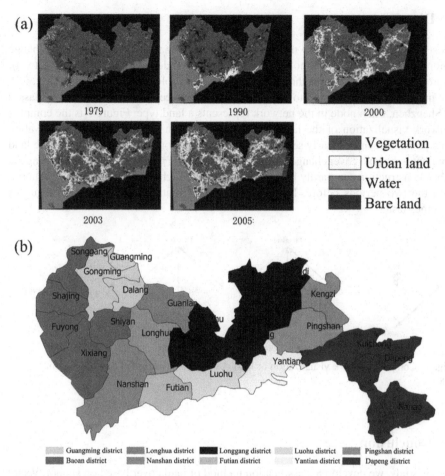

**Fig. 2.** The land cover maps in Shenzhen from 1979 to 2005 (a) and administrative map (b).

Then, we apply a statistical analysis on the land cover change. It is found there is an acceleration in the rate of urban land growth from 1979 to 2003. The vegetation drops the most rapidly from 2000 to 2003. It shows the urbanization in Shenzhen is implemented at a cost of vegetation reduction. The water area types in our study region include reservoir, river and sea. Figure 2 shows that the water areas exhibit the trend of falling down from 1979 to 2003. It indicates that the reclamation activity is the main factor to affect the water area in Shenzhen. Between 2000 and 2003, the increase of urban land mainly happened on Baoan and Longgang districts. In 2005, the land use change is still dominated by urbanization in outlying districts and reclamation in central districts.

## 4  Discussions

Most studies use the transfer matrix to describe the characters of land cover change among classes, but it is still not intuitive and easily understood, the complex network approach is an important method for the study of sociology and informatics [22, 23]. We make use of this method to describe the shift between different land cover classes in Shenzhen. Each node in the network represents a land type. Figure 3 is the complex network visualization of the land cover transformation between the five observation time points. It can be clearly seen that the transformation from vegetation to bare land dominates the land cover change during 1979–1990. The medium of bare land during the urban construction can hardly be witnessed, even though the time intervals are shorter than the two periods before. It reveals the transformation from vegetation to urban becomes faster.

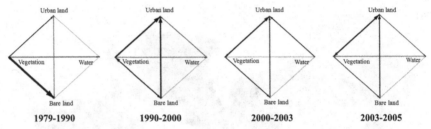

**Fig. 3.** Complex network visualization of land cover transformation in Shenzhen from 1979 to 2005.

## 5  Conclusion

In this study, we develop a classification method of land cover based on support vector machines by use of Landsat images. As a case study, we retrieve Shenzhen land cover maps in five timing points from 1979 to 2005, and then take use of the complex network approach to analyze the character of land use-cover change from an overall perspective. The complex network visualization shows that the main changes of land use-cover are different over time. The medium of bare land during the urban construction can hardly be witnessed, even though the time intervals are shorter than the two periods before. It reveals the transformation from vegetation to urban becomes faster. The transformation from vegetation to bare land is hard to be witnessed in the late stage. As bare land is the medium for transforming vegetation to urban land in Shenzhen during the past years from 1979 to 2005.

**Acknowledgments.** This work was supported in part by the Basic and Applied Basic Research Funding Program of Guangdong Province of China (Grant No. 2019A1515110303), the Natural Science Foundation of Guangdong Province (Grant No. 2018A030313014), the research team project of Dongguan University of Technology (Grant No. TDY-B2019009), the Guangdong University Key Project (2019KZDXM012).

# References

1. Benning, T.L., LaPointe, D., Atkinson, C.T., Vitousek, P.M.: Interactions of climate change with biological invasions and land use in the hawaiian islands: modeling the fate of endemic birds using a geographic information system. Proc. Natl. Acad. Sci. **99**, 14246–14249 (2002)
2. Reid, R., et al.: Land-use and land-cover dynamics in response to changes in climatic, biological and socio-political forces: the case of southwestern ethiopia. Landscape Ecol. **15**, 339–355 (2000)
3. Kreuter, U.P., Harris, H.G., Matlock, M.D., Lacey, R.E.: Change in ecosystem service values in the san antonio area, texas. Ecol. Econ. **39**, 333–346 (2001)
4. Moran, E.F.: News on the land project. Glob. Change Newsl. **54**, 19–21 (2003)
5. Turner, B.L., Meyer, W.B., Skole, D.L.: Global land-use/land-cover change: towards an integrated study. Ambio **23**, 91–95 (1994)
6. Dewan, A.M., Yamaguchi, Y.: Using remote sensing and gis to detect and monitor land use and land cover change in dhaka metropolitan of bangladesh during 1960–2005. Environ. Monit. Assess. **150**, 237–249 (2009)
7. Yin, J., et al.: Monitoring urban expansion and land use/land cover changes of shanghai metropolitan area during the transitional economy (1979–2009) in China. Environ. Monit. Assess. **177**, 609–621 (2011)
8. Lambin, E.F., et al.: The causes of land-use and land-cover change: moving beyond the myths. Glob. Environ. Change **11**, 261–269 (2001)
9. Schaldach, R., et al.: An integrated approach to modelling land-use change on continental and global scales. Environ. Model Softw. **26**, 1041–1051 (2011)
10. Seto, K.C., Kaufmann, R.K.: Modeling the drivers of urban land use change in the pearl river delta, China: integrating remote sensing with socioeconomic data. Land Econ. **2003**(79), 106–121 (2003)
11. Seto, K.C., Fragkias, M.: Quantifying spatiotemporal patterns of urban land-use change in four cities of China with time series landscape metrics. Landscape Ecol. **20**, 871–888 (2005)
12. Li, X., Yeh, A.G.-O.: Analyzing spatial restructuring of land use patterns in a fast growing region using remote sensing and gis. Landscape Urban Plan. **69**, 335–354 (2004)
13. Sui, D.Z., Zeng, H.: Modeling the dynamics of landscape structure in Asia's emerging desakota regions: a case study in Shenzhen. Landscape Urban Plan. **53**, 37–52 (2001)
14. Yang, Q., Li, X., Shi, X.: Cellular automata for simulating land use changes based on support vector machines. Comput. Geosci. **34**, 592–602 (2008)
15. Farr, T.G., et al.: The shuttle radar topography mission. Rev. Geophys. **45**(RG2004), 1–33 (2007)
16. Nordkvist, K., Granholm, A.-H., Holmgren, J., Olsson, H., Nilsson, M.: Combining optical satellite data and airborne laser scanner data for vegetation classification. Remote Sens. Lett. **3**, 393–401 (2011)
17. Friedl, M.A., Brodley, C.E.: Decision tree classification of land cover from remotely sensed data. Remote Sens. Environ. **61**, 399–409 (1997)
18. Franco-Lopez, H., Ek, A.R., Bauer, M.E.: Estimation and mapping of forest stand density, volume, and cover type using the k-nearest neighbors method. Remote Sens. Environ. **77**, 251–274 (2001)
19. Yanfei, Z., Liangpei, Z.: An adaptive artificial immune network for supervised classification of multi-/hyperspectral remote sensing imagery. IEEE Trans. Geosci. Remote Sens. **50**, 894–909 (2012)
20. Mountrakis, G., Im, J., Ogole, C.: Support vector machines in remote sensing: a review. ISPRS J. Photogram. Remote Sens. **66**, 247–259 (2011)

21. Pal, M., Mather, P.: Support vector machines for classification in remote sensing. Int. J. Remote Sens. **26**, 1007–1011 (2005)
22. Pengfei, W., Huili, G., Demin, Z.: Land use and land cover change in watershed of guanting reservoir based on complex network. Acta Geogr. Sin. **67**, 113–121 (2012)
23. Yang, C., et al.: Spatiotemporal evolution of urban agglomerations in four major bay areas of US, China and Japan from 1987 to 2017: evidence from remote sensing images. Sci. Total Environ. **671**(25), 232–247 (2019)

# Temporal Consistency Based Deep Face Forgery Detection Network

Chunlei Peng[1]([⊠]), Wenbo Zhang[1], Decheng Liu[2], Nannan Wang[3],
and Xinbo Gao[2,4]

[1] State Key Laboratory of Integrated Services Networks,
School of Cyber Engineering, Xidian University, Xi'an 710071, China
`clpeng@xidian.edu.cn`
[2] Video and Image Processing System Laboratory, School of Electronic Engineering,
Xidian University, Xi'an 710071, China
[3] State Key Laboratory of Integrated Services Networks,
School of Telecommunications Engineering, Xidian University, Xi'an 710071, China
[4] Chongqing Key Laboratory of Image Cognition, Chongqing University of Posts
and Telecommunications, Chongqing 400065, China

**Abstract.** With the rapid development of deep learning techniques as
well as increasingly more visual information being made publicly available on the Internet, image translation methods have achieved great
progress and encouraging performance. The manipulation and fabrica-
tion of visual information has become accessible and difficult to distinguish by the naked eye, which will have adverse effects on cloud and
communication security. Thus, face forgery detection techniques have
recently attracted increasing attention. Most recent works regard the
face forgery detection problem as the typical image classification task,
ignoring the exploration of inherent properties of forgery visual information itself. In this paper, we first explore the inherent limitation of fake
videos, and find that the temporal consistency could help distinguish
fake faces from real faces. A temporal consistency based deep face forgery
detection network is proposed to directly detect fake videos when given
multiple consistent video frames. The proposed method effectively considers the frame consistency property and achieves promising detection
performance. Experimental results on the face forgery detection dataset
demonstrate the superior performance of the proposed method.

**Keywords:** Cloud security · Visual information analysis · Face forgery
detection

## 1 Introduction

In cloud and communication security, facial visual information has drawn
increasing attention because of its convenience and safety. Especially for face
recognition tasks, many recent works [1,2] have achieved encouraging superior

© Springer Nature Switzerland AG 2020
X. Chen et al. (Eds.): ML4CS 2020, LNCS 12488, pp. 55–63, 2020.
https://doi.org/10.1007/978-3-030-62463-7_6

performance, even when facial visual information is captured with different sensors [3–7]. It is due to the rapid development of deep learning techniques and the availability of a large amount of cloud visual information on the Internet. In addition, image translation tasks are also becoming an important topic in computer vision and machine learning. Generative adversarial network methods [8–10] have yielded fine realistic textures and have further improved the quality of generated images. Because of the wide application of these image translation techniques, manipulating facial videos and images has become easier, and the results are becoming harder to distinguish by human eyes. These image forgery manipulation techniques can not only generate fake images or videos, but also create fake news and scams. There is no doubt that social media will make the propagation of fake visual information more convenient. Thus, it is indeed an important task to develop an effective forgery detection method for cloud and communication security.

In the early stage, the manipulation of visual content in the media requires complex sophisticated editing tools, and high image manipulation expertise. Additionally, the forgery of these videos is always time consuming, and the degree of realism is limited. For example, [11] utilized existing footage to create a new video of a person mouthing words, which tracked points on the mouth of the speaker. However, due to the increasing high computing power and rapid progress of machine learning applications, visual content manipulation is becoming easier. The end-to-end deep learning technique reduces the computational time. Existing face forgery detection methods can be roughly categorized into two categories: the traditional classification based methods [12–15] and deep learning based methods [16–19]. These methods mostly consider the face forgery detection task as the common binary classification task, and their aim is usually to train a strong and robust classifier to accurately distinguish fake images from real images.

With the development of deep learning, researchers have found that the convolutional neural networks can learn extremely powerful image features for classification, which indeed further promotes the face forgery detection field. Although recent deep learning based methods have achieved superior detection performance, these works do not consider the inherent properties of forgery media. Thus, the performance of these face forgery detection techniques is not good enough to be widely applied in the real world.

To address this challenge, we propose a temporal consistency based deep face forgery detection network. Our method first analyzes the frame consistency of forgery media, and finds the difference between fake videos and real videos. Then we design a temporal consistency based deep face forgery detection network to directly distinguish fake videos with an end-to-end network model. A novel objective function is designed to integrate the consistency information for better detection performance.

The main contributions of this paper are summarized as follows:

1. The proposed temporal consistency based framework considers the inherent characteristics of fake face videos, and experimental results prove its efficiency.

2. We design a temporal consistency based deep face forgery detection network that can effectively integrate frame temporal consistency to boost the detection performance. Furthermore, the backbone network of our method can be extended for better performance.
3. Experimental results on the UADFV dataset illustrate the superior performance of the proposed method.

We organize the rest of this paper as follows. Section 2 gives a review of face forgery detection works. Section 3 presents the temporal consistency based deep face forgery detection network. Section 4 shows the experimental results and provides an analysis of the algorithm, and the conclusion is drawn in Sect. 5.

## 2   Related Work

In this section, we review the face forgery detection methods in the aforementioned categories: traditional classification based methods and deep learning based methods.

In the early stage, researchers always focused on the biometric feature extraction to detect face forgeries [12,20]. [13] presented a holistic liveness detection paradigm that collaborated with face biometrics. Considering the lighting differences in optical flow fields generated by movements, [21] utilized the degree of differences between the two fields to distinguish a real face from a photograph. [14] proposed a method for masked fake face detection that utilized reflectance disparity. [15] captured facial physiological patterns with the bioheat information contained in the thermal images for face forgery analysis.

Recent works in the computer vision filed prove that convolutional neural networks can effectively extract strong and robust visual features for downstream tasks. Thus, deep learning based forgery detection methods have attracted increasing attention. [16] presented the incremental learning based method for the classification of GAN generated images, where multiple binary classifiers are utilized for the detection task. [22] found that fake videos are always created by splicing synthesized face regions into holistic regions, and proposed a novel method to estimate 3D head poses as features to distinguish fake videos. [17] directly targeted the artifacts in affine face warping to distinguish real and fake images, without using any generated images as training data. [18] presented a novel method to expose fake face videos generated and utilized the detection of eye blinking to detect generated videos. [19] combined a recurrent convolutional neural network model and face alignment approach to improve detection performance.

Inspired by previous works, we find that the key is to explore the inherent property of the generated fake face videos and train a strong binary classifier. In this work, we aim to train an end-to-end deep network to directly distinguish fake face videos, where temporal fame consistency information is considered as the clue.

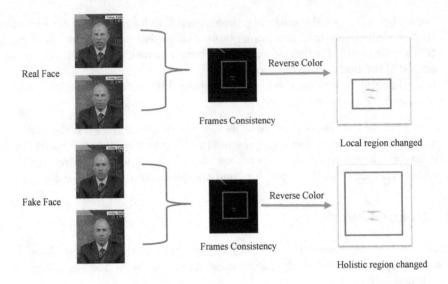

**Fig. 1.** The temporal consistency analysis results of face forgery videos.

# 3   Proposed Method

In this section, we present a novel temporal consistency based deep face forgery network to distinguish fake face videos. In the subsection, we will first describe the motivation, and then give a detailed explanation of our proposed method.

## 3.1   Motivation

In real-world scenarios, fake face videos are always generated from a deep network based image translation model. Naturally, our assumption is that the difference in temporal consistency will help to detect fake videos because these fake face videos are usually created with consistently generated with consistent generated frames from the image translation model, where the temporal relationship is ignored. In this work, we regard the temporal consistency of generated videos as the inherent characteristic for face forgery detection.

For convenience, we directly calculate the difference between the previous frame and the subsequent frame to find the distinct feature. As shown in Fig. 1, when we analyze the frames consistency of both real and fake videos, we find that there exists only a local changed region in the real videos, but a holistic appearance changed region is found in fake videos because the fake videos are generated by combining individual fake frames, which makes it hard to control the temporal consistency of face poses, expressions and countenances. Thus, we regard the temporal consistency of face videos as the distinguishing feature for forgery detection. In addition, the inherent features of the original videos themselves also provide discriminative information. Next, we provide more details of our proposed temporal consistency based deep face forgery detection network.

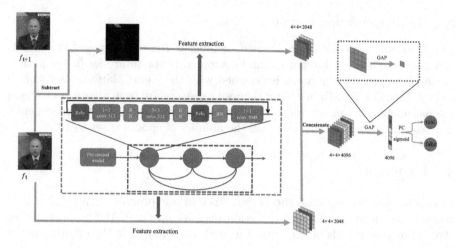

**Fig. 2.** Overview of the proposed temporal consistency detection based deep face forgery detection algorithm.

## 3.2 Temporal Consistency Based Deep Face Forgery Detection Network

In this subsection, we provide a detailed description of the proposed deep face forgery network. Here, we choose ResNet50 as the backbone network to introduce our method. Figure 2 shows the framework of the proposed algorithm. We consider that the input face video contains consistent frames $\{f_t\}_{t=1}^{T}$, where $T$ is the number of all frames in one video. To avoid overfitting problem, we utilize images from ImageNet dataset to pre-train the backbone network to improve the generalization ability of the extracted features. Inspired by related works, we set the parameters of former layers to effectively extract low-level common discriminative information.

We separately input the original frame $f_t$ and the difference in consistent frames $f_t - f_{t-1}$ into two network branches at time $t$. Next, the pre-trained backbone network extracts the discriminative image features as shown in Fig. 2. We directly concatenate frame feature $fea_t$ with the consistent frame feature $fea_t^{cons}$ as the final fused distinguishing feature $[fea_t, fea_t^{cons}]$ for face forgery detection. A global average pool layer is utilized to fuse different features, and then a fully connected layer is designed for binary classification.

The objective function of our network is designed as follows:

$$L = E[-log(p(label|fea_t, fea_t^{cons}))], \tag{1}$$

where $label \in \{0, 1\}$. Here $p(label|fea_t, fea_t^{cons})$ is the predicted probability of the consistent frames fused feature at time $t$.

## 3.3  Implementation Details

The backbone network is pre-trained by images from the ImageNet dataset. Different backbone architectures can be adopted in the future for better performance. The proposed network is designed with the Keras platform and run on a GTX 1660Ti GPU. Here we use the root mean square prop algorithm to train the parameters in our network. The batch size is set as 10 in our experiments. The learning rate is set to 1e–3, and the number of epochs is set to 10.

# 4  Experiments

In this section, we evaluate the performance of our proposed temporal consistency based deep face forgery detection network on the UADFV dataset [22]. We will present details of experimental results and illustrate the effectiveness of the proposed method.

## 4.1  Dataset and Evaluation Metric

The UADFV dataset [22] contains 98 videos in total, with 49 real face videos and 49 fake face videos. The average length of the videos is 11.14 s, and the size of most frames is $294 \times 500$. With the same protocol as that proposed in [22], we select 35 real face videos and 35 fake videos as training data which contains 23,981 frames in total, and we use the remaining 14 real videos and 14 fake videos as testing data, which contain 10,241 frames in total.

To further evaluate the face forgery detection performance to mimic real-world scenarios, we utilize the evaluation metrics mentioned in [22]: individual frames based metric. Here, we consider the face forgery detection task a binary classification task. Naturally, the individual frames of videos could be evaluated as units with the area under the ROC curves (AUC) as the evaluation metric. In the following experiment, we all choose the AUC to analyze individual frames as the performance metric.

## 4.2  Experimental Results

As mentioned before, we evaluate four different backbones of our proposed method on the UADFV dataset. Here we choose four common pre-trained models VGG16, VGG19, ResNet50 and ResNet101 as the backbone networks to prove the generalization ability and efficiency of our method. As shown in Fig. 3, the VGG16, VGG19, ResNet101 and ResNet152 models achieveed AUCs of 87.9%, 83.6%, 97.9% and 98.1% respectively. It is noted that the recent state-of-the-art method [22] achieves an AUC of only 89.0%. Compared with the state-of-the-art algorithm [22], our proposed method increases the AUC by 9.1%. It is because our method considers the inherent temporal consistency property of face forgery videos, and the deep networks could indeed extract strong discriminative features for forgery detection.

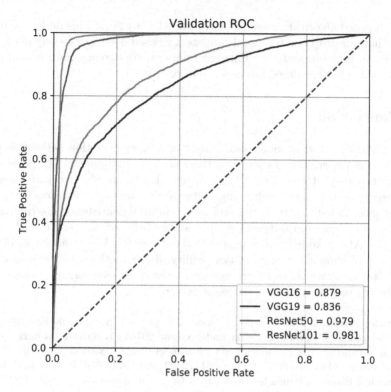

**Fig. 3.** ROC curves of the different backbone networks of our proposed method on the UADFV dataset.

## 4.3 Ablation Study

In this subsection, we further analyze the proposed temporal consistency based deep face forgery detection network. In the ablation study, we compare our network with the networks after removing the temporal consistency branch. As shown in Table 1, AUCs of all the CNN models apparently decrease after removing the temporal consistency network branch. Using a VGG16 backbone

**Table 1.** The experimental results (AUC) of the proposed temporal consistency based deep face forgery detection network with different backbones on the UADFV database.

| Models | W/O temporal consistency | Final performance |
|---|---|---|
| Method [22] | / | 89.0% |
| Proposed with VGG16 | 85.3% | 87.9% |
| Proposed with VGG19 | 81.5% | 83.6% |
| Proposed with ResNet50 | 96.6% | 97.9% |
| Proposed with ResNet101 | 97.0% | 98.1% |

in our proposed algorithm even decreases the AUC by 26%. This finding demonstrates that our proposed temporal consistency based deep face forgery detection network can help integrate the inherent temporal consistency information, and boost the face forgery detection performance.

## 5    Conclusion

A novel temporal consistency based deep face forgery detection network is proposed in this paper. The proposed method explores the inherent temporal consistency property of face forgery videos, and design an effective two-branches deep forgery detection network to fuse different discriminative features for better performance. Benefiting from the proposed temporal consistency discriminative features, our algorithm outperforms the state-of-the-art method by achieving a superior AUC. Additionally, we explore different backbone networks of our framework to show the generalization ability of our method. In the future, we will try to integrate more inherent characteristics of face forgery videos, and evaluate our proposed method on more datasets.

**Acknowledgement.** This work was supported in part by the National Key Research and Development Program of China under Grant 2016YFB0801001, in part by the National Natural Science Foundation of China under Grant 61806152, Grant 61876142, Grant 61432014, Grant U1605252, Grant 61772402, and Grant 61671339, in part by the Fundamental Research Funds for the Central Universities under Grant JB190117 and JB191502, in part by Natural Science Basic Research Plan in Shaanxi Province of China under Grant 2019JM-289, in part by the National Key Research and Development Program of China under Grant 2016QY01W0200, in part by the Key Industrial Innovation Chain in Industrial Domain under Grant 2016KTZDGY04-02, in part by the National High-Level Talents Special Support Program of China under Grant CS31117200001, in part by the Young Elite Scientists Sponsorship Program by CAST under Grant 2016QNRC001, in part by the Young Talent fund of University Association for Science and Technology in Shaanxi, China, in part by the CCF-Tencent Open Fund, in part by the China 111 Project under Grant B16037, in part by the Key Research and Development Program of Shaanxi under Grant 2020ZDLGY08-08, in part by the China Post-Doctoral Science Foundation under Grant 2018M631124 and 2019T120880, and in part by the Xidian University-Intellifusion Joint Innovation Laboratory of Artificial Intelligence.

## References

1. Wang, H., et al.: Large margin cosine loss for deep face recognition. In: IEEE Conference Computer Vision Pattern Recognition, pp. 5265–5274 (2018)
2. Deng, J., Guo, J., Xue, N., Zafeiriou, S.: Arcface: additive angular margin loss for deep face recognition. In: IEEE Conference Computer Vision Pattern Recognition, pp. 4690–4699 (2019)
3. Liu, D., Li, J., Wang, N., Peng, C., Gao, X.: Composite components-based face sketch recognition. Neurocomputing **302**, 46–54 (2018)

4. Liu, D., Gao, X., Wang, N., Li, J., Peng, C.: Coupled attribute learning for heterogeneous face recognition. IEEE Trans. Neural Netw. Learn. Syst. (2020)

5. Peng, C., Wang, N., Li, J., Gao, X.: Dlface: deep local descriptor for cross-modality face recognition. Pattern Recogn. **90**, 161–171 (2019)

6. Peng, C., Gao, X., Wang, N., Li, J.: Graphical representation for heterogeneous face recognition. IEEE Trans. Pattern Anal. Mach. Intell. **39**(2), 301–312 (2017)

7. Peng, C., Gao, X., Wang, N., Li, J.: Sparse graphical representation based discriminant analysis for heterogeneous face recognition. Signal Process. **156**, 46–61 (2019)

8. Isola, P., Zhu, J.Y., Zhou, T., Efros, A.A.: Image-to-image translation with conditional adversarial networks. In: IEEE Conference Computer Vision Pattern Recognition, pp. 1125–1134 (2017)

9. Zhu, J.Y., Park, T., Isola, P., Efros, A.A.: Unpaired image-to-image translation using cycle-consistent adversarial networks. arXiv preprint (2017)

10. Choi, Y., Choi, M., Kim, M., Ha, J.W., Kim, S., Choo, J.: Stargan: unified generative adversarial networks for multi-domain image-to-image translation. arXiv preprint, 1711 (2017)

11. Bregler, C., Covell, M., Slaney, M.: Video rewrite: driving visual speech with audio. In: Proceedings of the 24th Annual Conference on Computer Graphics and Interactive Techniques, pp. 353–360 (1997)

12. Galbally, J., Marcel, S., Fierrez, J.: Biometric antispoofing methods: a survey in face recognition. IEEE Access **2**, 1530–1552 (2014)

13. Kollreider, K., Fronthaler, H., Bigun, J.: Verifying liveness by multiple experts in face biometrics. In: IEEE Computer Society Conference on Computer Vision and Pattern Recognition Workshops, pp. 1–6. IEEE (2008)

14. Kim, Y., Na, J., Yoon, S., Yi, J.: Masked fake face detection using radiance measurements. JOSA A **26**(4), 760–766 (2009)

15. Buddharaju, P., Pavlidis, I.T., Tsiamyrtzis, P., Bazakos, M.: Physiology-based face recognition in the thermal infrared spectrum. IEEE Trans. Pattern Anal. Mach. Intell. **29**(4), 613–626 (2017)

16. Marra, F., Saltori, C., Boato, G., Verdoliva, L.: Incremental learning for the detection and classification of gan-generated images. In: International Workshop on Information Forensics and Security (2019)

17. Li, Y., Lyu, S.: Exposing deepfake videos by detecting face warping artifacts. arXiv preprint arXiv:1811.00656 (2019)

18. Li, Y., Chang, M.C., Lyu, S.: In ICTU oculi: exposing AI created fake videos by detecting eye blinking. In: International Workshop on Information Forensics and Security (WIFS), pp. 1–7. IEEE (2018)

19. Sabir, E., Cheng, J., Jaiswal, A., AbdAlmageed, E., Masi, I., Natarajan, P. Recurrent convolutional strategies for face manipulation detection in videos. Interfaces (GUI) **3**, 1 (2019)

20. Hadid, A., Evans, N., Marcel, S., Fierrez, J.: Biometrics systems under spoofing attack: an evaluation methodology and lessons learned. IEEE Signal Process. Mag. **32**(5), 20–30 (2015)

21. Bao, W., Li, H., Li, N., Jiang, W.: A liveness detection method for face recognition based on optical flow field. In: International Conference on Image Analysis and Signal Processing, pp. 233–236. IEEE (2009)

22. Yang, X., Li, Y., Lyu, S.: Exposing deep fakes using inconsistent head poses. In: IEEE International Conference on Acoustics, Speech and Signal Processing (ICASSP), pp. 8261–8265. IEEE (2019)

# CBOW and LSTM Based User Mobile Trajectory Prediction

Geng Sun[1] [ID] and Ming Tao[2,3]([envelope]) [ID]

[1] School of Computer, Guangdong University of Technology, Guangzhou,
People's Republic of China
iamsungeng@163.com
[2] School of Computer Science and Technology, Dongguan University of Technology,
Dongguan, People's Republic of China
ming.tao@mail.scut.edu.cn
[3] Institute of Artificial Intelligence and Blockchain,
Guangzhou University, Guangzhou 510006, People's Republic of China

**Abstract.** Trajectory prediction of mobile object has been a hotspot in current research. Many current mobile object trajectory models treat locations as isolated points, which do not suitable for scenes with a large number of locations and ignore the relationship between the locations. In order to reduce the dimension of location and obtain the movement pattern, we apply the Continuous-valued word representations (CBOW) word embedding method in natural language processing to represent the urban position. At the same time, the Long Short Term Memory (LSTM) model is established in combination with the historical position of the user to predict the next position of the user. Experiments show that the investigations in this paper can improve the prediction accuracy.

**Keywords:** CBOW · LSTM · Urban mobility · Trajectory prediction

## 1 Introduction

With the widespread use of wearable devices and position services, a large amount of trajectory data recording human movements has been generated. For example, GPS track data recorded by GPS sensors of mobile phones or other electronic devices, and call detail records (CDR) recorded by telecom operators. These data can provide important support for location-related services. How to better use user position information for location-related services has become the focus of current attention.

Mobile object trajectory prediction is the basic research of location-related services. Because various location-based information services are more and more concerned by research scholars, trajectory prediction technology is one of the hottest issues. The trajectory prediction technology has high research value and wide application fields. If the future of urban residents can be dynamically predicted in real time, higher quality services can be provided. For example: network providers can provide a seamless network switching service according to the user's historical movement trajectory and real-time

© Springer Nature Switzerland AG 2020
X. Chen et al. (Eds.): ML4CS 2020, LNCS 12488, pp. 64–73, 2020.
https://doi.org/10.1007/978-3-030-62463-7_7

predicting the user's next location by reserve network resources for it. Navigation tools can recommend high-quality roads and parking lots to drivers by predicting the next movement of vehicles in real time inferring road traffic conditions and parking lot states. Advertising companies or other information service providers can push more accurate information to users based on their next location.

The current prediction methods for mobile individuals are MM [1, 2], IO-HMM [3], LSTM [4, 5]. And obtain good results in corresponding scenarios. However, the relationship between the positions is ignored, and when the number of positions is large, the amount of calculation is large, and prediction is difficult. In order to make predictions for mobile individuals with a large number of positions, we use the CBOW embedding model to obtain the relationship between positions, and reduce the position representation dimension. And LSTM is employed to establish the corresponding prediction model.

## 2 Related Work

In order to predict the next position of moving objects, some researchers consider that the object's movements are related to the current position, Chen M et al. [1] used the Markov model, taking the current position into consideration, and predicting the next movement of the object by transition probability. Tang J et al. [6] obtained the transition probabilities of different transition modes through non-negative integer factorization to predict the object's movement. Kang C et al. [7] used a gravity model which thinks the transfer between object positions is affected by other surrounding positions. Predicting the moving objects' next position by calculating the transition probability of the current position to other positions. The model calculates the transition probability of every two positions, taking the distance relationship between the positions into account, but the amount of calculation is large. Predicting the next movement of the object based on the current position is that abstract the position as discrete points. The next position is only related to the current position and ignoring the object's movement is related to historical sequence and historical position. It is simply abstracted the object's movement as the transition probability, resulting in lower accuracy problem.

Some researchers consider that object's next movements are related to historical sequences. Qiao Y et al. [4] used hotspot position transfer sequences as moving sequences to predict object's movements through statistical movement patterns. Song L J et al. [2] based on the Markov model and use historical trajectories to predict the next position of moving objects. Rossi A et al. [5] expressed locations as discrete points through clustering, and together with the influence factors such as time, taxi ID constitute embedding vectors to predict the next position of the objects. Feng J et al. [8] consider the problem of information loss caused by long sequences, taking the position that the user may be at the same time of history day and the position predicted by the recurrent neural network as candidate points, then using the attention mechanism to screen the candidate points to assist predict the next movement of the object. The trajectory-based prediction finds the laws from the historical sequence of object's movement and improves prediction accuracy.

In addition, some scholars consider that the next movement of the object is related to social semantics. Yin M et al. [3] used IO-HMM to model the daily behavior of the work group. They take whether it is Sunday, time in one day and other influencing factors as input, and the social activity of the object as hidden states to infer the possible position of the objects. Tian S et al. [9] further used the LSTM model to predict the position of the objects and generate sequences of moving positions by the social activity. Yao D et al. [10] found that position is related to time and social semantics. The position, time and content are combined to form an embedded vector to predict the next position. Liao D et al. [11] change the LSTM Cell to input activities and trajectory at the same time, to predict the next position of the object. Ying et al. [12] establish the STP-Tree to predict the object position by mining the social semantic information represented by the building at the moving location, and convert the moving trajectory into a building trajectory sequence. Karatzoglou et al. [13] predict the next position of objects in different level by the multi-level abstraction of activities. Activity-based prediction has achieved good results, which can effectively reflect social semantics.

In the above research methods, all of them have achieved good results in some areas. However, when the number of positions is large, these methods cannot obtain good prediction results. Meanwhile the position is expressed as a discrete point, the relationship between the positions is ignored. As a useful technology in many natural language processing applications, CBOW is used to represent words with embedding vector [14]. In this paper, we use CBOW to represent the position of movement object to extract the position relationship and the movement pattern, and use LSTM model to predict the user's next position.

## 3   Model

The overall framework of our prediction model is shown in Fig. 1. The model consists of three parts: data preprocessing part, the CBOW embedding model part and the LSTM prediction model part. The following three sections will be used to explain these three parts in detail.

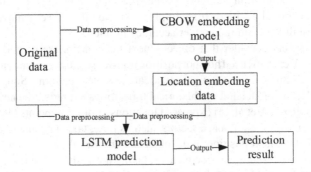

**Fig. 1.** The overall framework of our prediction model.

### 3.1   Data Preprocessing

Due to the errors in GPS data collection and data format problems, we need to preprocess the data.

The original GPS trajectory data can be abstractly expressed as $T = \{\langle l_1, t_1 \rangle, \langle l_2, t_2 \rangle, \ldots, \langle l_k, t_k \rangle\}$, where $l_i$ is the GPS data, $t_i$ is the time when GPS data was collected. To get precise prediction result. We did the following data preprocessing.

**Filter.** The data used in this article includes data outside Beijing. Because the number of trajectories outside Beijing is small, but the distance span is large. When the position granularity is small, the amount of position data will be large, which will cause a large number of locations to affect the results. Therefor we filter out trajectory outside Beijing. Trajectory likes $T = \{ \langle l_i, t_i \rangle | \exists l_i\ is\ outside\ of\ Beijing \}$ will be filtered

**Segment.** Because GPS data has a fixed time interval when it is collected, when the time interval is large, there may no data was collected in the middle, so we segment the trajectory with a threshold $\delta_t$. Trajectory $T = \{\langle l_1, t_1 \rangle, \langle l_2, t_2 \rangle, \ldots, \langle l_k, t_k \rangle\}$ will be split as $T_1 = \{\langle l_1, t_1 \rangle, \langle l_2, t_2 \rangle, \ldots, \langle l_i, t_i \rangle\}$ and $T_2 = \{\langle l_{i+1}, t_{i+1} \rangle, \langle l_{i+2}, t_{i+2} \rangle, \ldots, \langle l_k, t_k \rangle\}$, when $t_{i+1} - t_i > \delta_t$.

**Gridding.** Because the GPS data is too trivial and redundant, the trajectory outside the study area cannot directly and effectively extract the user's movement laws. So the data needs to be abstracted. Common abstract methods include hotspot extraction, conversion to activities, and gridding methods. Extracting hotspots and converting to activities both convert the user's trajectory into the trajectory of the cluster point, ignoring the intermediate information, which makes the position prediction granularity of the moving object larger. Therefore, we use the grid method to abstract the trajectory to obtain fine-grained prediction results. Geohash coding is commonly used in geographic position representation. The principle of Geohash coding is to process the earth into a two-dimensional plane, divide the plane into grids according to the latitude and longitude coordinates, and recursively divide these grids into grids. And encode these grids in a binary string. It is then converted to a string, which is Geohash encoded. For the original data $T = \{\langle l_1, t_1 \rangle, \langle l_2, t_2 \rangle, \ldots, \langle l_k, t_k \rangle\}$, we grid it and filter out the repeated codes, the result is $T' = \{\langle L_1, t_1 \rangle, \langle L_1, t_2 \rangle, \ldots, \langle L_m, t_m \rangle\}$.

**One-Hot Encoding.** After geohash coding, due to its discrete nature, it cannot be directly used in the neural network. In the previous related literature, the discrete position data was converted into one-hot encoding. That is, the position data is converted into a vector whose length is the number of positions, and the corresponding position is 1, the remaining positions are 0.

One-hot encoding can solve the input problem of discrete data well. But there are two problems based on one-hot encoding: (1) When the geohash encoding length is long, the data volume is large, and the one-hot vector length is large, which will cause data sparseness and dimensional explosion. (2) The position is expressed as an isolated point, and the mutual relationship between the positions is ignored.

In order to obtain the movement pattern from trajectory and reduce the one-hot dimension, we use the word embedding method CBOW to represent the position.

## 3.2 CBOW Embedding Model

The CBOW Embedding model is shown in the Fig. 2. The main idea for CBOW is that the current position is represented by the nearby location. We use the previous and next positions of trajectory to predict the vector of the current position. That is, for the trajectory sequence segment, the previous vector and the following vector are used to predict the intermediate vector and generate the corresponding embedding vector to represent the current position.

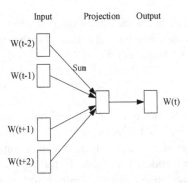

**Fig. 2.** CBOW architecture for predicting the current position based on the nearby one.

Let $l_{around}$ represent the length before and following positions numbers for the current position. The object of CBOW is learn the context info form trajectory to derive the movement info. When we know the surround location, and infer the current location. The probability calculation formula of prediction current position is formula (1), our goal is to maximize $p$.

$$p = p(w(t)|w(t - l_{around}), \dots .w(t - 1), \dots, w(t + l_{around})) \tag{1}$$

In order to ensure each position has its embedded vector, we add another position before and after the trajectory. And adopt a convenient method to construct the training set. Generate the embedding vector corresponding to each position. The specific algorithm is shown in Algorithm 1.

---

**Algorithm 1** CBOW train algorithm

---

**Input:** Beijing trajectory set $S_{T'}$.

**Output:** the geohash to embedding vector dictionary $S_{geo2emb} = \{< L_1, e_1 >\}$.

1: **for** $T_i'$ in $S_{T'}$ **do**

2:    Add padding to trajectory before and after trajectory $T_i'$ ;

3: **end for**

4: $S_{c2fb} = []$;

5: /* Build training set. */

6: **for** $T_i'$ in $S_{T'}$ **do**

7:    **for** slide window $T_s'$ in $T_i'$ **do**

8:       Extract $\{data, label\}$;

9:       Add $\{data, label\}$ to $S_{c2fb}$;

10:    **end for**

11: **end for**

12: $S_{geo2emb} = []$;

13: **while** $batchindex < batchCount$ **do**

14:    **for** $< data, label >$ in $S_{c2fb}$ **do**

15:       Input $< data, label >$ into Neural Network;

16:       Calculate $e_i$;

17:       /* Save the pair of geohash encode and embedding vector. */

18:       Add $< label, e_i >$ to $S_{geo2emb}$;

19:       Calculate $error$;

20:       Backpropagation parameters;

21:    **end for**

22: **end while**

---

### 3.3  LSTM Based Prediction Model

Trajectory data is a kind of time series data, which contains a strong spatiotemporal correlation. Individual movement prediction is defined as the use of historical movement position data to predict the position of the future moment. A recurrent neural network model can be used. The recurrent neural network allows the previous information to be used later by memory of historical information. Ordinary recurrent neural networks will have gradient disappearance or gradient explosion when the time scale increases. It is difficult to describe the characteristics of long-term sequence data. And the memory gate and the forget gate in the LSTM hidden layer, can greatly solve this problem. So we can Use LSTM to extract pattern information in personal mobile time series data and make predictions.

---

**Algorithm 2** LSTM prediction model

---

**Input:** Move sequence $S_{T}{}'$.

**Output:** prediction result $L$.

1:  $Dataset = []$;

2:  **for** $T_i{}'$ in $S_{T}{}'$ **do**

3:    **for** slide window $T_s{}'$ in $T_i{}'$ **do**

4:      Extract $data = \{e_a, e_{(a+1)}, ..., e_{(b-1)}\}, label = \{< e_b, l_{onehot} >\}$;

5:      Add $\{data, label\}$ to $Dataset$; // Save data and label.

6:    **end for**

7:  **end for**

8:  /* Disturb the data set and divide it into training set and test set. */

9:  $TrainSet, TestSet = dividedSet(suffer(Dataset))$;

10: **while** $batchindex < batchCount$ **do**

11:   **for** each {data, label} in $DataSet$ **do**

12:     Input $\{data, label\}$ into neural Network;

13:     Calculate $error$;

14:     Backpropagation parameters;

15:   **end for**

16: **end while**

17: Input $TestSet$ into LSTM calculate predict $L$;

---

Compared with the one-hot vector, the embedding vector calculated by CBOW embedding model has the advantages of low dimensionality and containing context information. Therefore, the position prediction algorithm based on LSTM can better deal with the problem of individual movement position prediction.

In order to facilitate training and testing, data and labels need to be constructed for LSTM network. Here, a sliding window is used to collect input data, and the data behind the window is the label. That is, the input data $x = \{e_1, e_2, \ldots, e_{sl}\}$, label $y = L_{sl+1}$, where $sl$ is the length of slide widow.

After inputting the data and labels into the LSTM neural network for training, the LSTM neural network outputs prediction results. The process is shown in Algorithm 2.

## 4  Experiments and Analysis

The data used in our experiment is Geolife [15, 16], the public data set come from Microsoft Asia Research Institute. This data set contains 182 individuals and GPS mobile data with a time span of 4 years.

We segment the original trajectory with a threshold of 20 min. At the same time, Geohash encoding is used to mesh the trajectory. The Geohash encoding length is 6 (every grid's size is about 1200 m * 600 m). After filtering out the trajectories outside Beijing, a total of 8806 position points are generated.

To conducted CBOW train set, we set $l_{around}$ equal to 1, 3, 5, 7 and 9, the length of embedding vector is 50, 100, 200, 300 and 400 to get good representation. The accuracy of embedded vector training is shown in the Table 1. We can see that the CBOW model get good performance when $l_{around}$ equal to 1 and the embedding length equal to 100. When the length of embedding vector is more than 100 does not improve. So we use the embedding vector of CBOW embedding model, which parameter is that $l_{around}$ equal to 1 and the embedding length equal to 100, as location representation vector.

**Table 1.** The accuracy of embedded vector representation.

|   | 50 | 100 | 200 | 300 | 400 |
|---|---|---|---|---|---|
| 1 | 0.890 | 0.900 | 0.900 | 0.904 | 0.906 |
| 3 | 0.880 | 0.888 | 0.888 | 0.894 | 0.886 |
| 5 | 0.852 | 0.850 | 0.860 | 0.860 | 0.864 |
| 7 | 0.796 | 0.814 | 0.854 | 0.832 | 0.832 |
| 9 | 0.746 | 0.778 | 0.810 | 0.806 | 0.814 |

*The first row of table is the length of embedding vector. The first column of table is $l_{around}$, which is the number of forward and after relative positions about current position.

In order to facilitate the observation of the spatial distribution of the embedding vectors, the position embedding vector is reduced to two dimensions through principal component analysis (PCA), and the distribution of some position embedding vectors in the two-dimensional vector space is shown in Fig. 3. The labels at each position point are their respective Geohash codes. The closer the two space vectors in the figure are, the more relevant they are in the trajectory.

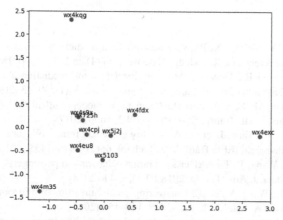

**Fig. 3.** Schematic diagram of spatial distribution of position embedding vector.

After constructing the embedding vector, we replace the geohash sequence with the embedding vector and train the LSTM neural network to get prediction results. And compared with the existing HMM, LSTM based on one-hot, MC. And the result is shown in Table 2.

**Table 2.** Prediction result for different model.

| HMM | LSTM | MC | CBOW-LSTM |
| --- | --- | --- | --- |
| 0.569 | 0.556 | 0.475 | 0.617 |

Experiments show that our work can effectively improve the prediction accuracy. Compared with the one-hot based LSTM and other model, it not only improves the training efficiency, but also improves the accuracy.

## 5  Conclusion

Urban mobility prediction can provide support for various location-based services. The traditional method is difficult to achieve the prediction result when the number of positions is large. In this paper, we use CBOW to combine the information before and after the moving point to represent the current position. At the same time use LSTM to predict the user's next location. Experiments show that the proposed embedding method improves the prediction accuracy by comparing with one-hot based LSTM.

**Acknowledgments.** This work was supported in part by the Natural Science Foundation of Guangdong Province (Grant No. 2018A030313014), the Guangdong University Key Project (2019KZDXM012), and the research team project of Dongguan University of Technology (Grant No. TDY-B2019009).

## References

1. Chen, M., Liu, Y., Yu, X.: NLPMM: a next position predictor with markov modeling. In: Pacific-Asia Conference on Knowledge Discovery and Data Mining, pp. 186–197 (2014)
2. Song, L.J., Meng, F.R., Yuan, G.: Moving object position prediction algorithm based on Markov model and trajectory similarity. J. Comput. Appl. **36**(1), 39–43 (2016)
3. Yin, M., Sheehan, M., Feygin, S., et al.: A generative model of urban activities from cellular data. IEEE Trans. Intell. Transp. Syst. **19**(6), 1682–1696 (2017)
4. Qiao, Y., Cheng, Y., Yang, J., et al.: A mobility analytical framework for big mobile data in densely populated area. IEEE Trans. Veh. Technol. **66**(2), 1443–1455 (2016)
5. Hu, X., An, S., Wang, J.: Taxi driver's operation behavior and passengers' demand analysis based on GPS data. J. Adv. Transp. 2018(PT.1), 1–11 (2018)
6. Tang, J., Liu, F., Wang, Y., et al.: Uncovering urban human mobility from large scale taxi GPS data. Physica A Stat. Mech. Appl. **438**, 140–153 (2015)
7. Kang, C., Ma, X., Tong, D., et al.: Intra-urban human mobility patterns: an urban morphology perspective. Physica A Stat. Mech. Appl. **391**(4), 1702–1717 (2012)
8. Feng, J., Li, Y., Zhang, C., et al.: DeepMove: predicting human mobility with attentional recurrent networks. In: International World Wide Web Conference, pp. 1459–1468 (2018)
9. Tian, S., Zhang, X., Zhang, Y., et al.: Spatio-temporal position prediction model for mobile users based on LSTM. In: International Conference on Parallel and Distributed Systems, pp. 967–970 (2019)

10. Yao, D., Zhang, C., Huang, J., et al.: SERM: a recurrent model for next position prediction in semantic trajectories. In: ACM on Conference on Information and Knowledge Management, pp. 2411–2414 (2017)
11. Liao, D., Liu, W., Zhong, Y., et al.: Predicting activity and position with multi-task context aware recurrent neural network. In: International Joint Conferences on Artificial Intelligence, pp. 3435–3441 (2018)
12. Ying, J.J.C., Lee, W.C., Weng, T.C., et al.: Semantic trajectory mining for position prediction. In: ACM SIGSPATIAL International Conference on Advances in Geographic Information Systems, pp. 34–43 (2011)
13. Karatzoglou, A., Köhler, D., Beigl, M.: Semantic-enhanced multi-dimensional Markov chains on semantic trajectories for predicting future locations. Sensors **18**(10), 3582 (2018)
14. Imani, A., Vakili, A., Montazer, A., Shakery, A.: Deep neural networks for query expansion using word embeddings. In: Azzopardi, L., Stein, B., Fuhr, N., Mayr, P., Hauff, C., Hiemstra, D. (eds.) ECIR 2019. LNCS, vol. 11438, pp. 203–210. Springer, Cham (2019). https://doi.org/10.1007/978-3-030-15719-7_26
15. Zheng, Y., Zhang, L., Xie, X., Ma, W.: Mining interesting locations and travel sequences from GPS trajectories. In: International World Wide Web Conference, pp. 791–800 (2009)
16. Tao, M., Wei, W., Huang, S.: Location-based trustworthy services recommendation in cooperative-communication-enabled internet of vehicles. J. Netw. Comput. Appl. **126**, 1–11 (2019)

# Framework Optimization for Face Recognition

Chao Chen[1] , Xin Wang[1]([⊠]) , and Yong-Xin He[2]

[1] Jiangsu Vocational Institute of Architectural Technology, Xuzhou, China
920658920@qq.com, 54520081@qq.com
[2] China University of Mining and Technology, Xuzhou, China
719788101@qq.com

**Abstract.** In recent times, with the increasing interest in face recognition for smart homes. However, most of these studies are focused on the individual modules of such a system, and there is an evident lack of research on a face recognition system framework that can integrate and manage the entire face recognition system. Therefore, in this study, we propose a framework that enables the user to effectively develop an face recognition system in different data volume applications. This paper designs an edge computing architecture and a cloud architecture. The edge computing architecture is designed with a Centralized-Edge and Peer-to-Peer Edge. At the same time, the face screening, face disguise, training timing and dynamic adjustment of training samples in face recognition are analyzed to give a feasible solution. In particular, the face screening rules are designed to reduce unnecessary training and repeated training. This paper has important application value for the intelligentization of the Internet of Things.

**Keywords:** Raspberry Pi · DNN · Face recognition · Framework · OpenCV

## 1 Introduction

Face recognition is one of the research hotspots in computer vision, image processing and neural networks in recent years. It is widely used in the fields of public safety, verification systems and human-computer interaction [5]. The development of face recognition has mainly gone through three stages: based on structural features, based on statistical features, based on big data and complex models [19]. From the earliest methods of geometric features and template matching to a scheme based on artificial features and classifiers, face recognition technology has begun to enter the automatic machine recognition stage. In recent years, with the continuous development of deep learning technology, face recognition technology has also begun to transform from traditional machine learning methods to deep neural networks. However, the calculation of neural networks often involves a large number of matrix operations, which puts high requirements on the computing power of hardware devices [16, 18]. The emergence of powerful GPU hardware devices has greatly reduced the model calculation time, which has promoted the widespread application of face recognition technology. At present, researchers have trained complex deep neural network models with tens or even hundreds of millions of undetermined parameters, and constantly refresh the highest record of face recognition

© Springer Nature Switzerland AG 2020
X. Chen et al. (Eds.): ML4CS 2020, LNCS 12488, pp. 74–85, 2020.
https://doi.org/10.1007/978-3-030-62463-7_8

accuracy. The DeepFace [14] released by Facebook, the DeepID [6–8] series researched by the Chinese University of Hong Kong, and the FaceNet [12] released by Google have reached or surpassed human recognition capabilities.

The breakthrough progress of artificial intelligence has prompted facial recognition technology to integrate into people's lives, which not only promotes the construction of smart cities, but also improves the efficiency of social operations. However, the deployment of most face recognition systems relies on cloud computing resources. With the rapid growth of the number of connected devices, in order to meet the fast and accurate identification effect, high requirements are placed on the network bandwidth and computing processing capabilities of the centralized physical data center that reflects the "cloud" [11]. In addition, the openness of the network environment and the sensitivity of identity data lead to the risk of privacy leakage in practical applications of face recognition technology [17]. The cloud-based method transmits sensitive data of a large number of users to the cloud. If attacked, it is likely to cause user privacy to leak. Therefore, in terms of protecting users' private data, there is a certain degree of insecurity in cloud-based deployment [3]. As a supplement to cloud computing, in recent years, edge computing has attracted great interest from researchers. Unlike cloud computing, edge computing provides end-to-end services. Data can be processed directly on the edge device without being transferred to the cloud, so it shows excellent performance in reducing communication delay and reducing bandwidth load. Considering the superiority of edge computing, some face recognition systems based on edge computing have been proposed. He et al. [2] proposed a lightweight and fast face detector (LFFD) for edge devices. Prentice et al. [9] developed a set of Raspberry Pi-based end-to-end smart office applications. The developed solution can monitor various environmental conditions and can use facial recognition to identify users.

At present, although face recognition technology has achieved some important results, there is still a lot of work to be done to design and implement a practical face authentication system. For the research based on Raspberry Pi face recognition at home and abroad, most of them focus on the Raspberry Pi [1, 10] software and hardware optimization and face recognition algorithm optimization, applied to specific smart home scenarios, and the multi-Raspberry pi collaboration constitutes a whole architecture research. It remains to be seen, and it has important research value for architecture optimization. This paper studies the face recognition in smart family as the application scenario. For the sake of family privacy protection, this paper studies the face training on the local Raspberry Pi, and uses the Raspberry Pi as the terminal and recognition for collecting face images. Through its continuous training, the end can make it possible to identify family members in almost any dress, hairstyle and face occlusion. By comparing several face recognition system architectures designed in this paper, the face recognition performance is continuously optimized.

This paper mainly studies the following two aspects:

1. In the Raspberry Pi collection terminal, how to choose a suitable screening image method to reduce unnecessary repetitive training.
2. To meet the needs of different data volumes, design non-cloud architecture and cloud architecture for face recognition, and study and design many details.

## 2  Raspberry Pi System Configuration

First, we will embody the scene. In the Raspberry Pi terminal, the face image and the person's identity are matched one by one through RFID. If the face reaches the training condition, the next step of training is performed; if not, only the door opening operation is performed. In the Raspberry Pi, the basic configuration is shown below (see Fig. 1):

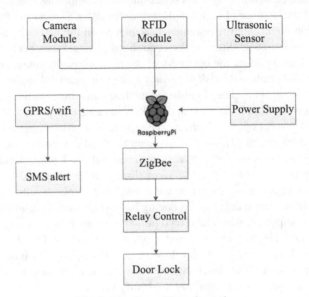

**Fig. 1.** Raspberry Pi configuration

### 2.1  Hardware Section

The Camera Module functions to capture a face image.

The role of RFID is to make the face image and the identity of the person correspond.

The Ultrasonic Sensor detects the distance of the person. When the distance is less than the preset distance, the command is sent to allow the camera to continuously capture 6 face images.

After we collect the image of the character, we send the image to the Raspberry Pi for processing. We will first screen it. The screening rules will be further explained below. If the training conditions are met, the model will be retrained.

In this architecture, the control module and communication module use the ZigBee platform. This technology is energy efficient, self-configuring, low cost, and provides high precision transmission. After we confirm the identity and determine that we can perform the operation such as opening the door, the Raspberry Pi sends a command to the relay through ZigBee, allowing the relay to control the corresponding hardware to perform the opening operation.

A whitelist is a collection of information about all legally identifiable people, and a blacklist is opposed to it. If the face image detection result is not in the white list but the RFID information is in the white list, or the existence of face disguise, the Raspberry Pi will not perform the action such as opening the door. At the same time, the Raspberry Pi sends a warning message to the administrator via GPRS or wifi, including the warning time and place.

## 2.2  Software Part

**Face Recognition Algorithm.** We installed the OpenCV platform in the Raspberry Pi and implemented the face recognition using the DNN algorithm. There are many platforms for implementing DNN, such as Caffe, TensorFlow, OpenCV, Caffe, etc. We refer to the article by Delia Velasco-Montero [15], who is based on Accuracy, Throughput, and Power Consumption.), FoM (Figure of Merit) and other aspects of the assessment. Define a variable FoM with the following formula:

$$FoM = Accuracy \cdot \frac{Throughput}{Power} \tag{1}$$

The experimental results show that OpenCV and SqueezeNet are the best performing components; for high throughput, OpenCV and TensorFlow are the best choices; for low energy budgets, OpenCV and Caffe2 are the most suitable tools. In summary, we see that OpenCV has a good performance in performance, throughput and power consumption, so this article decided to using OpenCV platform in the Raspberry Pi.

**Face Screening Algorithm.** When the system size and the amount of data are small, the old data and the new data can be trained together each time new face images needs training. However, as the scale of the system expands, more and more face images need to be newly identified. If the old and new data are still trained together, the training cost will be greatly improved and the training efficiency will be lower. Improve, remove some unnecessary training face images, and filter the face images. Face screening is considered in two aspects. One is to screen out face images such as unclear and blacklists to avoid wasting training time and resources; The second is to screen out the necessary training faces and not to train those faces that have been accurately identified.

Screening the face image that needs to be trained according to the preset face recognition accuracy threshold. For example, when the recognition rate is less than 50%, the image is discarded; when the recognition rate is greater than 90%, the image is not trained, but is stored; when identifying the image was trained at a rate between 50% and 90%. The specific flow chart is as follows (see Fig. 2):

In this paper, the threshold is used as the screening rule. In practical applications, when collecting human faces, we can require the examiner not to bring decorations such as hats and glasses. The increase of these rules can greatly reduce unnecessary face collection and training.

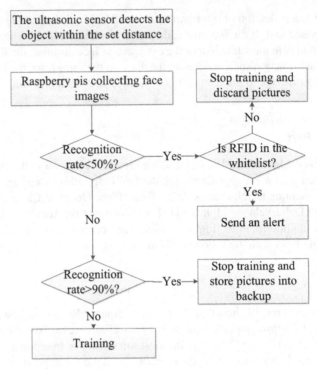

**Fig. 2.** Face image screening flow chart.

**To Prevent Face Deception.** There are many ways to prevent face deception. In the method proposed by Piyush Devikar et al., a temperature sensor based detection method is proposed because the surface temperature of the mask is close to the ambient temperature and is not as high as the real surface. At ambient temperature. The system takes images from a camera connected to the Raspberry Pi and then detects the faces in the image by OpenCV. The face temperature captured by the camera is obtained by an infrared temperature sensor. If a face is detected in the image and its temperature is greater than the threshold (skin) temperature, the face is true, otherwise it is false. This method can already block the fake face, face photo and the like formed by the high-end silicone mask.

In order to further prevent the occurrence of fraud, this article is supplemented by blink detection. We use the blink detection method of Tereza Soukupova and Jan Cech [13], using haar features to locate faces, and shape predictor 68 face landmarks.dat to mark face structures with 68 points to monitor human eyes, As shown in Fig. 3. The change in the distance between the upper and lower eyelids determines whether or not the eye is blinking. The principle is as shown below (see Fig. 4):

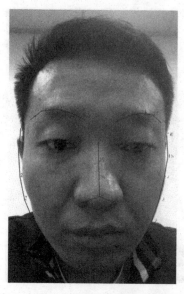

**Fig. 3.** Mark face structures with 68 points to monitor human eyes.

**Fig. 4.** Human eye mark.

As you can see in the image above, one eye will mark six points and define a variable EAR using the following formula:

$$\text{EAR} = \frac{\|P_2 - P_6\| + \|P_3 - P_5\|}{2\|P_1 - P_4\|} \tag{2}$$

The person blinks once for about 0.2–0.4 s, and when it detects that the EAR is less than 0.3, it is considered to be blinking. For face recognition, we need to label it, including the name, identity and age, as shown below (see Fig. 5):

**Fig. 5.** Face recognition effect.

As you can see, the recognition result Name is chenchao, the accuracy rate is 64.33%, the identity is administrator, and the age is 27.

## 3   Edge Computing Architecture Design

First, we design the edge computing architecture [4]. The face recognition and model training are all performed on the edge. The edge can be the Raspberry Pi or the NVIDIA TX2. Keeping the data locally saves user privacy to a large extent.

Use SQLite Studio to store information in any particular sequence. This article is used to store training data, trained models, whitelists and other information, and can be displayed during the detection process. For any new user registration, you need to add his/her information to this database.

In the edge computing architecture, there are two design approaches, one is a Centralized-Edge and the other is a Peer-to-Peer Edge. Centralized-Edge we use one or several Raspberry Pi as a training server. All the face images that need to be trained are sent to this training server. After training, the model is synchronized to each Raspberry Pi. In the Peer-to-Peer Edge, each Raspberry Pi is an independent training terminal. If any Raspberry Pi is trained and the model is updated, the latest models will be synchronized to other Raspberry Pis.

The first one adopts a Centralized-Edge, as shown in the following figure (see Fig. 6):

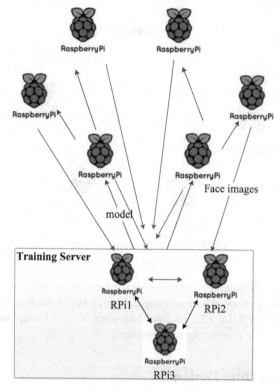

**Fig. 6.** Centralized architecture.

One or more Raspberry Pis are used as training servers, and each other Raspberry Pi is used as a face collection terminal. The collected faces are filtered and submitted to the Raspberry Pi server for training. After the trained, the model is distributed to each tree. Raspberry pi terminal. When the server has multiple Raspberry Pis, face recognition is performed by means of loop recognition.

Suppose now that the server consists of three Raspberry Pis. After receiving the identification request, it first asks if RPi1 can be identified. If it cannot be sent to RPi2, if RPi2 can't identify it, it will send it to RPi3. If RPi3 can recognize it, it will return recognition result. As a result, if it is not identifiable, training is performed at RPi3. At the same time, if the three Raspberry Pis can recognize different face sets different from each other, and exchange training data that cannot be recognized for training, the advantage of this is that on the one hand, the server performance can be improved, and on the other hand, the server side can be guaranteed. After a problem occurs in any Raspberry Pi, it doesn't affect the normal operation of the system.

The second adopts the Peer-to-Peer Edge. When any Raspberry Pi collects the face, it first filters and sends the filtered photos to the edge for training and updating the model. The structure is as follows (see Fig. 7):

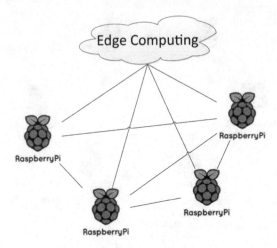

**Fig. 7.** Edge computing structure.

In the Raspberry Pi training, due to the limitations of its hardware performance, it can't meet the training work of a large sample. Therefore, on the Raspberry Pi training server. We can add NVIDIA TX2 or Intel Movidius Neural Computing Stick to the Raspberry Pi to improve computing performance.

## 4   Cloud Architecture Design

The architecture of the previous section is suitable for application scenarios where the model is small and the training data is small. As the amount of data continues to increase, the architecture of the previous section is clearly unable to meet the needs of big data processing. This chapter designs the cloud architecture and puts the training of preprocessing and model into the cloud.

The following Table 1 is a raspberry pi information table that records the id of the raspberry pi, the ip address of the Raspberry Pi, the version of the Raspberry Pi model, and the number of photos that need to be trained and the time the request was sent.

**Table 1.**  Raspberry pi information.

| ID | Time | Raspberry Pi ID | Number of images | Raspberry Pi address |
|----|------|-----------------|------------------|----------------------|
| 1 | 2018-12-10-12:10:10 | 3203001 | 5 | 120.109.0.111 |
| 2 | 2018-12-10-13:20:10 | 3203002 | 6 | 120.109.0.112 |

Here is the cloud architecture diagram (see Fig. 8):

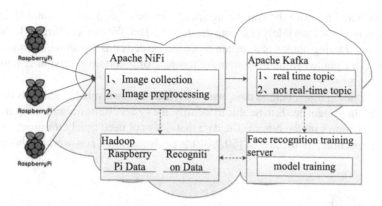

**Fig. 8.** Face recognition in cloud architecture.

Apache NiFi is a web project, we can use nifi to do a lot of pre-processing work, to reduce the computing burden of the cloud server. Specific to this article, we can add time and other attributes to the Raspberry Pi data, perform some pre-processing operations before the image training, for some data that needs to be stored, you can send it to Hadoop for storage.

Apache Kafka will be used as a messaging service because it provides high throughput, reliable delivery, and horizontal scalability. Kafka classifies messages according to Topic when they are saved. The sender becomes the Producer and the message receiver becomes the Consumer. In addition, the Kafka cluster consists of multiple Kafka instances, and each instance becomes a broker. Specific to this article, we can divide the data processed by nifi into two parts that need to be processed in time and not processed in time according to the training timing, and sent to different consumers for processing.

The cloud training host is responsible for training the data sent by Kafka. We can use cluster-based organization or only set up a high-performance training host. For priority training of data that needs timely training, meet the needs of real-time identification, and complete the training. Then send the model to the corresponding Raspberry Pi and let Hadoop store it.

Hadoop implements a distributed file system, referred to as HDFS. The core design of Hadoop's framework is HDFS and MapReduce. HDFS provides storage for massive amounts of data, while MapReduce provides calculations for massive amounts of data. This article is mainly used to store data such as Raspberry Pi and some trained models.

We can see that all parts of the cloud work together. First, nifi preprocesses the data sent from the Raspberry Pi, including image preprocessing, image categorization, and basic information storage. After that, nifi sends the pre-processed data to Kafka. Kafka further classifies it according to Topic, then passes the data to storm and spark for model training and recognition, and finally passes some model information to Hadoop for storage.

As time increases, the training data will become larger and larger. In order to maintain the efficiency of system training, on the one hand, we can increase the hardware configuration and improve the computing speed. On the other hand, we can reduce the training amount by reasonablely reducing the training data. We can set a threshold. When the number of training photos of a certain precision exceeds this threshold, we randomly delete the extra photos so that the training samples of a certain precision remain the same.

For the training time of cloud data, we are divided into two situations: one is that for the first time the system needs to be able to identify, the system starts training immediately after the data comes in; the other is the data that does not need timely training, such as the training accuracy is located at [50, 90] Photo, we arrange non-working hours for training.

## 5 Conclusions

In this study, we proposed a face recognition system framework. We designs an edge computing architecture and a cloud architecture. For each part of the architecture, such as face recognition, face screening, prevention of face disguise, and the design and connection of various components in the cloud architecture, solutions are given. The experimental results indicate that the proposed framework is effective and easy-to-use.

In the future, a face recognition system framework that is more useful can be expected if it can provide a guide that will help developer-s correct their own errors, or if it can correct errors automatically. In addition, for the face screening rule, the determination of the threshold needs further to be improved. We need further experimentation to find the most appropriate threshold. It can let us further improve efficiency and accuracy.

## References

1. Gunawan, T.S., Gani, M.H.H., Rahman, F.D.A., Kartiwi, M.: Development of face recognition on raspberry pi for security enhancement of smart home system. Indonesian J. Electr. Eng. Informatics (IJEEI) 5(4), 317–325 (2017)
2. He, Y., Xu, D., Wu, L., Jian, M., Xiang, S., Pan, C.: LFFD: a light and fast face detector for edge devices. arXiv preprint arXiv:1904.10633 (2019)
3. Li, J., Kuang, X., Lin, S., Ma, X., Tang, Y.: Privacy preservation for machine learning training and classification based on homomorphic encryption schemes. Inf. Sci. 526, 166–179 (2020)
4. Marjanovic, M., Antonic, A., Zarko, I.P.: Edge computing architecture for mobile crowdsensing. IEEE Access 6, 10662–10674 (2018)
5. Masi, I., Wu, Y., Hassner, T., Natarajan, P.: Deep face recognition: a survey. In: 2018 31st SIBGRAPI Conference on Graphics, Patterns and Images (SIBGRAPI), pp. 471–478. IEEE (2018)
6. Ouyang, W., et al.: Deepid-net: multi-stage and deformable deep convolutional neural networks for object detection. arXiv preprint arXiv:1409.3505 (2014)
7. Ouyang, W., et al.: DeepID-Net: deformable deep convolutional neural networks for object detection. In: Proceedings of the IEEE Conference on Computer Vision and Pattern Recognition, pp. 2403–2412 (2015)
8. Ouyang, W., et al.: DeepID-Net: object detection with deformable part based convolutional neural networks. IEEE Trans. Pattern Anal. Mach. Intell. 39(7), 1320–1334 (2016)

9. Prentice, C., Karakonstantis, G.: Smart office system with face detection at the edge. In: 2018 IEEE SmartWorld, Ubiquitous Intelligence & Computing, Advanced & Trusted Computing, Scalable Computing & Communications, Cloud & Big Data Computing, Internet of People and Smart City Innovation (SmartWorld/SCALCOM/UIC/ATC/CBDCom/IOP/SCI), pp. 88–93. IEEE (2018)

10. Sajjad, M., et al.: Raspberry pi assisted face recognition framework for enhanced law enforcement services in smart cities. Future Gener. Comput. Syst. **108**, 995–1007 (2017)

11. Satyanarayanan, M.: The emergence of edge computing. Computer **50**(1), 30–39 (2017)

12. Schroff, F., Kalenichenko, D., Philbin, J.: FaceNet: a unified embedding for face recognition and clustering. In: Proceedings of the IEEE Conference on Computer Vision and Pattern Recognition, pp. 815–823 (2015)

13. Soukupova, T., Cech, J.: Eye blink detection using facial landmarks. In: 21st Computer Vision Winter Workshop, Rimske Toplice, Slovenia (2016)

14. Taigman, Y., Yang, M., Ranzato, M., Wolf, L.: DeepFace: closing the gap to human level performance in face verification. In: Proceedings of the IEEE Conference on Computer Vision and Pattern Recognition, pp. 1701–1708 (2014)

15. Velasco-Montero, D., Fernandez-Berni, J., Carmona-Galan, R., RodrıguezVazquez, A.: Performance analysis of real-time DNN inference on raspberry pi. In: Real-Time Image and Video Processing 2018, vol. 10670, p. 106700F. International Society for Optics and Photonics (2018)

16. Wang, X., Kuang, X., Li, J., Li, J., Chen, X., Liu, Z.: Oblivious transfer for privacy preserving in VANET's feature matching. IEEE Trans. Intell. Transp. Syst. (2020)

17. Wang, X., Li, J., Kuang, X., Tan, Y.A., Li, J.: The security of machine learning in an adversarial setting: a survey. J. Parallel Distrib. Comput. **130**, 12–23 (2019)

18. Wang, X., Li, J., Li, J., Yan, H.: Multilevel similarity model for high-resolution remote sensing image registration. Inf. Sci. **505**, 294–305 (2019)

19. Yang, J., et al.: Neural aggregation network for video face recognition. In: Proceedings of the IEEE Conference on Computer Vision and Pattern Recognition, pp. 4362–4371 (2017)

# Machine Learning Assisted OSP Approach for Improved QoS Performance on 3D Charge-Trap Based SSDs

Zongwei Zhu[1] , Chao Wu[2]([⊠]) , Cheng Ji[2] , and Xianmin Wang[3]

[1] Suzhou Institute for Advanced Study, University of Science and Technology of China, Suzhou 215123, China

[2] Department of Computer Science, City University of Hong Kong, Hong Kong 999077, China
chaowu6-c@my.cityu.edu.hk

[3] Institute of Artificial Intelligence and Blockchain, Guangzhou University, Guangzhou 510006, China

**Abstract.** 3D charge-trap based SSDs have become an emerging storage solution in recent years. One-shot-programming in 3D charge-trap based SSDs could deliver a maximized system I/O throughput at the cost of degraded Quality-of-Service performance. This paper proposes RLOSP, a reinforcement learning based approach to improve the QoS performance for 3D charge-trap based SSDs. By learning the I/O patterns of the workload environments as well as the device internal status, the proposed approach could properly choose requests in the device queue, and allocate physical addresses for these requests during one-shot-programming. In this manner, the storage device could deliver an improved QoS performance. Experimental results reveal that the proposed approach could reduce the worst-case latency at the $99.9^{th}$ percentile by 37.5–59.2%, with an optimal system I/O throughput.

**Keywords:** 3D charge-trap based SSD · One-shot-programming · Reinforcement learning · I/O throughput · QoS performance

## 1  Introduction

With the rapid development of high-density NAND flash technology, 3-dimensional (3D) SSDs have been employed as the prevalent storage solution in the market. There are two mainstream 3D SSD technologies, floating-gate (FG) technology and charge-trap (CT) technology [21]. CT-based technology has been deemed as a pre-dominant candidate since it eliminates the cell-to-cell interference [21]. Thus, multiple pages in a word line could be programmed simultaneously, which is called one-shot-programming (OSP), e.g., 3 pages for TLC-based SSDs. This technique can maximize the system I/O throughput, thus improving the user experience significantly.

---

The original version of this chapter was revised: Subfigures (g) and (h) in Fig. 7 have been removed. The correction to this chapter is available at https://doi.org/10.1007/978-3-030-62463-7_51

© Springer Nature Switzerland AG 2020 corrected publication
X. Chen et al. (Eds.): ML4CS 2020, LNCS 12488, pp. 86–99, 2020.
https://doi.org/10.1007/978-3-030-62463-7_9

While OSP technique is recommended for superior system I/O throughput, this study identifies that OSP could significantly degrade the quality-of-service (QoS) performance. QoS demands all I/O latency to be constrained within a specific range [20]. Although SSDs provide a faster accessing speed comparing with HDDs, they suffer from large I/O performance variation and long worst-case latency owing to various reasons, e.g., garbage collection, read/write interference, process variation [6,20]. The enlarged I/O performance variation and prolonged worst-case latency could violate the QoS demand and impact the user experience. As all programmed pages share one completion time, OSP could spread one prolonged worst-case latency to the latency of all programmed pages. Moreover, as the storage device is oblivious to the process information of I/O requests, I/O from different processes might be programmed simultaneously. The prolonged worst-case I/O latency could be spread to different processes by OSP. As a result, OSP could significantly enlarge the I/O performance variation, thus incurring unpredictable I/O latency. Among these I/Os, some might have to experience long worst-case latency, thus violating the QoS demand and impacting the user experience.

Prior works fail to address the QoS performance incurred by OSP on CT-based SSDs. Some works propose to improve the system I/O performance by enhanced internal parallelism [15,21], which does not consider the QoS performance. There are also works proposed to improve the QoS performance by revised garbage collection [2,4,14], or mitigate the impaired storage lifetime incurred by process variation [1,5,17]. In addition, there are ways to optimize HDFS I/O performance in deep learning cloud computing platforms [25] or reduce system I/O latency by offloading computing power at the edge [24]. Since these works do not take account of the unique feature of OSP technique, there still exists a demanding need to enhance the efficacy when leveraging the OSP operations on 3D SSDs.

In this paper, a reinforcement learning based approach, RLOSP, is proposed. Through learning the I/O patterns and the device internal status, RLOSP properly decides which requests in the device queue to be programmed with OSP and the target physical addresses of these requests at each time. The considered I/O patterns consist of process ID and host I/O information of requests in the device queue, and the considered device status include the busy/free status of each flash chip and the space utilization of each plane. In this manner, the storage device could deliver reduced worst-case latency and improved QoS performance. Experimental results show that RLOSP could reduce the standard deviation of I/O latency by 51.8%, the worst-case I/O latency at the $99.9^{th}$ percentile by 37.5–59.2%, meanwhile delivering an optimal system I/O throughput (1.3% lower) comparing with existing approach. To the best of the author's knowledge, this is the first work proposed to improve the QoS performance of the 3D CT-based SSDs by OSP optimization.

In summary, this paper makes the following contributions:

- Identified that OSP in 3D CT-based SSDs could degrade the QoS performance of the storage device;

- Proposed RLOSP, a reinforcement learning assisted technique to make programming decisions adaptive to various I/O patterns and device internal status;
- Evaluated the proposed approach and verified that the proposed approach manages to effectively improve the QoS performance of the storage devices.

The remainder of this paper is as follows. Section 2 reviews related background and prior works. Section 3 states the problem of OSP technique in 3D CT-based SSDs. Section 4 presents the proposed RL assisted OSP approach. Section 5 evaluates the proposed approach and analyzes the experiment results. Finally, Sect. 6 concludes this paper.

## 2    Related Work

3D flash storage technology has attracted much attentions during past years. Liu et al. [15] propose to enhance the device internal parallelism by implementing the block-level parallelism and partial page accessing in the storage hardware architecture. Wu et al. [21] propose a distance-aware round robin page allocation scheme for improving the utilization of internal parallelism and thus improving the read performance. There are also works proposed to mitigate the detrimental effects of garbage collection activities [2,4,14], or mitigate the retention error and degraded system performance incurred by the process variation [1,3,6]. However, few of these works target on the degraded QoS performance incurred by the OSP in 3D CT-based SSDs.

There are also works proposed to improve the QoS performance of SSDs. Some works propose to decrease the I/O performance variation and reduce the worst-case latency by garbage collection oriented optimizations [11,19,22,23]. Gugnani et al. [8] propose a set of strategies for providing QoS performance guarantee with NVMe SSDs on virtual environment. Wu et al. [20] propose a reinforcement learning-assisted I/O merging technique for improved QoS performance on SSDs. However, few considers the QoS violation incurred by OSP in 3D CT-based SSDs.

## 3    Motivation

To study the advantage and disadvantage of One-Shot-Programming (OSP) on system performance of CT-based SSDs, experiments are evaluated with SSD-sim [9]. The simulator is configured to simulate a new TLC SSD and a TLC SSD device aged by 70%. The experimental results are compared with a device without OSP. The configuration of the simulator is described in Sect. 5.

Figure 1 shows the system throughput and I/O performance variation results. By writing three sub-requests into three consecutive pages, OSP can benefit system throughput significantly, as shown in Fig. 1(a). On average, system throughput is improved by 25.1% for a new device with OSP compared with a new device without OSP. For SRC1_2, the improvement of system throughput is 56.3%. However, the benefit of OSP on system throughput decreases as the aging status of the device increases. For a device aged by 70%, system throughput is improved by OSP by only 14.1% compared with a device without OSP. The reason is three-fold. First, the enlarged programming page number caused by OSP could incur garbage collection and block I/O requests, especially for aged devices with less physical space [10,12]. Then, the I/O requests grouped by OSP operation increases the possibility of read/write interference [13,18]. Finally, the grouped I/O requests may spread the prolonged latency of all blocked I/O requests, which degrades system performance further.

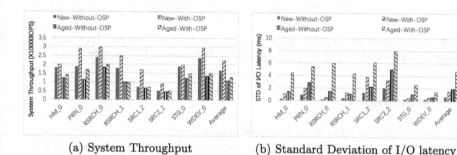

(a) System Throughput          (b) Standard Deviation of I/O latency

**Fig. 1.** Comparison Results of 3D SSD with and without One-Shot-Programming (OSP). New-Without-OSP means a new SSD without OSP, New-With-OSP means a new SSD with OSP, Aged-Without-OSP and Aged-With-OSP mean a SSD with and without OSP with 70% physical space used respectively.

The blocked I/O requests may experience a long I/O latency, which incurs enlarged I/O performance variation [20]. As shown in Fig. 1(b), the standard deviation of I/O latency for a SSD with OSP is 55.3% larger than a device without OSP on average for a new device, while the value is 58.9% for a device aged by 70%. This enlarged I/O performance variation makes I/O latency unpredictable, while several I/O may experience a worst-case latency. The unpredictable and extremely long I/O latency could affect user experience and violates the quality-of-service (QoS) requirement.

Figure 2 shows the comparison results of worst-case latency of example traces between SSDs with and without OSP operation. At the 99.9% percentage, the worst-case latency for a device with OSP is 45.4% for HM_0 and 61.8% for PRN_0 higher than a device without OSP for a new device. On average among all 8 adopted traces, the worst-case latency is prolonged by OSP by 44.2–67.8%, 55.3% on average at the 99.9% percentage for a new device. For a device aged by 70%, the worst-case latency is prolonged by OSP by 26.6–80.3%, 56.1% on average at the 99.9% percentage. Experimental results verified that OSP operation significantly prolongs worst-case latency and degrades QoS performance.

In summary, OSP in 3D charge-trap based SSD could improve system I/O throughput at the cost of prolonged worst-case latency and degraded QoS performance. The key observation here is that there is a need to propose a new approach to mitigate the worst-case latency while delivering a maximum I/O throughput during the adoption of OSP.

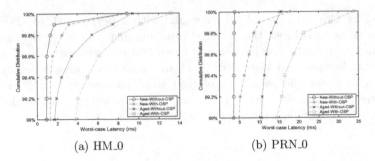

(a) HM_0                              (b) PRN_0

**Fig. 2.** Worst-case Latency of Example Traces. The horizontal axis is I/O latency in millisecond, the vertical axis is the cumulative distribution of I/O latency.

## 4    Methodology

This work proposes RLOSP, a reinforcement learning based approach to mitigate the worst-case latency issue incurred by one-shot-programming in CT-based SSDs. RLOSP learns from the I/O pattern and the storage internal status, and make OSP decisions accordingly. To avoid the worst-case I/O latency from spreading to different host I/Os and processes, the process ID (PID) and host I/O information of requests in the device queue is considered. The PID of requests is considered to avoid the worst-case I/O latency from spreading to various processes. Although current storage device is oblivious to this information, it is convenient to acquire this information from the host side with open-channel SSD technology [16] with trivial overhead. The host I/O information is considered to avoid the worst-case I/O latency from spreading to different host I/Os. To record the PID and host I/O information of each request in the device queue, a FIFO-list is maintained which is synchronized with the maintenance of all requests in the device queue. The considered storage internal status consists of busy/free status of flash chips and the space utilization of each plane. Requests programmed to the busy chips by OSP could be blocked until the chips are free, thus incurring a long I/O latency. Moreover, programming I/O requests to the planes with high space utilization might incur garbage collection, which could significantly degrade the I/O latency. The enlarged number of programmed pages in OSP increases the possibility of triggering garbage collection. Therefore, the busy/idle status of flash chips and the space utilization of each plane are considered in RLOSP.

Figure 3 describes the architecture of RLOSP. When a new request is inserted to the device queue (RQ1), the device controller splits this request into several one-page transactions. Then, the agent in RLOSP collects the PID and host I/O information of requests in queue from the storage device controller, as well as the busy/idle status of each flash chip and the space utilization of each plane ②. Accordingly, the agent identifies the current state $S_i$. After that, the agent refers to the $\varepsilon - Greedy$ policy to decide to perform exploration or exploitation ③. An action $A_i$ is selected according to the current state in the Q-table ④,

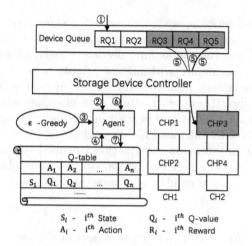

**Fig. 3.** Architecture of the proposed framework. $CHP$ is the flash chips, $CH$ is the device channel. $RQ$ is the request in device queue.

which chooses requests in queue to be programmed with OSP (RQ3, RQ4 and RQ5, assume all are one-page requests) and assigns physical addresses for these requests. These requests are then programmed to the assigned addresses in the target flash chip (CHP3) with OSP ⑤. After the latency of the programmed requests is perceived by the storage device controller, the agent acquires the latency from the storage device controller ⑥ and calculates the reward $R_i$ of the action $A_i$. Finally, the Q-value $Q_i$ is calculated and updated into the corresponding state-action entry in the Q-table ⑦.

## 4.1   Model Construct

**State Space.** The state space definition should provide sufficient information of the environment, so that the agent could fully learn from the environment and make OSP decisions accordingly. In RLOSP, the state definition considers storage internal status, including busy/idle status of each flash chip and the space utilization of each plane. In this manner, the OSP node could be assigned to a free chip and a plane with low space utilization, so as to boost the OSP node and improve the QoS performance. The space utilization of each plane is classified in to several regions (3 by default) in state definition.

The number of the state is illustrated as Eq. 1 shown.

$$S_n = 2^{CHP} \times SO^P \tag{1}$$

In which $CHP$ is the number of flash chips. $2^{CHP}$ is the number of busy/idle status of all flash chips. $P$ is the number of planes in the back-end storage device. $SO^P$ means the space utilization region distribution of all planes.

**Action Space.** The action space is defined by choosing requests in the device queue to be programmed by OSP and assigning physical addresses for these requests. There are two rules in action selection. Firstly, transactions from the same host I/O are programmed with highest priority in the same OSP node to avoid the worst-case latency from spreading to different host I/O. Then, transactions with the same PID are programmed with moderate priority in the same OSP node to avoid the worst-case latency from spreading to various processes. Finally, transactions with different PID are programmed with lowest priority in the same OSP node. In this manner, the action is defined by choosing $A$ transactions from I/O requests in the device queue, and assigning physical address for these transactions. $A$ is the number of transactions programmed by OSP each time.

**Reward.** The reward in RL is defined to represent the correctness of the last action. In RLOSP, improper OSP operations could incur long worst-case latency thus violating the QoS demand and impacting the user experience. The reward is defined by the I/O latency of the transactions programmed by last OSP action.

Figure 4 describes the reward definition in RLOSP. The I/O latency of each OSP operation is classified into three regions, as shown in the figure. For OSP actions with low latency, a bonus is assigned as the reward. For actions with medium latency, the reward is none. A minus reward is assigned for actions with long latency.

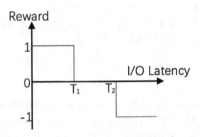

**Fig. 4.** Reward definition.

### 4.2   Algorithm

Algorithm 1 shows the algorithm of RLOSP. Four groups of elements are input in RLOSP, including the PID of each request in the queue $G(PID_1, .., PID_n)$, the host I/O information of each request in the device queue $G(HI_1, .., HI_n)$, busy/idle status of each chip $G(S_1, .., S_{CHP})$ and the space utilization of each plane $G(SO_1, .., SO_P)$. $n$ is the number of requests in queue, $P$ is the number of planes in the flash storage and $CHP$ is the number of flash chips. When the device controller prepares for an OSP operation, the agent picks an action

---

**Algorithm 1.** RL-Based OSP Approach

---

**Require:**
$\quad G(PID_1, .., PID_n), G(HI_1, .., HI_n),$
$\quad G(S_1, .., S_{CHP}), G(SO_1, .., SO_P)$
**Ensure:** $A_t$
$\quad$ **if** $Data/Eviction$ **then**
$\quad\quad S_t = state\_identify(G(PID_1, .., PID_n),$
$\quad\quad G(HI_1, .., HI_n), G(S_1, .., S_{CHP}), G(SO_1, .., SO_P));$
$\quad\quad A_t = \varepsilon - Greedy(S_t);$
$\quad\quad Perform\ OSP;$
$\quad\quad Acquire\ I/O\ latency;$
$\quad\quad Reward = Cal\_Reward(T_{Lat});$
$\quad\quad Q(s, a) = Q(s, a) + \gamma * Reward;$
$\quad$ **end if**

---

referring to $\varepsilon - Greedy$, which means deciding which requests in queue will be programmed with OSP and which physical address will be assigned for these requests. After that, the I/O latency of the last OSP operation is perceived and the Q-value $Q(s, a)$ is calculated and updated to the corresponding state-action $(s, a)$ entry in the Q-table.

Figure 5 describes a walk-through example of the comparison between the current OSP technique and RLOSP. In the current OSP technique, the device controller simply selects three requests in the tail of the device queue (RQ3, RQ4 and RQ5) to perform OSP operation. As the storage controller is oblivious to the busy/idle status of each flash chip and the space utilization of each plane in physical address allocation, the selected requests might been issued to busy chips (CHP3) or planes with high utilization, thus incurring garbage collection. In this case, RQ3, RQ4 and RQ5 could be blocked thus incurring a long worst-case latency. When these requests are from different processes (PID1, PID2 and PID3), the long worst-case latency could be spread to different processes, thus blocking the latency of all involved processes and impacting the user experience significantly. As shown in the figure, the wait time of PID1, PID2 and PID3 blocks the latency of all processes, the latency of PID1 is prolonged to T5, and the latency of PID2 and PID3 are prolonged to T4.

### 4.3 Walk-Through Example

For RLOSP, the agent considers the process ID and host I/O information of requests in queue, as well as the busy/idle status of each flash chip and space utilization of each plane in physical address allocation. The device controller firstly selects RQ1, RQ2 and RQ3 to serve, which are all from the same process (PID1). Then, an idle chip (CHP1) and plane with low space utilization will be assigned for these requests. Finally, other requests in queue (RQ4, RQ5) will be served simultaneously with OSP operation. In this manner, the wait time of PID1 process is eliminated, which could significantly reduce the worst-case latency of all processes and improve the QoS performance for SSDs. As shown

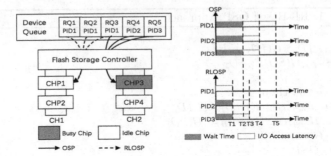

**Fig. 5.** Walk-through example of the comparison between the current OSP technique and RLOSP.

in the figure, the latency of PID1 for RLOSP is reduced into T1, the latency of PID2 and PID3 for RLOSP is reduced into T2.

## 5　Results and Analysis

### 5.1　Experiment Environment

**Experimental Setup.** In this work, experiments are conducted with a trace-driven simulator SSDsim [20] to verify the proposed approach. The simulator is configured as a new 3D SSD device and a 3D SSD device aged by 70% respectively. OSP operation is implemented in the simulator. The access latency is configured following a SAMSUNG NAND-Flash product [7]. The configuration of the simulator is described in Table 1.

**Table 1.** Configuration of SSDsim.

| Channel number | 2 | Chip per channel | 2 |
|---|---|---|---|
| Die per chip | 1 | Plane per die | 1 |
| Block per plane | 2048 | Page per block | 576 |
| Page capacity | 4 KB | Over provisioning ratio | 10% |
| Garbage collection scheme | Greedy | Page read latency | 90 us |
| Page write latency | 900 us | Block erase | 10 ms |

**Workload Environment.** In the experiments, MSRC traces [20] are adopted, which are published by Microsoft Cambridge collected from production servers. The characteristics of the adopted traces are shown in Table 2. The size distribution of the adopted traces shows the diversity among all traces collected from various workload environments. Interval denotes the average arrival time interval between each two consecutive I/O requests. Write ratio means the percentage of write I/O requests in all I/O requests.

**Table 2.** Characteristics of I/O traces.

| Trace | Size distribution | | | Write ratio | Interval (ms) |
|---|---|---|---|---|---|
| | 4 KB | 8–16 KB | >16 KB | | |
| HM_0 | 11.1% | 1.1% | 87.8% | 5.2% | 1.45 |
| PRN_0 | 72.2% | 11.1% | 16.7% | 85.7% | 2.14 |
| RSRCH_0 | 69.6% | 23.6% | 6.8% | 91.8% | 1.63 |
| RSRCH_1 | 5.3% | 1.6% | 93.1% | 2.1% | 1.60 |
| SRC1_2 | 43.8% | 3.0% | 53.2% | 12.7% | 1.40 |
| SRC2_2 | 74.5% | 16.2% | 9.3% | 79.4% | 1.56 |
| STG_0 | 70.0% | 20.0% | 10.0% | 50.0% | 0.09 |
| WDEV_0 | 70.0% | 20.0% | 10.0% | 50.0% | 0.09 |

## 5.2 Experimental Results

In the experiments, the evaluation results of RLOSP is compared with the performance of the current OSP technique in CT-based 3D SSDs. In addition, the performance of SSDs without OSP is also evaluated to show the advantage and disadvantage of OSP technique, and show the optimizing efficiency of RLOSP further.

**System I/O Throughput.** Figure 6(a) compares the system I/O throughput of RLOSP with existing approaches. OSP technique could significantly improve the system I/O throughput, while RLOSP could maintain the benefit brought by OSP. Compared with the performance of the current OSP technique, RLOSP reduces the system I/O throughput by only 1.3% on average on a new device. For a device aged by 70%, RLOSP reduces the system I/O throughput by 2.0% on average compared with the current OSP technique. Oriented at OSP optimization, RLOSP chooses proper requests to program with OSP in the device queue instead of canceling the OSP operation. Therefore, RLOSP could improve the QoS performance of the storage device without influencing the benefit gained by OSP technique.

**I/O Performance Variation.** The current OSP technique could enlarge the I/O performance variation, and part I/O might have to experience long worst-case latency thus degrading the QoS performance of the storage device. This issue is successfully mitigated by RLOSP. Figure 6(b) compares the standard deviation of I/O latency of RLOSP with existing approaches. On average, RLOSP could decrease the I/O performance variation by 51.8% compared with the current OSP technique on a new device. On a device aged by 70%, RLOSP could decrease the standard deviation of 56.2%. For HM_0 trace in the aged device, the I/O performance variation is decreased by RLOSP by 73.4%. RLOSP considers PID and host I/O information of requests in the device queue, thus avoiding the

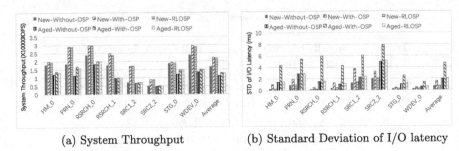

(a) System Throughput               (b) Standard Deviation of I/O latency

**Fig. 6.** Evaluation results of various approaches.

worst-case I/O latency from spreading to multiple I/O requests and different processes. In addition, the consideration of busy/idle status of flash chips and space utilization of planes in RLOSP could prevent the I/O requests from being blocked by busy chips or garbage collection. In this manner, RLOSP manages to eliminate the detrimental effects of OSP technique on QoS performance.

**Worst-Case I/O Latency.** Figure 7 describes the worst-case I/O latency of various approaches at the tail 1% percentile. OSP technique could significantly prolong the worst-case I/O latency, thus degrading the QoS performance of the device. For a new device, the worst-case latency is prolonged by OSP 26.0–54.7%, 35.0% on average at the 99% percentile, prolonged by 44.2–67.8%, 55.3% on average at the 99.9% percentile compared with a device without OSP. For a device aged by 70%, the worst-case latency is prolonged by OSP by 19.5–82.5%, 47.2% on average at the 99% percentile, by 26.6–80.3%, 56.1% on average at the 99.9% percentile. This detrimental effects of OSP technique is successfully eliminated by the proposed RLOSP approach. Among all adopted traces, RLOSP reduces the worst-case latency on a new device by 22.4–48.6%, 30.2% on average at the 99% percentile, by 37.5–59.2%, 46.7% on average at the 99.9% percentile compared with a device with OSP technique. For a device aged by 70%, RLOSP reduces the worst-case latency by 11.7–77.0%, 42.5% on average at the 99% percentile, by 23.5–78.7%, 52.0% on average at the 99.9% percentile compared with a device with OSP technique. By considering the I/O patterns and the storage internal status, RLOSP manages to choose the proper requests in the device queue and allocate right physical addresses for these requests, thus boosting the worst-case I/O latency significantly.

In summary, the proposed RLOSP approach could eliminate the detrimental effects of the OSP technique on the QoS performance, while maintaining the benefits gained by OSP on system I/O throughput. In this manner, the 3D CT-based SSDs could deliver an optimal system throughput with significantly improved QoS performance as well as the user experience.

**Overhead Analysis.** RLOSP incurs trivial overheads. First, the computation overhead of RLOSP is trivial. The storage system acquires an action at the cost

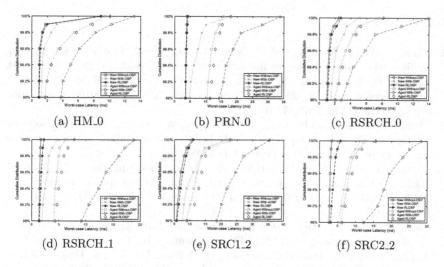

| (a) HM_0 | (b) PRN_0 | (c) RSRCH_0 |
| (d) RSRCH_1 | (e) SRC1_2 | (f) SRC2_2 |

**Fig. 7.** Evaluation results of the worst-case I/O latency. The horizontal axis is I/O latency in millisecond, the vertical axis is the cumulative distribution of I/O latency.

of a table traversing operation, whose computation overhead is negligible. Moreover, the memory overhead of RLOSP is trivial. The major memory overhead of RLOSP is incurred by the maintenance of the Q-table. According to Eq. 1, there are 1296 states. In RLOSP, the action is allocating a plane for selected requests in the device queue. There are 4 actions in the experiments in this work. Therefore, the Q-table is a $1296 \times 4$ vector, which takes around 10 KB memory overhead. This memory overhead is trivial over current SSDs with a DRAM larger than 1 GB.

## 6 Conclusion

This paper proposes RLOSP, a reinforcement learning assisted approach for the optimization of the OSP technique on 3D CT-based SSDs. First, evaluation results reveal that OSP could maximize the system I/O throughput at the cost of degraded QoS performance. Then, RLOSP considers the I/O patterns and the storage internal status in each OSP operation. In this manner, RLOSP manages to eliminate the detrimental effects of OSP while keeping the benefits of OSP on system I/O performance. We expect that this work could help the designers in 3D SSDs oriented optimizations.

# References

1. Chen, J., Wang, Y., Zhou, A.C., Mao, R., Li, T.: PATCH: process-variation-resilient space allocation for open-channel SSD with 3D flash. In: Teich, J., Fummi, F. (eds.) Design, Automation & Test in Europe Conference & Exhibition, DATE 2019, Florence, Italy, 25–29 March 2019, pp. 216–221. IEEE (2019). https://doi.org/10.23919/DATE.2019.8715197

2. Chen, S., Chang, Y., Liang, Y., Wei, H., Shih, W.: An erase efficiency boosting strategy for 3D charge trap NAND flash. IEEE Trans. Comput. **67**(9), 1246–1258 (2018). https://doi.org/10.1109/TC.2018.2818118

3. Chen, S.H., Chen, Y.T., Wei, H.W., Shih, W.K.: Boosting the performance of 3D charge trap nand flash with asymmetric feature process size characteristic. In: 2017 54th ACM/EDAC/IEEE Design Automation Conference (DAC), pp. 1–6. IEEE (2017)

4. Chen, T., Chang, Y., Ho, C., Chen, S.: Enabling sub-blocks erase management to boost the performance of 3D NAND flash memory. In: Proceedings of the 53rd Annual Design Automation Conference, DAC 2016, Austin, TX, USA, 5–9 June 2016, pp. 92:1–92:6. ACM (2016). https://doi.org/10.1145/2897937.2898018

5. Ji, C., et al.: Inspection and characterization of app file usage in mobile devices. ACM Trans. Storage (TOS) **16**(4), 1–25 (2020)

6. Du, Y., Zhou, Y., Zhang, M., Liu, W., Xiong, S.: Adapting layer RBERs variations of 3D flash memories via multi-granularity progressive LDPC reading. In: Proceedings of the 56th Annual Design Automation Conference 2019, DAC 2019, Las Vegas, NV, USA, 02–06 June 2019, p. 37. ACM (2019). https://doi.org/10.1145/3316781.3317759

7. Samsung Electronics: K9F8G08UXM Flash Memory Datasheet, March 2007

8. Gugnani, S., Lu, X., Panda, D.K.: Analyzing, modeling, and provisioning QoS for NVMe ssds. In: Sill, A., Spillner, J. (eds.) 11th IEEE/ACM International Conference on Utility and Cloud Computing, UCC 2018, Zurich, Switzerland, 17–20 December 2018, pp. 247–256. IEEE Computer Society (2018). https://doi.org/10.1109/UCC.2018.00033

9. Hu, Y., Jiang, H., Feng, D., Tian, L., Luo, H., Zhang, S.: Performance impact and interplay of SSD parallelism through advanced commands, allocation strategy and data granularity. In: Proceedings of the International Conference on Supercomputing, pp. 96–107 (2011)

10. Jung, M., Choi, W., Srikantaiah, S., Yoo, J., Kandemir, M.T.: HIOS: a host interface I/O scheduler for solid state disks. In: ACM/IEEE 41st International Symposium on Computer Architecture, ISCA 2014, Minneapolis, MN, USA, 14–18 June 2014, pp. 289–300. IEEE Computer Society (2014). https://doi.org/10.1109/ISCA.2014.6853216

11. Jung, S., Song, Y.H.: Garbage collection for low performance variation in NAND flash storage systems. IEEE Trans. CAD Integr. Circ. Syst. **34**(1), 16–28 (2015). https://doi.org/10.1109/TCAD.2014.2369501

12. Kang, W., Shin, D., Yoo, S.: Reinforcement learning-assisted garbage collection to mitigate long-tail latency in SSD. ACM Trans. Embedded Comput. Syst. **16**(5s), 134:1–134:20 (2017). https://doi.org/10.1145/3126537

13. Lee, H., Lee, M., Eom, Y.I.: Mitigating write interference on SSD in home cloud server. In: IEEE International Conference on Consumer Electronics, ICCE 2018, Las Vegas, NV, USA, 12–14 January 2018, pp. 1–3. IEEE (2018). https://doi.org/10.1109/ICCE.2018.8326216

14. Liu, C., Kotra, J., Jung, M., Kandemir, M.T.: PEN: design and evaluation of partial-erase for 3D NAND-based high density SSDs. In: Agrawal, N., Rangaswami, R. (eds.) 16th USENIX Conference on File and Storage Technologies, FAST 2018, Oakland, CA, USA, 12–15 February 2018, pp. 67–82. USENIX Association (2018). https://www.usenix.org/conference/fast18/presentation/liu

15. Liu, C., Kotra, J.B., Jung, M., Kandemir, M.T., Das, C.R.: SOML read: rethinking the read operation granularity of 3D NAND SSDs. In: Bahar, I., Herlihy, M., Witchel, E., Lebeck, A.R. (eds.) Proceedings of the Twenty-Fourth International Conference on Architectural Support for Programming Languages and Operating Systems, ASPLOS 2019, Providence, RI, USA, 13–17 April 2019, pp. 955–969. ACM (2019). https://doi.org/10.1145/3297858.3304035

16. Lu, Y., Shu, J., Zhang, J.: Mitigating synchronous I/O overhead in file systems on open-channel SSDs. TOS **15**(3), 17:1–17:25 (2019). https://doi.org/10.1145/3319369

17. Luo, Y., Ghose, S., Cai, Y., Haratsch, E.F., Mutlu, O.: HeatWatch: improving 3D NAND flash memory device reliability by exploiting self-recovery and temperature awareness. In: IEEE International Symposium on High Performance Computer Architecture, HPCA 2018, Vienna, Austria, 24–28 February 2018, pp. 504–517. IEEE Computer Society (2018). https://doi.org/10.1109/HPCA.2018.00050

18. Nguyen, D.T., Zhou, G., Xing, G.: Poster: towards reducing smartphone application delay through read/write isolation. In: Campbell, A.T., Kotz, D., Cox, L.P., Mao, Z.M. (eds.) The 12th Annual International Conference on Mobile Systems, Applications, and Services, MobiSys 2014, Bretton Woods, NH, USA, 16–19 June 2014, p. 378. ACM (2014). https://doi.org/10.1145/2594368.2601458

19. Wang, X., Li, J., Li, J., Yan, H.: Multilevel similarity model for high-resolution remote sensing image registration. Inf. Sci. 505 (2019). https://doi.org/10.1016/j.ins.2019.07.023

20. Wu, C., et al.: Maximizing I/O throughput and minimizing performance variation via reinforcement learning based I/O merging for SSDs. IEEE Trans. Comput. **69**(1), 72–86 (2020). https://doi.org/10.1109/TC.2019.2938956

21. Wu, F., Lu, Z., Zhou, Y., He, X., Tan, Z., Xie, C.: OSPADA: one-shot programming aware data allocation policy to improve 3D NAND flash read performance. In: 36th IEEE International Conference on Computer Design, ICCD 2018, Orlando, FL, USA, 7–10 October 2018, pp. 51–58. IEEE Computer Society (2018). https://doi.org/10.1109/ICCD.2018.00018

22. Xie, W., Chen, Y.: A cache management scheme for hiding garbage collection latency in flash-based solid state drives. In: 2015 IEEE International Conference on Cluster Computing, CLUSTER 2015, Chicago, IL, USA, 8–11 September 2015, pp. 486–487. IEEE Computer Society (2015). https://doi.org/10.1109/CLUSTER.2015.75

23. Yan, S., et al.: Tiny-tail flash: near-perfect elimination of garbage collection tail latencies in NAND SSDs. TOS **13**(3), 22:1–22:26 (2017). https://doi.org/10.1145/3121133

24. Zhu, Z., Han, G., Jia, G., Shu, L.: Modified DenseNet for automatic fabric defect detection with edge computing for minimizing latency. IEEE Internet Things J. **7**(10), 9623–9636 (2020)

25. Zhu, Z., Tan, L., Li, Y., Ji, C.: PHDFS: optimizing i/o performance of HDFS in deep learning cloud computing platform. J. Syst. Arch. **109**, 101810 (2020)

# A Location Method for Reliability-Critical Paths Based on Monte Carlo Search Tree

Ziwen Sun$^{(\boxtimes)}$ (iD), Qiou Ji, Zecheng Wu, and Aizhu Liu

Zhejiang University of Technology, Hangzhou 310023, China
sziwen@foxmail.com

**Abstract.** The development of nanoscale technology for integrated circuit chips highlights the reliability of circuit reliability. The path formed by the basic unit of the circuit has a significant influence on the overall reliability of the circuit. Therefore, it is necessary to quickly and effectively locate the path to understand the difference in reliability between logic circuits of different structures in time. Combined with the Monte Carlo tree search strategy and based on relaxation operations, this paper realizes the rapid location of sensitive paths in the combinational circuit while ensuring the accurate location of sensitive paths.

**Keywords:** Reliability evaluation · Sensitive unit · Monte Carlo tree search

## 1 Introduction

With the continuous updating and extensive application of circuit integration technology, the size of transistors keeps shrinking in very large-scale integrated circuits. The reduction in the size of the transistor leads to a reduction in the total power of the circuit and the supply voltage [1, 2]. The situation caused a significant reduction in charge of the circuit nodes and ultimately reduced the reliability of the circuit [3–5]. The reinforcement strategy for the basic unit of the circuit can effectively improve circuit reliability. As we all know, fault simulation is expensive in terms of computation time and storage space. Therefore, improving the efficiency of fault simulation is of great significance.

M. Abramovic et al. proposed critical path tracing–a top-down fault simulation method. The Algorithm (CPT) was first introduced in [6], which is an alternative to the original traditional fault simulation method. Initially, the algorithm does not consider the criticality of the fan-out source, and can only be used to detect fixed faults, that is, fixed 0 Fault (SA 0) and fixed one fault (SA 1). In [7], the effect of fan-out re-convergence is considered, and the accuracy of the original CPT is improved. The accuracy version still has a high time complexity, and the fan-out analysis in its improved method results in additional time consumption. [8] Heuristic method is used to analyze the re-convergent fan-out, but it is still an approximate method. Then in [9], the approach of label transfer is used to determine the criticality of the fan-out source. At the same time, based on the 6-value symbol simulation, the critical path tracing algorithm is

X. Chen et al. (Eds.): ML4CS 2020, LNCS 12488, pp. 100–114, 2020.
https://doi.org/10.1007/978-3-030-62463-7_10

applied to the detection of transition delay faults, which is falling delay faults (SF 0) and rising delay fault (SR 1). This method uses symbolic values to represent logical value transitions. Although avoiding the need to consider the size of the delay in the traditional process in timing simulation, still, it does not contemplate the X value and Z value, and the self-shielding phenomenon of the fan-out source is ignored. [10] A series of rules are given to determine the criticality of the fan-out source, and considered the X value. However, the application of the rules is not suitable for multi-input logic gates. [11] An accurate and effective critical path tracing algorithm is proposed for fault simulation.

Analysis found that the above method usually only considers the influence of single input, and ignores the influence of self-shielding and multi-path sensitization on the criticality of the fan-out source. Meantime, it does not consider the difference of the sensitivity levels of different units in the path. In order to overcome the shortcomings of the existing methods, the sensitivity of each unit is obtained based on the benchmark method. Then use topological sorting [12] and Monte Carlo tree search [13] two strategies to locate sensitive paths. This paper implements a reliability-oriented circuit sensitivity path location, which helps to quickly grasp the circuit the overall reliability level and the corresponding central impact unit.

The structure of this paper is as follows. Section 2 introduces the development of CPT. Section 3 introduces the research method. Section 4 introduces the performance evaluation and discusses the experimental results. Section 5 summarizes the full paper.

## 2   Development of CPT

We will use the concept of a Criticality as defined in [14].

**Definition 1:**  A line $L$ has a critical value $v$ in the test (vector) $t$ iff $t$ detects the fault L $s$-$a$-$v$. A line with a critical value in $t$ is said to be critical in $t$.

The traditional test evaluation method includes determining the path of the critical line through the backtracking process starting from the PO, which is called the critical path. By finding the critical line in a test (vector) $t$, we can know the fault detected by (vector) $t$.

**Definition 2 [6]:**  The gate input $I$ is sensitive if supplementing the value of $I$ changes the value of the gate output.

**Definition 3:**  If at least one input of the gate is sensitive, the gate is sensitive; otherwise, the gate is insensitive.

From the Definitions 1, 2, and 3, the principle of traditional CPT can be easily explained. For each test vector t, a fault-free simulation is fulfilled to determine the value held by each network in the circuit. Then, start the backtracking process from each Primary Output (PO). It tracks each critical network input through the sensitive gate and then stops when it reach-es the Primary Input (PI) or insensitive gate. Nevertheless, the method can only guarantee accurate results in the fanout-free regions (FFR) of the circuit.

Later CPT considered the effect of self-masking and proposed an ac-curate and effective CPT algorithm for fault simulation in digital circuits. This method reduces the number of gates to be processed. The algorithm uses an 18-valued algebra, including values 'X' and 'Z' (three-state and high-impedance), and performs both transition and stuck-fault simulations in one pass. Using this type of algebra can distinguish between fixed faults and transient faults.

## 3   Research Methods

Combining circuit and graph theory, and applying the shortest path strategy to locate the sensitive path, a method is proposed in this paper to eliminate the influence of self-shielding and multi-path sensitization on the criticality of the fan-out source. The circuit unit sensitivity is given by the benchmarking strategy locating method. The weights between the inter-connected units are initialized to solve the effect of a multi-input vector driving on a sensitive path locating. First, calculate the sensitivity of the circuit unit through the benchmarking strategy. The specific details are shown in the Sensitivity Calculation () [15]. Finally, this paper proposes two algorithms to locate the sensitive path accurately.

---

Sensitivity Calculation ()

1.$\forall t \in \{1, 2, ..., Ng\}$, set $Nrg_t = 0$ and set $CGR_t = 0$
2.$\forall j \in D$, set $Nrs_j = 0$
3.create a sample set $siv$ for input vectors by Input Vectors ()
4.create a sample set $srg$ for reliability-critical gates by Reliability-critical Gates ()
5.FOR $j = 1$ to D
6.   $\forall t \in \{1, 2, ..., k\}$: extract a sample $srg_i$ from $srg$ and encode the gates by the method presented, then perform $Nrs_j = \text{sum}(srg_i)$
7.   extract a sample $siv_j$ from $siv$ and set it as the input vector of the circuit
8.   calculate the sensitivity $sen_j$ for the combination
           samples of $siv_j$ with $srg_i$
9.   FOR $t = 1$ to $Ng$
10.      IF $srg_i(t) = 1$
11.         $Nrg_t = Nrg_t + 1$
12.            $CGR_t = CGR_t + sen_j/Nrs_j$
13.      END IF
14.   END FOR
15.END FOR
16.FOR $t = 1$ to $Ng$
17.   set $CGR_t = CGR_t/Nrg_t$
18.END FOR

---

### 3.1 Obtain the Sensitivity of Circuit Unit

The sensitivity of each circuit unit is obtained according to the benchmarking strategy. The object of benchmarking analysis is the sensitive unit returned by the pruning technology based on the input vector. The purpose is to establish a benchmarking comparative analysis model to quickly and effectively distinguish the sensitivity of these sensitive units, and accurately obtain it by combining with the SCA algorithm Its final sensitivity.

1. Input vector produced by the uniform non-Bernoulli sequence
2. Design of sampling method for sensitive unit
3. Construction of the sensitivity calculation method based on the benchmarking model
4. Design adaptive convergence strategies
5. Method integration

The simulation results show the advantages of the proposed method in terms of locating accuracy, computational stability, and space-time overhead.

### 3.2 Algorithm 1: Topological Sorting and Relaxation

Considering that the circuit has no directed loop and negative loop, the circuit can be regarded as a directed acyclic graph (DAG) [12]. Therefore, the sensitive path can be quickly and accurately located by the relevant strategies in graph theory. This paper uses topological sorting and relaxation operations to locate sensitive paths. The detailed steps are as follows.

1. Netlist analysis and related quantity initialization
2. Topological sorting of circuit units
3. locate the circuit's sensitive path based on relaxation operation

In step 1, read the netlist and extract the necessary unit information of the circuit and its interconnection with other similar parts. Furthermore, build the corresponding integrity list LC. The details are presented in Netlist Analysis (). Then in step 2, extract the unit in-degree information in the integrity linked list LC to sort the circuit units topologically. Finally, use the sequence obtained by the topological sorting of circuit units to perform the relaxation operation in step 3. The idea of the algorithm is to select a vertex with a degree of 0 from the circuit network to output, then delete this vertex, and delete an arc with a vertex as a tail, continue to repeat this step until all vertices are output or there is no degree of 0 in the circuit to the apex. The concept of "relaxation" comes from the analogy of the estimation of the shortest path and the length of the spiral tension spring, which is not designed for compression. Initially, the cost of the shortest path was overestimated, like an extended spring. When a shorter path is found, the cost is estimated to decrease, and the spring relaxes. Finally, the shortest path is found, and if it exists, the spring will relax to the length of the rest. The result of the relaxation step can be regarded as the constraint relaxation $d[n] \leq d[m] + w\ (m, n)$, According to the triangle inequality, it must be satisfied if and only if $d[m] = \delta\ (s, m)$ and $d[n] = \delta\ (s, n)$. That is, if $d[n] \leq d[m] + w\ (m, n)$, there is no pressure to satisfy this constraint, then the constraint is relaxed.

| Netlist Analysis () |
| --- |
| Input: ISC format netlist |

Output: Original output linked list ($LO$), integrity linked list ($LC$)

1.   Compile the ISC format netlist, initially related variables.
    1.1)     Initialize the variable $j$, let $j = 1$.
    1.2)     Traverse the ISC netlist from the beginning to determine the traversal location. If it does not reach the tail, extract the $j_{th}$ data and generate the corresponding basic unit $g_j$, find its input information in the $LC$; if it has traversed to the end, go to Step 1.5.
    1.3)     Determine whether the information of $g_j$ is known, if it is not known, store it in the non-integrity list $LI$, and jump to step 1.5; otherwise, go to step 1.4.
    1.4)     Store the unit at the end of the $LC$ to determine whether it is the output of the circuit. If it is, store it with the information in the $LC$ and store it in the $LO$.
    1.5)     Let $j = j + 1$ and go to step 1.2.
2.   Improve the $LC$ and process the nodes in the non-integrity list $LI$
3.   End the algorithm.

Topological sequence: Let $T = (V, E)$ be a directed graph with $n$ vertices, and the sequence of vertices $V_1, V_2, \ldots, V_n$ in V is satisfied if there is a path from the vertex $V_i$ to $V_j$. Then in the vertex sequence, the vertex $V_i$ must precede the vertex $V_j$. Such a vertex sequence is called a topological sequence, and the topological sequence is not unique. Topological sorting is the process of constructing a topological sequence. When there is no cycle in the circuit network, all vertices are output.

The running time complexity of the line 1 of the algorithm is $O\ (V + E)$, and the running time complexity of the 2 to 6 lines is $O\ (V)$. Line 7 iteration $|\ V\ |$ times, Line 8 iterates over each edge of the node, the relaxation operation from line 9 to line 13 takes linear time, so the total running time complexity is $O\ (V + E)$. Therefore, Algorithm 1 can complete sensitive path location in linear time. The details are shown in Algorithm 1.

| Algorithm 1: DAG-Sensitive-Path $(T, w, s)$ |
|---|
| Input: *LO* and *LC* related data |
| Output: Sensitive-Path |
| 1. Topological sorting of the vertices of $T$ |
| 2. Initialization- single source $(T, s)$ |
| 3.    FOR each vertex $n \in T.V$ |
| 4.        $n.d = \infty$ |
| 5.        $n.\pi = NIL$ |
| 6.    EDN FOR |
| 7. FOR each vertex $m$, taken in Topological sort order |
| 8.    FOR each vertex $n \in T.\text{Adj}[m]$ |
| 9.        Relax $(m, n, w)$ |
| 10.          IF $n.d > m.d + w(m,n)$ |
| 11.              $n.d = m.d + w(m,n)$ |
| 12.              $n.\pi$ |
| 13.          END IF |
| 14.      END FOR |
| 15. END FOR |

Reading the netlist to extract basic gates, original input terminals, original output terminals, and other information shows that the circuit can be treated as a directed acyclic graph, and the shortest path Strategy in graph theory can be used to generate a sensitive path based on the reliability of the circuit unit. At present, the classic algorithms on the shortest path in graph theory: Dijkstra's algorithm, Floyd algorithm, and A * algorithm, and others. However, the time complexity of these algorithms is relatively high. For example, the complexity of Dijkstra's algorithm is $O(n^2)$, and the complexity of Fibonacci heap is $O(E + V \lg V)$; the time complexity of Floyd's algorithm is $O(n^3)$, and the space complexity is $O(n^2)$. This paper proposes a locating method for linear complexity $O(V + E)$ in conjunction with the specific situation of the circuit.

### 3.3    Algorithm 2: Monte Carlo Tree Search

Adopting the Monte Carlo tree search strategy idea (MCTS) in Alpha Go can locate more accurate sensitive paths in a faster time. Alpha Go is an engineering masterpiece that integrates several methods. One of the core components of Alpha Go is the Monte Carlo tree search-a specific variant of tree traversal using the PUCT function [16].

The rapid attention of MCTS is mainly due to the success of the Alpha Go program and its potential application in many problems. It combines the generality of random simulation and the accuracy of the tree search. Theoretically, in addition to the game itself, MCTS can be used in any field where {state, action} is used to define and use simulation to predict the output. MCTS can be seen as a general technique for making decisions in a perfect information game scenario. When the MCTS strategy is applied to the combinational circuit, it is not so complicated. In the combinational circuit, this paper uses the Monte Carlo tree search strategy to locate the sensitive path obtained by sorting the sensitive unit every time. This strategy is a method for making the best decisions in artificial intelligence problems, usually in the form of a combination of mobile plans in the game.

The search tree is constructed according to the fan-out structure of the combinational circuit and the circuit unit. The process can be divided into the following steps, as shown in Fig. 1:

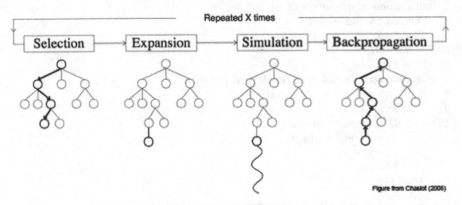

**Fig. 1.** Search tree construction process.

1. Netlist analysis and initialization source
2. Selection: Go down from the root node T and select an optimal child node in the subsequent node according to the following formula *UCT*. The PUCT algorithm is defined in the Alpha paper, so we can directly select the optimal child node according to the formula.
3. Expansion: Take the current node as the parent node and add an unexpanded child node. The full expression is shown in Fig. 2.

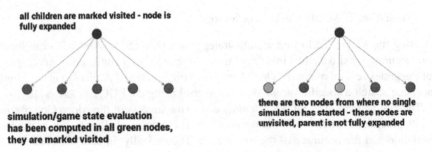

**Fig. 2.** Explanation of node not fully expanded during the expansion.

4. Simulation: Through the expansion process recursively to the termination node, the final score of the path is obtained.

5. Backpropagation: Recursively update all parent nodes from the last node. That is, all parent nodes need to add the current simulated path score. When the simulation of a node is completed, the result is propagated back to the root node T of the current tree. Backpropagation ensures that the statistical information of each node can reflect the simulation results of all descendants of that node. This process ensures accurate updates of sensitive paths.
6. Locate the sensitive path: Locate the sensitive path based on the backpropagation score.

In the construction of the search tree, since we have not performed any simulations, we first select the nodes that are not traversed during the expansion phase. The node that has never been traversed starts the simulation, and the simulation result is propagated back to the root node $T$. In order to select the next node on our path to start the next simulation, we need to consider the information all the child nodes of $v_1, v_2, ..., v_k$ and $v$ itself. This information forms a part: Upper Confidence Bound applied to Trees ($UCT$). $UCT$ is a function that allows us to select the next node from the visited nodes to traverse and is also the core function of MCTS.

$$UCT(v_i, v) = Q(v_i)/N(v_i) + c\sqrt{\log(N(v))/N(v_i)}$$

During MCTS traversal, the current node maximizes the value of $UCT$. First, the $UCT$ function is a function of the node $v$ and its child nodes of $v_1, v_2, ..., v_k$. It consists of two parts. The previous part is also called the exploitation component, which can be regarded as the priority estimate of the child node $v_1, v_2, ..., v_k$ (total score /total number of times = average revenue per time). The central concept of MCTS is searching. The search is a set of traversals along the search tree. A single traversal is a path from the root node $T$ (current state) to an incompletely expanded node. An incompletely expanded node means that it has at least one unvisited child node. When encountering a node that is not fully expanded, select one of the child nodes of the node that has not been visited for a simulation. The result of the simulation is then backpropagation to the root node $T$ of the current tree, and the node's statistics are updated. When the search is over, we can decide which path is sensitive based on the collected statistical information.

Selection phase starts from the root node $T$. It recursively selects the optimal child node until reaching the leaf node $N$. Then in the expansion phase, if the child node is not the termination node (that is, it will not cause the process to terminate), then create one or more child nodes and select one of $v_1, v_2, ..., v_k$. Simulation phase: starting from termination to run a simulated output until the root node of the tree, judging the criticality of the identified sensitive path based on the final output. Finally, at the backpropagation selection stage, the simulation results are used to update the current sensitive path sequence. The specific operation is shown in Algorithm 2.

| Algorithm 2: MCTS Locating Sensitive Paths |
|---|

Input: Current strategy function and circuit status information

Output: sensitive path

1.  Netlist analysis and initialization source
2.  Choose according to the current state
3.  Calculate the value of the current node
4.  Select the next best child node according to the strategy function
5.  FOR each child node $t$
6.      IF Reaching the end of the circuit
7.          Output Sensitive Path
8.      ELSE Expand the child node
9.      END IF
10. END FOR
11. Expand the tree by generating new child nodes
12. The simulation process generates a complete tree
13. Score each path
14. FOR every current node
15.     Update the score of the current child node
16.     Recursively update the scores of all predecessor nodes of the current child node
17. END FOR
18. Count node scores on the path
19. Output sensitive path

The output of the search tree is a score where each node represents a circuit unit. Moving from a node to a child node of that node (if it exists) and the number of children of a node is called the branching factor. The end node of the Search tree is a node that has no child nodes. The root node T of the Search tree represents the initial state of the circuit. The heuristic of MCTS is that MCTS does not require any knowledge of a given domain strategy or specific practical knowledge to make reasonable decisions. This algorithm can work effectively without any knowledge about the circuit except basic rules. An advantage of Algorithm 2 is that it can locate multiple sensitive paths, while Algorithm 1 can only locate one.

## 4 Performance Evaluation

This chapter verifies the performance of the proposed method by constructing relevant simulation experiments. The experimental results constructed the calculation formulas of accuracy and stability according to the size of the circuit scale and established a complete evaluation standard. First, the data verifies the accuracy of the proposed method by the MC method, and then the stability of the proposed method is verified. Finally, based on the proposed method, an experimental analysis of some relevant validity is carried out.

As we all know, except for the MC method, there is no recognized effective method that can accurately locate the sensitive path in the circuit. However, the MC model is only suitable for small circuits, and accurate results cannot be obtained in large-scale circuits. This section selects some typical 74 series and ISCAS85 series circuits and uses the proposed method to locate these circuits with sensitive paths. The sensitive path location results of some circuits are shown in Table 1.

It can be seen from Table 1 that the proposed Algorithm 1 (topological sorting and relaxation operation) and the accurate MC method's sensitive path location results are consistent, and from the perspective of the fitness of the location results, it is better than the traditional accurate strategy-MC method; the proposed Algorithm 2 (Monte Carlo tree search) location results are basically consistent with the MC method. The location of the sensitive path based on the method proposed in this paper is close to the accuracy of the MC model. Among the five small and medium circuits tested, the Algorithm 1 achieves an accuracy of 100% on all five circuits; Algorithm 2 The accuracy of 100% is achieved on 4 of these circuits, and the total average accuracy on other circuits is 99%.

In order to further verify the performance of the method proposed in this paper, we selected more classic circuits. The time consumption and Memory consumption of the three methods on different circuits are shown in Table 2. The location time of the two methods mentioned in this paper is faster than the traditional method. The fastest is Algorithm 1, and Algorithm 2 is slightly slower than Algorithm 1. From the average results, the proposed Algorithm 1 and Algorithm 2 are better than the MC method is 121.350797 times faster and 85.803935 times faster, respectively. The memory overhead of the three methods is similar, and the smallest is Algorithm 1, followed by Algorithm

**Table 1.** Sensitive Paths Simulation Results of Circuits.

| Method | Circuit | Sensitive path |
| --- | --- | --- |
| MC model | 74182_63 | [9, 12, 62, 63] |
| | c17_22 | [2, 16, 20, 22] |
| | c2670_1971 | [224, 394, 1823, 1971] |
| | c3540_3987 | [50, 398, 950, 1244, 2133, 2325, 2344, 2755, 2973, 3201, 3484, 3531, 3835, 3885, 3925, 3987] |
| | c7552_10025 | [367, 581, 871, 880, 9893, 10025] |
| Algorithm 1 | 74182_63 | [9, 12, 62, 63] |
| | c17_22 | [2, 16, 20, 22] |
| | c2670_1971 | [224, 394, 1823, 1971] |
| | c3540_3987 | [50, 398, 950, 1244, 2133, 2325, 2344, 2755, 2973, 3201, 3484, 3531, 3835, 3885, 3925, 3987] |
| | c7552_10025 | [367, 581, 871, 880, 9893, 10025] |
| Algorithm 2 | 74182_63 | [9, 12, 62, 63] |
| | c17_22 | [2, 16, 20, 22] |
| | c2670_1971 | [224, 394, 1823, 1971] |
| | c3540_3987 | [116, 456, 763,1059, 1490, 1707, 2138, 2334, 2345, 2756, 2977, 3204, 3384, 3528, 3697, 3710, 3777, 3885, 3925, 3987] |
| | c7552_10025 | [367, 581, 871, 880, 9893, 10025] |

**Table 2.** Comparison of Time and Memory Consumption of different methods.

| Circuit | MC model | | Algorithm 1 | | Algorithm 2 | |
|---|---|---|---|---|---|---|
| | Time (s) | Memory (MB) | Time (s) | Memory (MB) | Time (s) | Memory (MB) |
| 74157_40 | 0.508728 | 44.359374 | 0.004996 | 43.968749 | 0.009990 | 43.996093 |
| 74182_63 | 1.242812 | 44.363750 | 0.010994 | 44.093749 | 0.012990 | 44.304686 |
| c17_22 | 0.743576 | 44.429686 | 0.006995 | 44.050781 | 0.008994 | 44.335936 |
| c17_23 | 0.689605 | 44.398436 | 0.010988 | 43.992187 | 0.011993 | 44.093749 |
| c2670_1971 | 1.323362 | 44.417969 | 0.012987 | 44.000001 | 0.014993 | 44.023437 |
| c3540_1713 | 0.560660 | 44.421875 | 0.010993 | 44.183594 | 0.012992 | 44.363281 |
| c3540_3987 | 14.855019 | 44.554688 | 0.122930 | 44.156251 | 0.133925 | 44.468750 |
| c5315_4737 | 1.445177 | 44.375001 | 0.009996 | 43.953124 | 0.012994 | 44.062499 |
| c5315_4739 | 1.422883 | 44.335938 | 0.007996 | 44.062499 | 0.010994 | 44.296874 |
| c7552_10025 | 3.264456 | 44.414063 | 0.013992 | 44.292968 | 0.036977 | 44.355469 |

2 and MC method. It can be seen that the two methods mentioned in this paper have higher accuracy and faster running time when the memory overhead is similar. Although the time and memory overhead of Algorithm 2 is slightly larger than that of Algorithm 1, the heuristic of Algorithm 2 is greater than the former, and it is more conducive to use in extensive circuits.

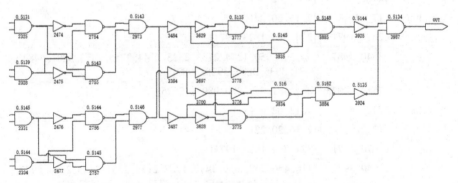

**Fig. 3.** c3540_3987 circuit.

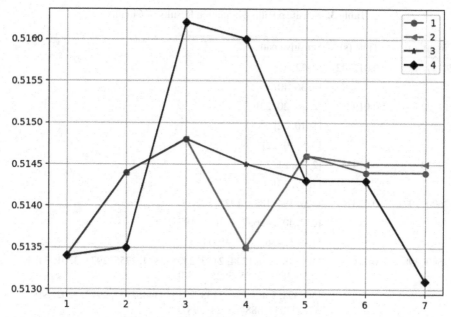

**Fig. 4.** c3540_3987 gate sensitivity changes.

Algorithm 2 is different from the MC method in the c3540_3987 circuit in Table 1 because of the fan-out structure of the circuit. The sensitivity relationship of the circuit node and part of the circuit structure of c3540_3987 are shown in Fig. 3, because the sensitivity of the leading node is low, which causes the location result to be different from the MC method. Some of the circuit gates that are not marked with sensitivity are because these units are buff gates and have no effect on the simulation results. Shown in Fig. 4 are the changes in the sensitivity of the circuit unit before the original output in the four paths, which graphically illustrates the reasons for the different simulation results. In the subsequent improvement, the better node selection strategy was changed, and now the result of the sensitive path location of the Algorithm 2 has been consistent with the MC method.

The time and memory overhead of Algorithm 2 is slightly higher than Algorithm 1. The advantage of Algorithm 2 is that it can locate multiple sensitive paths, but Algorithm 1 can only locate one. Table 3 shows the sensitive path location results of some circuits. The importance of the sensitive paths in the circuit is from top to bottom. The sensitive path of the c17 circuit includes c17_22 and c17_23, as shown in Fig. 5.

The locating method of the sensitive circuit unit based on the benchmarking strategy ensures the validity of its locating results by using operations such as pruning, benchmarking, and uniform non-Bernoulli sequence. The sensitive path location method (Algorithm 1) based on topological sorting and relaxation operations ensures the effectiveness of its locating results and the complexity of linear time through depth-first search operations. The application of topological sorting speeds up the convergence

**Table 3.** Sensitive Paths Simulation Results of Circuits.

| Circuit | Time (s) | Sensitive path |
|---|---|---|
| 74157_40 | 0.012742 | [2, 32, 40] |
| | | [6, 36, 40] |
| c17_22 | 0.019983 | [2, 16, 20, 22] |
| | | [1, 10, 22] |
| | | [3, 8, 10, 22] |
| c17_23 | 0.014331 | [2, 16, 21, 23] |
| | | [7, 19, 23] |
| c2670_1971 | 0.02376 | [224, 394, 1823, 1971] |
| | | [237, 402, 496, 795, 1823, 1971] |
| | | [1, 405, 486, 1823, 1971] |
| c3540_3987 | 0.416934 | [50, 398, 950, 1244, 2133, 2325, 2344, 2755, 2973, 3201, 3484, 3531, 3835, 3885, 3925, 3987] |
| | | [50, 398, 950, 1244, 2133, 2325, 2344, 2755, 2973, 3202, 3484, 3530, 3834, 3884, 3924, 3987] |
| | | [116, 456, 763, 1059, 1490,1702, 2137, 2331, 2346,2757,2977,3204,3384,3528, 3697, 3710, 3777, 3885, 3925, 3987] |
| c7552_10025 | 0.041974 | [367, 581, 871, 880, 9893, 10025] |
| | | [367, 581, 871, 1108, 1115, 9740, 10025] |

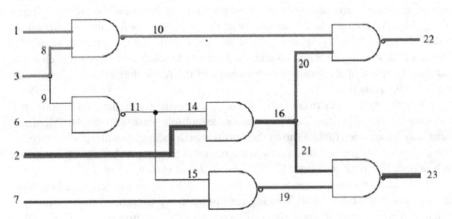

**Fig. 5.** c17 Sensitive Path-c17_22, c17_23.

process of sensitive unit analysis because it avoids an unnecessary comparison of sensitive units with similar sensitivity. The sensitive path location method (Algorithm 2) based on Monte Carlo Tree Search (MCTS) ensures the fast and enlightening results of its locating results through operations such as optimal decision; MCTS provides a better method than traditional tree search. The MC method will inevitably lead to additional memory overhead. From the experimental data, it can be seen that the MC method memory overhead is slightly larger than the two algorithms proposed in this paper.

## 5 Conclusion

Locating sensitive paths in circuits is an essential step in improving circuit reliability at low cost, and is also an essential goal in evaluating the reliability of circuit structures. In this paper, the proposed method is experimentally verified, and some related application examples are briefly given. Algorithm 1 has a linear advantage in terms of complexity. Algorithm 2 can locate multiple sensitive paths and heuristic advantages. The simulation results on the benchmark circuits prove the advantages of the pro-posed method in terms of accuracy and efficiency. Therefore, this meth-od can play an important role in improving circuit reliability at a low cost.

## References

1. Xiao, J., Jiang, J.H., Li, X.X., et al.: A novel trust evaluation method for logic circuits in IoT applications based on the E-PTM model. IEEE Access 6(7), 35683–35696 (2018)
2. Sridharan, K.: A transistor-level probabilistic approach for reliability analysis of arithmetic circuits with applications to emerging technologies. IEEE Trans. Reliab. 66(2), 440–457 (2017)
3. Xiao, J., Lee, W., Jiang, J.H., et al.: Sensitivity evaluation of input vectors with masking effects in digital circuits. Chin. J. Comput. 41(10), 2282–2294 (2018)
4. Bickford, J., Habib, N., Li, B., et al.: Integrated circuit chip reliability qualification using a sample-specific expected fail rate. USA Patent (2018)
5. Han, J., Hao, C., Liang, J.H., et al.: A stochastic computational approach for accurate and efficient reliability evaluation. IEEE Trans. Comput. 63(6), 1336–1350 (2014)
6. Abramovici, M., Menon, P.R., Miller, D.T.: Critical path tracing: an alternative to fault simulation. IEEE Des. Test Comput. 1(1), 83–93 (1984)
7. Menon, P.R., Levendel, Y., Abramovici, M.: Critical path tracing in sequential circuits. In: IEEE International Conference on Computer-aided Design, ICCAD-88 Digest of Technical Papers. IEEE (1988)
8. Favalli, M., Olivo, P.: A novel critical path heuristic for fast fault grading. IEEE Trans. Comput. Aided Des. Integr. Circuits Syst. 10(4), 544–548 (1991)
9. Girard, P., Landrault, C., Pravossoudovitch, S.: A reconvergent fan-out analysis for the CPT algorithm used in delay-fault diagnosis. In: European Test Conference. IEEE (1993)
10. Wu, L., Walker, D.M.H.: A fast algorithm for critical path tracing in VLSI digital circuits. In: 20th IEEE International Symposium on Defect and Fault Tolerance in VLSI Systems, 2005. DFT 2005. IEEE (2005)
11. Bosio, A., Girard, P., Pravossoudovitch, S.: An exact and efficient critical path tracing algorithm. In: Fifth IEEE International Symposium on Electronic Design, Test and Application, 2010. DELTA 2010. IEEE (2010)

12. Pang, Chaoyi., Junhu Wang, Yu., Cheng, Haolan Zhang, Li, Tongliang: Topological sorts on DAGs. Inf. Process. Lett. **115**(2), 298–301 (2015)
13. Bayer, K., Koch, S., Klein, R.: Monte carlo tree search. Wirtschaftswissenschaftliches studium **47**(12), 11–18 (2018)
14. Wang, D.T.: An algorithm for the generation of test sets for combinational logic networks. IEEE Trans. Comput. **C-24**(7), 742–746 (1975)
15. Xiao, J., Shi, Z.H., Jiang, J.H., et al.: A locating method for reliability-critical gates with a parallel-structured genetic algorithm. J. Comput. Sci. Technol. **34**(5), 1136–1151 (2019)
16. Silver, D., Schrittwieser, J., Simonyan, K., et al.: Mastering the game of Go without human knowledge. Nature **550**(7676), 354–359 (2017)

# Research of Expressway Vehicle Detection Based on Faster R-CNN and Domain Adaptation

Meng Yu, Yong Hu, Guanxiang Yin, Fei Du, Meng Wang, and Yuejin Zhang[✉]

School of Information Engineering, East China Jiaotong University, Nanchang 330013, China
zyjecjtu@foxmail.com

**Abstract.** For the traditional object detection model affected by weather, illumination and occlusion factors, the application scene is single, high missing rate of small vehicle and poor detection results problems in the expressway scene, this paper put forward an improved domain adaptive Faster R-CNN algorithm, by adding image-level, instance-level domain classifier, consistency loss component, and improvement RPN network. Adopting multi-scale training during the training and using mining the hard samples fine tuning the train model. The improved model can get the gain by 4.8%. The experimental results show that the adaptive domain is effective to deal with the domain transfer between different domains, and the improved model can significantly raise the detection performance of small-scale object detection. It is proved that the improved method effectively enhance the accuracy and robustness of the model, and also has certain generalization ability.

**Keywords:** Faster R-CNN · Expressway · Domain adaption · Small-scale object · Hard sample mining

## 1 Introduction

With increase of the vehicle number, the traffic problem will be suffered the more challenge. In the practical application process, the traditional vehicle detection algorithm based on video has problems such as low detection accuracy and slow speed in complex environment. With the research on the deep learning, founding convolutional neural network is very suitable for detection. Compared with the traditional detection model, the detection accuracy of convolutional neural network (CNN) model can be significantly improved. The detection algorithm based on CNN has been achieved excellent performance in the benchmark dataset [1], but it still faces great challenges when applied to the expressway vehicle detection. If we use the classical detection algorithm directly applied to the expressway vehicle detection, the following problems will occur:

(1) The traditional model is easily affected by illumination, bad weather, occlusion and out of view factors, and its robustness and real-time performance are poor. Because of the visual differences, it is difficult to apply the detection model learned from one domain to another [2]. Therefore, when faced with large and complex domain shifts, how to preserve image-objects and maintaining structure consistency is of great significance [3].

© Springer Nature Switzerland AG 2020
X. Chen et al. (Eds.): ML4CS 2020, LNCS 12488, pp. 115–124, 2020.
https://doi.org/10.1007/978-3-030-62463-7_11

(2) It is generally considered that training data and test data are extracted from similar or the same distribution. In the actual scene, the distribution mismatch between the source and the target domain may lead to significant degradation in performance of model [4].

(3) For small-scale object, there are some problems for model such as false, missing detection and poor detection accuracy. Multi-scale problems caused by local occlusion and object distance changes in the collected image [5]. The small-scale object contains little information, it is difficult to mark or mark inaccuracy the training data leads to the bad results.

Base on the above problems, this article put forward an improved domain adaptive Faster R-CNN model [6] by adding to three domain adaptive components and improved the structure of RPN network, and establish end-to-end model. Construct expressway datasets containing multiple weather and scale change scenes. Using residual network and multi-scale training, hard sample mining strategy raise the performance of the model. Finally, experiments with multiple datasets.

## 2    Improved Domain Adaptive Faster R-CNN Model

As shown in Fig. 1, three components are introduced into the Faster R-CNN [7] model. Image-level domain classifier is added at the end of the convolution layer, and instance-level domain classifier is added after the ROI feature. The classifier keeps the RPN domain unchanged through connection loss of consistency. Based on the H-divergence theory, a domain adaptive component is proposed, it is learned from adversarial training. Two domain classifiers strengthen RPN by uniform regularization to learn a domain invariant model. Experiments show that for small-scale object, the features extracted from the lower layer of CNN are more useful than the upper layer. In this view, through improving the RPN network structure, a multi-scale detection model is obtained. The vehicle detection process of the model is as follows:

(1) Input the image, extract the vehicle contour and color features in Faster R-CNN, feature can be shared by CNN for RPN layer and fully connected layers (FC) layer. (2) Generate regional proposal through RPN network. Output layer determined image category is foreground or background, and obtain the accurate region suggestion by the bounding box. (3) ROI collects step2 feature maps and region proposals, outputs the feature maps of different scales to a uniform scale, and extracts the region proposal feature map sent to FC layer to determine the category. (4) In image-level layer by gradient reversal layer (GRL) connect a domain classifier make model can't distinguish samples of source and target domain, reduce the sensitivity of the features to the domain, and learn the common features. (5) At instance-level layer, perform step4 on the bounding box of the object. Image level classifies each position of the feature map, last regularizes it by loss of RPN.

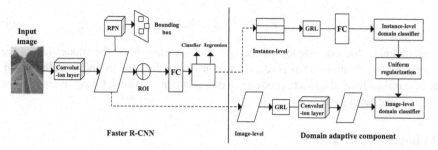

**Fig. 1.** Network architecture

# 3 Vehicle Detection Network Architecture

## 3.1 Domain Adaptive Component

H-divergence [8] measures the divergence between two groups of samples with different distributions. The definitions are as follows: feature vector n, source domain $N_S$, target domain $N_T$, hypothesis domain classifier h:n $\rightarrow$ {0, 1}, $N_S = 0$, $N_T = 1$. Suppose H is a set of domain classifiers, and the H-divergence defines the distance between two domains, such as Formula (1).

$$D_S(S, T) = 2\left(1 - \min_{h \in H}\left(E_{err_S}(h(x)) + E_{err_T}(h(x))\right)\right) \qquad (1)$$

$$\min_l D_S(S, T) \Leftrightarrow \max_l \min_{h \in H}\left\{E_{err_S}(h(x)) + E_{err_T}(h(x))\right\} \qquad (2)$$

Respectively, $E_{rrs}$ and $E_{rrT}$ are the prediction errors of the samples in the source and the target domain samples $h(x)$. Assuming that the error rate of the best domain classifier $D_S(S, T)$ is high, it shows that the distribution of the two domains are very close. So, if we want to align the two domains, we should make the network $l$ output feature vector $x$ to minimize the domain distance $D_S(S, T)$ [9]. This can be optimized in adversarial training.

It is found that the detector is consistent between two domains: to detect any image, no matter which domain the image belongs to, the detection result should be the same. In Faster R-CNN model, the image representation is the feature mapping output of the base convolution layer. So, in order to solve the problem of domain shift, the distribution of the image representations of the two domains is forced to be the same. Formula (3) is called image-level adaptation. It helps to reduce offsets caused by global image differences (image style, image scale, lighting, etc.). Similarly, the semantic consistency between two domains, also the distribution of instance representations is forced to be the same. Formula (4) is called instance-level alignment. It helps reduce local instance differences in object appearance, size, viewpoint, etc., and trains a domain classifier through feature vectors to align instance-level distribution.

$$P_S(I) = P_T(I) \qquad (3)$$

$$P_S(B, I) = P_T(B, I) \tag{4}$$

By maintaining the consistency of the two domain classifiers, the model learns the robustness of the bounding box predictor, so consider adding consistency regularization. Because image-level classifiers produce an output when activated, taking average value of it as image-level probability.

## 3.2 Improvement of RPN

According to the principle of Faster R-CNN algorithm model, we know the feature of region proposal generated by RPN, only accept the feature images of the last of convolution layer through ROI pooling. The high-level features obtained in this way will cause problems for small-scale object detection. The solution is to use multi-scale feature map in CNN network to obtain multi-scale detection. As shown in Fig. 2, through CNN extracted feature maps of different scales from the input image are sent to the RPN of each layer to generate region proposal, then map the generated regions proposal to the feature map. Finally, through ROI pooling layer output uniform size features, and then make it classification, so make full use of low-level features for small object detection.

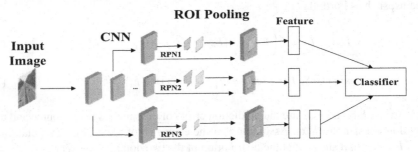

**Fig. 2.** Improvement of RPN

## 3.3 Network Loss Function

The network framework is shown in Fig. 1. In this paper, the domain adaptive component is used to extend the Faster R-CNN infrastructure. Three new components are introduced into Faster R-CNN. The final training loss of the proposal network is the sum of each component part, which can be written as follow. Among $L_0$ is the loss sum of training loss RPN and ROI, $L_1$ is the image level adaptive loss, $D_i$ is the training domain, $D_0$ and $D_1$ represent the source domain and target domain respectively, $p_i^{(m,n)}$ represent the output of the domain classifier, $L_2$ is the case level adaptive loss, $L_3$ is the consistency regularization, $|K|$ represents the total number of activation in the feature mapping, and $l_2$ is the distance.

$$L = L_{RPN} + L_{ROI} + \lambda \left\{ -\sum\nolimits_{i,m,n} \left[ D_i log p_i^{(m,n)} + (1 - D_i) log \left( 1 - p_i^{(m,n)} \right) \right] \right.$$

$$-\sum\nolimits_{i,j}\left[D_i log p_{i,j} + (1 - D_i)log\left(1 - p_i^{(m,n)}\right)\right]$$
$$+\sum\nolimits_{i,j}\left\|\frac{1}{|K|}\sum\nolimits_{m,n}p_i^{(m,n)} - p_{i,j}\right\|_2\right\}$$

(5)

### 3.4 Network Model Training Method

The process of model building is divided into two parts: model training and model testing. In the model training stage, the original ImageNet model downloaded from the network is used, and the data set pre-training model is used to initialize the Faster R-CNN training model. In order to improve the efficiency of the training model, the training is first carried out with the VOC data set, the samples in different scenarios and the vehicle samples from different angles. In order to improve the training speed and efficiency of the model, the end-to-end training mode is selected to replace the alternate training mode. Model training and testing process, as shown in Fig. 3. According to the structure of pre-training model, the vehicle detection network is constructed in the training stage. This method is used to solve the problem of training sample class imbalance and the effective gradient dilution of the model. In the model testing stage, the model is verified by vehicle test samples with different density in different scenarios.

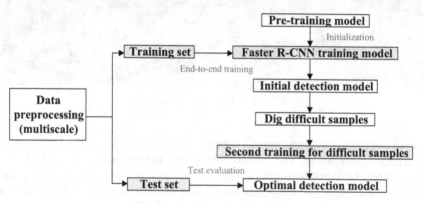

**Fig. 3.** Model training flowchart

In the process of image processing and analysis, in order to obtain as much visual information as possible, different scales of image analysis are taken. Therefore, multiple scales of images are used to train the model. In this paper, three scales (600,850,1100) are trained. It enables the model to learn a series of features of the target and achieve the purpose of improving the detection performance of the model.

## 4 Experiment and Result Analysis

The original Faster R-CNN model was taken as the baseline, the evaluation threshold was 0.5, and $\lambda = 0.1$ was set for all experiments. According to [9] setting super parameters.

**Fig. 4.** VOC dataset detection results

**Fig. 5.** COCO dataset detection results

There are two parts in the experimental dataset: (1) VOC2007, 2012, COCO dataset; (2) Jiangxi Expressway dataset. The dataset contains six types of vehicles: sedan, SUV, light truck, light bus, large bus, heavy truck. Five weather scenes: sunny, rainy, foggy, cloudy and snowy. Select about 10000 pictures as training and test dataset and 500 pictures as incremental test set.

Experiments in multiple datasets VOC2007, VOC2012 and COCO, detection results as shown in Fig. 4. Both VOC2007 and VOC2012 are large-scale datasets with simple object types. It can be seen from the detection results of the algorithm proposed are good, and almost all vehicles can be detected without missing or false detection. For the more difficult COCO dataset, because of the high resolution of the sample images, the moving speed of the vehicle is relatively slow, and there is no background clutters. Most vehicle objects can be detected well as shown in Fig. 5.

As shown in Fig. 6, there are great differences between the expressway scene test set and the training set. Although the domain adaptive component is added to the algorithm before the model is improved, the result of small-scale object detection is still not very good. In the Faster R-CNN model, although the RPN is introduced to extract multi-scale vehicle feature map, it only receive the features extracted from the last layer of the

**Fig. 6.** Detection results in expressway scene

convolutional layer network, which seriously reduces the detection accuracy of the model for small-scale object. It can be seen from Fig. 6 the detection accuracy of small-scale vehicle is not high, and even there is a case of missing detection.

**Fig. 7.** Detection results in different weather scenes in improvement model

In Fig. 7, the detection results shows that the improved model can basically detect small-scale vehicle object in different scenes, but there is overfitting problem, which may be due to the over training of the model. The results indicate that the model has obvious performance improvement for the problem of small-scale object missing detection in the complex environment, and the improved model can effectively solve the problem and meet the detection requirements in the expressway scene.

On Table 1, the mAP results of the VOC2007 dataset is about 1–3% higher than the VOC2012 with VGG16 as the baseline. Because the object size of the VOC07 dataset is larger than the VOC12 and it is less difficulty for detection, the detection result is relatively better. Not surprisingly, it is much better to use VOC07+12 to train than alone, because with the expansion of the sample data set, after training the model can learn more data to make its judgment closer to the correct result, shortcomings are the training time will be longer and expensive cost of label more dataset.

**Table 1.** mAP results of various object detection algorithms in multiple dataset

| Method | Network | mAP(07) | mAP(12) | mAP(07+12) | mAP(COCO) |
|---|---|---|---|---|---|
| R-CNN | VGG16 | 66.2 | 63.1 | 67.1 | |
| Fast RCNN | VGG16 | 69.9 | 67.9 | 70.2 | |
| Faster R-CNN | VGG16 | 76.7 | 74.8 | 78.1 | |
| YOLO | VGG16 | 64.1 | 61.3 | 66.2 | |
| SSD(512) | VGG16 | 75.9 | 74.8 | 76.5 | |
| OUR | VGG16 | 78.0 | 76.5 | 79.8 | |
| Faster R-CNN | ResNet50 | 81.3 | 80.4 | 81.9 | |
| YOLOv2 | ResNet50 | 78.5 | 73.3 | 79.1 | |
| R-FCN | ResNet50 | 77.5 | 76.1 | 79.6 | |
| OUR | ResNet50 | 83.1 | 82.0 | 83.9 | |
| R-FCN | ResNet50 | | | | 49.1 |
| Faster R-CNN | ResNet50 | | | | 55.6 |
| SSD | ResNet50 | | | | 50.1 |
| YOLOv2 | ResNet50 | | | | 43.9 |
| OUR | ResNet50 | | | | 56.8 |
| OUR | ResNet101 | | | | 58.9 |

In COCO dataset, the detection accuracy of mAP is about 20–30% lower than VOC dataset. Because the object scale of COCO is much smaller than VOC, and the background is more complicated, the mAP value is much lower. The mAP of R-CNN [10] algorithm is the lowest of all algorithms, only 66.2%, 19% lower than OUR. Compared with other two stage algorithms, it is 3.7% lower than Fast R-CNN [11], and 10.5% lower than Faster R-CNN. Among the one stage algorithm, the min mAP is YOLO [12] algorithm is 64.1%, which is 11.8% lower than that SSD [13]. YOLOv2 [14] is 78.5% lower 4.6% than OUR, and mAP of R-FCN [15] is 77.5% lower 5.6% than OUR. From Table 1, it show that on ResNet50 baseline network the mAP of R-FCN algorithm is also the lowest only 49.1%, 7.7% lower than OUR, and 6.5% lower than Faster R-CNN. In one stage algorithm, the mAP of YOLOv2 is the lowest only 43.9%, 6.2% lower than SSD algorithm.

It is conclusion that the network model from VGG16 to ResNet50 will increase by 4–5%, and then to ResNet101 will increase by 2%. The detection accuracy has been significantly raise. Finally, it draw a conclusion that mAP of the proposed algorithm in this paper is the best of all the algorithms.

From the results in Table 2, the object detection method in this paper effectively enhance the accuracy of the detection for small-scale vehicles and the discriminative performance of the model about the Faster R-CNN model has a domain transition problem. The method proposed in this paper improves the accuracy of the detection results of Faster R-CNN increase by 4.8%. On the Table 2, through multi-scale training strategy

**Table 2.** Quantitative analysis of domain adaptive results

| Method | Image | Instance | Hard | Multi-scale | S → T | T → S | Recall % |
|---|---|---|---|---|---|---|---|
| Faster R-CNN | | | | | 51.9 | 73.7 | |
| DAF | | | | | 55.1 | 76.1 | |
| OUR | ✓ | | | | 53.5 | 74.8 | |
| | | ✓ | | | 55.8 | 76.3 | |
| | ✓ | ✓ | | | 57.4 | 78.5 | |
| | | | | | | | 84.7 |
| | | | ✓ | | | | 86.6 |
| | | | | ✓ | | | 85.5 |
| | | | ✓ | ✓ | | | 88.9 |

the model has certain robustness for different vehicle sizes and improves the detection performance of the model. The results show that is increased by about 1% through three scales are used for training. The background clutters in the scene is complex, when the vehicle color is similar to the background color, the complex background especially makes the vehicle recognition less obvious, which undoubtedly makes the vehicle location more difficult and easily lead to miss detection result. The strategy of hard sample mining is an effective method to improve the ability of model recognition. The gain of about 2% can be obtained by properly increasing the ratio of hard samples.

## 5  Conclusion

This paper improves the framework of the original Faster R-CNN model and adds the corresponding domain adaptive component. In the training stage use multi-scale image training and hard samples fine-tuning the model, so that the model can learn more complex features and enhance the recognition ability of the model. The experimental results show that the detection method proposed in this paper effectively solves the problems of traditional vehicle object detection methods, improves the detection accuracy of small object and the discrimination of detection model. The reliability of the algorithm is proved, and the test results achieve the expectation.

## References

1. Everingham, M., Van Gool, L., Williams, C.K., Winn, J., Zisserman, A.: The Pascal visual object classes (VOC) challenge. Int. J. Comput. Vision **88**(2), 303–338 (2010). https://doi.org/10.1007/s11263-009-0275-4
2. Gopalan, R., Li, R., Chellappa, R.: Domain adaptation for object recognition: an unsupervised approach. Int. J. Comput. Vis. (2011)
3. Lin, C.-T.: Cross domain adaptation for on-road object detection using multimodal structure-consistent image-to-image translation. In: IEEE International Conference on Image Processing (ICIP), Taipei, Taiwan, pp. 3029–3030 (2019)

4. Lin, T.-Y., et al.: Microsoft COCO: common objects in context. In: Fleet, D., Pajdla, T., Schiele, B., Tuytelaars, T. (eds.) ECCV 2014. LNCS, vol. 8693, pp. 740–755. Springer, Cham (2014). https://doi.org/10.1007/978-3-319-10602-1_48
5. Hu, X., Xu, X., Xiao, Y., et al.: SINet: a scale-insensitive convolutional neural network for fast vehicle detection. Trans. Intell. Transp. Syst. **20**(3), 1010–1019 (2019)
6. Wang, T., Zhang, X., Yuan, L.: Few-shot adaptive faster R-CNN. In: 2019 IEEE/CVF Conference on Computer Vision and Pattern Recognition (CVPR), Long Beach, CA, USA, pp. 7166–7175 (2019)
7. Chen, Y., Li, W., Sakaridis, C., et al.: Domain adaptive faster R-CNN for object detection in the wild. In: IEEE Conference on Computer Vision and Pattern Recognition (2018)
8. Ben-David, S., Blitzer, J., Crammer, K., Kulesza, A., Pereira, F., Vaughan, J.W.: A theory of learning from different domains. Mach. Learn. **79**(1-2), 151–175 (2009). https://doi.org/10.1007/s10994-009-5152-4
9. Ren, S., He, K., Girshick, R., Sun, J.: Faster R-CNN: towards real-time object detection with region proposal networks. IEEE Trans. Pattern Anal. **39**(6), 1137–1149 (2015)
10. Girshick, R., Donahue, J., Darrell, T., et al.: Rich feature hierarchies for accurate object detection and semantic segmentation. In: Conference on Computer Vision and Pattern Recognition, OH, pp. 580–587 (2014)
11. Girshick, R.: Fast R-CNN. In: 2015 IEEE International Conference on Computer Vision (ICCV), Santiago, pp. 1440–1448 (2015)
12. Redmon, J., Divvala, S., Girshick, R., et al.: You only look once: unified, real-time object detection. In: 2016 IEEE Conference on Computer Vision and Pattern Recognition (CVPR), Las Vegas, NV, pp. 779–788 (2016)
13. Liu, W., et al.: SSD: single shot multibox detector. In: Leibe, B., Matas, J., Sebe, N., Welling, M. (eds.) ECCV 2016. LNCS, vol. 9905, pp. 21–37. Springer, Cham (2016). https://doi.org/10.1007/978-3-319-46448-0_2
14. Redmon, J., Farhadi, A.: YOLO9000: better, faster, stronger. In: 2017 IEEE Conference on Computer Vision and Pattern Recognition (CVPR), Honolulu, HI, pp. 6517–6525 (2017)
15. Dai, J., Li, Y., He, K., et al.: R-FCN: object detection via region-based fully convolutional networks. In: Advances in Neural Information Processing Systems (NIPS), Barcelona, Spain, pp. 379–387 (2016)

# A Short-Term Traffic Flow Forecasting Method Based on Support Vector Regression Optimized by Genetic Algorithm

Aiyun Zhan[1], Fei Du[1], Guanxiang Yin[1,2], Meng Wang[1], and Yuejin Zhang[1(✉)]

[1] School of Information Engineering, East China Jiaotong University, Nanchang 330013, China
zyjecjtu@foxmail.com
[2] College of Computer and Control Engineering, Nankai University, Tianjin 300071, China

**Abstract.** This paper uses support vector regression to predict short-term traffic flow, and studies the feasibility of support vector regression in short-term traffic flow prediction. The short-time traffic flow has many influencing factors, which are characterized by nonlinearity, randomness and periodicity. Therefore, support vector regression algorithm has advantages in dealing with such problems. In order to improve the prediction accuracy of the support vector regression, this paper uses genetic algorithm to optimize the support vector regression and other parameters to obtain the global optimal solution. The optimal parameters are used to construct the support vector regression prediction model. This paper selects the traffic flow data of the California Department of Transportation (PEMS) database to verify the feasibility and effectiveness of the model proposed in this paper.

**Keywords:** Machine learning · Support vector machine · Genetic algorithm · Parameter optimization · Penalty factor

## 1 Introduction

With the increasing number of cars in the country, the problem of traffic congestion is becoming more and more serious. Traffic congestion not only delays people's travel, but also reduces the efficiency of economic development and wastes a lot of resources. In order to solve the problem of traffic congestion, Intelligent Transport System (ITS) has gradually attracted more and more people's attention, and traffic flow prediction is the core function of ITS [1]. Traffic flow prediction is an important bridge to build intelligent traffic system, and also an important basis for traffic management departments to effectively manage traffic congestion and other problems [2].

At present, traffic flow prediction methods are mainly divided into two categories: mathematical statistics method and intelligent prediction method [3]. At present, the statistical theoretical models used by many scholars mainly include Historical Average Model, Kalman Filter model and Autoregressive Integrated Moving Average model (ARIMA). The Historical Average Method is the primary method for short-term traffic flow prediction. In this method, the data are averaged in the whole period and the average

© Springer Nature Switzerland AG 2020
X. Chen et al. (Eds.): ML4CS 2020, LNCS 12488, pp. 125–134, 2020.
https://doi.org/10.1007/978-3-030-62463-7_12

value is used as the prediction result. The Historical Average Method is simple, the calculation speed is fast, but the prediction accuracy is low, and it cannot deal with emergencies and complex traffic conditions, so it is suitable for the static prediction system with low accuracy [4]. Iwao et al. applied the Kalman filter model to the short-term traffic flow prediction for the first time, and achieved good prediction results [5]. Ahmaed et al. first used the time series theory in the study of traffic flow prediction. The method used in this study is autoregressive-integrated moving average (ARIMA), which is the most widely used model [6].

Due to the non-linear and stochastic characteristics of traffic flow, the mathematical statistics method is only applicable to the prediction of a single object point, and the prediction accuracy cannot meet the actual needs. In order to better capture the characteristics of traffic flow data, some intelligent prediction methods are widely used in the short-term traffic flow prediction. Smith et al. applied the neural network model to the prediction of short-time traffic flow for the first time and proved the superiority of neural network by comparing with other models [4]. Baher Abdulhal et al. carried out short-time traffic flow prediction based on the time-delay neural network model optimized by genetic algorithm without exploring the temporal and spatial relationship of detection points [5]. Liu et al. proposed a KNN-SVR model based on Non-parametric Regression and Support Vector Regression (SVR), used the K-Nearest Neighbor Algorithm to reconstruct the historical traffic flow sequence similar to the current traffic flow, and then used the SVR for traffic flow prediction [6]. Tian et al. found that the input length of most time series prediction models is static, and the optimal time lag cannot be accurately determined. Therefore, the Long Short-Term Memory (LSTM) is used to predict the traffic flow sequence, which overcomes the problem of determining the optimal time lag. Moreover, LSTM is compared with several well-known models such as Support Vector Machine (SVM), Single-Layer Feedforward Neural Network and Cascaded Autoencoder [7].

Based on the above research, the mathematical statistical prediction method represented by regression analysis and time series cannot deal with the sudden and random situations of traffic flow prediction. Although the artificial neural network has a strong nonlinear fitting characteristic, it is easy to fall into local optimization and output instability [8]. Different from artificial neural network, SVR can obtain the global optimal solution and map the nonlinear regression problem to the linear regression problem by applying kernel function [9]. Genetic Algorithm (GA) is a global optimization search algorithm that borrows from natural selection and genetic theory in the process of biological evolution [10]. It has the characteristics of group search and intrinsic heuristic random search, and is not easy to fall into local optimization, so it is very suitable for large-scale parallel computing [11]. Based on this, GA and SVR are combined to give full play to the advantages of the two, and the key parameters in the SVR model are obtained by using the characteristics of GA optimization, and then the traffic flow prediction model is established to obtain accurate results.

## 2    Prediction Model Construction

### 2.1    SVR

In 1963, the research group of ATE - T Bell LABS, led by Vanpik, first proposed the theoretical method of SVM [12]. SVM itself is proposed for binary classification problems, and SVR is an important branch of SVM (support vector machine). When SVM is applied to regression fitting analysis, its basic idea is no longer to find an optimal classification surface to separate the two types of samples, but to find an optimal classification surface to minimize the error of all training samples from the optimal classification surface [13]. Vapnik et al. introduced the $\gamma$ insensitive loss function and established the SVR model.

SVR models are used to model large-scale nonlinear data in high dimensional space. Given the training set $T = \{(x_1, y_1), (x_2, y_2), \cdots, (x_n, y_n)\}$, where $x_i(x_i \in R^d)$ is the input column vector of the *i-th* training sample and $x_i = [x_i^1, x_i^2, \cdots, x_i^d]^T, y_i \in R$ is the corresponding output value. The linear regression function established in the high latitude feature space is:

$$y = f(x) = \omega^T \phi(x) + b; i = 1, 2, \cdots, n \tag{1}$$

Where, $\phi(x)$ is the nonlinear mapping function, and $\omega(\omega \in R^{nh})$ and $b(b \in R)$ are the setting parameters of the SVR model. These two parameters can be minimized through the following function:

$$R_{SVR} = \frac{1}{N} \sum_{i=1}^{N} \Theta\left(y_i, W^T \varphi(x_i) + b\right) \tag{2}$$

Where $\Theta(y_i, f(x_i))$ can be given by the following function:

$$\Theta(y_i, f(x_i)) = \begin{cases} |f(x) - y| - \varepsilon, & |f(x) - y| \geq \varepsilon \\ 0, & \text{Else} \end{cases} \tag{3}$$

Define the linear $\gamma$ insensitive loss function:

$$L(f(x), y, \gamma) = \begin{cases} 0, |y - f(x)| \leq \gamma \\ |y - f(x)| - \gamma, |y - f(x)| > \gamma \end{cases} \tag{4}$$

Where, $f(x)$ is the predicted value returned by the regression function; $y$ is the corresponding true value. If the difference between $f(x)$ and $y$ is less than or equal to $\gamma$, the loss is equal to 0; if the difference between $f(x)$ and $y$ is greater than $\gamma$, the loss is equal to $|y - f(x)| - \gamma$. According to the principle of risk minimization, relaxation factors $\xi_1$ and $\xi_2$ are introduced, and the above problem of finding $\omega$ and $b$ is described in mathematical language, and the SVR problem is translated into:

$$\min \frac{1}{2} \|\omega\|^2 + C \sum_{i=1}^{1} (\xi_i, \xi_i^*)$$

$$s.t. \begin{cases} y_i - (\omega^T x_i + b) < \varepsilon + \xi_i \\ (\omega^T x_i + b) - y_i < \varepsilon + \xi_i^* \\ \xi_i, \xi_i^* \geq 0 \end{cases} \tag{5}$$

Where, $C$ is the penalty factor. The larger $C$ is, the greater the punishment for the samples whose training error is greater than $\gamma$ is. $\gamma$ specifies the error requirements of the regression function, and the smaller $\gamma$ is, the smaller the error of the regression function is. In order to improve the generalization ability, the $\varepsilon$ pipeline needs to be expanded. This minimizes the likelihood that the unknown point will exceed the region. However, when the training set is nonlinear, the generalization performance of the obtained regression function is very poor even after the optimization is completed. Therefore, by introducing the kernel function $K(x_i, x_j)$ and Lagrange multiplier $\alpha$ and $\alpha^*$, the low-dimensional nonlinear problem is transformed into a high-dimensional linear problem, and finally the regression problem is transformed into the following optimization problem:

$$\max = \begin{cases} -\frac{1}{2} \sum_{i,j=1}^{l} (\alpha_i - \alpha_i^*)(\alpha_j - \alpha_j^*) K(x_i \cdot x_j) \\ -\sum_{i=1}^{l} \alpha_i(\varepsilon - y_i) - \sum_{i=1}^{l} \alpha_i^*(\varepsilon + y_i) \end{cases}$$

$$s.t. \begin{cases} \sum_{i=1}^{l} (\alpha_i - \alpha_j^*) = 0 \\ 0 \leq \alpha_i, \alpha_i^* \leq C (i = 1, 2, \cdots, l) \end{cases} \tag{6}$$

Among them, the $K(x_i, x_j) = \phi(x_i)\phi(x_j)$ as the kernel function. So we can get the optimal nonlinear regression function:

$$f(x) = \sum_{i=1}^{l} (\alpha_i - \alpha_i^*) K(x_i \cdot x) + b \tag{7}$$

Generally, the radial basis function is selected as the kernel function, which is expressed as:

$$K(x_i, x_j) = \exp\left(-\frac{\|x_i - x_j\|^2}{2\sigma^2}\right) \tag{8}$$

## 2.2 GA

GA is a randomized search method that borrows from the evolutionary law of biology (survival of the fittest, genetic mechanism of survival of the fittest). Professor J. Halolland first proposed it in 1975 in the United States [14]. Its main feature is that it directly operates on structural objects, without the limitation of derivation and function continuity. It has inherent implicit parallelism and better global optimization ability. The probabilistic optimization method can automatically obtain and guide the optimized search space and adaptively integrate the search direction without any definite rules. These properties of genetic algorithm have been widely used in combinatorial optimization, machine learning, signal processing, adaptive control and artificial life [15]. Only the fittest chromosomes in the population have the chance to interact (reproduce or mate) and create a second population. The mating process uses three different techniques:

- Cloning: a parent chromosome is copied exactly to the next generation. 50% of the new population is generated by Cloning.
- Crossover: two parents exchange a part of their chromosome at one or more randomly selected breakpoints to create two new children. 47% of the new population is generated by Crossover.
- Creation: a completely new chromosome is generated randomly, 3% of the new population is generated by Creation.

## 2.3  Optimization of SVR Parameters by GA

GA is a search heuristic algorithm for solving optimization problems. With scalability as its main feature, this algorithm can be combined with other algorithms to form a high-quality hybrid algorithm combining the advantages of both sides. Based on the defects of SVR prediction method itself, GA is used to carry out optimization, which can effectively optimize the threshold and weight under its initialization state, and can better carry out actual training and prediction for SVR [16]. The key parameter of RBF kernel function is $\varepsilon$, namely the width of the kernel, which controls the radial range of the function. The penalty parameter $C$ plays a balancing role between model complexity and training error. The larger the $C$ value is, the greater the penalty for the data beyond the loss function is, which affects the generalization ability of the model. The embedded dimension $E$ and time delay $\tau$ determine the quality of the samples, so it has a great impact on the accuracy of the prediction model. At present, there is no definite theory to guide the parameter selection of regression machine. The basic idea of genetic algorithm for SVR parameter optimization is to introduce the principle of biological evolution into the coding serial population formed by the optimization parameter $(C, \sigma, E, \tau)$. Iterate the individuals through the Cloning, Crossover and Creation in the inheritance according to the selected fitness function until the termination condition is satisfied, to achieve the purpose of intelligent optimization. The realization process of SVR prediction model optimization by genetic algorithm is as follows:

(1) Define the necessary data. Including training data set, appropriate features, the number of SVR input and output, the size of GA population, termination conditions and GA parameter (population size, crossover probability, mutation probability, etc.) settings.
(2) Standardized traffic data. Select 90% of the traffic data as the training data set, and the remaining 20% of the traffic data as the test data set.
(3) Set GA parameters and encode SVR parameters. Chromosomes are expressed in the form of $\{x_1, x_2, x_3, x_4\}$, where $x_1, x_2, x_3$ and $x_4$ represent $C, \sigma, E$, and $\tau$, respectively, and are encoded in real numbers, that is, each chromosome Represented by a string of real numbers.

(4) Calculate the fitness of each individual. At the same time, appropriate criteria need to be introduced to rank the individuals in the population. In this paper, Mean Absolute Percentage Error (MAPE) is selected as the fitness function:

$$MAPE\% = \frac{1}{N} \sum_{i=1}^{N_{es}} \sigma_i \tag{9}$$

$$\sigma_i \% = \frac{|\overline{y_i} - y_i|}{y_i} \times 100, \ i = 1,2,\cdots, N_{es} \tag{10}$$

Where $N_{es}$ is the number of prediction points.

(5) Determine whether the iteration conditions are met. If not satisfied, clone, crossover and creation operations are performed. The clone operation uses the roulette method to screen the population with a high fitness value into the next generation. The fitness value of the individual $X_i$ is $f(X_i)$, and the probability of being selected as the next generation is:

$$P(X_i) = \frac{N \times f(X_i)}{\sum\limits_{j=1}^{N} f(X_i)} \tag{11}$$

The crossover operation uses a linear combination to crossover two chromosomes with a certain probability $p(p \in [0, 1])$, namely:

$$\begin{cases} X_1 = pX_1 + (1-p)X_2 \\ X_2 = (1-p)X_1 + pX_2 \end{cases} \tag{12}$$

The creation operation adopts uniform creation mode, that is, for the individual to be mutated $X = (x_1, x_2, \cdots x_n)$, randomly generate a random number $k \in [1, n]$, generate a new descendant $X = \left(x_1, x_2 \cdots, x_k', \cdots x_n\right)$, where $x_k'$ is a random number subject to uniform distribution in $[l_k, u_k]$.

(6) The fitness value of each generation was calculated, the survival of the fittest was eliminated, and the offspring were obtained through cloning, crossover and creation after determining the parent individuals. The optimal individual was selected from the sub generation and the optimal combination of four parameters $(C, \sigma, E, \tau)$ was determined after decoding.

(7) Construct the SVR prediction model and substitute the parameters obtained by genetic algorithm optimization into the SVR prediction model for simulation prediction.

The specific flow chart is shown in Fig. 1.

## 3 Experiment and Result Analysis

### 3.1 Data Sources

Traffic flow data from Performance Measurement System (PeMS) database are selected in this paper. Since the traffic flow data follow certain stability and regularity, as shown

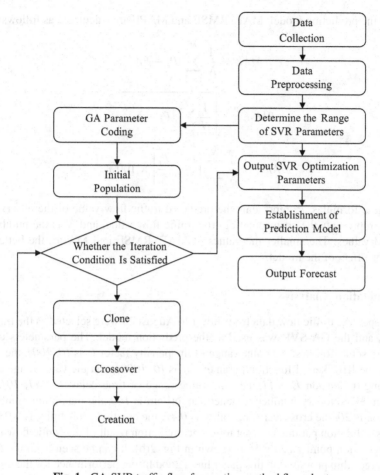

**Fig. 1.** GA-SVR traffic flow forecasting method flow chart

in Fig. 2(a), especially the time of the weekend has different characteristics, in order to make full use of the regularity of the collected data, the data used in this experiment does not include the traffic flow of the weekend, and the experimental data is finally verified. In this paper, 20 observation points east of highway 91 in Los Angeles area are selected as the target stations. Taking the data of 50 working days from July 1, 2019 to September 6, 2019 as experimental data, taking the data of 40 working days from July 1, 2019 to August 23, 2019 as the training data, the data of 10 working days from August 26, 2019 to September 6, 2019 were used as the test data.

### 3.2 Error Analysis and Comparison

To better explain the prediction accuracy of this optimization model, this paper introduces three error evaluation indexes, namely Mean Absolute Error (MAE), Root Mean Square Error (RMSE) and Mean Absolute Percentage Error (MAPE), to evaluate the prediction effect of the prediction model. MAE, RMSE and MAPE are calculated as follows:

$$MAE = \frac{1}{N} \sum_{i=1}^{N} |P_i - F_i| \tag{13}$$

$$RMSE = \sqrt{\frac{1}{N} \sum_{i=1}^{N} (P_i - F_i)^2} \tag{14}$$

$$MAPE = \frac{1}{N} \sum_{i=1}^{N} \left| \frac{P_i - F_i}{P_i} \right| \times 100\% \tag{15}$$

In these formulas, $P_i$ represents the predicted traffic flow value of the observation point, $F_i$ represents the corresponding true traffic flow value, and $N$ is the number of predicted values. The smaller the values of MAE, RMSE and MAPE, the better the prediction effect of the model.

### 3.3 Simulation Analysis

In this paper, the traffic flow data from July 1 to August 23 were selected as the training data set, and the GA-SVR was used as the prediction model. The parameters of the model are set as follows: the value range of the penalty factor $C$ is *[0, 100]*, the value range of the RBF kernel function parameter $\sigma$ is *[0, 1000]*, and the value range of the embedding dimension $E$ is *[1, 10]*, the value range of time delay $\tau$ is *[1, 10]*. The maximum evolutionary number of genetic algorithm is *100*, the maximum number of population is *20*, the crossover probability is *0.47*, the creation probability is *0.03*, and the cross-validation parameter is set to *5*. The prediction result of the traffic flow of the fifth observation point on July 15 is shown in Fig. 2(b). It can be seen from the figure that the algorithm proposed in this paper has a good fitting performance for the training set.

In order to verify the prediction performance of the model proposed in this paper and observe the prediction effect of the model more intuitively, three classical prediction models with good test performance and the un-optimized SVR model are selected as the benchmark comparison models. The three classical models are Artificial Neural Network (ANN), BP Neural Network and ARIMA. MAE, RMSE, and MAPE were used as evaluation indicators. The comparison results of predictions from 17:00 to 18:00 on September 6, 2019 are shown in Fig. 2(c). It can be seen from the figure that the difference between GA-SVR model and the real value is the smallest.

Figure 2(d) shows the comparison of MAPE indexes of all models to each observation point. It can be seen from the figure that the prediction error of the model proposed in this paper is lower than that of other models on 10 test data sets, showing its superiority in performance.

Table 1 lists the MAE, RMSE and MAPE values of the four prediction models. MAE, RMSE and MAPE are used to measure the prediction results of the model. The smaller the values of the two indicators are, the more accurate the prediction results will be. Table 1 shows that the GA-SVR prediction model MAE, RMSE and MAPE are

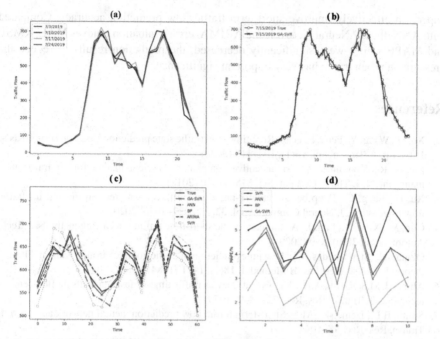

**Fig. 2.** (a) Hourly traffic flow on wednesdays in July; (b) The fitting effect of GA-SVR model on training set; (c) Comparison of model prediction results; (d) Performance evaluation.

better than the ANN, BP Neural Network and un-optimized SVR, with better prediction accuracy, among which the un-optimized SVR has the worst performance.

**Table 1.** Model prediction performance comparison

| Model | MAE | RMSE | MAPE/% |
|---|---|---|---|
| GA-SVR | 18.7 | 25.4 | 6.9 |
| ANN | 22.4 | 31.8 | 8.4 |
| BP | 26.5 | 30.7 | 8.3 |
| SVR | 30 | 34.1 | 10.5 |

## 4   Conclusion

SVR has powerful nonlinear fitting characteristics of artificial intelligence algorithm and overcomes the shortcoming of unstable output results. It has excellent characteristics when dealing with complex nonlinear laws in the case of small samples, but it is difficult to determine the parameter values in practical applications. By combining GA and SVR, using the optimized parameters of SVR forecasting model is established and the actual traffic flow data to verify, results show that the prediction model is proposed in this paper can effectively improve short-term traffic flow prediction accuracy. Compared with ANN, the BP Neural Network and ARIMA, error evaluation indexes MAE, RMSE and MAPE values were significantly decreased, the prediction results to reduce the pressure on urban traffic has very important significance.

# References

1. Xu, D., Wang, Y., Peng, P., et al.: Real-time road traffic state prediction based on kernel-KNN. J. Transp. A-Transp. Sci. **16**(1), 104–118 (2020)
2. Mehta, R., Vazirani, V.V.: An incentive compatible, efficient market for air traffic flow management. J. Theor. Comput. Sci. **818**, 41–50 (2018)
3. Xu, H., Jiang, C.: Deep belief network-based support vector regression method for traffic flow forecasting. J. Neural Comput. Appl. **32**(7), 2027–2036 (2020)
4. Guo, M., Xiao, X., Lan, J.: A summary of the short-time traffic forecasting methods. J. Tech. Autom. Appl. **28**(6), 8–9 (2009)
5. Okutani, I., Stephanedes, Y.J.: Dynamic prediction of traffic volume through Kalman filtering theory. J. Transp. Res. Part B Methodol. **18**(1), 1–11 (1984)
6. Ahmaed, M.S., Cook, A.R.: Analysis of freeway traffic time-series data by using Box-Jenkins technique. J. Transp. Res. Rec. **722**, 1–9 (1979)
7. Smith, B.L., Demetsky, M.J.: Short-term traffic flow prediction: neural network approach. J. Transp. Res. Rec. 98–104 (1984)
8. Moazenzadeh, R., Mohammadi, B., Shamshirband, S., Chau, K.W.: Coupling a firefly algorithm with support vector regression to predict evaporation in northern Iran. J. Eng. Appl. Comput. Fluid Mech. **12**(1), 584–597 (2018)
9. Shi, H., Xu, M., Li, R.: Deep learning for household load forecasting-a novel pooling deep RNN. J. IEEE Trans. Smart Grid **9**(5), 5271–5280 (2018)
10. Son, B., Kim, J.W., Lee, D., Jung, S.Y.: Genetic algorithm with species differentiation based on kernel support vector machine for optimal design of wind generator. IEEE Trans. Magn. **55**(9), 1–4 (2019)
11. Sukawattanavijit, C., Chen, J., Zhang, H.S.: GA-SVM algorithm for improving land-cover classification using SAR and optical remote sensing data. IEEE Geosci. Remote Sens. Lett. **14**(3), 284–288 (2017)
12. Vapnik, V.N.: Statistical learning theory. In: Encyclopedia of the Sciences of Learning, vol. 41, no. 4, p. 3185 (2012)
13. Abdulhai, B., Porwal, H., Recker, W.: Short-term traffic flow prediction using neuro-genetic algorithms. Intell. Transp. Syst. J. **7**(1), 3–41 (2002)
14. Liu, Z., Du, W., Yan, D., et al.: Short-term traffic flow forecasting based on combination of K-nearest neighbor and support vector regression. J. Highw. Transp. Res. Dev. (Engl. Ed.) **12**(1), 89–96 (2018)
15. Tian, Y., Pan, L.: Predicting short-term traffic flow by long short-term memory recurrent neural network. In: 2015 IEEE International Conference on Smart City (Smart City), Chengdu, pp. 153–158. IEEE Press (2015)
16. Zhang, Y.J., Wang, M., et al.: Research on adaptive beacon message broadcasting cycle based on vehicle driving stability. Int. J. Netw. Manag. Spec. Issue Paper (2020)

# Short-Term Traffic Flow Prediction Based on Hybrid Model

Yong Hu, Meng Yu, Guanxiang Yin, Fei Du, Meng Wang, and Yuejin Zhang[✉]

East China Jiaotong University, Nanchang 330013, China
zyjecjtu@foxmail.com

**Abstract.** Accurate and reliable short-term traffic flow prediction can provide effective help for people's travel and road planning. In order to improve the accuracy of short-term traffic flow prediction, this paper proposes a hybrid model of improve long-term short-term memory (LSTM) and radial basis function neural network (RBFNN). Firstly, according to the temporal and spatial characteristics of traffic flow, LSTM and RBFNN models are constructed. Then, by adding the percentage error term to balance the loss function of the LSTM, and an improved LSTM (ILSTM) is proposed. Finally, the prediction results of these models are weighted by the Entropy method to obtain the final result. The experimental results show that the ILSTM-RBFNN model can achieve higher prediction accuracy compared with traditional models.

**Keywords:** Traffic flow prediction · LSTM · RBFNN · Entropy method

## 1 Introduction

With the continuous increase of highway mileage and car ownership, road traffic congestion, frequent traffic accidents and environmental pollution are becoming more and more serious. The development of Intelligent Transportation provides new ideas for solving these transportation problems. A core technology and key component of Intelligent Transportation is traffic flow prediction (e.g. short-term traffic flow prediction). Short-term traffic flow prediction refers to the use of history and real-time traffic flow data to analyze traffic flow status and predict the traffic flow at the next period (5 to 15 min) [1].

Due to the complexity, randomness, and uncertainty of actual traffic flow, accurate and reliable prediction methods have always been a research hotspot in the field of intelligent transportation [2]. During the past few decades, many scholars have proposed many prediction models, such as Kalman filter model, support vector regression model, artificial neural network model and so on. Among these models, it is difficult for a single model to adapt to the time-varying characteristics of short-term traffic flow, and the hybrid model combines the advantages of each single model, fully considers the basic characteristics of traffic flow, and further improves the prediction accuracy [3].

Based on the above discussion, this paper proposes a hybrid prediction model based on improved LSTM and RBFNN. This model uses the LSTM and RBFNN to model

© Springer Nature Switzerland AG 2020
X. Chen et al. (Eds.): ML4CS 2020, LNCS 12488, pp. 135–143, 2020.
https://doi.org/10.1007/978-3-030-62463-7_13

the temporal and spatial characteristics of traffic flow, respectively, and weights each single model by the Entropy method to obtain the final prediction result. At the same time, considering that the LSTM produce high-flow data tendencies when using gradient descent to optimize parameters, this paper adds a balance term to the loss function to improve prediction accuracy.

The rest of this paper is organized as follows. Section 2 introduces the related work. Section 3 gives the hybrid model framework. Section 4 discusses and analyzes the experimental results. Finally, Sect. 5 concludes the paper.

## 2  Related Work

Generally, traffic flow prediction models are divided into three categories: parametric models, non-parametric models, and hybrid models. The parametric model mainly includes the Autoregressive Integrated Moving Average (ARIMA) model and Kalman filter (KF) model. Scholars made many improvements on the traditional ARIMA model, and proposed a series of variant models such as adaptive ARIMA [4], spatial-temporal ARIMA [5], seasonal ARIMA [6] and so on. In 2018, Huang et al. [7] proposed an online seasonal adjustment factors plus adaptive Kalman filter (OSAF+AKF) model to predict traffic flow. Although the above improved model improves prediction performance, these simple linear models still cannot accurately extract the characteristics of traffic flow due to the nonlinearity and randomness of traffic flow.

Non-parametric models mainly include Support Vector Machine (SVM) models, K-nearest neighbor (KNN) models, and deep neural network models (ANN). In order to improve the prediction accuracy, scholars have proposed many different improved models. In 2016, Kong et al. [8] proposed a multi-core SVM model, which used linear kernel, polynomial kernel and Gaussian kernel for SVM training respectively. In 2018, Liu et al. [9] proposed a KNN model based on manifold distance, and averaged data with high manifold similarity to obtain the prediction result. In 2020, Du et al. [10] proposed the deep irregular convolution residuals LSTM prediction model for deep spatial-temporal feature (DST-ICRL), which used DST-ICRL to learn the traffic flow features and get the final prediction result. In general, although non-parametric models can better fit non-linear traffic flow data, most of them have complex structures and cannot be adapted to multiple environments.

The hybrid model mainly uses a combination algorithm to combine multiple single prediction models to obtain the optimal prediction result. In 2018, Duan et al. [11] proposed a CNN-LSTM model based on greedy algorithm, in which used CNN to extract spatial characteristics and LSTM to process time series. At the same time, training the network by a greedy policy to shorten the prediction time. In 2019, Guo et al. [12] proposed a short-term anomaly passenger flow prediction method based on SVR and LSTM, which used historical data and real-time data to train two models, respectively, and finally adaptively weight the prediction results of the two models. Feng et al. [13] proposed a new short-term traffic flow prediction algorithm based on spatial-temporal characteristics and adaptive multi-core support vector machine (AMSVM-STC). However, most of the hybrid models only consider the structural optimization of the model and do not analyze the characteristics of traffic flow data.

Different from the above work, this paper focuses on analyzing the basic characteristics and spatial-temporal characteristics of traffic flow, and selecting the appropriate time scale and space station to predict the traffic flow. At the same time, this paper improved the LSTM network structure and combined the single model by using the Entropy method, so the prediction results will always be biased towards the side with higher accuracy, and even got a better fitting effect.

## 3 Hybrid Model

Short-term traffic flow is a complex nonlinear system. The current traffic flow at the target station is not only related to the data of the neighboring historical moments, but also related to the traffic flow data of the neighboring stations. In this paper, ILSTM and RBFNN are used to predict the traffic flow from the perspective of temporal correlation and spatial correlation, respectively, and the prediction value of each model is fused using the Entropy method to obtain the final prediction result. The hybrid model framework is shown in Fig. 1. Data preprocessing includes data filling and normalization. The details will be illustrated in Sect. 4.1. Historical and real-time traffic flow data are defined as temporal data, the traffic flow data of the neighboring stations are defined as spatial data.

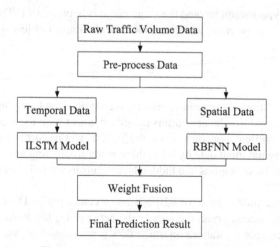

**Fig. 1.** The framework of the hybrid model

### 3.1 ILSTM Model

The LSTM is an improved Recurrent Neural Network (RNN), which overcomes the gradient disappearance and explosion problems of the RNN model, and is suitable for processing and predicting time series events with long time intervals and high correlation. Each LSTM unit adds a "cell". The "cell" includes three gates, namely the forget gate, input gate and output gate. When the information enters the network, these gates can determine whether it is useful. The more useful the information, the greater the weight, and the completely unrelated information will be forgotten [14].

The LSTM model uses the Mean Square Error (MSE) as the loss function to measure the difference between the true value and the estimated value, which is

$$l = \frac{1}{N} \sum_{i=1}^{N} (Y_i - \hat{Y}_i)^2 \tag{1}$$

Where $Y_i$ and $\hat{Y}_i$ represent real value and predicted value of sample, $N$ represents the sample numbers.

It can be seen from the above formula that when the LSTM model evaluates the traffic flow error, the error value at the high traffic flow is larger than the error value at the low traffic flow. That is, MSE has a tendency to high-flow data, and the model cannot learn low-flow data features during training, which leads to a decline in prediction accuracy. In order to solve this problem, this paper adds a balance term to the loss function, which is

$$l = \frac{1}{N} \sum_{i=1}^{N} \left[ (1 - \zeta)(Y_i - \hat{Y}_i)^2 + \zeta(\frac{Y_i - \hat{Y}_i}{Y_i})^2 \right] \tag{2}$$

Where $\zeta$ is a hyperparameter, and the value range is between 0–1. The balance term is essentially a percentage error, which can increase the impact of low-flow data on the model.

## 3.2 RBFNN Model

RBFNN is a typical three-layer neural network, including input layer, hidden layer and output layer. First, the input layer neurons directly connect with the hidden layer neurons to transfer the original input data. Then, the hidden layer performs high dimensional mapping on the original input data by RBF, thereby approximating any nonlinear function. Finally, output layer neurons and hidden layer neurons are fully connected through a weight matrix [15].

Each node in the hidden layer of RBFNN has a center point. The closer the input information is to the center point, the greater the network feedback. At the same time, the influence of neighboring stations on traffic flow is not consistent, and its influence is affected by multiple factors such as spatial distance and random events. Therefore, the use of RBFNN to learn the spatial correlation of traffic flow can better simulate the traffic environment of the road network and filter out suitable neighboring stations. That is, the higher the spatial correlation between the neighboring stations and the target station, the greater the weight.

## 3.3 Weight Fusion

This paper combine the ILSTM and RBFNN predicted values, and output the final predicted results of the target station I at the next period, which is

$$Y_I(t + 1) = \lambda_1 Y_I^1(t + 1) + \lambda_2 Y_I^2(t + 1) \tag{3}$$

Where $Y_I^1(t+1)$ and $Y_I^2(t+1)$ represent ILSTM model and RBFNN model predicted values of $I$ at the next period, $\lambda_1, \lambda_2$ are the fusion weights, which can be calculated by Entropy method.

The specific steps of the Entropy method to calculate the fusion weights of each model are as follows [13]:

1. Calculate the relative error rate $p_{jt}$ by the prediction error $e_{jt}$ at the $j$th prediction model at the $t$th period.

$$p_{jt} = \frac{e_{jt}}{\sum_{t=1}^{n} e_{jt}}, \quad t = 1, 2, \ldots, n; \quad j = 1, 2, \ldots, \mu \tag{4}$$

$$e_{jt} = \begin{cases} 1 & \left|(y_t - \hat{y}_{jt})/y_t\right| \geq 1 \\ \left|(y_t - \hat{y}_{jt})/y_t\right| & 0 \leq \left|(y_t - \hat{y}_{jt})/y_t\right| < 1 \end{cases} \tag{5}$$

2. Calculate the entropy value $h_j$ of the $j$th prediction model according to the relative error rate.

$$h_j = -k \sum_{t=1}^{n} p_{jt} \ln(p_{jt}), \quad j = 1, 2, \ldots, \mu \tag{6}$$

Where $k$ is a constant, generally take $k = \frac{1}{\ln n}$.

3. According to the principle that the size of entropy is opposite to the change degree of the error, the coefficient of change degree of the $j$th prediction model can calculated.

$$d_j = 1 - h_j, \quad j = 1, 2, \ldots, \mu \tag{7}$$

4. Finally, calculate the fusion weights $\lambda_j$ of the $j$th prediction model.

$$\lambda_j = \frac{1}{\mu - 1}\left[1 - \frac{d_j}{\sum_{j=1}^{\mu} d_j}\right] j = 1, 2, \ldots, \mu \tag{8}$$

# 4  Experiments

## 4.1  Data Description

In this paper, the experimental data were obtained from the official traffic electronic database provided by Portland Oregon Regional Transportation Archive Listing (POR-TAL). The traffic volume collected from multiple stations on the U.S. WA I-5 and SR500 highways, with station IDs are 1581, 1509, 1553, 1515, 1626, 1503 and 1618. Among them, 1581 is the target station, and others are neighboring stations. The data collection interval is 15 min, and the collection period is 58 days, from October 1 to November 27, 2019. For each station, the first 51 days of data are used as the model train sets, and the rest 7 days of data are used as the test sets.

## 4.2  Data Pre-processing

Data pre-processing is mainly divided into two parts: one is the filling of missing data; the other is the normalization of data. For the filling of missing data, this paper uses linear interpolation, and the formula is as follows:

$$y(t) = \frac{t - t_1}{t_0 - t_1} \cdot y(t_0) + \frac{t - t_0}{t_1 - t_0} \cdot y(t_1) \tag{9}$$

Where $y(t)$ represents the missing data, $t_0$ and $t_1$ are the adjacent periods of $t$ period.

For the normalization of data, this paper uses standard deviation normalization, that is, z-score standardization, which is

$$x^* = \frac{x - \mu}{\sigma} \tag{10}$$

Where $x^*$ is the normalized data, $x$ is the raw traffic volume data, $\mu$ is the original data mean, $\sigma$ is the raw data standard deviation.

## 4.3  Experiment Results and Analysis

In order to evaluate the prediction performance of the hybrid model, this paper selects Root Mean Square Error (RMSE) and Mean Absolute Percentage Error (MAPE) as the evaluation metrics. They are defined as follows:

$$RMSE = \sqrt{\frac{1}{N} \sum_{t=1}^{N} (y_t - \hat{y}_t)^2} \tag{11}$$

$$MAPE = \frac{1}{N} \sum_{t=1}^{N} \frac{|y_t - \hat{y}_t|}{y_t} \tag{12}$$

Where $y_t$ and $\hat{y}_t$ represent the real value and predicted value of the traffic flow, $N$ represents the number of samples.

According to multiple experiments, the structure of the hybrid model are set. The number of layers of the ILSTM model is set to 5, including an input layer, two hidden layers, and an output layer, the number of nodes is set to 1, 20, 10, and 1, respectively. The other hyperparameters are set as follows: the learning rate was set at 0.01, and the number of training times was set to 500. For the RBFNN model, the number of input layer neurons is set to 6, the number of hidden layer neurons is set to 25, the output neurons are set to 1, and the algorithm parameter propagation is set to 1.2.

This paper calculate the evaluation metrics of LSTM and ILSTM with different iteration times. The results are shown in Fig. 2. It can be seen that the metrics of both models show a downward trend, indicating that within a certain range, as the number of iterations increases, the prediction performance of the model will become better. In addition, the RMSE and MAPE values of the ILSTM are lower than the LSTM, indicating that the improved loss function proposed in this paper can effectively optimize the LSTM structure and improve the prediction accuracy.

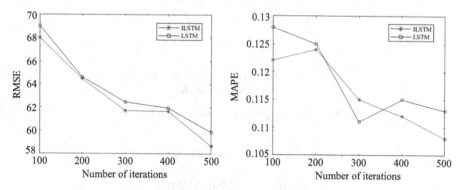

**Fig. 2.** The evaluation metrics of different model

In order to verify the superiority of the model proposed in this paper (ILSTM-RBFNN), we compare ILSTM-RBFNN with the traditional ARIMA, SVR, LSTM models. Table 1 presents the prediction results of different models and Fig. 3 shows randomly selected partial prediction results of ILSTM-RBFNN. It can be seen from Table 1 that the RMSE and MAPE values of LSTM are lower than those of ARIMA and SVR, indicating that LSTM is more suitable to deal with nonlinear traffic flow prediction problems. In addition, the RMSE and MAPE values of ILSTM are 2 and 0.48% lower than LSTM, respectively. More importantly, the RMSE and MAPE values of ILSTM-RBFNN is the lowest, only 52 and 9.86%, respectively, which is 8 and 1.43% lower than a single LSTM, indicating that comprehensively considering the spatial-temporal characteristics of traffic flow can get better prediction results than considering only temporal characteristics, the hybrid model proposed in this paper has achieved better prediction accuracy.

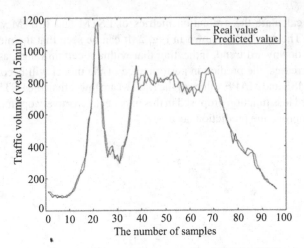

**Fig. 3.** Traffic flow prediction results of ILSTM-RBF

**Table 1.** Prediction results of different model

| Model | RMSE | MAPE |
|---|---|---|
| ARIMA | 121 | 20.24 |
| SVR | 70 | 13.16 |
| LSTM | 60 | 11.29 |
| ILSTM | 58 | 10.81 |
| ILSTM-RBF | 52 | 9.86 |

## 5 Conclusion

This paper proposes a method of using improve LSTM and RBFNN hybrid model to predict short-term traffic flow. Compared with traditional models, the experimental results show that the proposed model in this paper has better prediction performance and more accurate prediction results. In the future, we will consider using swarm intelligence optimization algorithm to optimize the structural parameters of the neural network to reduce the time of model training and further improve the accuracy of prediction.

## References

1. Zhang, L., Alharbe, N.R., Luo, G., Yao, Z., Li, Y.: A hybrid forecasting framework based on support vector regression with a modified genetic algorithm and a random forest for traffic flow prediction. Tsinghua Sci. Technol. **23**(4), 113–126 (2018)
2. Zheng, Z., Pan, L., Pholsena, K.: Mode decomposition based hybrid model for traffic flow prediction. In: 3rd International Conference on Data Science in Cyberspace, Guangzhou, China, pp. 521–526. IEEE (2018)

3. Zhang, M., Fei, X., Liu, Z.H.: Short-term traffic flow prediction based on combination model of Xgboost-Lightgbm. In: 2018 International Conference on Sensor Networks and Signal Processing (SNSP), Xi'an, China, pp. 322–327. IEEE (2018)
4. Han, C., Song, S., Wang, C.H.: A real-time short-term traffic flow adaptive forecasting method based on ARIMA model. Acta Simulata Systematica Sinica 16(7), 043 (2004). No. 1530–1456
5. Ding, Q.Y., Wang, X.F., Zhang, X.Y., et al.: Forecasting traffic volume with space-time ARIMA model. Adv. Mater. Res. 11(5), 156–157 (2010)
6. Williams, B.M., Hoel, L.A.: Modeling and forecasting vehicular traffic flow as a seasonal ARIMA process: theoretical basis and empirical results. J. Transp. Eng. 129(6), 664–672 (2003)
7. Huang, W., Jia, W., Guo, J., et al.: Real-time prediction of seasonal heteroscedasticity in vehicular traffic flow series. IEEE Trans. Intell. Transp. Syst. 19(10), 3170–3180 (2018)
8. Kong, X., Xu, Z., Shen, G., et al.: Urban traffic congestion estimation and prediction based on floating car trajectory data. Future Gener. Comput. Syst. 61, 97–107 (2016)
9. Liu, Q., Cai, Y., Jiang, H., et al.: Traffic state spatial-temporal characteristic analysis and short-term forecasting based on manifold similarity. IEEE Access 6(2), 9690–9702 (2018)
10. Du, B., Peng, H., Wang, S., et al.: Deep irregular convolutional residual LSTM for urban traffic passenger flows prediction. IEEE Trans. Intell. Transp. Syst. 21(3), 972–985 (2020)
11. Duan, Z., Yang, Y., Zhang, K., et al.: Improved deep hybrid networks for urban traffic flow prediction using trajectory data. IEEE Access 6(2), 31820–31827 (2018)
12. Guo, J., Xie, Z., Qin, Y., et al.: Short-term abnormal passenger flow prediction based on the fusion of SVR and LSTM. IEEE Access 7, 42946–42955 (2019)
13. Feng, X., Ling, X., Zheng, H., et al.: Adaptive multi-kernel SVM with spatial-temporal correlation for short-term traffic flow prediction. IEEE Trans. Intell. Transp. Syst. 20(6), 2001–2013 (2019)
14. Xie, Z., Liu, Q.: LSTM networks for vessel traffic flow prediction in inland waterway. In: 2018 IEEE International Conference on Big Data & Smart Computing, Shanghai, China, pp. 418–425. IEEE (2018)
15. Esfetanaj, N.N., Kazemzadeh, R.: A novel hybrid technique for prediction of electric power generation in wind farms based on WIPSO, neural network and wavelet transform. Energy 149, 662–674 (2018)

# Uneven Clustering Routing Protocol for Geological Disaster Monitoring Sensor Network

Yicheng Wei[1], Xun Wang[2], Xu Gou[1], Zhaoma Pan[2], and Zhi Li[1(✉)]

[1] College of Electronic Information Engineering, Sichuan University, Chengdu 610065, China
lizhi@scu.edu.cn

[2] China Railway Eryuan Engineering Group Co., Ltd., Chengdu, China

**Abstract.** Geological disasters frequently occur in the mountainous area which seriously threaten the safety of railways and human life. Although wireless sensor network (WSN) technology can be used in the geological disaster remote monitoring system by deploying sensor nodes in the mountainous area, the difficulties faced in long-term and large-area monitoring, such as limited battery energy of sensor nodes and high energy consumption of long-distance transmission. In order to prolong the network life and extend monitoring range, this paper takes the distribution and the residual energy of sensor nodes into account to propose an uneven clustering multi-hop routing protocol for geological disaster monitoring sensor network which server can use to realize optimal clustering formation and establish energy efficient routing. In our protocol, particle swarm optimization (PSO) algorithm is used to realize dynamic uneven clustering and minimum spanning tree (MST) algorithm is used to establish energy efficient inter-cluster multi-hop routing that will reduce the energy consumption of long-distance transmission. Compared with the previous routing protocols, the simulation results show the superiority of the proposed protocol in balancing energy consumption of sensor nodes, prolonging network life and long-distance low-power transmission.

**Keywords:** Wireless sensor network · Uneven clustering · Routing protocol · Multi-hop transmission · Particle swarm optimization · Minimum spanning tree

## 1 Introduction

The railway network in southwest China is usually located around mountainous areas in which the geological disasters, such as debris flow, landslides and rockfalls [1], often occur due to the complex geological structure and the changeable climate. Therefore, it is necessary to build a long-term monitoring system of mountainous environments to give early warning before geological disasters occur and reduce the losses of life and property.

Although wired monitoring instruments, such as Trip wires [2] and geophones [3], are conventionally used to detect geological disasters, it cannot provide any further information once the wire is broken. Moreover, the limited monitoring regions and the high

© Springer Nature Switzerland AG 2020
X. Chen et al. (Eds.): ML4CS 2020, LNCS 12488, pp. 144–159, 2020.
https://doi.org/10.1007/978-3-030-62463-7_14

cost of these instruments cannot achieve the requirement in practical applications [4]. Recently, WSN technology has been widely used in outdoor large-area environmental monitoring which constitutes spatially dispersed sensor nodes meant to record and monitor various physical and environmental conditions. WSN technology has the capability to capture quickly, process, and transmit of critical high-resolution data for real-time monitoring [5, 6]. However, the large-area and long-term of the geological disaster monitoring in mountainous area and the limited energy of sensor batteries, designing an energy-efficient routing protocol to prolong the network life is a vital issue in WSN.

The clustering scheme has better flexibility and scalability which has been considering to be one of the most effective solutions. A number of the current researches on routing protocols are based on the clustering scheme. In routing protocols LEACH [7] and LEACH-C [8], although random cluster head (CH) selection strategy can balance energy consumption of CH nodes in some case, it may cause the hotspot problem [9, 10], the single-hop transmission scheme, in addition, will consume more energy in long-distance transmission scenarios. Reference [11] takes distance of intra-cluster and residual energy of CH nodes into account and uses PSO algorithm to achieve optimal clustering formation. However, the protocol does not consider the distribution of clusters which may cause imbalance energy consumption of inter-cluster. Reference [12] considers various parameters such as intra-cluster distance, BS distance and residual energy of sensor nodes to select CH nodes. Due to several factors are repeatedly considered both in the CH selection and cluster formation, some cluster member nodes far from CH nodes will die prematurely when the CH node selected by the nodes has more residual energy and is far away from these nodes.

Most of the aforementioned routing protocols only consider the CH nodes selection, they do not well consider the uneven distribution of sensor nodes in mountainous area, which cannot well balance the energy consumption of sensor nodes. Moreover, star topology adopted by previous routing protocols will consume more energy of sensor nodes in large-area monitoring sensor network. In order to balance the energy consumption of sensor nodes and improve existing clustering routing protocol, this paper proposes a PSO-based uneven clustering multi-hop routing protocol (PSO-UCRP), which can be used by server to optimize the network topology, balance the energy consumption of sensor nodes and prolong the network life. The major contributions of this paper can be summarized as follows.

- Aiming at the issues of imbalance energy consumption of WSN in geological disaster monitoring caused by uneven distribution of sensor nodes, we propose uneven clustering algorithm based on PSO, which distribute more CH around BS to form unequal size clusters. To do so can reduce the tasks of CH nodes close to BS, achieve load balancing and prolong the network life.
- Aiming at the issues of high energy consumption of long-distance data transmission, we propose an energy efficient inter-cluster multi-hop routing algorithm based on MST. The algorithm takes communication distance and the residual energy as weight and uses Prim algorithm to establish a minimum spanning tree between BS and CHs as the initial route and optimizes the initial route according to the distribution of sensor nodes. Hence, the transmission energy consumption of CH nodes far away from BS is reduced.

The rest of the paper is organized as follows. Section 2 proposes the network and energy models. The overview of PSO is presented in Sect. 3. The design details of PSO-UCRP protocol are introduced in Sect. 4. Simulation experiment results are shown in Sect. 5. The conclusion is presented in Sect. 6.

## 2 Network and Energy Models

This paper assumes that there are $N$ sensor nodes randomly deploy in a square area (represented by symbol $A$) of size $M \times M$ and BS is positioned at the boundary of $A$ to form an initial network system model. The assumptions of the initial network model parameters are as follows.

- All sensor nodes are homogeneous and have the same initial energy. BS have infinite Energy.
- All sensor nodes can be selected as CH node and cluster member.
- Each sensor node has a unique ID and can be aware of its own residual energy and location.
- Each sensor node collects data periodically and can adjust the transmit power according to the distance to the receiver.
- After all the sensor nodes are deployed, they are fixed.

In order to establish the energy consumption model, this paper mainly considers energy consumption of sending data, power amplification and the receiver to run the radio electronics. The model adopts the free space and the multi-path fading channel depending on the distance between transmitter and receiver [13, 14]. The energy consumption ($E_T$) of transmitting a $l$-bit packet from the transmitter to its receiver over link of distance $d$ can be calculated by Eq. (1).

$$E_{Tx}(l, d) = \begin{cases} l \times E_{elec} + l \times \varepsilon_{fs} \times d^2, d < d_0 \\ l \times E_{elec} + l \times \varepsilon_{fs} \times d^2, d \geq d_0 \end{cases} \tag{1}$$

where $E_{elec}$ is the energy consumed by a sensor node transmit 1-bit data. $\varepsilon_{fs}$ and $\varepsilon_{mp}$ are two amplifier coefficients of free-space model and multi-path fading model respectively and $d_0$ is threshold distance which can be calculated by Eq. (2). When $d < d_0$, the energy consumption model adopts the free space model, and the amplifier coefficient is $\varepsilon_{fs}$. When $d \geq d_0$, the energy consumption model adopts the multi-path fading model, and the amplifier coefficient is $\varepsilon_{mp}$.

$$d_0 = \sqrt{\varepsilon_{fs}/\varepsilon_{mp}} \tag{2}$$

## 3 Overview of PSO Algorithm

Particle Swarm Optimization (PSO) algorithm [15] was proposed by Eberhart et al. in 1995. PSO algorithm is based on the foraging behavior of birds in nature, the algorithm abstract bird as a particle in search space, and the food is abstracted as solution of the problem, the process of birds searching for food can be expressed as the process of particle search for the optimal solution of the problem [16]. The PSO algorithm flow is described as following steps.

**Step1.** Assume that there are $N_P$ particles deployed in D dimensional space, each particle $P_i (1 \leq i \leq N_P)$ has its position $X_{i,d} (1 \leq d \leq D)$ and velocity $V_{i,d} (1 \leq d \leq D)$ in the $d$th dimension of the multidimensional space, and the particle $P_i$ can be represented by Eq. (3).

$$P_i = (X_{i,1}, X_{i,2}, \ldots, X_{i,D}) \tag{3}$$

**Step2.** Set a fitness function and calculate the current fitness of $P_i$. The fitness function is used to evaluate each particle quality of the solution of the problem. The personal best ($Pbest_i$) is the best position of each $P_i$. The global best ($Gbest$) is the best position of all particles.

**Step3.** In order to find the optimal solution of the problem throughout the iteration, each particle $P_i$ follows its own best, i.e., $Pbest_i$ and $Gbest$ to update its own velocity and position. In each iteration, Eq. (4) and (5) are the update formulas for velocity and position of the particle respectively.

$$V_{i,d}(t+1) = w \times V_{i,d}(t) + c_1 \times r_1 \times (Pbest_{i,d} - X_{i,d}(t)) + c_2 \times r_2 \times (Gbest - X_{i,d}(t)) \tag{4}$$

$$X_{i,d}(t+1) = V_{i,d}(t+1) + X_{i,d}(t) \tag{5}$$

where $t$ is current iterations, $1 \leq t \leq T_{max}$, $T_{max}$ is the maximum number of iterations. $c_1$ and $c_2 (0 \leq c_1, c_2 \leq 2)$ are the acceleration coefficients. $r_1$ and $r_2 (0 \leq r_1, r_2 \leq 2)$ are random values between 0 and 1. $w$ is self-adapting weight factor and $w = 0.8$ in this paper.

**Step4.** Use the new position of the particle $P_i$ to calculate the fitness value and update $Pbest_i$ and $Gbest$ for the minimization problem by Eqs. (6) and (7).

$$Pbest_i = \begin{cases} P_i, & if \ (Fitness(P_i) < Fitness(Pbest_i)) \\ Pbest_i, & otherwise \end{cases} \tag{6}$$

$$Gbest = \begin{cases} P_i, & if \ (Fitness(P_i) < Fitness(Gbest)) \\ Gbest, & otherwise \end{cases} \tag{7}$$

According to Eq. (4–7), each particle $P_i$ updates its own velocity and position by the influences of previous velocity ($V_{i,d}(t)$), previous position ($X_{i,d}(t)$), $Pbest_i$ and $Gbest$. After a number of iterations, all particles move toward to $Gbest$ and find the optimal solution in the searching space.

## 4   Proposed Algorithm

In order to achieve low energy consumption and long-distance transmission in geological disaster monitoring WSN. The proposed algorithm runs in server, which according to the distribution of sensor nodes in mountainous area and the status of each nodes per round to dynamically form the optimal clusters and establish the energy efficient inter-cluster routing. Figure 1 is the completed workflow of PSO-UCRP.

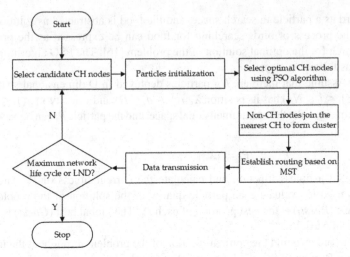

**Fig. 1.** The workflow of PSO-UCRP

## 4.1 Particle Representation

In each round of PSO-UCRP, different particles represent different clustering formations, the optimal solution of the particles represents the optimal clustering formation. We select V nodes with higher energy than the average energy of network as the candidate CH nodes at first, then define $K(K < V)$ clusters to the network, thus, there are $C_V^K$ results of clustering in total. According to [8], we have considered only 5% of sensor nodes as CH nodes.

## 4.2 Initialize Particle Swarm

We randomly select $N_P$ sensor nodes' position as the initial position of particle $P_i$ ($1 \le i \le N_P$) in 2-dimensional network area $A$, thus, each particle $P_i = (X_{i,1}, X_{i,2}) = (x_i, y_i)$ has its own position in $A$. The distance $d(i, j)$ ($1 \le j \le V$) between $V$ candidate CH nodes and particle $P_i$ can be calculated by Eq. (8).

$$d(i,j) = \sqrt{\left(x_{CHj} - x_i\right)^2 + \left(y_{CHj} - y_i\right)^2} \tag{8}$$

where $(x_{CHj}, y_{CHj})$ is the position of $j$th candidate CH node in $A$.

After generating the particles and calculating the distances between candidate CH nodes and particle $P_i$, we use the following Algorithm 1 to select $K$ CH nodes in $V$ candidate CH nodes and establish $K$ clusters for each particle.

**Algorithm 1.** Select CH nodes and form clusters.

**Input:**

Particle position: $P_i = (x_i, y_i)$ /* $1 \leq i \leq N_P$ */

The position of candidate CH nodes: $(x_{CHj}, y_{CHj})$ /* $(1 \leq j \leq V)$ */

The number of CH nodes: $K$

**Output:**

The nodes' ID for CH nodes: $\{\forall k \in K,\ \forall i \in N_P | CHID_{P_i,k}\}$

Clustering formation: $\{\forall k \in K,\ \forall i \in N_P | Cluster_{P_i,k}\}$

1:  **for** $i = 1$ to $N_P$ **do**
2:    **for** $j = 1$ to $V$ **do**
3:      Calculate $d(i, j)$ by using Equation (8)
4:    **end for**
5:    Sort the array $d(i, j)$ in ascending order and divide them into $K$ groups equally
6:    **for** $k = 1$ to $K$ **do**
7:      Randomly select a group member as CH node
8:      $CHID_{P_i,k}$ = selected sensor node's ID
9:    **end for**
10:   Non-CH nodes join the nearest CH to form $K$ clusters
11: **end for**

## 4.3 Derivation Fitness Function

Each particle selects $K$ CH nodes by using Algorithm 1, the position and the residual energy of CH nodes can be obtained by CH node's ID, thus, we can use the following fitness function and then use PSO algorithm to achieve the optimal clustering formation.

The fitness function is the bridge between PSO algorithm and optimization problem, different fitness functions can be defined according to the different optimization problems. This paper mainly focuses on the issues of balancing the energy consumption and avoiding the premature death of nodes to define the fitness function. Equation (9) is the fitness function in this paper, it contains four sub-functions.

$$Fitness(P_i) = \alpha_1 \times f_1 + \alpha_2 \times f_2 + \alpha_3 \times f_3 + \alpha_4 \times f_4 \tag{9}$$

where $\alpha_1, \alpha_2, \alpha_3$ and $\alpha_4$ ($\alpha_1 + \alpha_2 + \alpha_3 + \alpha_4 = 1$) are the weight factors of sub-function respectively. $f_1$ is defined by CH nodes residual energy. $f_2$ and $f_3$ are defined by the distribution of clusters. $f_4$ is defined by the intra-cluster distance.

### 4.3.1 Fitness Sub-function of Energy

Due to CH nodes undertake more tasks than non-CH nodes, such as data sending and receiving, routing and forwarding, the energy consumption of them is larger than that of the cluster member nodes in the cluster. In other words, the more the residual energy of CH nodes, the more reliable routing transmission of data in the network can be achieved.

The CH node's ID, obtained by Algorithm 1, can help us to calculate the proportion of CH nodes' energy in network energy to make sure that CH nodes have enough residual energy without dying soon. Equation (10) is sub-function of energy.

$$f_1(P_i) = \frac{\sum_{j=1}^{N_{alive}} E(n_j)/N_{alive}}{\sum_{k=1}^{K} E(CH_{P_i,k})/K} \tag{10}$$

where $N_{alive}$ is the number of alive sensor nodes. $E(n_j)$ is the energy of the $j$th sensor node. $E(CH_{P_i,k})$ is the energy of CH node $k$ which is selected by particle $P_i$. $K$ is the number of the clusters.

$f_1$ denotes the ratio of the average energy of the whole network to the average energy of CH nodes, the smaller the $f_1$ is, the more the residual energy of CH nodes.

### 4.3.2  Fitness Sub-functions of Distribution of Clusters

The distribution of clusters directly affects CH nodes life, two factors determine the distribution of clusters: the distance between CH nodes and BS and the inter-cluster distance. In order to reduce the energy consumption of forwarding tasks of CH nodes which are close to BS, more CH nodes should be distributed around BS. Moreover, the longer the inter-cluster distance, the more transmission energy the CH nodes consume. Thus, two sub-functions of distribution of clusters are proposed as following Eq. (11) and (12).

$$f_2(P_i) = \frac{\sum_{i=1}^{K} d(BS, CH_{P_i,k})/K}{d(BS, NC)} \tag{11}$$

where $d(BS, CH_{P_i,k})$ is the distance between BS and $k$th CH node of particle $P_i$. $d(BS, NC)$ is the distance between BS and network area center.

Sub-function $f_2$ denotes the ratio of the average distance between CH nodes and BS to the distance between BS and network area center. The smaller the $f_2$ is, the more the CH nodes around the BS.

$$f_3(P_i) = max\{\forall k \in K | Overlap(Cluster_{P_i,k})\} \tag{12}$$

where $Overlap(Cluster_{P_i,k})$ is the overlapping ratio of cluster $k$, and according to the following Fig. 2, we can calculate the overlapping rate of cluster by Eq. (13).

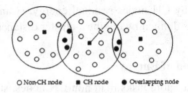

O Non-CH node     ■ CH node     ● Overlapping node

**Fig. 2.**  The overlapping ratio of cluster

In Fig. 2, $d$ is the average distance between cluster member nodes and CH node.

$$Overlap(Cluster_{P_i,k}) = \left(\sum_{j=1}^{C_{overlap}} \frac{N_{overlap,C_j}}{N_{intra,C_j}}\right)/C_{overlap} \tag{13}$$

where $C_{overlap}$ is the number of overlapping clusters. $N_{overlap,C_j}$ and $N_{intra,C_j}$ are the number of overlapping nodes and the number of sensor nodes in cluster $j$ respectively.

Sub-function $f_3$ denotes the maximum overlapping ratio of clusters of particle $P_i$, a smaller value of $f_3$ means more reasonable inter-cluster distance which can prevent the imbalance energy consumption of CH nodes caused by too dense or too sparse distribution.

### 4.3.3 Fitness Sub-function of Intra-cluster Distance

The intra-cluster distance determines the energy consumption of cluster member nodes. The CH nodes should be distributed as closely as possible to the center of the cluster to balance the transmission energy consumption of cluster members. The sub-function of intra-cluster distance in is shown as Eq. (14).

$$f_4(P_i) = max\left\{ \forall k \in K \mid \frac{\sum_{j=1}^{N_{intra,C_k}} d\left(CH_{P_i,k}, n_j\right)}{N_{intra,C_k} - 1} \right\} / d_0 \tag{14}$$

where $d\left(CH_{P_i,k}, n_j\right)$ is the distance between cluster member $j$ and CH node. $N_{intra,C_k}$ is the number of sensor nodes in cluster $k$.

$f_4$ denotes the maximum average distance of intra-cluster, a smaller value of $f_4$ means CH nodes are distributed closely at the center of clusters.

### 4.4 Optimal Clustering Formation

The optimal clustering formation process based on PSO algorithm is as follows.

---

**Algorithm 2.**  Optimal Clustering formation based on PSO.

---

**Input:**
Set of sensor nodes: $S = \{s_1, s_2, \ldots, s_N\}$
The number of particles: $N_P$
The number of CH nodes: $K$
Maximum iterations of PSO: $T_{max}$

---

**Output:**
The nodes' ID for optimal CH nodes: $\{\forall k \in K | CHID_k\}$
Optimal clustering formation: $\{\forall k \in K | Cluster_k\}$

---

1:    Select $V$ sensor nodes with higher energy than the average energy of network as candidate CH nodes
2:    **for** $i = 1$ to $N_P$ **do**
3:        $P_i = (x_i, y_i)$ /* $(x_i, y_i)$ is the position of sensor node */
4:        Get $CHID_{P_i,k}$ using Algorithm 1 and establish $K$ clusters
5:        Calculate $Fitness(P_i)$ /* Using Equation (9) */
6:        $Pbest_i = P_i$
7:        **if** $Fitness(P_i) < Fitness(Gbest)$ **then**
8:            $Gbest = P_i$
9:            $\{\forall k \in K | CHID_k\} = \{\forall k \in K, \forall i \in N_P | CHID_{P_i,k}\}$
10:      **end if**
11: **end for**
12: **for** $t = 1$ to $T_{max}$ **do**
13:    **for** $i = 1$ to $N_P$ **do**
14:        Update velocity and position of particle $P_i$ using Equation (4) and (5)
15:        **if** $Fitness(P_i) < Fitness(Pbest_i)$ **then**
16:            $Pbest_i = P_i$
17:            **if** $Fitness(P_i) < Fitness(Gbest)$ **then**
18:                $Gbest = P_i$
19:                $\{\forall k \in K | CHID_k\} = \{\forall k \in K, \forall i \in N_P | CHID_{P_i,k}\}$
20:            **end if**
21:        **end if**
22:    **end for**
23: **end for**
24: Form $K$ optimal clusters according to $CHID_k$

---

### 4.5  Establish Inter-cluster Routing

After determining the position of CH nodes, the distribution of clusters is also determined, it is necessary to establish an energy-efficiency routing in network before transmit data to BS. This paper takes transmission distance and the residual energy of CH nodes as edge weight then uses Prim algorithm to create a minimum spanning tree (MST) as inter-cluster routing. The following steps are the establishment of routing.

**Step1.** Add CH nodes and BS as vertexes to set $V = \{BS, CH_1, CH_2, \ldots, CH_K\}$.
**Step2.** Calculate weight of edges by following Eq. (15).

$$Weight_{i,j} = \frac{d_{vertex(i,j)}/d_0}{E_i + E_j}(1 \leq i \leq K+1) \tag{15}$$

where $d_{vertex(i,j)}$ is the distance between vertex $i$ to vertex $j$, $d_{vertex(i,j)}$ divided by $d_0$ is normalized to adjust its value in the range [0, 1]. $E_i$ and $E_j$ are the residual energy of $CH_i$ and $CH_j$ respectively.

**Step3.** Add vertex BS as root to set $U$ ($U = \{BS\}$).

**Step4.** In set $V - U$, select the vertex with minimum weight to the vertexes in set U and add it to set $U$.

**Step5.** Repeat Step4 until $V = U$.

**Step6.** If the distances between CH node and BS are less than $d_0$, so these CHs transmit data to BS directly. If the distances from cluster member nodes to CH node are less than that distance to BS, so these cluster member nodes transmit data to BS directly.

Figure 3 is the result of optimal clustering formation and eventual routing in simulation environment with 200 sensor nodes deployed in 200 m × 200 m network area. It can be seen that CH nodes are distributed around BS as close as possible and are not distributed too dense or too sparse and the transmission distances between clusters are as short as possible. The residual energy of the sending nodes and receiving nodes is considered in routing establishment process to make sure the CH nodes to have enough energy to forward packets. Hence, the consumption of energy in data transmission is significantly reduced and balanced.

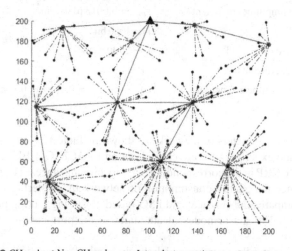

▲ BS   ● CH node   • Non-CH node   —— Inter-cluster routing   ····· Intra-cluster routing

**Fig. 3.** Clustering and routing diagram in simulation environment

## 5   Simulation Experiments and Results

To evaluate the performance of PSO-UCRP, MATLAB is used to perform simulations. All comparison protocols adopt the following simulation principles.

- The operation of protocol is broken up into rounds.
- In each round, non-CH nodes send a packet to their own CH, then CH aggregates the received packets and sends it to the next hop.

- In each round, we recorded and compared the number of packets received by BS, the number of alive nodes and the residual energy of network. Moreover, the round in which the first dead node occurs, the round in which half of nodes die and the round in which 80% nodes die are also recorded and compared.
- The sensor nodes consume energy in transmitting packets and only the sensor nodes with enough energy can send packets successfully. The adopted network and energy models show in Sect. 2.

**Table 1.** Simulation parameters.

| Parameter | Value |
|---|---|
| Area of network | $A = M \times M\,(M = 200\,\text{m}, 400\,\text{m})$ |
| Base station position | $(M/2, M)$ |
| Number of sensor nodes | $n = 100, 200$ |
| Initial energy of sensor nodes | $E_0 = 0.5\,\text{J}$ |
| Energy consumption on circuit | $E_{elec} = 50\,\text{nJ/bit}$ |
| Free-Space amplifier coefficient | $\varepsilon_{fs} = 10\,\text{pJ/bit/m}^2$ |
| Multi-path fading amplifier coefficient | $\varepsilon_{mp} = 0.0013\,\text{pJ/bit/m}^4$ |
| Packet size | $l = 4000\,\text{bit}$ |
| Maximum life cycle of network | $r = 2500\,\text{rounds}$ |
| The acceleration coefficients | $c_1 = c_2 = 2$ |
| The weight factors of fitness function | $\alpha_1 = 0.3, \alpha_2 = 0.2, \alpha_3 = 0.3, \alpha_4 = 0.2$ |

The simulation parameters of network are shown in Table 1.

The simulation experiments were carried out on LEACH-C, HEED, PSO-C, PSO-ECHS and PSO-UCRP in the corresponding network circumstances. The proposed algorithm is to balance the energy consumption of network by clustering. Therefore, this paper mainly compares the classical and PSO-based clustering routing protocols.

Table 2 shows 100 sensor nodes' death states of the protocols in 200 m × 200 m network area.

where FND denotes the round in which the first node dead occurs. HND denotes the rounds in which half of nodes die. LND denotes the round in which 80% nodes die.

Table 2 indicates that the PSO-UCRP protocol has better performance in balance energy consumption of sensor nodes and in prolong network life than the other four

**Table 2.** The death states of sensor nodes.

| Death state | LEACH-C | HEED | PSO-C | PSO-ECHS | PSO-UCRP |
|---|---|---|---|---|---|
| FND | 326 | 92 | 623 | 787 | 945 |
| HND | 1007 | 1453 | 1102 | 1374 | 1751 |
| LND | 1371 | 1640 | 1826 | 1512 | 1901 |

protocols under the same network conditions. Compared with LEACH-C, HEED, PSO-C and PSO-ECHS, the number of running rounds of PSO-UCRP is increased by 38.6%, 15.9%, 4.1% and 25.7% respectively.

Table 3 shows the average (AVE) and standard deviation (STD) of residual energy of nodes in the 500th round in the 200 m × 200 m network area with 100 sensor nodes.

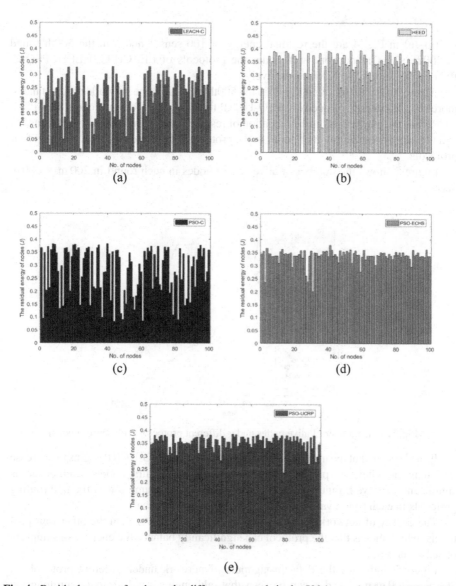

**Fig. 4.** Residual energy of nodes under different protocols in the 500th round: (a) LEACH-C; (b) HEED; (c) PSO-C; (d) PSO-ECHS; (e) PSO-UCRP

**Table 3.** Average and standard deviation of residual energy of nodes in 500th round.

|     | LEACH-C | HEED   | PSO-C  | PSO-ECHS | PSO-UCRP |
| --- | ------- | ------ | ------ | -------- | -------- |
| AVE | 0.2253  | 0.2874 | 0.2775 | 0.3361   | 0.3517   |
| STD | 0.1052  | 0.1231 | 0.0926 | 0.0266   | 0.0263   |

(a)–(e) in Fig. 4 are the residual energy of 100 sensor nodes in the 500th round in 200 m × 200 m network area under the protocols of LEACH-C, HEED, PSO-C, PSO-ECHS and PSO-UCRP respectively.

As we can see in (a)–(e) in Fig. 4, in the 500th round in same network circumstances, more residual energy of nodes in PSO-UCRP than the other four protocols, and PSO-UCRP has the smallest standard deviation of residual energy in Table 3 which turns out that the energy consumption of nodes in the protocol is more balance than the other four protocols.

Figure 5 shows the number of alive sensor nodes in each round in 200 m × 200 m network area.

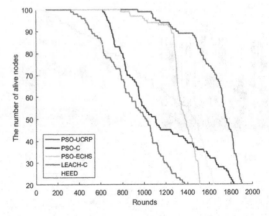

**Fig. 5.** The number of alive nodes under different protocols (100 sensor nodes)

It can be seen that the number of alive sensor nodes of PSO-UCRP begins to decrease later than the other four protocols and the round when the first node's death occurs is significantly delayed. Figure 6 shows how the energy consumption of the five routing protocols in each round varies in 200 m × 200 m network area.

The energy of network in PSO-UCRP is decreased slower than the other four protocols, which means that the protocol can significantly balance the energy consumption of sensor nodes.

Figure 7 indicates the data throughputs of network under different protocols in 200 m × 200 m net. And it can be seen that, although several protocols prolong the network life, the packets received by BS in PSO-UCRP protocol is the highest of the five compared protocols. In other words, the protocol can collect more data in each round.

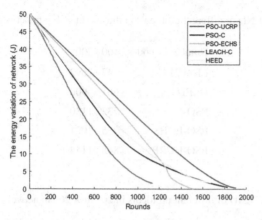

**Fig. 6.** Energy consumption of network under different protocols (100 sensor nodes)

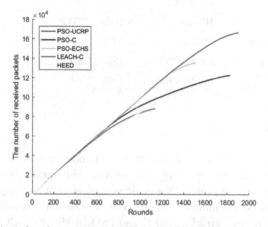

**Fig. 7.** The number of packets received by BS under different protocols (100 sensor nodes)

We also compared the scalability of the network nodes number and network areas of the five protocols, the LNDs of LEACH-C, HEED, PSO-C, PSO-ECHS and PSO-UCRP were tested in the 200 m × 200 m and 400 m × 400 m network areas with different number of sensor nodes, respectively. Table 4 shows the LNDs of the five protocol in 400 m × 400 m network area with different number of sensor nodes. Table 5 shows the LNDs of the five protocol in 200 m × 200 m network area with different number of sensor nodes.

Table 4 and 5 show that in the network area with different number of nodes, the LNDs of the PSO-UCRP occurred later than the other four protocols. The results show that due to the balanced energy consumption of network, PSO-UCRP protocol has better scalability for network area with different number of sensor nodes and different sizes.

The experiment results show that, although several protocols prolong the network life, the packets received by BS in PSO-UCRP protocol is the highest one, which means the energy consumption of sensor nodes in our protocol are more balanced and have enough energy to transmit packets.

**Table 4.** LND of different protocols in the 400 m × 400 m network area

| Number of nodes | 100 | 200 |
|---|---|---|
| LEACH-C | 547 | 693 |
| HEED | 661 | 900 |
| PSO-C | 733 | 1130 |
| PSO-ECHS | 735 | 1021 |
| PSO-UCRP | 848 | 1144 |

**Table 5.** LND of different protocols in 200 m × 200 m network area

| Number of nodes | 100 | 200 |
|---|---|---|
| LEACH-C | 1371 | 1755 |
| HEED | 1640 | 1770 |
| PSO-C | 1826 | 2019 |
| PSO-ECHS | 1512 | 1803 |
| PSO-UCRP | 1901 | 2152 |

## 6  Conclusions

Considering the large-area of monitoring in the mountainous area, the uneven distribution of sensor nodes and the battery energy are limited. This paper proposes an uneven clustering multi-hop routing protocol for geological disaster monitoring sensor network to balance the energy consumption of sensor nodes. Simulation experiments showed that compared with LEACH-C, HEED, PSO-C and PSO-ECHS, PSO-UCRP significantly delays the time of first dead node occurs, prolongs the network life and improves the success rate of packets received by BS. In other words, PSO-UCRP can achieve long-term and large-area requirements of geological disaster monitoring in the mountainous area and can be adopted by other outdoor wireless monitoring systems.

**Acknowledgements.** The work is supported by the Research Project of China Railway Eryuan Engineering Group CO. LTD. (No. KYY2019033(19-20)) and the Support project of Key Research and Development of Chengdu, China (No. 2019-YF08-00160-GX).

## References

1. Kotta, H.Z., Rantelobo, K., Tena, S., Klau, G.: Wireless sensor network for landslide monitoring in nusa tenggara timur. TELKOMNIKA (Telecommun. Comput. Electron. Control) **9**(1), 9–18 (2011)
2. Huang, Y.M., Chen, W.C., Fang, Y.M., Lee, B.J., Chou, T.Y., Yin, H.Y.: Debris flow monitoring - a case study of Shenmu area in Taiwan. Disaster Adv. **6**(11), 1–9 (2013)

3. Hsiao, T., Lee, B., Chou, T., Lien, H., Chang, H.L.: Debris flow monitoring system and observed event in Taiwan: a case study at Aiyuzi river. Wuhan Univ. J. Nat. Sci. **12**, 610–618 (2007). https://doi.org/10.1007/s11859-006-0298-4
4. Lee, H., Ke, K., Fang, Y., Lee, B., Chan, T.: Open-source wireless sensor system for long-term monitoring of slope movement. IEEE Trans. Instrum. Meas. **66**(4), 767–776 (2017)
5. Behera, T.M., Mohapatra, S.K., Samal, U.C., Khan, M.S., Daneshmand, M., Gandomi, A.H.: I-SEP: an improved routing protocol for heterogeneous WSN for IoT-based environmental monitoring. IEEE Internet Things J. **7**(1), 710–717 (2020)
6. Ramesh, M.V.: Design, development, and deployment of a wireless sensor network for detection of landslides. Ad Hoc Netw. **13**(Part A), 2–18 (2014)
7. Heinzelman, W.R., Chandrakasan, A., Balakrishnan, H.: Energy-efficient communication protocol for wireless microsensor networks. In: Proceedings of the 33rd Annual Hawaii International Conference on System Sciences, Maui, HI, USA, vol. 2 (2000). 10 pp.
8. Heinzelman, W.B., Chandrakasan, A.P., Balakrishnan, H.: An application-specific protocol architecture for wireless microsensor networks. IEEE Trans. Wireless Commun. **1**(4), 660–670 (2002)
9. Jannu, S., Jana, P.K.: A grid based clustering and routing algorithm for solving hot spot problem in wireless sensor networks. Wireless Netw. **22**(6), 1901–1916 (2015). https://doi.org/10.1007/s11276-015-1077-y
10. Singh, S.K., Kumar, P., Singh, J.P.: An energy efficient protocol to mitigate hot spot problem using unequal clustering in WSN. Wireless Pers. Commun. **101**(2), 799–827 (2018). https://doi.org/10.1007/s11277-018-5716-3
11. Latiff, N.M.A., Tsimenidis, C.C., Sharif, B.S.: Energy-aware clustering for wireless sensor networks using particle swarm optimization. In: 2007 IEEE 18th International Symposium on Personal, Indoor and Mobile Radio Communications, Athens, pp. 1–5 (2007)
12. Rao, P.C.S., Jana, P.K., Banka, H.: A particle swarm optimization based energy efficient cluster head selection algorithm for wireless sensor networks. Wireless Netw. **23**(7), 2005–2020 (2016). https://doi.org/10.1007/s11276-016-1270-7
13. Shen, J., Wang, A., Wang, C., Hung, P.C.K., Lai, C.: An efficient centroid-based routing protocol for energy management in WSN-assisted IoT. IEEE Access **5**, 18469–18479 (2017)
14. Hamida, E.B., Chelius, G.: A line-based data dissemination protocol for wireless sensor networks with mobile sink. In: 2008 IEEE International Conference on Communications, Beijing, pp. 2201–2205 (2008)
15. Eberhart, R., Kennedy, J.: A new optimizer using particle swarm theory. In: MHS 1995, Proceedings of the Sixth International Symposium on Micro Machine and Human Science, Nagoya, Japan, pp. 39–43 (1995)
16. Liang, Y., Yu, H.: PSO-based energy efficient gathering in sensor networks. In: Jia, X., Wu, J., He, Y. (eds.) MSN 2005. LNCS, vol. 3794, pp. 362–369. Springer, Heidelberg (2005). https://doi.org/10.1007/11599463_36

# Clustering Analysis Method of Ethnic Cultural Resources Based on Deep Neural Network Model

Mingjing Tang⑩, Chao Sun(✉) ⑩, and Li Liang⑩

Yunnan Normal University, Kunming, China
1052962641@qq.com

**Abstract.** This paper proposes a method of clustering analysis of ethnic cultural resources based on deep neural network model. Firstly, the feature word extraction and vectorization of ethnic cultural resources texts are realized by doc2vec document vectorization tool. Then K-means clustering algorithm is used to cluster the ethnic cultural resources texts after vectorization, and the Elbow method is used to determine the best aggregation. So as to obtain the correlation between the texts of ethnic cultural resources, which is used for the collection, storage and intelligent service of massive ethnic cultural resources provides technical support. At the end of the paper, the ethnic cultural resources in the specific ethnic website are taken as an example to analyze the above methods.

**Keywords:** Deep neural network model · Ethnic cultural resources · Clustering analysis · Elbow method

## 1 Introduction

Ethnic culture is an indispensable part of the culture that has been deposited in China of five thousand years of history. However, with the continuous development of modernization and internationalization process of China, the protection and inheritance of ethnic culture, especially minority culture, faces a huge crisis. At present, the latest research on the protection of ethnic culture has the digital protection mentioned in the reference [1] and the reference [2]. For example, reference [3] have invented an algorithm that can effectively mine a large number of rock carving patterns; also the network and information protection mentioned in the reference [4] and the reference [5], such as the construction of various ethnic cultural websites. However, these methods all rely on the continuous collection of minority cultures. While in the face of massive ethnic cultural resources, they can't be quickly classification, collection and sharing, which is not conducive to the protection and inheritance of ethnic culture. Therefore, new theoretical guidance and tool support are urgently in need. The most ideal state of ethnic cultural resources management is to realize intelligent management of data resources [6], such as automatic collection, classification and sharing of cultural resources. Using distributed web crawling technology, natural language processing technology, text mining technology [7] to collect, parse, preprocess and other operations on the ethnic cultural resources text. And then the obtained text of ethnic cultural resources is analyzed by clustering so

© Springer Nature Switzerland AG 2020
X. Chen et al. (Eds.): ML4CS 2020, LNCS 12488, pp. 160–170, 2020.
https://doi.org/10.1007/978-3-030-62463-7_15

as to have a better understand on the deep semantics of ethnic cultural resources texts and obtain the association between ethnic cultural texts. All of these will help the automatic collection, identification and sharing of massive ethnic cultural resources, and provide technical support for the development of ethnic cultural resources in the direction of intelligent management.

The current text mining is mainly based on text feature extraction of deep learning. For example, the reference [8] introduces the related research of deep learning text extraction, and the reference [9] introduces the text mining technology that combines deep learning features. Text mining is widely used, For example, reference [10] analyzes the differences between Chinese and American science and technology policies Based on text mining and visual analysis; Reference [11] studies how the big data service of scientific and technological literature develops towards intelligent question answering, based on text mining technology; reference [12] research the automatic classification of product description based on text mining. Inspired by the reference [10] and the reference [12], this paper is based on the text data of ethnic cultural resources, for the purpose of to enhance the identifiability and comprehensibility of massive ethnic cultural resources and to facilitate the intelligent collection and sharing of massive ethnic cultural resources. This paper propos a clustering analysis method of ethnic cultural resources based on neural network probabilistic language model [13]. Firstly, the ethnic cultural resources are crawled through distributed web crawling technology. The ethnic cultural resources data are processed by natural language processing technology to preprocess and segment Chinese word. And based on document vectorization tool, the vectorization of ethnic cultural resources data is realized. Then, using appropriate clustering algorithm to cluster the vectorized ethnic cultural resources data, and using the Elbow Method to select the optimal cluster number; finally, analyze the clustering result to identify and discover the deep semantics and associations between the ethnic cultural resources Texts. Thus provide support for the mining of massive ethnic cultural resources. The specific workflow is shown in Fig. 1.

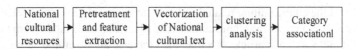

**Fig. 1.** Ethnic cultural resources clustering process based on deep neural network.

The organization structure of this paper is as follows: Sect. 2 crawls on ethnic cultural resources and performs data preprocessing and vectorization representation; Sect. 3 describes the clustering process of ethnic cultural resources data and analyzes the clustering results; Sect. 4 takes the ethnic cultural resources data in the specific website as an example to verify the method of this paper. Finally, Sect. 5 summarizes the full text and future work prospects.

## 2 Ethnic Cultural Resources Data Vectorization

At present, the main mode of dissemination of minority cultures is the various cultural websites of ethnic minorities. The content of ethnic cultural resources of such websites is relatively scattered and difficult to be discovered and utilized. Using distributed network crawler technology, natural language processing technology and data mining technology, we can collect, analyze and preprocess the ethnic cultural resources.

## 2.1 Feature Extraction

The ethnic cultural resources crawled from the webpage are stored in the format of the HTML document. In order to extract the useful ethnic cultural resources texts in the HTML webpage, the relevant document parsing library needs to be called to delete the head and other unrelated areas of the obtained ethnic cultural resources data text and conduct preprocessing operations such as unlabeling, so as to extract the text content of the webpage. Then use natural language processing tools to remove prepositions, adverbs and other meaningless words, retaining entity words such as verbs and nouns. The specific process is described as follows:

---

**Algorithm 1.** Feature word extraction algorithm for ethnic cultural resources

---

**Input:** ethnic cultural resources text
**Output:** feature word set $\varphi(s)$
1.    Set $\varphi(s)=\emptyset$;
2.    Scrapy (html);    // *Climb the ethnic culture related pages;*
3.    BeautifulSoup(P);    // *Extract node element text content;*
4.    for(p in P) {    // *traverse for each text content;*
5.        wordFilter(p);    //*Remove stop words with less meanings such as prepositions, adjectives and adverbs, and retain vocabulary such as verbs and nouns;*
6.        reductWord(p);    //*Convert the vocabulary of various tenses into a general form, and make a part of speech reduction;*
7.        add($\varphi(p)$);    //*Add to feature set*
8.    }
9.    **end**

First, Crawl the ethnic culture related webpage from the ethnic culture related website and save it, then extract the ethnic culture related articles in the node from the saved webpage text and save each article as a line, then traverse all the ethnic culture articles separately. Remove prepositions, adjectives, adverbs and other stop words in the article, retain verbs, nouns and other entity words. Finally, the tense of the vocabulary is transform into a general form and add to the feature set.

## 2.2 Ethnic Cultural Resources Text Vectorization

After completing the extraction of ethnic cultural resources, in order to measure the similarity between texts and then realize the cluster analysis of ethnic cultural resources texts [14]. It is necessary to vectorize each document information of ethnic cultural resources. Doc2vec is an unsupervised learning algorithm [15], which is used to predict a vector to represent different documents, it mainly adopts two models: Distributed Memory (DM) and Distributed Bag of Words (DBOW). The Distributed Memory Model works by predicting the current word based on its context information. Each paragraph is represented by a vector that represents a column vector in the paragraph matrix, each word is represented by a vector that represents a column vector in the word matrix, the paragraph vector and the word vector are averaged or connected to predict the next word in the context. The working principle of Distributed Bag of Words model is the same as that of Distributed Memory model, but predicts the context probability based on the target word, ignoring the input context word, Let the model predict a random word in the paragraph. Specifically, in each iteration of the random gradient descent, a window is sampled from the text, and then a word is randomly sampled from the window, and a classification task [16] is formed according to the paragraph vector. In this paper, the doc2Vec DBOW model is used to characterize the ethnic cultural resources characterized by feature words. The specific process is shown in Fig. 2.

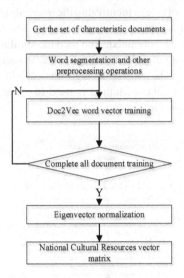

**Fig. 2.** Vectorization process of ethnic cultural resources.

As can be seen from Fig. 2, the process of ethnic cultural resources vectorization is as follows:

Firstly, the data crawling operation of the above section can obtain a feature document set representing the ethnic cultural resources, and perform data pre-processing operations such as deleting stop words and word segments.

Then, based on the doc2vec document vector tool, the Distributed Bag of Words Model is constructed to train the document vector of ethnic cultural resource document set. In order to avoid the impact of content size in the text of ethnic cultural resources on the value of feature vectors, it is necessary to normalize the feature vectors of each document:

$$\bar{\Delta}x(i,j) = \Delta x(i,j)/\|\Delta x(i)\|2 \tag{1}$$

The ethnic cultural resources feature vector can be normalized to the [0, 1] by formula (1).

Finally, the characterization of the ethnic cultural resources vector matrix can be obtained. Assume that the entire ethnic cultural resources have $n$ texts, and after vectorization, the following ethnic cultural resources vector matrix is obtained:

$$\delta i = [\delta i(1), \delta i(2), \cdots \delta i(n)] \tag{2}$$

According to Formula (2), $\delta i \in R^{n \times m}$, $n$ is the number of texts of ethnic cultural resources, and $m$ is the number of characteristic words of ethnic cultural resources.

## 3  Ethnic Cultural Resources Clustering

Clustering is one of the important research fields in data mining and pattern recognition and so on. It plays an important role in identifying the internal structure of data [17]. After the above ethnic cultural resources texts are vectorized, the feature vectors corresponding to each ethnic cultural text can be obtained. And then the similarity between the texts can be measured by the clustering algorithm to realize the association and differentiation of ethnic cultural resources. The clustering algorithm used in this paper is K-means clustering algorithm, and the Elbow Method is used to evaluate the clustering effect. The specific workflow is shown in Fig. 3.

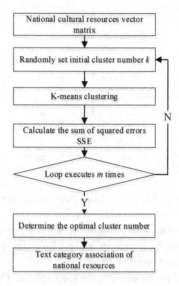

**Fig. 3.** Cluster analysis process of ethnic cultural resources text.

### 3.1  K-Means Clustering Algorithm

K-means clustering algorithm [18, 19] as a classical clustering analysis algorithm, which has the advantages of simple implementation, fast clustering and so on. The main principle of this algorithm is to divide samples into $k$ clusters according to the distance between samples. Taking the ethnic cultural resources sample set as an example, the specific steps are as follows: First, $k$ points are randomly selected from the vector set space as the initial cluster center. Then, the vector of ethnic culture and its nearest clustering center are classified into one category, and Then calculate the average of all vectors of each cluster to update the values of each cluster center. Finally, the above two steps are continuously iterated until the cluster center no longer changes to obtain the final clustering result [20]. The specific algorithm is described as follows:

---

**Algorithm 2 .** Ethnic cultural resources text vector clustering algorithm

---

**Input:** ethnic cultural resources text vector set, cluster number $k$

**Output:** $k$ clusters

1. Randomly select $k$ vectors in the text vector set of ethnic cultural resources as the initial clustering center of $k$ clusters;
2. For any vector, calculate its similarity with $k$ cluster centers by Euclidean distance method, and mark the category of vector as the cluster corresponding to the nearest cluster center;
3. Update the cluster centers corresponding to the $k$ clusters by calculating the average value of each cluster;
4. Iterate over the above two steps until the center point no longer changes.
5. **end**

## 3.2 Evaluation of Optimal Cluster Number Based on Elbow Method

Since the K-means clustering algorithm is an unsupervised learning task, the iterative method is used, and only the local optimal solution can be obtained. The selection of $k$ is not easy to grasp, so the clustering effect needs to be evaluated. In this paper, the Elbow Method is used to evaluate the clustering effect of ethnic cultural resources [21].

**Calculate the Sum of Squared Errors (SSE).** The sum of squared errors of the text vector of ethnic cultural resources is calculated by formula (3):

$$SSE = \sum_{i=1}^{k} \sum_{p \in c_i} |p - m_i|^2 \tag{3}$$

Where, $k$ represents the number of clusters, $c_i$ is all elements in the i-th cluster, $m_i$ represents the cluster center of the i-th cluster, and $p$ represents each element in the cluster.

**The Relationship Between K Value and SSE Value (Elbow Shape).** As shown in Fig. 4, in order to obtain the best clustering effect, the k-means clustering algorithm needs to be repeated $n$ times and the error sum of squares needs to be calculated each time, and obtain the correspondence between $K$ values and $SSE$ values. Draw the relationship between $K$ value and $SSE$ value (elbow shape), the $K$ value point corresponding to the elbow portion ($k = 4$) in the relationship graph is selected as the optimal cluster number.

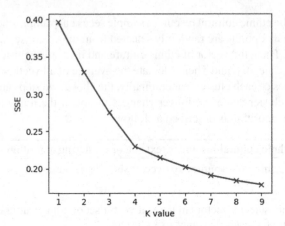

**Fig. 4.** K value and SSE value.

## 4  Experimental Analysis

In the experimental part of this paper, five virtual machines will be created by using virtualization technology in the OpenStack environment, one of which is the Master node and the other four are the slave nodes. Install the crawler module Scrapy on each slave node to implement the main function of resource crawling; The Redis database is installed in the master node to realize URL queue management and maintenance of multiple distributed crawlers. Scrapy is a screen grabbing and web grabbing framework developed by python, which is used to grab web sites and extract structured data from pages [22].

### 4.1  Experimental Data and Steps

The experimental data in this paper were obtained by crawling articles related to ethnic culture from the Ethnic Affairs commission of the People's Republic of china (http://www.seac.gov.cn) and Ethnic network (http://www.minzu56.net), and a total of 6830 ethnic cultural resources web pages have been obtained. The specific experimental process includes: ethnic cultural resources data crawling, ethnic cultural resources text preprocessing, ethnic cultural resources text vectorization, ethnic cultural resources text clustering, identifying and discovering the deep semantic and relationship between the texts of ethnic cultural resources according to the clustering results.

The programming language used in the experimental part of this article is the Python language, and all experimental steps are completed using a Python-based development library. For example: The crawler module Scrapy is used to realize the crawling of national cultural resources; Use Python's natural language development library NLTK to complete the ethnic cultural resources text preprocessing. In the vectorization stage of ethnic cultural resources text, the doc2vec development interface is called by Python's gensim library to construct a text vector training model, model training and feature extraction for ethnic cultural resources texts, and automatic mapping to $k$ resource category clusters through unsupervised learning. In the text processing stage, a total of

6,830 ethnic cultural resources webpage texts were generated, which resulted in 6830 feature vectors in the text vectorization stage. The K-means clustering algorithm was implemented by Python's sklearn library to clustering the ethnic cultural resources text vector matrix.

### 4.2   Analysis of Experimental Results

The relationship between the sum of the squared errors (SSE) and the cluster number $k$ is shown in Fig. 5. For the more accurate evaluation, we also draw the change of the Silhouette Coefficient [23] as a reference. We can see from the figure that when $k = 7$, the SSE polyline chart shows a larger inflection point (elbow). At the same time, the Silhouette Coefficient also reaches the maximum at this point. Therefore, $k = 7$ is the choice of the optimal cluster number $k$ value. As shown in Fig. 6, 6830 texts in the ethnic cultural resources data are mapped to 7 ethnic culture types, of which type 7 contains the most text, accounting for 1782 texts; type 1 contains the least text, accounting for 423 texts. So far, the discovery and identification of the relationship between the texts of ethnic cultural resources has been completed.

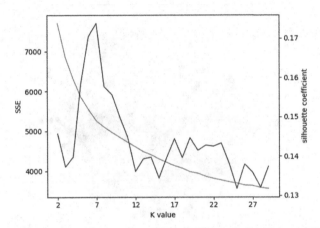

**Fig. 5.**  The relationship between SSE and Silhouette Coefficient and K.

Due to the high dimension of text vector matrix of ethnic cultural resources, it is not convenient for data visualization analysis [24]. In this paper, Principal Component Analysis (PCA) is adopted to decrease the dimension of the clustering results of the text of ethnic cultural resources, and calls Python's painting library matplotlib, Visualize the effect of clustering. The clustering effect is shown in Fig. 7. It is better to divide the ethnic cultural resources into seven different types.

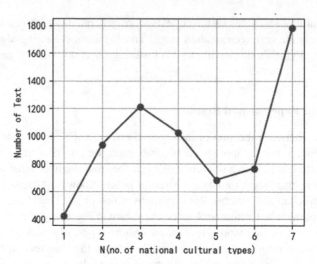

**Fig. 6.** Number of text and cultural type.

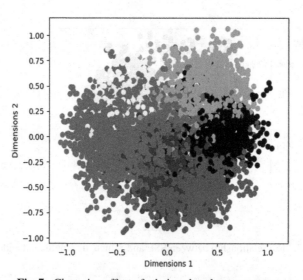

**Fig. 7.** Clustering effect of ethnic cultural resources text.

## 5   Conclusion

In this paper, a clustering analysis method of ethnic cultural resources based on deep neural network model is proposed, which adopts the unsupervised text vector representation method (doc2vec) based on deep neural network, using cosine distance formula to calculate the topic similarity between vectors effectively reduces the dimension of vector space and improves the efficiency of training, which is conducive to the mining of massive ethnic cultural resources texts. At the same time, the K-means clustering algorithm is used for text clustering, and the clustering effect evaluation method combining

Elbow Method and Silhouette Coefficient is adopted to select the optimal clustering number $k$ value, it solves the problem that the number of clusters $k$ in K-means algorithm is difficult to be determined. Finally, through experimental analysis, the method has high accuracy and efficiency, and effectively divides ethnic cultural resources into different categories. And realizes the differences and associations between ethnic cultural resources, revealing the deep semantics of ethnic cultural resources, and providing support for the massive ethnic cultural resources mining and intelligent services. In the next step, we can apply the clustering analysis method of ethnic culture based on the deep neural network model to the construction of the intelligent management system of ethnic culture, and realize the automatic collection, classification and sharing of ethnic culture resources.

**Acknowledgment.** This work was supported by the National Natural Science Foundation of China (Grant No. 61662085), Natural Science Foundation of the Department of Education of Yunnan Province of China (Grant No. 2017ZZX073), and Program for innovative research team (in Science and Technology) in University of Yunnan Province.

# References

1. Xiao, W., Lu, Y.: Digital protection analysis of national traditional culture based on big data. Inf. Commun. **2019**(05), 177–178 (2019)
2. Liu, X.C., Song, W.: Research on digital protection of national traditional culture under the condition of big data. J. Cent. Univ. Natl. (Nat. Sci. Ed.) **25**(03), 44–49 (2016)
3. Zhu, Q., Wang, X., Keogh, E., Lee, S.-H.: An efficient and effective similarity measure to enable data mining of petroglyphs. Data Min. Knowl. Disc. **23**(1), 91–127 (2011). https://doi.org/10.1007/s10618-010-0200-z
4. Mei, H.: Research on the information service mode and document information resource construction of national libraries under the network environment. Commun. Res. **1**(08), 176 (2017)
5. Kummer, T.-F., Leimeister, J.M., Bick, M.: On the importance of national culture for the design of information systems. Bus. Inf. Syst. Eng. **4**(6), 317–330 (2012). https://doi.org/10.1007/s12599-012-0236-2
6. Manjunath, T.N., Ravindra, S.H., Umesh, I.M., Ravikumar, G.K.: Realistic analysis of data warehousing and data mining application in education domain. Int. J. Mach. Learn. Comput. **2**(04), 419–422 (2012)
7. Sun, Y.: A text mining approach to analyze public media science coverage and public interest in science. Int. J. Mach. Learn. Comput. **4**(06), 496–500 (2014)
8. Liang, H., Sun, X., Sun, Y., Gao, Y.: Text feature extraction based on deep learning: a review. EURASIP J. Wirel. Commun. Network. **2017**(1), 211 (2017). https://doi.org/10.1186/s13638-017-0993-1
9. Chen, X.: The feature extraction of the text based on the deep learning. In: Advanced Science and Industry Research Center. Proceedings of the 2014 International Conference on Network Security and Communication Engineering (NSCE 2014). Advanced Science and Industry Research Center: Science and Engineering Research Center, 2014, no. 5 (2014)
10. Wu, Y., Yuan, Y., Gong, Z.D.: A comparative study of Sino-US science and technology policies under the background of artificial intelligence——based on text mining and visualization analysis. J. China Acad. Electron. Sci. **14**(08), 891–896 (2019)

11. Wen, Y.K., Wen, H., Qiao, X.D.: Let knowledge generate wisdom——study on text mining and question answering technology based on artificial intelligence. J. Inf. **38**(07), 722–730 (2019)
12. Lee, H., Yoon, Y.: Engineering doc2vec for automatic classification of product descriptions on O2O applications. Electron. Commer. Res. **18**(3), 433–456 (2017). https://doi.org/10.1007/s10660-017-9268-5
13. Xiong, F.L., Deng, Y.H., Tang, X.W.: The core architecture of word2vec and its application. J. Nanjing Normal Univ. (Eng. Technol. Ed.) **15**(01), 43–48 (2015)
14. Rajhans, M., Pradeep, K.: Clustering web logs using similarity upper approximation with different similarity measures. Int. J. Mach. Learn. Comput. **2**(03), 219–221 (2012)
15. Wang, C.C.: A model of chinese sentiment analysis with more general applicability. In: Information Engineering Research Institute (USA), Asia Pacific Human-Computer Interaction Research Center (Hong Kong). Proceedings of 2018 4th ICMSMA International Conference on Advances in Intelligent Information Technologies (ICAIIT 2018). Information Engineering Research Institute (USA), Asia Pacific Human-Computer Interaction Research Center (Hong Kong): Intelligent Information Technology Application Society (2018)
16. Le, Q.V., Mikolov, T.: Distributed representations of sentences and documents (2014)
17. Mohamed, N.E., Monzer, M.Q.: Analysis of some algorithms for clustering data objects. Int. J. Mach. Learn. Comput. **4**(01), 99–105 (2014)
18. Sun, J.G., Liu, J., Zhao, L.Y.: Research on clustering algorithms. J. Softw. **2008**(01), 48–61 (2008)
19. Liu, P., Teng, J.Y., Ding, E.J., et al.: Spark-based large-scale text k-means parallel clustering algorithm. Chin. J. Inf. Sci. **2017**(04), 150–158 (2017)
20. Yogi, W.R., Devi, F.: The comparative study on clustering method using hospital facility data in Jakarta District and surrounding areas. Int. J. Mach. Learn. Comput. **9**(06), 749–755 (2019)
21. Kanungo, T., Mount, D.M., Netanyahu, N.S., et al.: An efficient k-means clustering algorithm: analysis and implementation. IEEE Trans. Pattern Anal. Mach. Intell. **24**(7), 881–892 (2002)
22. Ryan, M.: Python Network Data Collection. People's Posts and Telecommunications Press, Beijing (2016). 2016.3
23. Sarunya, K.: A novel outlier detection applied to an adaptive k-means. Int. J. Mach. Learn. Comput. **9**(05), 569–574 (2019)
24. Supaporn, B., Thuttaphol, I., Nittaya, K., Kittisak, K.: Text-independent speaker identification using deep learning model of convolution neural network. Int. J. Mach. Learn. Comput. **9**(02), 143–148 (2019)

# Optimal Group Consensus of Second-Order Multi-agent Systems

Yize Yang[1,2], Hongyong Yang[1,3(✉)], Yuling Li[1,3], and Yuanshan Liu[1,3]

[1] School of Information and Electrical Engineering, Ludong University, Yantai 264025, China
hyyang@yeah.net
[2] School of Electrical and Electronic Engineering, University of Adelaide, Adelaide, Australia
[3] Key Laboratory of Cyber-Physical System and Intelligent Control in Universities of Shandong, Ludong University, Yantai 264025, China

**Abstract.** The optimal control problems play an important role in modern control theory. This paper focuses on the optimal problem for group flocking movement of multi-agent systems (MAS). Two new cost functions are proposed with distributed optimal cooperative control. By using modern control theory and algebraic graph theory, optimal control of group consensus trajectory for dynamic MAS is studied. Moreover, a properly chosen value of the optimal scaling factor is presented, where the optimal cost functions for group consensus of MAS can be achieved by choosing the appropriate scaling factor. Numerical simulations are provided to illustrate the effectiveness of the theoretical results.

**Keywords:** Optimal control · Group consensus · Multi-agent systems · Coordinated motion

## 1 Introduction

### 1.1 A Subsection Sample

In the distributed automatic control fields, cooperative control of multi-agent systems (MAS) is currently a critical research topic. It has many applications for cooperative control of MAS, including driverless cars, unmanned aerial vehicles and unmanned submarine detectors. These applications of distributed MAS provide the convenience of social life and promote the development of scientific research.

Consensus is an important issue in the cooperative control of MAS, which means the agents achieve the agreement of the position, velocity or phase by designing a communication protocol. In early research, centralized control of MAS is explored with the system structural features. Compared with centralized control of MAS, distributed control of MAS has its unique advantages. Examples of the advantages include lower cost, faster response speed and more flexible structure. Distributed MAS have been studied extensively in the recent research works. For distributed cooperative control of MAS, the consensus control algorithms in different settings are studied in [1–5]. The consensus of linear MAS in different backgrounds is studied in [6–13].

© Springer Nature Switzerland AG 2020
X. Chen et al. (Eds.): ML4CS 2020, LNCS 12488, pp. 171–182, 2020.
https://doi.org/10.1007/978-3-030-62463-7_16

As a class of special condition in consensus control problem, the competition mechanism is introduced into MAS, group consensus control on the competition and cooperation mechanism has been proposed and studied in recent years [14–19]. In [14], group consensus for the first-order MAS with nonlinear input constraints is investigated. Group consensus of MAS with switching topologies and communication delays are addressed in [15–17]. In [18], the dynamics group consensus problem of heterogeneous multi-agent systems with time delays is investigated, in which agents' dynamics are modeled by single integrators and double integrators. To achieve group consensus, a novel group consensus protocol is proposed for MAS with a time-varying estimator of the uncertain parameters in [19]. Based on the different requirement, group consensus of MAS with cooperative relationship and the competition mechanism are studied in practical applications.

In the process of the cooperative control of the dynamic MAS, how to reduce the cost and energy has become more and more important. The optimal control problem of cooperative control of MAS has attracted more and more the attention. Optimality issues in consensus algorithms have been studied in [20, 21], where the global optimal consensus problem for MAS with bounded controls is studied. Instead of studying global optimal algorithms, the optimal consensus control laws for agent-based models are discussed in [22]. The optimal consensus problem of continuous-time MAS with a common state set constraint is reported in [23]. In [24], the optimal control of a multi-agent consensus problem in an obstacle-laden environment is investigated. The optimal coordination control for nonlinear MAS based on event-triggered adaptive dynamic programming method is concerned in [25].

In this paper, the distributed optimal group consensus control of dynamic MAS is investigated. The contribution of this paper is the optimal cost function for the optimal control problem is proposed with distributed communication protocol. Based on graph theory and optimal control method, the group consensus algorithm of dynamic MAS with leaders and without leaders are discussed. Based on the optimal scaling factor in group control algorithm, the optimal group motion can be achieved for the distributed MAS.

The remainder of this paper is organized as follows. In Sect. 2, the graph theory and definitions are introduced. A new optimal cost function is proposed, and group consensus control algorithm of dynamic MAS is obtained in Sect. 3. Based on the proper value of the optimal scaling factor in optimal group control protocol, the optimal group consensus is analyzed in Sect. 3. Numerical simulations are used to verify the optimal group consensus algorithms of MAS with leaders and leaderless in Sect. 4. A short conclusion is given in Sect. 5.

## 2  Graph Theory and Definitions

Let $G = (V, \omega, \Lambda)$ be a networked topology of order $n + m$ with the set of nodes $V = \{v_1, v_2, \ldots, v_{n+m}\}$, set of edges $\omega \subseteq v \times v$, and the symmetrical weighted adjacency matrix $\Lambda = [a_{ij}] \in R^{(n+m) \times (n+m)}$ with real adjacency elements $a_{ij}$. An edges of $G$ is denoted by $e_{ij} = (v_i, v_j)$, which starts from $i$ and ends on $j$. The adjacency elements associated with the edges of the graph are nonzero, i.e., $e_{ij} \in \omega$ if and only if $a_{ij} \neq 0$. For all nodes $i$, it is assumed that $a_{ii} = 0$. There are two ways to describe the communication link between two nodes. If there is a communication link between $v_i$ and $v_j$, the sending and receiving of information can be represented by $v_i \rightarrow v_j$ and $v_j \rightarrow v_i$ respectively. In the set of neighbors of node $v_i$ is denoted by $N_i = \{v_j | v_j \in V : e_{ij} \in \omega\}$. The set of a node subset neighbors $\varsigma \subset V$ is defined by $N_\varsigma = \{v_j | v_j \in V, v_i \in \varsigma : e_{ij} \in \omega\}$. The Laplacian matrix $L(G) = [l_{ij}]_{(n+m)\times(n+m)}$ of networked topology $G$ is defined by

$$
l_{ij} = \begin{cases} -a_{ij}, & j \neq i \\ \sum_{k=1, k\neq i}^{n+m} a_{ik}, & j = i \end{cases}
$$

The in-degree and out-degree of node $i$ are defined as

$$
Deg_{in}(i) = \sum_{j=1}^{n+m} a_{ij}, \quad Deg_{out}(i) = \sum_{j=1}^{n+m} a_{ji}
$$

$D = diag\left\{\sum_{j=1}^{n+m} a_{1j}, \cdots, \sum_{j=1}^{n+m} a_{n+m,j}\right\}$ is defined as degree matrix. The Laplacian matrix satisfies $L = D - A$.

**Definition 1.** A networked topology $G_1 = \{V_1, \omega_1, A_1\}$ is said to be a sub-network of a networked topology $G = \{V, \omega, A\}$, if $V_1 \subseteq V$ and $\omega_1 \subseteq \omega$. Furthermore, if $V_1 \subset V$ and $\omega_1 = \{(v_i, v_j) : i, j \in V_1, e_{ij} \in \omega\}$, the graph $G_1$ is a proper sub-graph of $G$.

We suppose a group $G = \{V, \omega, A\}$ can be partitioned into two bipartite graphs $G_1 = \{V_1, \omega_1, A_1\}$ and $G_2 = \{V_2, \omega_2, A_2\}$ with $V_1 \cup V_2 = V$ and $V_1 \cap V_2 = \emptyset$. Without loss of generality, a network graph $G$ with $n + m$ $(n, m > 1)$ agents indexed by $1, 2, \ldots, n + m$ is consisted with nodes $1, 2, \ldots, n$ in sub-graph $G_1$ and nodes $n + 1$, $n + 2, \ldots, n + m$ in sub-graph $G_2$.

Consider a second-order multi-agent system, the $i$ th agent with double-integrator given by

$$
\begin{aligned} \dot{x}_i(t) &= v_i(t) \\ \dot{v}_i(t) &= u_i(t) \end{aligned} \tag{1}
$$

where $x_i(t) \in R$, $v_i(t) \in R$ and $u_i(t) \in R$ are the position, velocity and control input of the $i$ th agent, respectively.

**Assumption 1.** There is a balance of effect between two sub-graph: $\sum_{j=n+1}^{n+m} a_{ij} = 0$, $\forall i = G_1$; $\sum_{j=1}^{n} a_{ij} = 0$, $\forall i = G_2$.

**Remark 1.** The weighting factor $a_{ij}$ in the networked topology could be negative, which provides the competition relationships between two sub-graph and more complex dynamic behavior of agents.

**Definition 2.** The multi-agent system described by (1) is considered. The protocol $u_{ij}$ is said to achieve asymptotically a group consensus problem if for any initial state $x(0) \in R^n$, the states of agents satisfy

(I) $\lim_{t \to \infty} \|x_i(t) - x_j(t)\| = 0, \forall i, j \in G_1, \lim_{t \to \infty} \|v_i(t) - v_j(t)\| = 0, \forall i, j \in G_1$

(II) $\lim_{t \to \infty} \|x_i(t) - x_j(t)\| = 0, \forall i, j \in G_2, \lim_{t \to \infty} \|v_i(t) - v_j(t)\| = 0, \forall i, j \in G_2$

**Assumption 2.** There may be two agents in the undirected graph $G$ as leaders. Each leader is a globally reachable node in each sub-graph ($G_1$ or $G_2$).

**Remark 2.** In this paper, the MAS with leaders and without leaders are considered. The symmetric Laplacian matrix can be derived from the networked topology $G$ in two conditions. Both of them can get the same theoretical result.

## 3   Optimal Group Consensus of MAS

In this section, we propose a cost function for the second-order MAS. From the LQR perspective, an optimal control algorithm is derived, which guarantees the MAS to achieve group flocking movement.

The dynamic multi-agent system (1) can be rewritten as follows

$$\begin{pmatrix} \dot{x}_i(t) \\ \dot{v}_i(t) \end{pmatrix} = A \begin{pmatrix} x_i(t) \\ v_i(t) \end{pmatrix} + Bu_i(t) \qquad i = \{1, 2, \ldots, n+m\} \qquad (2)$$

where $A = \begin{pmatrix} 0 & 1 \\ 0 & 0 \end{pmatrix}, B = \begin{pmatrix} 0 \\ 1 \end{pmatrix}$ are the system matrix and input matrix. It is assumed that distributed dynamic systems are composed of two sub-graphs with $n+m$ agents randomly connection, where the networked topology can be represented by $G_1 = \{1, 2, \ldots, n\}$ and $G_2 = \{n+1, n+2, \ldots, n+m\}$.

The dynamics of the leader is

$$\begin{aligned} \dot{x}_{0k}(t) &= v_{0k}(t), \\ \dot{v}_{0k}(t) &= 0, \end{aligned} \qquad k = 1, 2 \qquad (3)$$

where $x_{0k}$ is the position of the leader, $v_{0k}$ is the velocity of the leader.

**Remark 3.** The leader of the dynamic MAS has the velocity with the acceleration zero, which means that if other agents receive information from the leader, they will adjust their speed under the control algorithm (10), eventually catch up with the speed and position of the leader.

The group consensus cost function for dynamic second-order system (2) is proposed as

$$J_f = \sum_{i=1}^{n} \int_0^{\infty} \left\{ \sum_{j=1}^{i-1} a_{ij}[x_j(t) - x_i(t)]^2 + \sum_{j=1}^{i-1} a_{ij}[v_j(t) - v_i(t)]^2 + ru_i^2(t) \right\} dt \qquad (4)$$

where $a_{ij} \in R$ is the element of weight adjacency matrix of structure graph $G$, and $r$ is the given constant, $r > 0$.

The dynamic system (2) can be written as

$$\dot{X}(t) = \tilde{A}X(t) + \tilde{B}U(t) \tag{5}$$

where $X(t) = [x_1(t), \ldots, x_{n+m}(t), v_1(t), \ldots, v_{n+m}(t)]^T$, $U(t) = [u_1(t), \ldots, u_{n+m}(t)]^T$, $\tilde{A} = A \otimes I_{n+m}$, $\tilde{B} = B \otimes I_{n+m}$.

Thus, the optimization problem for system (5) as

$$\min_{a_{i,j}} J_a = \int_0^\infty [X^T(t)QX(t) + U^T(t)RU(t)]dt, \tag{6}$$

$$\text{subject to} : \dot{X}(t) = \tilde{A}X(t) + \tilde{B}U(t), \; U(t) = U^*(t) \tag{7}$$

where $Q = \begin{bmatrix} Q_1 & 0 \\ 0 & Q_3 \end{bmatrix} = \begin{bmatrix} \beta^2 H^2 & 0 \\ 0 & \beta^2 H^2 - 2\beta H \end{bmatrix}$ is a positive definite matrix, $R = I_{n+m}$ is an identity matrix, the parameter $\beta$ is an optimal scaling factor. $H$ is the positive definition symmetric Laplacian matrix associated with the structure of MAS.

**Remark 4.** According to the definition of the performance indicator function, $Q$ should be a positive definite matrix, that is $Q > 0$. Therefore, the optimal scaling factor $\beta$ should be ensured that the matrix $\beta^2 H^2 - 2\beta H$ is a positive definite matrix.

**Theorem 1.** For dynamic multi-agent systems (5) with Assumption 1 and Assumption 2. By solving the optimal control problem (6), the optimal group consensus control can be obtained

$$U^*(t) = -\beta\big(H \; H\big)X(t) \tag{8}$$

where $H$ is the symmetric Laplacian matrix and $\beta > 2/\lambda_{\min}$ corresponds to the minimum eigenvalue of matrix $H$.

**Proof.** Since the matrix $\tilde{A}$ and $\tilde{B}$ is controllable matrix in Eq. (2), the dynamic MAS (5) with the system matrix $(\tilde{A}, \tilde{B})$ is controllable, which implies that there exists a matrix $P$ satisfying the algebraic Riccati equation

$$\tilde{A}^T P + P\tilde{A} - P\tilde{B}R^{-1}\tilde{B}^T P + Q = 0_{n \times n} \tag{9}$$

Let $P = \begin{pmatrix} P_1 & P_2 \\ P_2^T & P_3 \end{pmatrix}$, $R = I_{n+m}$, the Riccati Eq. (8) can be written as

$$\begin{cases} P_2 R^{-1} P_2^T = Q_1 \\ P_2 R^{-1} P_3 = P_1 \\ P_3 R^{-1} P_3 - P_2 - P_2^T = 0 \end{cases} \tag{10}$$

It follows that $P = \begin{pmatrix} \beta^2 H^2 & \beta H \\ \beta H & \beta H \end{pmatrix}$. Then, we can obtain the optimal consensus control $U^*(t) = -\beta(H\ H)X(t)$, where $H$ is the symmetric Laplacian matrix.

For dynamic multi-agent systems (5) with Assumption 1 and Assumption 2, the optimal group consensus control (7) can be written as

$$u_i(t) = \begin{cases} \beta\left\{\left[\sum\limits_{j\in N_{1i}} a_{ij}(x_j - x_i) + b_i(x_{01} - x_i)\right] + \sum\limits_{j\in N_{2i}} a_{ij}(x_j - x_i)\right\} \\ \quad + \beta\left\{\left[\sum\limits_{j\in N_{1i}} a_{ij}(v_j - v_i) + b_i(v_{01} - v_i)\right] + \sum\limits_{j\in N_{2i}} a_{ij}(v_j - v_i)\right\}, \quad \forall i \in G_1 \\ \beta\left\{\sum\limits_{j\in N_{1i}} a_{ij}(x_j - x_i) + \left[\sum\limits_{j\in N_{2i}} a_{ij}(x_j - x_i) + b_i(x_{02} - x_i)\right]\right\} \\ \quad + \beta\left\{\sum\limits_{j\in N_{1i}} a_{ij}(v_j - x_i) + \left[\sum\limits_{j\in N_{2i}} a_{ij}(v_j - v_i) + b_i(v_{02} - v_i)\right]\right\}, \quad \forall i \in G_2 \end{cases}$$

(11)

where $a_{ij}$ is the element of the adjacency matrix of MAS, $\forall i, j \in G_1$ or $\forall i, j \in G_2$. The set $N_{1i} = \{v_j \in V_1 : (v_i, v_j) \in \omega\}$ and $N_{2i} = \{v_j \in V_2 : (v_i, v_j) \in \omega\}$ is neighbor set of agent $i$. If followers $i$ can receive information from the leader $x_{0k}$, then $b_i > 0$, otherwise $b_i = 0$. Note that $H = L(G) + B = \begin{bmatrix} L(G_1) + B_1 & \Delta \\ \Delta^T & L(G_2) + B_2 \end{bmatrix}$ is positive definition

symmetric matrix, $\Delta = -\begin{bmatrix} a_{1,n+1} & a_{1,n+2} & \cdots & a_{1,n+m} \\ a_{2,n+1} & a_{2,n+2} & \cdots & a_{1,n+m} \\ \vdots & \vdots & \ddots & \vdots \\ a_{n,n+1} & a_{n,n+2} & \cdots & a_{n,n+m} \end{bmatrix}$, $B_1 = diag\{b_i, i = 1, \ldots, n\}$,

$B_2 = diag\{b_i, i = n+1, \ldots, n+m\}$. With the Assumption 1 and Assumption 2, the matrix $H$ is positive definite.

**Remark 5.** In the optimal group consensus control protocol (11) of dynamic MAS, the weight value $b_i$ may be zero, which means that there is the leaderless condition of the dynamic MAS, or agents do not receive information from the leader. If there is a globally reachable node in subgroups, MAS with the cooperative and competitive relationship between the two subgroups will achieve the group flocking motion without leaders.

**Theorem 2.** Considering the formation of dynamic MAS (2), the networked topology is composed of $n + m$ agents with undirected graph. If Assumption 1 and Assumption 2 are established for the dynamic MAS (2) with the optimal group consensus control (11), the optimal group consensus can be achieved for MAS.

**Proof.** Let $\bar{x}_i(t) = x_i(t) - x_{0k}(t)$, $\bar{v}_i(t) = v_i(t) - v_{0k}(t)$. The dynamic system (2) with the optimal control algorithm (11) be written as

$$\dot{\bar{x}} = \bar{v}$$
$$\dot{\bar{v}} = (-\beta H\bar{x}) + (-\beta H\bar{v})$$

(12)

where $\bar{x} = \begin{bmatrix} \bar{x}_1, \bar{x}_2, \ldots, \bar{x}_{n+m} \end{bmatrix}^{\mathrm{T}}$, $\bar{v} = \begin{bmatrix} \bar{v}_1, \bar{v}_2, \ldots, \bar{v}_{n+m} \end{bmatrix}^{\mathrm{T}}$. Let $z = H\bar{x}$, $w = H\bar{v}$. We definite the Lyapunov function

$$V = V_1 + V_2, \tag{13}$$

where $V_1 = \bar{v}^{\mathrm{T}} H \bar{v}$, $V_2 = z^{\mathrm{T}} \beta z$. The derivative of the Lyapunov function along the solution trajectory of the system (2) is

$$\begin{aligned} \dot{V}_1 &= 2\bar{v}^{\mathrm{T}} H \dot{\bar{v}} \\ &= 2w^{\mathrm{T}} \cdot [(-\beta z) + (-\beta w)] \\ &= -2w^{\mathrm{T}} \beta z - 2w^{\mathrm{T}} \beta w \end{aligned} \tag{14}$$

$$\dot{V}_2 = 2w^{\mathrm{T}} \beta z \tag{15}$$

It can be obtain that

$$\dot{V} = \dot{V}_1 + \dot{V}_2 = -2w^{\mathrm{T}} \beta w \leq 0. \tag{16}$$

Note that the equilibrium states of the dynamic system satisfy $\dot{\bar{x}} = 0$, $\dot{\bar{v}} = 0$. We can obtain that $\bar{v} = 0$ and $u_i(t) = 0$, that is $-\beta H \bar{x} = 0$. We can obtain the equilibrium states of the dynamic system satisfy $\bar{x} = 0$ and $\bar{v} = 0$ from the positive definite matrix $H$.

When $\dot{V} \equiv 0$, there exists $w = -H\bar{v} = 0$ such that $\bar{v} = 0$ and $\dot{\bar{v}} = 0$. And then it has $\bar{x} = 0$, $\bar{v} = 0$. Therefore, when $\dot{V} = 0$, there exists only the equilibrium point in this solution set. According to the principle of Lasalle's invariant set, the dynamic MAS (2) are asymptotically stable at the equilibrium point. The optimal group consensus of the dynamic MAS (2) with the optimal control algorithm (11) will be achieved.

## 4 Numerical Examples

In this section, simulations of MAS with leaders/leaderless are given to illustrate the theoretical results of this paper.

### 4.1 The Simulation of MAS with Leaders

It is assumed that the networked topology of MAS with leaders is shown in Fig. 1. Two sub-graph composed of followers 1–3 and 4–7. Agents L1 and L2 are the corresponding leaders in two sub-graph. Lines and numbers indicate that the information transfer and associated weigh.

Based on the networked topology of Fig. 1, we can get its Laplacian matrix, and then we can obtain the eigenvalue of Laplacian matrix of Fig. 1.

We choose the initial states and the initial speeds of agents are $x(0) =$ $[5, 8, 6, 12, 4, 2, 5]^{\mathrm{T}}$, $v(0) = [4, 7, 5, 11, 3, 1, 4]^{\mathrm{T}}$. Considering leaders in two sub-graph with the initial state and the initial speed $x_{01} = 5$, $v_{01} = 4$ and $x_{02} = 6$, $v_{02} = 5$, respectively. The simulation results of the system motion are shown in Fig. 2.

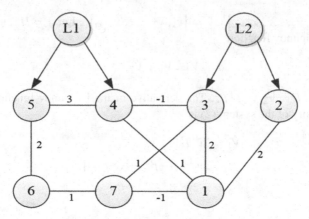

**Fig. 1.** Multi-agent system with leaders

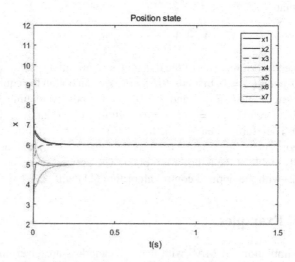

**Fig. 2.** Motion trajectory of MAS with leaders, $\beta = 18$

It has been revealed that the motion trajectories of each agent finally converge to two equilibrium states with the optimal control algorithm (11). The followers track the trajectory of leaders into two subgroups, and the two equilibrium states are the initial states of the two leaders ($x_{01} = 5$, $x_{02} = 6$).

Comparing Fig. 3 and Fig. 4, we can see that when $\beta = 0.8$ in Fig. 3, the convergence time $t$ required for the MAS to reach the equilibrium state is longer. Therefore, it is concluded that the optimal scaling factor $\beta$ has an effect on the convergence speed of the MAS. The larger the optimal scaling factor $\beta$, the faster the convergence speed.

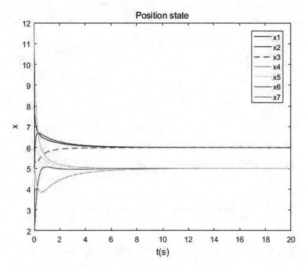

**Fig. 3.** Motion trajectory of MAS with leaders, $\beta = 0.8$

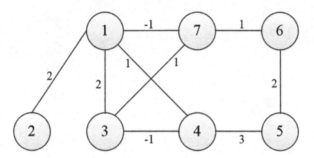

**Fig. 4.** Multi-agent system with leaderless

## 4.2 The Simulation of MAS with Leaderless

Now we assume the networked topology graph of MAS with leaderless is shown in Fig. 4. In the topology graph, the agents 1-7 are all substantive agents. There is information transfer between each agent.

The initial value of agents and the parameter are same with Subsect. 4.1. The optimal control algorithm (11) is applied in simulation, then the motion trajectory of MAS without leaderless is shown in Fig. 5. It can be seen that the agents move into two subgroups and the movement consensus of MAS with leaderless is realized.

Comparing Fig. 5 and Fig. 6, we can see that when $\beta = 0.8$ in Fig. 6, the convergence time $t$ required for the MAS to reach the equilibrium state is longer. Therefore, it is concluded that the relationship between optimal scaling factor $\beta$ and the convergence speed of the multi-agent system is proportional.

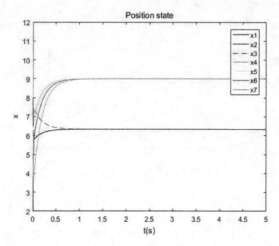

**Fig. 5.** Motion trajectory of MAS with leaderless $\beta = 33$

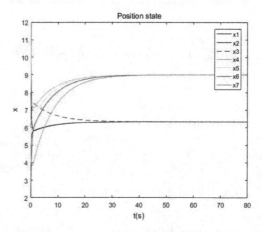

**Fig. 6.** Motion trajectory of MAS with leaderless

## 5  Conclusions

In the paper, the problem for optimal group consensus control of second-order MAS with/without leaders has been investigated. The group consensus cost function for dynamic system and the optimal consensus algorithm are proposed. By applying the LQR method, the symmetric Laplacian matrix associated with the undirected graph is derived. In addition, the optimal scaling factor for optimal control problem is studied. Based on algebraic graph theory and modern control theory, group flocking motions of second-order MAS are studied. Numerical examples are given to validate the theoretical results. One future research work will focus on the optimality issues for group consensus algorithms of high-order dynamic MAS.

**Acknowledgments.** The work is supported by the National Natural Science Foundation of China (61673200, 61771231), the Major Basic Research Project of Natural Science Foundation of Shandong Province of China (ZR2018ZC0438) and the Key Research and Development Program of Yantai City of China (2019XDHZ085).

# References

1. Yang, Y., Yang, H., Liu, F.: Group motion of autonomous vehicles with anti-disturbance protection. J. Netw. Comput. Appl. **162**, 102661 (2020)
2. Yang, H., Zhang, Z., Zhang, S.: Consensus of second-order multi-agent systems with exogenous disturbances. Int. J. Robust Nonlinear Control **21**(9), 945–956 (2010)
3. Yang, H., Zhu, X., Zhang, S.: Consensus of second-order delayed multi-agent systems with leader-following. Eur. J. Control **16**(2), 188–199 (2010)
4. Yang, H., Wang, F., Han, F.: Containment Control of Fractional Order Multi-Agent Systems With Time Delays. IEEE/CAA J. Autom. Sinica **5**(3), 727–732 (2018)
5. Cai, X., Wang, C., Wang, G., Liang, D.: Distributed consensus control for second-order nonlinear multi-agent systems with unknown control directions and position constraints. Neurocomputing **306**, 61–67 (2018)
6. Yang, H., Yang, Y., Han, F., Zhao, M., Guo, L.: Containment control of heterogeneous fractional-order multi-agent systems. J. Franklin Inst. **356**(2), 752–765 (2019)
7. Qin, J., Gao, H., Zheng, W.: Second-order consensus for multi-agent systems with switching topology and communication delay. Syst. Control Lett. **60**(6), 390–397 (2011)
8. Shang, Y.: Resilient consensus of switched multi-agent systems. Syst. Control Lett. **122**, 12–18 (2018)
9. Ma, T., Zhang, Z., Cui, B.: Adaptive consensus of multi-agent systems via odd impulsive control. Neurocomputing **321**, 139–145 (2018)
10. Liu, X., Zhang, K., Xie, W.: Consensus of multi-agent systems via hybrid impulsive protocols with time-delay. Nonlinear Anal. Hybrid Syst. **30**, 134–146 (2018)
11. Wang, Z., Xu, J., Song, X., Zhang, H.: Consensus problem in multi-agent systems under delayed information. Neurocomputing **316**, 277–283 (2018)
12. Yang, T., Zhang, P., Yu, S.: Consensus of linear multi-agent systems via reduced-order observer. Neurocomputing **240**, 200–208 (2017)
13. Yoon, M.: Consensus of adaptive multi-agent systems. Syst. Control Lett. **102**, 9–14 (2017)
14. Miao, G., Ma, Q.: Group consensus of the first-order multi-agent systems with nonlinear input constraints. Neurocomputing **161**, 113–119 (2015)
15. Gao, Y., Yu, J., Shao, J., Yu, M.: Group consensus for second-order discrete-time multi-agent systems with time-varying delays under switching topologies. Neurocomputing **207**, 805–812 (2016)
16. An, B., Liu, G., Tan, C.: Group consensus control for networked multi-agent systems with communication delays. ISA Trans. **76**, 78–87 (2018)
17. Yu, J., Wang, L.: Group consensus in multi-agent systems with switching topologies and communication delays. Syst. Control Lett. **59**(6), 340–348 (2010)
18. Wen, G., Yu, Y., Peng, Z., Wang, H.: Dynamical group consensus of heterogenous multi-agent systems with input time delays. Neurocomputing **175**, 278–286 (2016)
19. Hu, H., Yu, W., Xuan, Q., Zhang, C., Xie, G.: Group consensus for heterogeneous multi-agent systems with parametric uncertainties. Neurocomputing **142**, 383–392 (2014)
20. Yang, T., Wan, Y., Wang, H., Lin, Z.: Global optimal consensus for discrete-time multi-agent systems with bounded controls. Automatica **97**, 182–185 (2018)

21. Xie, Y., Lin, Z.: Global optimal consensus for multi-agent systems with bounded controls. Syst. Control Lett. **102**, 104–111 (2017)
22. Bailo, R., Bongini, M., Carrillo, J., Kalise, D.: Optimal consensus control of the Cucker-Smale model. IFAC-PapersOnLine **51**(13), 1–6 (2018)
23. Qiu, Z., Liu, S., Xie, L.: Distributed constrained optimal consensus of multi-agent systems. Automatica **68**, 209–215 (2016)
24. Chen, Y., Sun, J.: Distributed optimal control for multi-agent systems with obstacle avoidance. Neurocomputing **173**, 2014–2021 (2016)
25. Zhao, W., Zhang, H.: Distributed optimal coordination control for nonlinear multi-agent systems using event-triggered adaptive dynamic programming method. ISA Trans. **91**, 184–195 (2019)

# A Novel Game Machine Learning Method for Calculating Optimal Response for Edge Server

Rui Zhang, Hui Xia$^{(\boxtimes)}$, Ju-fu Cui, Yi-zhe Li, Shu-shu Shao, and Hang Ren

College of Computer Science and Technology, Qingdao University,
Qingdao 266100, China
xiahui@qdu.edu.cn

**Abstract.** Mobile edge computing extends traditional cloud services to the edge of the network and enables edge server to handle network requests with low latency requirements. However, the edge server is closer to the terminal device with relatively limited storage capacity and computing capacity, and is more vulnerable to the invasion of attackers. To solve this problem, we proposed a game machine learning method to determine the optimal response of edge server to attackers, so as to defend against attackers. First, we used Hidden Markov Model to fit the behavior model of the attacker; secondly, due to the payoff of edge server is closely related to the attacker's behavior model, we used the gradient ascent method to maximize the payoff of edge server; finally, the optimal response of edge server was determined. Detailed experimental results showed that the new scheme can improve the payoff of the edge server and defend against attackers.

**Keywords:** Mobile edge computing · Hidden Markov Model · Edge server · Optimal response

## 1 Introduction

Mobile edge computing technology [1] enables the network service environment and cloud computing technology to combine at the edge of the network, improves the computing and storage capacity of the edge network, and reduces the network operation and service delivery delay. The network structure of this technology is three layers [2], i.e., edge device layer, edge server layer and cloud server layer. The edge device layer usually deploys some low-level electronic equipment, which runs in the physical world to complete tasks such as sensing, driving, and control. The edge server layer consists of several sub-layers, which are composed

Supported by the National Natural Science Foundation of China (NSFC) under Grant No. 61872205, the Shandong Provincial Natural Science Foundation under Grant No. ZR2019MF018, and the Source Innovation Program of Qingdao under Grant No. 18-2-2-56-jch.

X. Chen et al. (Eds.): ML4CS 2020, LNCS 12488, pp. 183–192, 2020.
https://doi.org/10.1007/978-3-030-62463-7_17

of different edge servers. The cloud server layer includes the cloud server and the data processing center.

Mobile edge computing enables edge server to handle network requests with low latency requirements. However, the edge server is closer to the terminal equipment of the Internet of Everything. The openness and heterogeneity of the terminal equipment, as well as the relatively limited computing and storage resources, have greatly increased the difficulty of the edge server protection, which has led to widespread cyber threats in edge servers. For example, an attacker could tamper with communication data packets, inject spurious pressure measurements to trick the decision-maker, delay the action of control valve and cause equipment damage in the scenario of a smart manufacturing plant. Without appropriate safety precautions, not only the production process may be interrupted, but also the lives of workers will be threatened to a great extent. In mobile edge computing, the uav's operating system is attacked, which will generate simulated global positioning system signals, mislead the uav system components, and drive them to the target area for capture. Therefore, it is a prerequisite and necessary condition for the further development of mobile edge computing technology to guarantee the security of edge server and enhance its ability to resist various security threats.

At present, the security protection technology in the edge computing environment mainly includes four aspects: intrusion detection, access control, defense strategy and key management. Intrusion detection is mainly used to monitor and detect the abnormal data on the host side or network side. Zhou et al. [3] proposed a general IDS framework for fog computing and developed a cloud and fog hybrid intrusion detection scheme. Chaabouni et al. [4] reviewed existing NIDS implementation tools and data sets, as well as free and open source network sniffing software. However, such intrusion detection scheme did not give full play to the characteristics of mobile edge technology and failed to effectively utilize its advantages. Controlling the access of malware to the edge computing environment can effectively defend against attackers. Yu et al. [5] proposed a universal framework of functional encryption suitable for fog computing access control, which not only provided privacy and fine-grained access control in fog computing, but also ensured the security of fog computing under channel attack. Yang et al. [6] proposed an intelligent IoT medical big data storage system with adaptive access control capability. However, there are some problems in the access control scheme of mobile edge computing technology, such as heavy computation, complex model and difficulty in rewriting parameters. Zheng et al. [7] reviewed existing defense strategies for moving targets. Huang et al. [8] proposed a dynamic game framework to simulate the long-term interaction between stealth attackers and active defenders. However, the traditional security defense strategies do not take into account the random distribution of attackers or the overall network cost. Key management is an encryption technique for communication in edge computing networks. Anzani et al. [9] proposed an improved scheme for hybrid symmetric design based on the hybrid key predistribution method of symmetric design, which improved connectivity and durability. Bitansky et al.

[10] showed how to use secret key function encryption to obtain exponentially valid undistinguished obfuscation. However, traditional key management schemes have poor scalability and lack of lightweight implementation methods, so they are not suitable for edge computing networks with features such as resource sharing, scalability, and virtualization.

Inspired by the above scheme, this paper proposed a novel game machine learning method to determine the optimal response for the edge server to the attacker. The contribution of this paper is as follows:

(1) This paper used Hidden Markov Model to predict the attacker's observed action sequence in next T period.

(2) Based on the results of the first step, this paper used the gradient ascent method to maximize the payoff of edge server for determine its optimal response to the attacker.

## 2   Preface

This section describes the theory of Markov process.

### 2.1   Markov Process

Markov process is a kind of random process proposed by the Russian mathematician A.A. Markov in 1907, in which, given the current state, its future evolution does not depend on its past evolution. In the real world, the Brownian motion of papers in liquids, the number of people infected with infectious diseases, the number of people waiting at stations, the changes in the number of animals in the forest, etc., can all be regarded as Markov processes. A Markov process refers to the transition of each state in the process only depends on the previous $n$ states. The First-order Markov can be described as follows,

$$
\begin{aligned}
&\Pr(X_{n+1} = x | X_1 = x_1, X_2 = x_2, ..., X_n = x_n) \\
&= \Pr(X_{n+1} = x | X_n = x_n)
\end{aligned}
\tag{1}
$$

Similarly, the $m$-order Markov can be defined as follows,

$$
\begin{aligned}
&\Pr(X_n = x_n | X_{n-1} = x_{n-1}, X_{n-2} = x_{n-2}, ..., X_1 = x_1) \\
&= \Pr(X_n = x_n | X_{n-1} = x_{n-1}, X_{n-2} = x_{n-2}, ..., X_{n-m} = x_{n-m})
\end{aligned}
\tag{2}
$$

As can be seen from the above, the Markov model can be represented as a triple, $(S, \prod, A)$, where $S$ is a set of states, $\prod$ is the probability distribution of the initial state, and $A$ is the state transition probability. In practical applications, the Markov process is not sufficient to solve the existing problems. Therefore, the Hidden Markov Model is proposed. Hidden Markov model is a quintuple $\{N, M, n, A, B\}$, where $N$ is the number of hidden states; $M$ is the number of observable states, its value can be obtained from the training set; $n = \{n_i\}$ is the probability of the initial state, that is, the probability of each hidden state occurring in the initial state, $A = \{a_{ij}\}^{N*N}$ is the transfer matrix of the hidden state, which refers to the probability of transition from the first state to the second state, $B = \{b_{ij}\}^{N*M}$ is the confusion matrix, which refers to given the initial state $s$ the probability of the occurrence of an observation. Each probability in a state transfer matrix and a confusion matrix is time-independent, that is, as the system evolves, these matrices do not change over time.

# 3    Defense Strategy of Edge Server

This section describes the interaction process between the edge server and the attacker, and how to determine the optimal response of the edge server.

## 3.1    Two-Players Security Game

When the attackers attack the edge server, the attacker hopes to obtain the highest reward at the lowest cost, while the edge server hopes to defend attackers at the lowest cost, thus, the interaction between the attacker and the edge server can be modeled as a two-players security game. Assume that the attacker has two strategies: the non-attack strategy (i.e., $NA$) refers that the attacker doesn't launch an attack to the edge server; the attack strategy (i.e., $A$) refers that the attacker launches an attack to the edge server. Similarly, the edge server also has two strategies: the defense strategy (i.e., $D$) refers that the edge server would defense attackers; the non-defense strategy (i.e., $ND$) refers that the edge server would not defend the attacker. For attackers and edge servers, the payoff matrix is shown in Table 1.

**Table 1.** Payoff matrix

| Payoff CE\AT | A | NA |
|---|---|---|
| **D** | $p - c_D, u - c_A$ | $-c_D, 0$ |
| **ND** | $-r, u - c_A + r$ | $0, 0$ |

Where $AT$ refers to the attacker, $CE$ refers to the edge server, $r$ means the additional payoff obtained by the attacker when the attacker initiates the attack and the edge server does not defend, $c_A$ is the cost of the attacker, $c_D$ is the cost of the edge server to defend the attacker's attack, $u$ is the payoff from the attacker attacking the edge server. When the attacker plays the attack strategy $A$ and the edge server plays the defense strategy $D$, the edge server will obtain the payoff $p$ at the cost of $c_D$, and the attacker will gain the payoff $u$ at the cost of $c_A$; When the attacker plays the attack strategy $A$ and the computing center plays strategy $ND$, the payoff of the edge server is $-r$, the attacker will get the payoff $u$ and the additional payoff $r$ at the cost of $c_A$. When the attacker plays the attack strategy $NA$ and the edge server plays the defense strategy $D$, the payoff of the edge server is $-c_D$, and the attacker's payoff is 0; When the attacker plays the attack strategy $NA$ and the edge server plays the non-defense strategy $ND$, the payoff of the edge server and the attacker are 0.

Assuming $p - c_D > 0, u - c_A > 0$, from Table 1, the strategy profile (*attacker D, attacker A*) is a pure strategy *Nash equilibrium* of this game. However, in practical applications, this cannot be achieved because there is strong assumption: the information between the attacker and the edge server is known to each other, that is, the game is a perfect information game. This assumption cannot be realized when the edge server defends the attacker's intrusion, because the attack launched by the attacker is irregular and undirected, i.e. the edge computing is not certain that the attacker will be able to launch an attack properly, nor is it certain that the attack will be sufficient to launch an attack on it. To solve this problem, this paper proposes a game machine learning method to determine the edge server defense strategy.

## 3.2  Edge Computing and Attacker's Payoff

In order to determine the optimal defense strategy for edge servers, this paper defines the payoffs of the edge server as,

$$Payoff = \sum_{i \in attacker} \varphi_i(g, d) - \sum_{i \in attacker} c_{D_i} - \sum_{i \in attacker} r_i \qquad (3)$$

In order to clearly represent every element of the payoffs of the edge server, this paper gives the most primitive form of the payoff of edge server. Where $i$ is the number of attacks by an attacker, *attackers* is the set of attackers, $\varphi_i(g, d, s)$ is the edge server's payoff when it defends attackers, $r_i$ is the cost of edge server when it does not take a defense strategy while the attacker attacks, $g$ is the behavior model of the attacker, and $d$ is the state of the edge server (the strategy of the edge server). Similarity, the payoff of attackers can be defined as,

$$Utility = \sum_{i \in attacker} u_i - \sum_{i \in attacker} c_{A_i} + \sum_{i \in attacker} r_i \qquad (4)$$

Where $u_i$ is the payoff obtained by the attacker's successful attack, $c_{A_i}$ is the cost of the attacker launching the attack, and $r_i$ is the additional payoff obtained by the attacker. From (4), the payoff of the attacker is related to the payoff obtained by each attack and the cost of launching the attack. From (3), the payoff of the edge server is related to its own state, the attacker's behavior model, and the attacker's historical data. Therefore, in order to determine the optimal strategy of the edge server, we need to evaluate the attacker's behavior model. Before determining the optimal strategy of the edge server, we first define the optimal response of the edge server as follows,

**Optimal response**: a strategy which can maximize the payoff of the edge server calls the optimal response of the edge server to an attacker.

From the definition of optimal response, this paper can transform the process of solving the optimal strategy of the edge server into the process of optimizing the calculation center's payoff.

## 3.3   The Edge Server's Optimal Response

Assume that the attacker's next attack is only related to its current state. An attacker launches an attack, there may be multiple states, but the attacker has only two kinds of behaviors. The attacker's hidden state set can be defined as $S = \{s_1, s_2, \cdots s_n\}$, and the attacker's observable behavior set is $A = \{a_1, a_2\}$. The attacker's hidden state transition matrix $N$ and observable action transition matrix $M$ can be obtained,

$$N = \begin{pmatrix} p_{s_1,s_1} \cdots p_{s_1,s_n} \\ \vdots \quad \cdots \quad \vdots \\ p_{s_n,s_1} \cdots p_{s_n,s_n} \end{pmatrix} \qquad (5)$$

$$M = \begin{pmatrix} p_{s_1,a_1} \ p_{s_1,a_2} \\ \cdots \cdots \\ p_{s_n,a_1} \ p_{s_n,a_2} \end{pmatrix} \qquad (6)$$

Where $p_{s_i,s_j} = P(s_{t+1} = s_j | s_t = s_i)$, $p_{s_i,a_k} = P(a_t = a_k | s_t = s_i)$, the probability distribution of the original state is $\Pi = [\pi(i)]_n$, $\pi(i) = P(s_1 = s_i)$. The observable action $a_{t+1}$ can be generated according to the observable action transfer distribution $p_{s_i,s_j}$ of the attacker's hidden state $s_t$, and then the hidden state $s_{t+1}$ can be generated according to the state transfer distribution $p_{s_i,a_k}$ of the hidden state $s_t$. After $T$ rounds of iteration, we can the attacker's observation sequence $a = \{a_1, \cdots, a_T\}$.

With the help of Hidden Markov model, the attacker's observation sequence in the next T cycles can be obtained. We can maximize the payoff of the edge server with the gradient ascent method to determine the optimal response of the edge server. That is,

$$\max_{a} \arg \{ Pay = \sum_{z=t}^{T} \sum_{i \in attacker} (\varphi_{i,z}(g,a) - c_{D_{i,z}} - r_{i,z}) \} \qquad (7)$$

## 4   Experimental Stimulation

This paper uses anaconda integrated development tool to verify the game machine learning method. Firstly, this paper uses the Hidden Markov Model to predict the observable behavior sequence of attackers in 15 cycles under different initial states, and then based on this result determine the optimal response strategy of edge server. Secondly, based on the prediction results of Hidden Markov Model, this paper compares the changing trend of the edge server and the attacker's payoff in different initial states. Finally, to verify the efficiency of the proposed scheme, this paper compares and analyzes the OUR scheme with the Random scheme (randomly taking strategy), ALL-D scheme (always taking the defensive strategy) and ALL-ND (always taking the defensive strategy) to verify the OUR scheme can improve earnings at the edge of the computing center and defense the invasion of the attacker.

The parameter setting of the experiment is shown in Table 2. The hidden state transition matrix of the and the observable behavior state transition matrix of the attacker in Hidden Markov Model are shown in $N$ and $M$.

**Table 2.** Parameters setting

| Parameter | $n$ | $p$ | $u$ | $c_D$ | $c_A$ | $r$ | $a_0$ | $s_0$ | $T$ |
|---|---|---|---|---|---|---|---|---|---|
| Value | 3 | 0.6 | 0.3 | 0.2 | 0.2 | 0.5 | 0.5 | 0.5 | 0.5 |

$$N = \begin{pmatrix} 0.2 \ 0.3 \ 0.5 \\ 0.5 \ 0.2 \ 0.3 \\ 0.3 \ 0.5 \ 0.2 \end{pmatrix}, M = \begin{pmatrix} 0.5 \ 0.5 \\ 0.4 \ 0.6 \\ 0.7 \ 0.3 \end{pmatrix}$$

### 4.1   The Prediction of Observable Behavior of Attackers

Table 3 predicts the observable behavior sequence of attackers in the next 15 interaction cycles based on Hidden Markov Model. In Table 3, $AT$ is the attacker and $CE$ is the edge server. Assuming that the initial state of the attacker is $s_1$, the observable action $a_2$ can be generated according to the observable action transfer distribution $p_{s_i,s_j}$ of the attacker's hidden state $s_1$, and then the hidden state $s_3$ can be generated according to the state transfer distribution of the hidden state $s_1$. After 15 rounds of iteration, the observation sequence of the attacker can be obtained as $a = \{A, NA \cdots, A, NA\}$. Since the observable behavior of the attacker in the next 15 interaction cycles has been determined, the edge server can determine its own optimal response according to the result, maximize its own payoffs, and then resist the attack of the attacker.

### 4.2   Comparison of the Payoff of Attacker and Edge Server

Based on the prediction results of the Hidden Markov Model on the observable behavior sequence of attackers in the next 15 cycles, Fig. 1 compares the changing trends of the edge server and the attacker's payoff in the initial state of $s_1$ and $s_2$. It can be seen from Fig. 1(a) that the payoff of edge server is higher than that of attacker, which is determined by the payoff matrix of two players. According to the payoff parameters set in this paper, when the attacker plays strategy $A$, the optimal response of the edge server is strategy $D$, and the payoff of the edge server and the attacker is 0.4 and 0.1 respectively. When the attackers play strategy $NA$, the optimal response of the edge server is strategy $NA$, the payoff of the edge server and the attacker are 0.

From Fig. 1(a), the edge of computing center and the attacker's payoff is decreased in the second interaction cycle. The former yields decreased from 0.4 to 0, which yields decreased from 0.1 to 0, this is because the edge of computing center and the attackers are adjusted the strategy, that is, the edge server plays

**Table 3.** The observable behavior of attackers

| Number | State | Action | Strategy (AT) | Strategy (CE) |
|--------|-------|--------|---------------|---------------|
| 0 | $s_1$ | $a_2$ | A | D |
| 1 | $s_3$ | $a_1$ | NA | ND |
| 2 | $s_2$ | $a_2$ | A | D |
| 3 | $s_1$ | $a_1$ | NA | ND |
| 4 | $s_3$ | $a_1$ | NA | ND |
| 5 | $s_2$ | $a_2$ | A | D |
| 6 | $s_1$ | $a_2$ | A | D |
| 7 | $s_3$ | $a_1$ | NA | ND |
| 8 | $s_2$ | $a_2$ | A | D |
| 9 | $s_1$ | $a_2$ | A | D |
| 10 | $s_3$ | $a_1$ | NA | ND |
| 11 | $s_2$ | $a_2$ | A | D |
| 12 | $s_1$ | $a_1$ | NA | ND |
| 13 | $s_3$ | $a_2$ | A | D |
| 14 | $s_2$ | $a_2$ | A | D |
| 15 | $s_1$ | $a_1$ | NA | ND |

the strategy $D$ and the attacker plays the strategy $A$ in the first interaction cycle, but the computing center and the attacker's strategy adjustment for $ND$ and $NA$ respectively in the second interaction cycle. Similarly, we can know that the reason for the change of payoff curve of edge server and attacker. Comparing Fig. 1(a) and Fig. 1(b), it can be seen that the edge server can determine the optimal response even if the attacker's initial state is different.

### 4.3   Comparison Between Our Scheme and the Other Three Schemes

In order to verify the efficiency of the proposed scheme, Fig. 2 compares the payoff variation trend of the edge server in $OUR$ scheme, $Random$ scheme, $ALL$-$D$ scheme and $ALL$-$ND$ scheme in different initial states. According to Fig. 2(a), the results of $OUR$ scheme are the best, followed by $ALL$-$D$, and the worst effect of $ALL$-$ND$. This is because the OUR scheme uses the prediction results of Hidden Markov Model to determine the optimal response of edge server for the attacker's strategy.

Therefore, $OUR$ scheme has the highest payoff and the best effect. The reason for the worst effect of $ALL$-$ND$ is that no matter what strategy the attacker takes, the edge server always plays the strategy $ND$, so the edge server in this scheme has the lowest payoff and the worst effect. Similarly, it can be concluded that the effect of $Random$ and $ALL$-$D$ is lower than that of $OUR$ scheme. Contrast Fig. 2(a) and Fig. 2(b), the attacker in any initial state, the payoffs of the edge server are all the highest in the *OUR* scheme, this is because this paper uses Hidden Markov Model to predict the attacker in the observable behavior of the future interaction cycle, the edge server based on the prediction results defends against attackers.

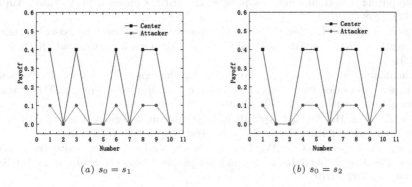

(a) $s_0 = s_1$        (b) $s_0 = s_2$

**Fig. 1.** Comparison of edge server and attacker's payoff.

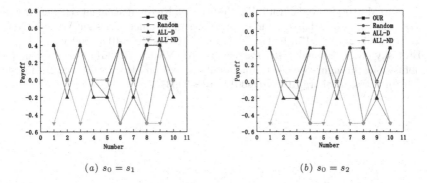

(a) $s_0 = s_1$        (b) $s_0 = s_2$

**Fig. 2.** Payoff comparison of edge server.

## 5    Conclusion

The key to promote the application of mobile edge computing technology is to improve the ability of edge server to resist attackers. In this paper, a game machine learning method is proposed to solve this problem. In the scheme, Hidden Markov Model is used to fit the behavior model of the attacker, and the gradient ascending method is used to maximize the benefits of the edge server, so as to determine the optimal response of the edge server to attacker. Detailed experimental results verify the effectiveness of the proposed scheme.

# References

1. Han, Y., Wang, X., Leung, V., Niyato, D.: Convergence of edge computing and deep learning: a comprehensive survey. arXiv preprint arXiv:1907.08349 (2019)
2. Mao, Y., You, C., Zhang, J., Huang, K., Letaief, K.: A survey on mobile edge computing: the communication perspective. IEEE Commun. Surv. Tutor. **19**(4), 2322–2358 (2017)
3. Zhou, X., Xing, L.: Sample selected extreme learning machine based intrusion detection in fog computing and MEC. Wirel. Commun. Mob. Comput. **2018**, 1–10 (2018)
4. Chaabouni, N., Mosbah, M., Zemmari, A., Sauvignac, C., Faruki, P.: Network intrusion detection for IoT security based on learning techniques. IEEE Commun. Surv. Tutor. **21**(3), 2671–2701 (2019)
5. Yu, Z., Man, H., Xu, Q.: Towards leakage-resilient fine-grained access control in fog computing. Future Gener. Comput. Syst. **78**(1), 763–777 (2018)
6. Yang, Y., Zheng, X., Guo, W., Liu, X., Chang, V.: Privacy-preserving smart IoT-based healthcare big data storage and self-adaptive access control system. Inf. Sci. **479**, 567–592 (2019)
7. Zheng, J., Namin, A.: A survey on the moving target defense strategies: an architectural perspective. J. Comput. Sci. Technol. **34**(1), 207–233 (2019)
8. Huang, L., Zhu, Q.: A dynamic games approach to proactive defense strategies against advanced persistent threats in cyber-physical systems. Comput. Secur. **89**, 101660 (2020)
9. Anzani, M., Haj Seyyed Javadi, H., Modirir, V.: Key-management scheme for wireless sensor networks based on merging blocks of symmetric design. Wireless Netw. **24**(8), 2867–2879 (2017). https://doi.org/10.1007/s11276-017-1509-y
10. Bitansky, N., Nishimaki, R., Passelègue, A., Wichs, D.: From cryptomania to obfustopia through secret-key functional encryption. J. Cryptol. **33**(2), 357–405 (2019). https://doi.org/10.1007/s00145-019-09337-9

# SDUW-Net: An Effective Retinal Vessel Segmentation Model

Hongkai Lin[1], Hongliang Kang[2], and Xinrong Cao[1(✉)]

[1] College of Computer and Control Engineering, Minjiang University, Fuzhou, China
cxrxmu@163.com
[2] Fujian Provincial Key Laboratory of Information Processing and Intelligent Control,
Minjiang University, Fuzhou, China

**Abstract.** Retinal vessel segmentation is the main step in the analysis of fundus images. However, gray-scales' uneven distribution, complex structure, and serious noise interference bring difficulties of automatic retinal vessels segmentation on fundus images. To solve these problems, we present an effective retinal vessel segmentation model, SDUW-Net, in this paper. The same scale dense connection is designed to improve U-Net's structure and remove different scales dense connections to accelerate the training speed. We use skip connections to merges the features between shallow layers and deep layers to retain more features that may be lost in the process of down sampling and convolution. Experimental results on the DRIVE dataset show that the retinal vessels can be effectively segmented out by our proposed SDUW-Net, which has AUC of 0.9811 with low computation and short training time.

**Keywords:** Retinal vessels segmentation · Same scale dense connection · Skip connection

## 1 Introduction

Retinal images are the common basis for the diagnosis and treatment of ocular diseases. Segmenting vessels accurately is a fundamental step in analyzing retinal images. However, the distribution of fundus blood vessels is complex, and fundus images are easily disturbed by various external factors. At the same time, retinal vessel segmentation is more likely to have different results due to different interpretation personnel. Therefore, even if it costs a lot of manpower and material resources, artificial segmentation of retinal vessels cannot meet the current clinical needs. Therefore, the automatic segmentation of retinal vessels has become a research focus.

With the development of science and technology, deep learning has achieved outstanding performance in computer vision tasks such as image classification, target detection, semantic segmentation, image description [1]. In 2015, the full convolution network (FCN) [2] changed the tedious process of convolution neural network in image segmentation at that time. It not only greatly improved the segmentation accuracy but also reduced the calculation amount. On this basis, there are many new network models,

© Springer Nature Switzerland AG 2020
X. Chen et al. (Eds.): ML4CS 2020, LNCS 12488, pp. 193–202, 2020.
https://doi.org/10.1007/978-3-030-62463-7_18

such as U-Net [3], SegNet [4], which are widely concerned in the field of medical image analysis.

In this work, we focused on a research of neural network model for retinal vessels automatic segmentation in retinal image. Combining the advantages of U-Net and dense connection, an effective retinal vessels segmentation model is proposed to be a useful computer-aided technology for the diagnosis and treatment of ocular diseases.

## 2   Related Works

U-Net is a network model of encoder-decoder structure with skip connection, which can strengthen the characteristic connection of different layers. Therefore, U-Net, which could perform well even with less data, has become one of the most widely used network structures in medical image analysis since it was proposed.

The structure of U-Net model consists of encoder and decoder, corresponding to the down-sampling and the up-sampling (see Fig. 1).

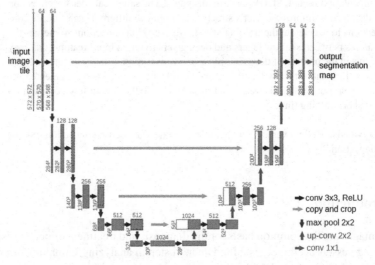

**Fig. 1.** U-Net structure [3].

The encoder is composed of several convolution layers and pooling layers alternately, which contracts the path of the input to captures the global information. The decoder is composed of several convolution layers and upper sampling alternately, which expands the feature map to locate each pixel accurately. The model does not have full connection layers, and uses the softmax to classify the images at the pixel levels.

Alom et al. [5] proposed a fundus vascular segmentation method combining recurrent neural network and ResNet [6] in U-Net, which used recursive residual convolution to complete feature accumulation to better represent the target features. Zhuang et al. [7] proposed a Ladder-Net, which combined residual network to form weight sharing residual block, and added multiple pairs of coding decoder to enhance feature information flow. Xu et al. [8] proposed a deep neural network model, combining the hierarchical symmetry structure and dense connection model [9], to achieve accurate segmentation of fundus vessels. Zhang et al. [10] utilized U-Net with residual connection to detect vessels, which payed more attention to the boundary areas of vessels and achieve a better performance, especially in tiny vessels detecting. Li et al. [11] developed a connection sensitive attention U-Net for accurate retinal vessel segmentation. This connection sensitive loss tightly integrates with the proposed attention U-Net to accurately segment retinal vessels and reserve the connectivity of thin vessels. Moreover, Wang et al. [12] proposed a new retinal vessel segmentation framework based on dense U-Net and the patch-based learning strategy to achieve competitive evaluation metrics.

It is found that if there is a shorter connection between the input layer and the output layer, the convolutional neural network can become deeper, more accurate and more efficient. Therefore, the dense connection is established to connect each layer in feedforward mode (see Fig. 2). That is to say, for each layer, all the feature maps of the front layers will be the input of the layer, and its own feature map will be the input of all subsequent layers. This method can effectively alleviate the gradient disappearance in the back propagation caused by the deepening of the network, and also strengthen the characteristic propagation.

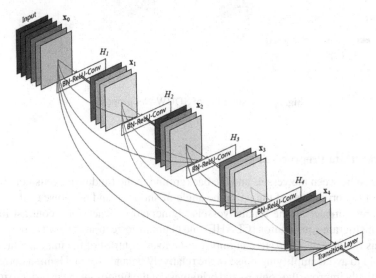

**Fig. 2.** The structure of dense connection [9].

Inspired by the U-Net structure and dense connection, a new model SDUW-Net (Sample scale Dense connection and Up-sampling W-Net) is constructed, which improves and modifies U-Net structure, removes the dense connections among different scales, and uses skip connections to merge the target's features. The experimental results show that the proposed SDUW-Net is suitable for retinal vessels segmentation with a comparable performance.

# 3 Methodology

According to the specific characteristics of fundus image, SDUW-Net is proposed for vessels segmentation in fundus image. SDUW's full name is Same-scale Dense-connection Up-front sampling W-type (see Fig. 3). It uses skip connections to connect the features among shallow layers and deep layers, and takes the same scale dense connections to strengthen the targets' features.

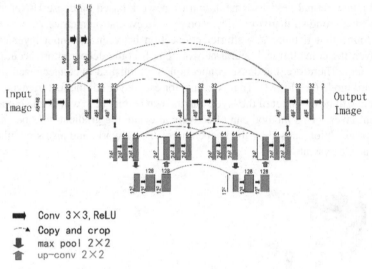

**Fig. 3.** The structure of SDUW-Net proposed.

## 3.1 Image Data Preprocessing

The contrast between the vessels and the background in the fundus image is not obvious because of the brightness of the light, the angle of shooting and the noise of the image. It has a great impact on the fundus vessels segmentation. Therefore, contrast limited adaptive histogram equalization (CLAHE) and Gamma correction are used to preprocess the fundus image data. CLAHE can improve the local contrast of the image, which sets a threshold to avoid amplifying noise in the relatively balanced area. Gamma correction can effectively improve the contrast and brightness of the fundus image in order to reduce the uneven illumination and local shadow in the fundus images (see Fig. 4).

Directly as training data, the performance of the model is affected by the limited quantity of fundus images. Therefore, the processing strategy of random block is adopted for the fundus images, that is, the pixels are randomly determined in the fundus image, and the sub images of 48 × 48 pixels are generated with the pixel as the center, (see Fig. 5). In this way, the generated sub images are different, but also contain similar vascular structure, which helps to train the network model, and reduce over fitting phenomenon. Compared with the original image, the sub image can also reduce the parameters and the network scale.

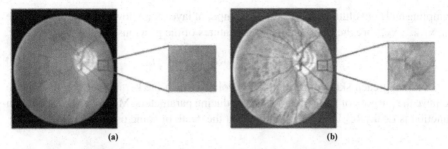

**Fig. 4.** Image preprocessing results: (a) gray retinal image; (b) preprocessing result.

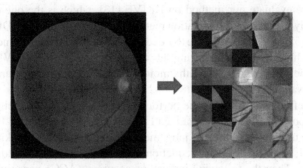

**Fig. 5.** Random image blocks

## 3.2 Building and Training

SDUW-Net inherits the advantages of U-Net and dense connection. It builds additional feature transfer paths by the same scale dense connection, and adds an up-sampling at the input.

The U-Net's structurer is reconstructed to enhance image details. Firstly, the U-Net is mirror extended to form a W-like structure, which can enhance the obvious features in the network. It is found that the details of target, especially small target, may be lost by encoding in U-net. Therefore, the up-sampling modular is added at the beginning of the model, which is used to enlarge image features so as to prevent key feature information from being encoded in the net.

Dense connections are established to connect each layer in feedforward mode to strengthen the characteristic propagation. Therefore, dense connection can be applied to U-net's structure. There may be some connections that only increase the computation of the network, but have no contribution to the overall performance, and even a negative impact.

The idea of the same scale dense connection is proposed, which only keeps the dense connection between same scales in dense block, and removes the dense connections between different scales. This reduces the number of connections in SDUW-Net from 55 to 13, which could simplify the model and reduce the computational complexity.

Moreover, skip connection can bring the shallow features with more details into the reconstruction process. It retains some features that may be lost in the process of down sampling and convolution. At this time, the input of layer N is shown in formula 1, where $x_0, x_1, \ldots, x_{N-1}$ are the same scale output features of the previous N-1 layers.

$$x_N = connect(x_{1,} \ldots x_{N-1}),  \tag{1}$$

Following each skip connection, a convolution with the kernel of $1 \times 1$ is done to achieve the purpose of fusing features and reducing parameters. Moreover, the activation function is LeakyReLU, which could inhibit the death of some useful neurons.

## 4 Materials and Performances

In this work, we evaluate our method on DRIVE [13], which is the one of the public datasets for studying vessels segmentation method in retinal images. The DRIVE dataset has forty $565 \times 584$ images, which are divided equally between training and testing sets. Specifically, both the training and testing sets are 20 images. For the DRIVE images, the vessels were labeled manually by ophthalmologists, and the binary segmentation masks were publicly available.

In order to effectively evaluate the performance of segmentation methods, we calculate the true positive (TP), true negative (TN), false positive (FP), and false negative (FN), which are used to show that the judgement of vessel pixels and background pixels is correct or not, respectively. This deduction is inconsistent with the judgment of the ophthalmologist. Then the following indicators, accuracy (ACC), sensitivity (SE), and specificity (SP), are used as shown in formula 2–4.

$$ACC = (TP + TN)/(TP + FP + TN + FN),  \tag{2}$$

$$SE = TP/(TP + FN),  \tag{3}$$

$$SP = TN/(FP + TN),  \tag{4}$$

Furthermore, we also calculate AUC to evaluate the comprehensive performance of retinal vessels segmentation. AUC is the area under ROC curve, which is drawn with false positive rate as the vertical axis and true positive rate as the horizontal axis. The higher AUC value, the better the classification result.

The retinal images of DRIVE dataset are processed by the proposed model on a computer configured as Intel (R) Xeon (R) gold 6129 CPU @ 2.30 GHZ.

### 4.1 Results of Improved Model

The structurer is reconstructed to improve the U-Net model. The U-Net is mirror extended, a W-like structure, to enhance image details and an up-sampling modular is added at the beginning so as to reduce the details that may be lost by encoding in U-net. Experimental result shows that the proposed improved U-Net with up-sampling and W-like structure is better than the original U-Net, as shown in Table 1.

Complete dense connection is established to connect each layer in feedforward mode to strengthen the characteristic propagation. But not every dense connection can play a

**Table 1.** Comparisons between original and improved U-Net's structures on DRIVE.

| Methods | AUC | ACC | SE | SP |
|---|---|---|---|---|
| Original U-Net | 0.9786 | 0.9557 | 0.7567 | **0.9847** |
| Improved U-Net | **0.9800** | **0.9565** | **0.7709** | 0.9835 |

role in the whole network. These connections from low scales to high scales are removed to form half dense connection, which does not affect the overall performance.

The experimental results show that half dense connection has the similar performance with complete dense connection, as shown in Table 2.

**Table 2.** Comparisons between complete and half dense connections on DRIVE.

| Methods | AUC | ACC | SE | SP |
|---|---|---|---|---|
| Complete dense connection | **0.9794** | 0.9562 | **0.7849** | 0.9812 |
| Half dense connection | **0.9794** | **0.9563** | 0.7754 | **0.9827** |

## 4.2 Results of Vessel Segmentation

The comparisons of performances among the SDUW-Net and other methods are shown in Table 3. The accuracy, sensitivity, and specificity of [8, 10] are slightly higher with their lager computation and lower training speed. However, in SDUW-Net, the data enhancement strategy of sub images is adopted to solves the problem of small data set of fundus image effectively, and the same scale dense connections are used to reduce computational complexity to cost less training time. In addition, the up-sampling and W-like structure help the model achieve better performance.

**Table 3.** Qualitative comparisons of segmentation results obtained by different methods on DRIVE.

| Methods | AUC | ACC | SE | SP | Time of training/s |
|---|---|---|---|---|---|
| Alom [5] | 0.9784 | 0.9556 | 0.7792 | 0.9813 | – |
| Zhuang [7] | 0.9793 | 0.9561 | 0.7856 | 0.9810 | – |
| Xu [8] | 0.9171 | **0.9707** | 0.7409 | **0.9929** | 1130 |
| Zhang [10] | 0.9799 | 0.9504 | **0.8723** | 0.9618 | 36000 |
| Wang [12] | 0.9740 | 0.9511 | 0.7986 | 0.9736 | – |
| SDUW-Net | **0.9811** | 0.9570 | 0.7976 | 0.9803 | **863** |

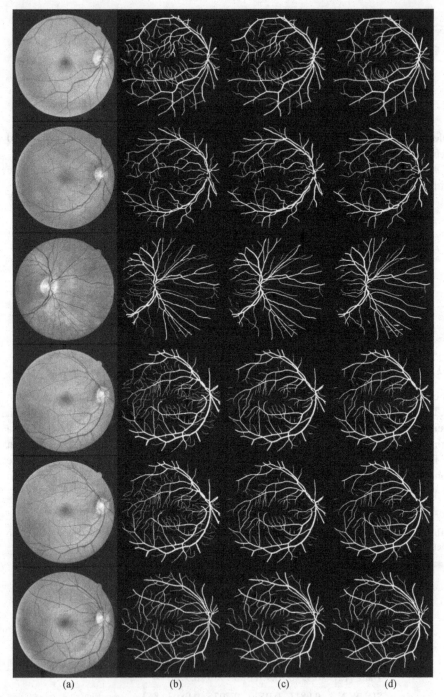

**Fig. 6.** Retinal images and vessel segmentation: (a) retinal images; (b) ground truths; (c) SDUW-Net's results; (d) original U-Net's results

The experimental results show that the proposed model SDUW-Net can effectively segment retinal vessels in fundus images (see Fig. 6). Compared with original U-Net's results, the vessels segmented by SDUW-Net are more complete and clearer. But through the lower sensitivity, it is found that tiny vessels cannot be detected completely because of some missing feature details in the model. Therefore, we will optimize the model to strengthen feature extraction and transmission in order to complete the accurate segmentation of tiny vessels in the future.

## 5 Conclusions

Automatic retinal vessel segmentation is a very difficult task, due to the complexity of retinal images. This paper presented an effective model, which is inspired by the U-Net and dense connections. The same scale dense connection is proposed and applied to the improved U-net's structure in order to overcome these problems of automatic retinal vessel segmentation. The experimental results show that the proposed SDUW-Net has good performances with low computation and short training time. It could assist ophthalmologists in retinal vessels segmentation for the diagnosis and treatment of ocular diseases.

**Acknowledgment.** This work was supported by Education and Scientific Research Project of Middle and Young Teachers in Fujian Province (JAT180394); National Natural Science Foundation of China (61972187); Fujian Provincial Leading Project (2019H0025); Natural Science Foundation of Fujian Province (2019J01756); Research Project of Minjiang University (MYK18048); Open Fund Project of Fujian Provincial Key Laboratory of Information Processing and Intelligent Control (Minjiang University) (MJUKF-IPIC201811).

## References

1. Lecun, Y., Bengio, Y., Hinton, G.: Deep learning. Nature **521**(7553), 436–444 (2015)
2. Long, J., Shelhamer, E., Darrell, T.: Fully convolutional networks for semantic segmentation. In: Proceedings on Computer Vision and Pattern Recognition, pp. 3431–3440. IEEE (2015)
3. Ronneberger, O., Fischer, P., Brox, T.: U-Net: convolutional networks for biomedical image segmentation. In: Navab, N., Hornegger, J., Wells, W., Frangi, A. (eds.) MICCAI 2015. LNCS, vol. 9351, pp. 234–241. Springer, Cham (2015). https://doi.org/10.1007/978-3-319-24574-4_28
4. Badrinarayanan, V., Kendall, A., Cipolla, R.: SegNet: a deep convolutional encoder-decoder architecture for image segmentation. IEEE Trans. Pattern Anal. Mach. Intell. **39**(12), 2481–2495 (2017)
5. Alom, M.Z., Hasan, M., Yakopcic, C., et al.: Recurrent residual convolutional neural network based on U-Net (R2U-Net) for medical image segmentation. arXiv preprint arXiv:1802.06955 (2018)
6. He, K., Zhang, X., Ren, S., et al.: Deep residual learning for image recognition. In: Proceedings on Computer Vision and Pattern Recognition, pp. 770–778. IEEE (2016)
7. Zhuang, J.: LadderNet: multi-path networks based on U-Net for medical image segmentation. arXiv preprint arXiv:1810.07810 (2018)

8. Xu, G.Z., Hu, S., Chen, S., et al.: Extraction of retinal vessels based on the combination of U-Net and dense net. Chin. J. Image Graph. **25**(9), 1569–1580 (2019)
9. Huang, G., Liu, Z., Van Der Maaten, L., et al.: Densely connected convolutional networks. In: Proceedings on Computer Vision and Pattern Recognition, pp. 4700–4708. IEEE (2017)
10. Zhang, Y., Chung, A.C.S.: Deep supervision with additional labels for retinal vessel segmentation task. In: Frangi, A., Schnabel, J., Davatzikos, C., Alberola-López, C., Fichtinger, G. (eds.) MICCAI 2018. LNCS, vol. 11071, pp. 83–91. Springer, Cham (2018). https://doi.org/10.1007/978-3-030-00934-2_10
11. Li, R., Li, M., Li, J., et al.: Connection sensitive attention U-Net for accurate retinal vessel segmentation. arXiv preprint arXiv:1903.05558 (2019)
12. Wang, C., Zhao, Z., Ren, Q., et al.: Dense U-Net based on patch-based learning for retinal vessel segmentation. Entropy **21**(2), 168 (2019)
13. Staal, J., Abramoff, M.D., Niemeijer, M., et al.: Ridge-based vessel segmentation in color images of the retina. IEEE Trans. Med. Imaging **23**(4), 501–509 (2004)

# A New Mobility Model Based on Markov Chain for MANETs

Tieyuan Liu$^{(\boxtimes)}$, Tianlong Gu, Liang Chang, and Hongzhuan Zhao

Guilin University of Electronic Technology, Guilin 541004, Guangxi, China
1379268391@qq.com

**Abstract.** Due to difficult and costly to build a real-world test-bed, simulation is still the main method in the research area of mobile ad hoc networks (MANETs). Depict a exactly mobile scene or select a proper mobility model is very important to the performance of the simulation. In this paper, we propose a new mobility model based on Markov chain, we call it Time-Speed Markov Random Path (TSMRP) mobility model, which can reflect more classic mobile scene and has better independent parameters controllability. The flexible movement manners of the nodes and the algorithm of TSMRP are derived and analyzed. In order to test the TSMRP effectively, we implement and compare it to the MRP model and the RWP model by NS2, which are widely used in the research area of MANETs. The simulation result shows that, according to choose different probability parameters can effectively simulate different scenarios, and also more stable than the other two models. The results also shows that, different mobility models have great effects on the performance of routing protocol in MANETs.

**Keywords:** Mobile ad hoc networks · Mobility model · Markov chain · Protocol evaluation

## 1 Introduction

MANETs is a temporary, dynamic, multi-hop, purpose-specific wireless network, it consisted by mobile nodes(MNs) with wireless transceiver devices, which depend on no default infrastructure and form in a self-organizing fashion. With the development of mobile Internet, MANET has been applied in more and more occasions, especially in impromptu meeting, battlefield communication and emergency rescue [1].

It is difficult and costly to conduct a real-world test-bed for depicting all kinds of dynamic and complicated scenarios, most of research in MANETs are done by simulation [2]. The rationality of a mobility model plays an important role to the accuracy and reliability of the network simulation results [3, 4].

The existing mobility models can be divided into two categories, one is entity movement model, the other is group movement model [5]. The entity movement model emphasizes the independent motion of each node in the network, such as Random Waypoint (RWP) model [7], Markov Random Path (MRP) model [8]. The group movement model

X. Chen et al. (Eds.): ML4CS 2020, LNCS 12488, pp. 203–214, 2020.
https://doi.org/10.1007/978-3-030-62463-7_19

emphasizes the correlation and relativity motion of each node, such as the Reference Point Group Mobility (RPGM) model [9]. For a detailed definition, classification and characteristics of mobile models, please refer to [5, 6]. In this work, we focus on the entity movement model study.

The MRP model alleviates the defects of RWP model by introducing probability mechanism, but it brings other problems, such as the node cannot be paused in the whole simulation process, the speed of a node is a constant in the whole simulation process, the moving direction of a node is limited, only can move forward, backward, left 90°, right 90°. To alleviates the defects in MRP, some improved model proposed [10, 11], but it still has only two optional speed values, the motion direction a node is still limited, it has only 8 moving directions.

In order to solve these problems, an mobility model based on Markov chain with pause time and variable speed (TSMRP) is proposed in this work. In this model, the node can be paused in the simulation process, the speed of the node can change smoothly and the moving direction can change arbitrarily through the probabilistic selection mechanism in the simulation process.

The rest of the paper is organized as follows. In Sect. 2, the RWP mobility model and the original MRP model are briefly described, which are closely related our work and widely used in Ad Hoc network simulation. In Sect. 3, we present our TSMRP model in details, including the definition, algorithm, simulation implementation and the analysis. In Sect. 4, we validated and compared our model with RWP and MRP by NS2. Conclusions and the future study are outlined in the last section.

## 2    Mobility Model of MANET

As the basis of simulation, How to design a mobility model more in line with actual mobile scenarios has been paid more and more attention by researchers in recent years [5, 6, 12]. In the next, we only briefly introduce the RWP mobility model and MRP mobility model, which are used in the following chapters.

### 2.1    RWP Mobility Model

Random Way Point (RWP) [7] is widely used because of its simplicity and easy implementation. Many typical simulation studies in MANETs are based on this mobility model [5, 6, 12, 13]. In this model, the moving style of nodes are generally described as follows: i) the nodes are stochastically suspended for a period of time; ii) a destination location is randomly selected in the simulation area and moved to the destination at a random speed v, where v obeys the uniform distribution between [minspeed, maxspeed]; iii) after arriving at the destination, repetition process i) and ii) until the end of the whole simulation. The initial position of each node is randomly distributed and the rebound boundary rule is adopted.

RWP model is a non-memory mobility model. because of the velocity attenuation, the random selection of velocity is very important in this model. Under the same average velocity, the larger velocity and the longer pause time will produce a more stable network topology. RWP model truly reflects the basic law of entity movement, and has a good reality. NS2 has its own implementation of this model. Therefore, RWP model has been used in many typical simulation studies of network protocols, and has been seemed as a benchmark mobility model [14].

## 2.2 MRP Mobility Model

The Markov Random Path Model (MRP) [8] uses a probability matrix to determine the location of the mobile node at the next moment. It uses three different states (states 0, 1, 2) to identify the location of the node in the $X$ direction and the $Y$ direction. State 0 denotes the current position of the node in the $X$ or $Y$ direction, State 1 denotes the front position of the node in the $X$ or $Y$ direction, and State 2 denotes the position of the node in the next $X$ or $Y$ direction that the node will move in the current direction. The state transition diagram is shown in Fig. 1. The state transition matrix obtained from the state transition diagram is shown in Formula (1).

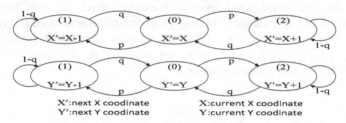

X':next X coodinate          X:current X coodinate
Y':next Y coodinate          Y:current Y coodinate

**Fig. 1.** State transition diagram of MRP model

$$P = \begin{bmatrix} p(0,0) \ p(0,1) \ p(0,2) \\ p(1,0) \ p(1,1) \ p(1,2) \\ p(2,0) \ p(2,1) \ p(2,2) \end{bmatrix} = \begin{bmatrix} 0 & p & p \\ q & 1-q & 0 \\ q & 0 & 1-q \end{bmatrix} \quad (1)$$

Each P (a, b) in the matrix represents the transition probability of the node from state a to state b. The positions of nodes in the $X$ and $Y$ directions are updated by the values of the elements in the matrix.

MRP model is a mobility model with memory. Compared with the nodes randomly moving in RWP mobility model, the nodes in MRP moving more smoothly by probabilistic. So it can reflect the real mobile mode much better. A notable feature of this model is that the probability of the node moving in the same direction is higher than changing direction. Therefore, it alleviates the sharp stopping and turning problem in RWP model.

# 3  A Markov Mobility Model with Pause Time and Variable Speed

As mentioned in Sect. 1, MRP model tries to achieve more realistic movement by intro-ducing probability selection mechanism, but it also brings other problems. The TSMRP model is proposed in this paper solves the above problems by matching the state of Markov chain with the motion behavior, instead of the original state and position.

## 3.1  TSMRP Model Definition

Similar to MRP mobility model, TSMRP model still uses two discrete-time Markov chains to represent the movement of nodes in the $X$ direction and $Y$ direction respectively, and the movement in the $X$ direction and the $Y$ direction are independent of each other. As shown in Fig. 2, the TSMRP model still uses three different states (states 0, 1, 2) to identify the positions of nodes in the $X$ and $Y$ directions. State 0 denotes the current position of the node in the $X$ or $Y$ direction, State 1 denotes the front position of the node in the $X$ or $Y$ direction, and State 2 denotes the position of the node in the next $X$ or $Y$ direction that the node will move in the current direction. Unlike the MRP model, we add the state transition from state 0 to state 0 itself, and change the original node's moving distance per step to a unit time per step. Thus, the distance between two adjacent positions in the same direction is the product of the random velocity $v$ and the unit time interval $T$. The state transition matrix obtained from the state transition diagram is shown in Formula (2). $P$ (a, b) denotes the transition probability of mobile nodes from state a to state B, $a, b \in \{1, 0, 2\}$. If all $P$ (a, b) are obtained, then the movement of nodes can be controlled by the state transition matrix $P$.

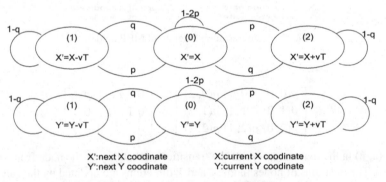

Fig. 2.  State transition diagram of TSMRP model

$$P = \begin{bmatrix} p(0,0) \; p(0,1) \; p(0,2) \\ p(1,0) \; p(1,1) \; p(1,2) \\ p(2,0) \; p(2,1) \; p(2,2) \end{bmatrix} = \begin{bmatrix} 1-2p & p & p \\ q & 1-q & 0 \\ q & 0 & 1-q \end{bmatrix} \qquad (2)$$

## 3.2 Description of TSMRP Model

As mentioned above, compared with the original MRP model, TSMRP adds state transition from state 0 to state 0 itself (transition probability is $1-2p$). In this way, we can express that the mobile node stays at the current position by the self-transfer of the mobile node in the current state 0, and the length of the pause time of the mobile node varies with the number of self-transfer of the state 0. At the same time, we can find that the distance of each movement is no longer a constant of unit distance, but the product of the random velocity $v$ and the time interval $T$. Because of the difference of the random velocity produced in $X$ and $Y$ directions, the composite velocity and direction will not be fixed. This means that the direction and velocity of the node can be changed arbitrarily, while the original MRP model has only four directions and one constant velocity; the improved SIMM model has only nine directions and two constant velocity values.

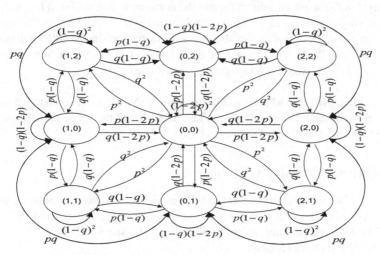

**Fig. 3.** Two-dimensional state transition diagram of TSMRP model

The three models of MRP, SIMM and TSMRP are based on Markov process. They use two discrete-time Markov chains to represent the movement of nodes in the $X$ direction and $Y$ direction respectively, and the movement in the $X$ direction and $Y$ direction is independent of each other. In this way, we can combine the two chains and get a two-dimensional Markov chain to describe the motion of nodes in the two-dimensional region. The specific state transition diagram is shown in Fig. 3. The former number in parentheses represents the state in the $X$ direction, and the latter number represents the state in the $Y$ direction.

As shown in Fig. 3, we can express the node's movement in the plane by state transition. Further, we can conclude that the TSMRP model has the following properties:

(1) In the current state $(0, 0)$, the probability of a time interval for a mobile node to stay in place is $(1-2p)^2$; the probability of moving one time interval in four directions,

namely, east, south, West and north, is $p(1-2p)$; and the probability of moving one time interval in east-south, east-north, west-south and west-north are all $p^2$. The larger the probability parameter $p$ is, the more mobile the node will be.

(2)  In the current state $(2, 0)$, the probability that the mobile node keeps moving eastward for one time interval is $(1-q)(1-2p)$; the probability of moving westward for one time interval is $p(1-2p)$; the probability of moving eastward South and eastward North for one time interval is $p(1-q)$; the probability of moving westward South and northwest for one time interval is $pq$. It can be seen that if $P$ is larger, with the decrease of $q$, the probability of sharp rotation (direction change 180°) will become smaller and smaller. When the current state is $(1, 0)$, $(0, 1)$, $(0, 2)$, the motion situation is similar to $(2, 0)$, and it is not elaborated one by one.

(3)  In the current state $(2, 2)$, the probability that the mobile node keeps moving one time interval in the northeast direction is $(1-q)^2$; the probability of moving one time interval in the south-east direction and the north-west direction is $q(1-q)$; and the probability of moving in the southwest direction is $q^2$. It can be seen that with the decrease of $q$, the probability of sharp rotation (direction change 180°) will become less and less. When the current state is $(1, 1)$, $(2, 1)$, $(1, 2)$, the motion situation is similar to $(2, 2)$, which is no longer elaborated one by one.

In particular, MRP model is a special case when TSMRP model sets p to 0.5 and the speed set to a constant. When setting p to less than 0.5 and the speed set to a constant, it is the SIMM model proposed in reference [10, 11].

### 3.3   TSMRP Node Mobility Algorithms

In view of the fact that node movement in MRP model does not have pause time and the moving speed is constant, and the limitation of node speed and direction in improved model SIMM, we propose an improved algorithm. The specific TSMRP model node movement algorithm is as follows:

**Step1.** *Initialize, set the number of nodes N, simulate the size of scene [xRange, yRange], simulate the t totalTime, the length of time Interval per unit timeinterval, the probability parameters p, q, the maximum speed of X and Y direction xMaxSpeed, yMaxSpeed.*

**Step2.** *Cycling N times, the initial distribution of N nodes is generated randomly. The initial states of X and Y directions of all nodes are 0, that is, xState = 0, yState = 0.*

**Step3.** *Location updates until the end of simulation time.*

*While (Current time <= simulation time)*
*{ for (i=0; i<Number of nodesN; i++)*
*{Random generation of X direction velocity xSpeed,*
*xSpeed∈ [0,xMaxSpeed] ;*
*Random generation of Y direction velocity ySpeed,*
*ySpeed∈ [0,yMaxSpeed];*
*Random generation of two probabilistic random numbers r1,r2 , r1,r2 [0,1],*
*control the movement of X and Y directions respectively.*
*Update the position in the X direction;*
*If (xState==0)*
*{If (r1<=p){xLoc[i]=xLoc[i]+xSpeed\*timeInterval;*
*xState=2;}*
*else if(r1>p&&r1<2p) {xLoc[i]=xLoc[i]-xSpeed\*timeInterval; xState=1;}*
*else {xLoc[i]=xLoc[i];}*
*}*

*If (xState==2)*
*{If(r1<=q) {xLoc[i]=xLoc[i]-xSpeed\*timeInterval; xState=0;}*
*Else {xLoc[i]=xLoc[i]+xSpeed\*timeInterval;*
*xState=2; }*
*}*
*If (xState==1)*
*{If (r1<=q) {xLoc[i]=xLoc[i]+xSpeed\*timeInterval; xState=0;}*
*else {xLoc[i]=xLoc[i]-xSpeed\*timeInterval;*
*xState=1;}*
*}*
*Update the position in the Y direction (using the same method as updating the X*
*direction);*
*The boundary of the simulation area is judged and processed according to the cor-*
*responding boundary rules;*
*Output in specified format;*
*}*
*Current time = current time+timeInterval;*
*}*

## 3.4 TSMRP Model Simulation Implementation

Compared with other entity mobility models, TSMRP model is more versatile and controllable. Different mobile scenarios can be simulated by setting different probability parameters simply. We implement the model in C++ under Linux, and use Gnuplot to plot the trajectory of a single node. The simulation scenario is 500 m × 500 m. The initial position of the node is (250, 250) and the total simulation time is 100 s. The time interval of state updating is 5 S. The maximum speed is 10 m/s. As shown in the following figures.

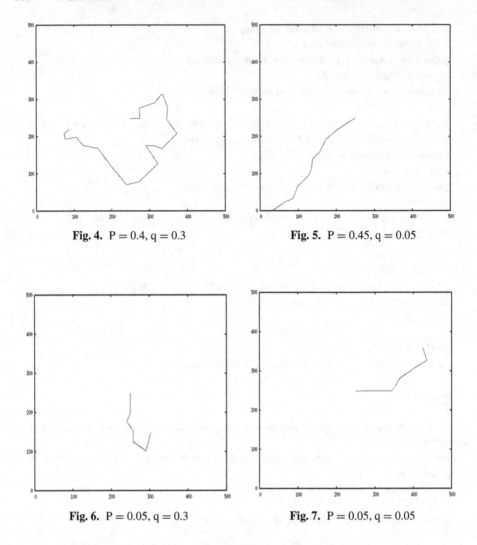

**Fig. 4.** P = 0.4, q = 0.3

**Fig. 5.** P = 0.45, q = 0.05

**Fig. 6.** P = 0.05, q = 0.3

**Fig. 7.** P = 0.05, q = 0.05

The probability parameters p = 0.4 and q = 0.3 set in Fig. 4. It can be seen from the figure that the pause time of nodes is shorter during the whole movement process, and there are few sudden stops. This can be used to simulate people's daily movement, because people tend to move in the same direction after completing their daily work, rarely turning suddenly and rarely repeating the road they take.

The parameters p = 0.45 and q = 0.05 set in Fig. 5. From the figure, it can be seen that there are almost no pauses and few changes in the direction of motion during the whole process, which can be used to simulate the real vehicle driving on the road.

The probability parameters p = 0.05 and q = 0.3 set in Fig. 6. It can be seen from the figure that the suspension time of nodes is longer during the whole movement process, which can be used to simulate the movement of disaster scene, meeting scene.

The parameters p = 0.05 and q = 0.05 set in Fig. 7. It can be seen from the figure that the suspension time of nodes is long and the direction of nodes is seldom changed during the whole process, which can be used to simulate the traffic situation of vehicles in urban streets.

# 4 Simulation Comparison

## 4.1 Building of Simulation Environment

In this section, we compared our model with the widely used RWP model and the original MRP model by NS2. The MRP model is implemented in C++ language under Linux, and the mobile scene of RWP model is generated by Setdest tool of NS2. The main simulation parameters are as follows: the size of simulation scenario is set to 1000 m × 1000 m; the number of nodes in the scenario is 50, and the transmission range of each node is set to 250 m; the maximum speed of node movement varies from 5 m/s to 30 m/s; 30 pairs of random data streams are generated by Cbrgen tool of NS2, in which the size of data packets is 512 bytes. The transmission interval is 0.5 s and the simulation time is 1210 s. The probabilistic parameters of MRP model are set to p = 0.5, q = 0.3 [4], the probabilistic parameters of TSMRP module are set to p = 0.3, q = 0.3, and the time interval of node update status is set to 2 s. In order to overcome the influence of initial distribution of nodes and clearing pipeline at the end of simulation, we only collected data between 600 s and 1200 s. Each data is averaged by running 10 mobile scenarios.

The routing protocol we choose here is Dynamic Source Routing (DSR) [7]. DSR is a reactive on-demand routing protocol. It does not periodically broadcast routing information to maintain the entire network topology, and only establishes routing when the source node needs data transmission.

This paper compares and analyses the impact of mobility model on DSR protocol performance from four main indicators of protocol evaluation, namely, packet delivery rate, average end-to-end delay, average hop count and routing load.

## 4.2 Comparison and Analysis of Simulation Results

Packet delivery rate refers to the ratio of the number of packets received by each node's application layer to the number of packets sent by each node's application layer in the whole simulation process, which is used to measure the data transmission and data loss of each node in the network. Figure 8 shows how the packet delivery rate varies with the node's moving speed under three mobile models: RWP, MRP and TSMRP. It can be seen that with the increase of node speed, the whole network topology changes faster and faster, and the packet delivery rate of nodes under each model decreases in varying degrees. The grouping delivery rate under MRP model is the lowest, while RWP and TSMRP are close. With the increase of speed, the grouping delivery rate under TSMRP model has a better trend than that of RWP model.

Average end-to-end time delay refers to the average time interval between the sending node's sending time and the receiving node's application layer's receiving time, which is used to measure the real-time performance of protocol data transmission. Figure 9 shows how the average end-to-end delay varies with the node's moving speed under three mobile models. It can be seen that the average end-to-end delay of TSMRP model

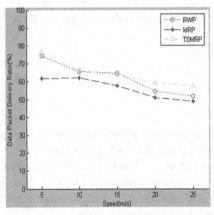
Fig. 8. Speed vs. data packets delivery ratio

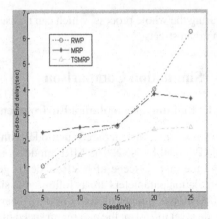
Fig. 9. Speed vs. end-to-end delay

is the shortest and the most stable, while that of RWP model changes dramatically with the change of speed.

The average hop count is the ratio of the total number of messages sent in the network (including those successfully sent and forwarded by intermediate nodes) to the total number of messages received successfully. Combined with packet delivery rate, this index can measure the effectiveness of routing protocols. If the packet transmission rate is high and the average hop count is low, it indicates that the protocol is only valid in the range of fewer hops, and the protocol will fail with the increase of hops. Figure 10 shows how the average hop number varies with the node's moving speed under three mobile models. It can be seen that DSR protocol is stable under three models. At the same speed, the average hops of MRP model and TSMRP model are higher than those of RWP model. At the same time, the joint Fig. 8 shows that the DSR protocol in RWP model is only valid in the range of fewer hops. This shows that the problem of nodes moving only near the initial position in RWP model can be alleviated by probabilistic mobility.

There are many definitions of routing load. Here we take the most accepted definition in academia, that is, the number of routing packets needed to send a packet. This index can be used to measure the efficiency of routing protocols. Figure 11 shows how the routing load varies with the moving speed of the nodes in three mobile models. It can be seen that under the three models, the routing load increases significantly with the increase of speed, because the increase of speed leads to the acceleration of network topology changes, resulting in DSR protocol need to send more routing requests, and the routing maintenance time is shortened. Because there is no pause mechanism in MRP model, its topology changes most frequently, which leads to the highest routing load in MRP model.

From Fig. 7 to Fig. 10, we can see that TSMRP model effectively alleviates the phenomena of sudden turning and short moving range of nodes in RWP model by introducing probabilistic mobility mechanism, and solves the problem that nodes in MRP model must move at the same speed until the end of simulation by introducing pause mechanism and variable speed mechanism. This also makes TSMRP more realistic

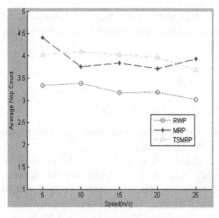

**Fig. 10.**  Speed vs. average hop count

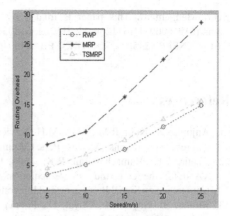

**Fig. 11.**  Speed vs. routing overhead

than the other two models. At the same time, we can see that the performance evaluation results of DSR protocol under these three different mobile models are very different. Therefore, it is particularly important to select a suitable mobile model in protocol evaluation.

## 5  Conclusion

This paper reviews the research progress of current entity mobility model. To overcome the shortcomings of MRP mobility model in which the node speed is single and the node cannot pause in the whole simulation process, by introducing probabilistic mobility, variable speed and suspend mechanism, a time-Speed Markov Random Path mobility model (TSMRP) is proposed, which can better reflect the real node movement law and has strong independent and controllable parameters. This model corrects the shortcomings of MRP model that nodes do not have pause time and nodes can only move at a uniform speed in the simulation process, and effectively alleviates the phenomenon of sudden stop in entity mobility model. The simulation implementation of the model shows that by adjusting the probability parameter p and q, it can effectively simulate various typical mobile scenarios, and has more versatility. The results of model comparison show that TSMRP model is more stable than RWP model and original MRP model, and it also shows that different mobile models have a greater impact on the performance of Ad Hoc network protocols. Therefore, in the process of network protocol design and network simulation, it is necessary to select or design appropriate mobile models according to different application scenarios and application requirements.

In the future, there are two directions to extend our work. We will give more mathematical derivation, such as the spatial probability distribution of the nodes in our model and the proof of our model is stable enough. We only evaluated the DSR protocol in this work, to valuate much more protocols and design suitable protocols are another interesting direction to extend our work.

**Acknowledgement.** This paper is funded by the National Natural Science Foundation of China (61966009, U1811264, U1711263, 61803113), the Guangxi Natural Science Foundation (2019GXNSFBA245059, 2019GXNSFBA245049).

# References

1. Anjum, S.S., Noor, R.M., Anisi, M.H.: Review on MANET based communication for search and rescue operations. Wireless Pers. Commun. **94**, 31–52 (2017)
2. Bhatia, T.K., Ramachandran, R.K., Doss, R., Pan, L.: A review of simulators used for VANETs: the case-study of vehicular mobility generators. In: 2020 7th International Conference on Signal Processing and Integrated Networks (SPIN), pp. 234–239. IEEE (2020)
3. Kumar, S., Goyal, M., Goyal, D., Poonia, R.C.: Routing protocols and security issues in MANET. In: 2017 International Conference on Infocom Technologies and Unmanned Systems (Trends and Future Directions) (ICTUS), pp. 818–824. IEEE (2017)
4. Xie, J., Wan, Y., Wang, B., Fu, S., Lu, K., Kim, J.H.: A comprehensive 3-dimensional random mobility modeling framework for airborne networks. IEEE Access **6**, 22849–22862 (2018)
5. Camp, T., Boleng, J., Davies, V.: A survey of mobility models for ad hoc network research. Wirel. Commun. Mob. Comput. **2**, 483–502 (2002)
6. Kanta, K., Sunil, M., Basant, S.: A brief survey of mobility models for FANET. In: Proceedings of the National Conference on Innovative Trends in Computer Science Engineering, Bahal, India, pp. 20–22 (2015)
7. Johnson, D.B., Maltz, D.A.: Dynamic source routing in ad hoc wireless networks. In: Imielinski, T., Korth, H.F. (eds.) Mobile Computing, pp. 153–181. Springer, Boston (1996). https://doi.org/10.1007/978-0-585-29603-6_5
8. Chiang, C.-C., Gerla, M.: Wireless network multicasting (1998)
9. Hong, X., Gerla, M., Pei, G., Chiang, C.-C.: A group mobility model for ad hoc wireless networks. In: Proceedings of the 2nd ACM International Workshop on Modeling, Analysis and Simulation of Wireless and Mobile Systems, pp. 53–60 (1999)
10. Campos, C.A.V., Otero, D.C., de Moraes, L.F.M.: Realistic individual mobility Markovian models for mobile ad hoc networks. In: 2004 IEEE Wireless Communications and Networking Conference (IEEE Cat. No. 04TH8733), pp. 1980–1985. IEEE (2004)
11. Campos, C.A.V., de Moraes, L.F.M.: A Markovian model representation of individual mobility scenarios in ad hoc networks and its evaluation. EURASIP J. Wirel. Commun. Netw. **2007**, 1–14 (2007)
12. Jo, Y.-I., Fathoni, M.F., Kim, K.: A new mobility model for multi-UAVs reconnaissance based on partitioned zone. Appl. Sci. **9**, 3810 (2019)
13. Susanto, B.M., Hariyanto, A.: Performance comparison of MANET routing protocol based on randomwaypoint mobility model. In: Proceedings of the 2017 International Conference on Computer Science and Artificial Intelligence, pp. 183–187 (2017)
14. Karthikeyan, P., Amudhavel, J., Infanta, E., Deepika, P., Saranya, M., Nandhini, C.: Impact on self-organization in mobile adhoc networks: an comprehensive review. In: Proceedings of the 2015 International Conference on Advanced Research in Computer Science Engineering & Technology (ICARCSET 2015), pp. 1–6 (2015)
15. Shruthi, S.: Proactive routing protocols for a MANET—a review. In: 2017 International Conference on I-SMAC (IoT in Social, Mobile, Analytics and Cloud) (I-SMAC), pp. 821–827. IEEE (2017)

# Image Dust Storm Synthetic Method Based on Optical Model

Jiayan Huang[1], Zuoyong Li[2(✉)], and Chuansheng Wang[3]

[1] College of Mathematics and Computer Science, Fuzhou University,
Fuzhou 350108, China
jyan_huang@163.com

[2] Fujian Provincial Key Laboratory of Information Processing and Intelligent
Control, College of Computer and Control Engineering, Minjiang University,
Fuzhou 350121, China
fzulzytdq@126.com

[3] School of Computer Science and Technology, Harbin University of Science
and Technology, Harbin 150080, China

**Abstract.** To address the problem that the dusty image dataset is small and difficult to collect, this paper presents a synthetic method for generating dusty image based on a classical optical model. The proposed method first learns the physical process of generating dusty image according to the classical optical model. Then, the transmission map is estimated and combined with the presupposed dust storm color map as inputs for obtaining a synthetic dusty image. Finally, considering the impact of image scene depth on the synthesis of dusty image, the proposed method selects an appropriate value of input parameter to obtain final synthetic dusty image. Experimental results on an image dataset captured in clear weather show that the synthetic dusty images obtained by the proposed method can be used as a good substitute for real dusty images.

**Keywords:** Dusty image synthesis · Optical model · Dedusting

## 1 Introduction

With the development of computer vision and digital image technology, monitoring system can capture clear scenes with more details. For example, car monitoring system can capture the accurate plate numbers and the basic information of other cars in clear weather, which can be considered as valid evidences of violations. However, a dusty weather not only affects the visibility of scene for people, but also the function of computer vision system which is responsible for outdoor monitoring tasks. A majority of existing algorithms focus on image defogging. Despite the fact that dust storm is one of the critical factors which may degrade the performance of outdoor monitoring devices, little attention has been paid to image dedusting, which has resulted in major difficulty in many post-processing

© Springer Nature Switzerland AG 2020
X. Chen et al. (Eds.): ML4CS 2020, LNCS 12488, pp. 215–226, 2020.
https://doi.org/10.1007/978-3-030-62463-7_20

works. Therefore, image dedusting research still has great application prospects in the field of image processing.

The availability of large labeled datasets can advance the research of computer vision. However, it is not yet feasible to annotate such datasets for every new research subject. Thus, focus has been put on learning synthetic data in recent years. Furthermore, synthetic image dataset becomes a good substitute for real dataset in many related experiments. For instance, Jeong et al. [1] adopted a fog synthesis method based on depth information and temporal filtering for road environment. Sakaridis et al. [2] presented a fog synthesis method to obtain a fog simulation image dataset which automatically inherits its true and clear semantic annotations. Besides, for detection of road object in rainy weather, Jeong's method [3] proposes a rainy image synthetic algorithm. And Iizuka et al. [4] tested their proposed image editing system with fog synthesis. Exactly, because of the fact that dusty image data is too small and difficult to collect in reality, the further development of this field has been seriously hindered.

Scattering of particles in atmosphere is the main cause of dust storm. Harald's method [5] finds that lower visibility of foggy images is caused by absorption and scattering of global light by particles in atmosphere. And McCartney et al. [6] explained that the scattering of particles leads to attenuated transmission of light between object and camera, and then adds a layer of atmospherically scattered light. For the problem of lower visibility of foggy image, Nayar et al. [7] explained the imaging process of foggy images with a mathematical model. The atmospheric scattering model provides a powerful basis to develop many algorithms focusing on defogging and even related image processing works, and advances research in this field. Motivated by the success of convolutional neural network (CNN) in feature extraction [8], Ren et al. [9] proposed an image enhancement method based on CNN.

Since similar physical generation process of dust storm and fog, we utilized a standard optical model [10] used in image defogging field to synthesize dusty images. The optical model has been used to model the impact of dust storm on a clear image [9,11,12], and its definition is,

$$I(x) = J(x)t(x) + L(1 - t(x)),  \tag{1}$$

where $I(x)$ denotes the observed dusty image at pixel $x$, $J(x)$ is a clear raw image, $L$ is atmospheric light, and transmission map $t(x)$ is defined as:

$$t(x) = e^{(-\beta l(x))},  \tag{2}$$

where $\beta$ denotes scattering coefficient of atmosphere, which controls the thickness of dust storm. The larger values of $\beta$ mean thicker dust storm. As an example, Fig. 1 shows different thickness of dusty scene corresponding to different $\beta$. And $l(x)$ is the distance between object and camera. Equation (1) provides a significance basis for simulating dust storm on images with clear-weather.

The main contributions of the paper are: 1) we learn the dusty image model according to classic foggy model, and evaluate the transmission map accurately; 2) we propose a new synthetic method based on classic optical model for dusty

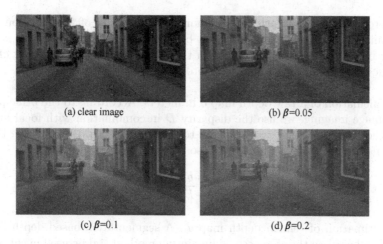

(a) clear image

(b) $\beta$=0.05

(c) $\beta$=0.1

(d) $\beta$=0.2

**Fig. 1.** Synthetic dusty images with different values of $\beta$.

images from a clear dataset; 3) we make subjective evaluation and comparison between synthesized dusty images and real dusty scenes, and the final experimental results show that the dusty images synthesized by our approach can be a substitute for real datasets in image dedusting tasks; 4) we extend the dusty image dataset for advancing related post-processing work to some extent.

## 2   The Proposed Method

Motivated by the classical optical model [10], this paper proposed an image dust storm synthetic method. The required inputs of the proposed synthetic method include an evaluation of the transmission matrix $t(x)$ at each pixel, a hypothetical dust storm color map $C$, and a clear-weather raw image $J(x)$ for dust storm synthesis. Since the $J(x)$ is simple and easy to be satisfied, we will dedicate to estimate the image transmission $t$, and selection of dust storm color map $C$.

**Estimation of Transmission:** To estimate a transmission map $t$ for the proposed synthetic method, the necessary inputs are:

- an original color image $J$ took in clear-weather as the left image of stereo pair;
- a right image $Q$ for stereo pair;
- the camera focal length $f$, and two camera baseline $b$;
- a dense, raw disparity $D$ with the same resolution as the clear image $J$;
- a set $M$ composed of the pixels without the value of $D$.

For these required inputs, we can use a stereo camera and standard matching algorithm [13] to acquire them easily.

Before using image depth as input for the calculation of transmission map $t$, we can improve the quality of depth by using an accurate structure served in color images of stereo pair. Based on this main idea, the main works of our synthetic method can be summarized into five steps:

*step1* Calculation of a raw depth map $d$ in meters. According to the principle of stereo image imaging, we use the disparity $D$ in combination with focal length $f$ and the baseline values $b$ of cameras to obtain $d$ by Eq. (3) directly. Those pixels missing disparity $D$ values are also missing depth values in $d$.

$$d(x) = \frac{b * f}{D(x)}. \tag{3}$$

*step2* Estimation of refined depth map $d'$. A segmentation-based depth filling method [2] based on the stereoscopic inpainting method [14] be used in this step. Specifically, we use a superpixel segmentation of the clear image $J$ to guide the denoising and completion of $d$ at the superpixel level, and make an assumption that each individual superpixel corresponding to a plane in the 3D scene.

Firstly, we check photo-consistency between $J$ and $Q$, all pixels in $J$ whose color deviation (measured in RGB color space) is greater $\epsilon = 12/255$ from the corresponding pixel in $Q$ are considered invalid in image depth, and are added to set $M$.

Secondly, we implement SLIC [15] to segment clear image $J$ into superpixels. $\widehat{K}$ denotes the expected number of superpixels and $m$ is the related range domain scale parameter, respectively (set $\widehat{K} = 2048$ and $m = 10$). The final output number of superpixels of SLIC is represented as $K$. According to the number of invalid or missing depth pixels in a superpixel, all superpixels are classify into reliable and unreliable. A surperpixel $T$ is regarded as a reliable superpixel if and only if the following condition [14] is satisfied:

$$T_{rel} = \begin{cases} 1, & card(T\backslash M) \geq max\{P, \lambda card(T)\} \\ 0, & otherwise \end{cases} \tag{4}$$

where $card(\cdot)$ denotes the number of pixels in a set, $T\backslash M$ is the set difference between $T$ and $M$, setting $P = 20$ and $\lambda = 0.6$.

In order to fit a depth plane for each superpixel, we run RANSAC (Random Sample Consensus) on its pixels with valid depth value, and account for differences in the range of depth values between distinct superpixels by an adaptive inlier threshold defined as:

$$\theta = 0.01 \underset{x \in T\backslash M}{median}\{d(x)\}, \tag{5}$$

where $d(x)$ denotes the estimated depth of superpixel $x$. By using adaptive RANSAC, the maximum number of iterations is set to 2000, and the bound of probability $p = 0.99$ for obtaining a pure sample.

We then use a greedy algorithm [9] to match unreliable superpixels to reliable ones pairwise, and assign the fitted depth plane of reliable superpixel to its matched unreliable superpixel. Different from the method [14], we apply a novel objective function [2] for matching of superpixel pair $(s, t)$, which is defined as:

$$E(s, t) = ||C_s + C_t||^2 + \alpha||x_s + x_t||^2. \tag{6}$$

The first term on right-hand side of Eq. (6) is used to describe the similarity between two superpixels in color space, $C_s$ denotes the average CIELAB color of the superpixel $s$. More specifically, the method [2] penalized the squared Euclidean distance of the average colors of two superpixels in CIELAB color space for increasing perceptual uniformity while another method [14] designs its cost of range domain by using cosine similarity of average colors of superpixels:

$$E_{cls}(s, t) = 1 - \frac{C_s}{||C_s||} \cdot \frac{C_t}{||C_t||}. \tag{7}$$

where $E_{cls}(s, t)$ is the color similarity between the two superpixels, $C_s$ and $C_t$ are the average color vectors of superpixels $s$ and $t$, respectively. In some cases, Eq. (7) exists a problem that it may assign zero matching penalty to dissimilar colors. For example, in the RGB color space, the pair of colors $(\delta, \delta, \delta)$ and $(1 - \delta, 1 - \delta, 1 - \delta)$, where $\delta$ is a small positive constant is assigned zero match cost, even though the former color is very dark grey and the latter is very light grey.

The second term on right-hand side of Eq. (6) calculates the squared Euclidean distance between two centroid $x_s$ and $x_t$ of superpixel pair $(s, t)$ in spatial space, which be used to denote the similarity of two superpixels. Due to the Eq. (7) gives zero matching penalty to adjacent superpixel and unit cost to non-adjacent superpixel, a superpixel $s$ will mismatch a far superpixel $t$ as long as the range domain of superpixel pair is minimum and has a different appearance with all adjacent superpixels of $s$. On the contrary, the method [2] handled well for the problem that existed in [14], i.e. other superpixels with less similar appearance yet smaller distance to $s$ will be the first choice if $t$ has a large spatial cost to matching $s$.

The parameter $\alpha > 0$ in Eq. (6) is used to balance the importance between spatial space and color space. Let $\alpha = m^2/S^2$, where $S = \sqrt{N/K}$, $N$ is the total number of pixels in the image, $m$ and $K$ are the same as for SLIC [15]. The matching target of Eq. (6) is similar to the distance defined in SLIC. In this paper, we use this distance to measure the similarity of superpixel pair.

After all superpixels have been assigned a depth plane, we use these planes to complete the missing depth values of pixels belonging to set $M$. Besides, the depth values of those pixels, which do not belong to $M$ but constitute large-margin outliers with their corresponding plane (deviation larger than $\hat{\theta} = 50$ m) are replaced with the values imputed by the plane, and finally results in a denoising and complete depth map $d'$.

*step3* Computation of scene distance $l(x)$ from camera. Each pixel value in the depth map $d'$ represents the distance between a point in the scene and the

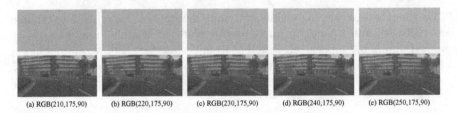

(a) RGB(210,175,90)    (b) RGB(220,175,90)    (c) RGB(230,175,90)    (d) RGB(240,175,90)    (e) RGB(250,175,90)

**Fig. 2.** Different hypothetical dust storm color maps and their corresponding synthetic dusty images. (Color figure online)

camera. Therefore, we use the coordinates of principal point that takes the center point of the image as the coordinate origin, and the focal length of camera to calculate the distance $l(x)$ of scene from camera at each pixel $x$ based on $d'(x)$.

*step4* Calculation of initial transmission map $\hat{t}$. We can substitute $l(x)$ into Eq. (2) to obtain an initial transmission map $\hat{t}$.

*step5* Guided filter of $\hat{t}$ using $J$ as guidance to compute the final transmission $t(x)$. In order to smooth transmission $t(x)$ while preserving boundaries of clear image $J$, a guided filtering [16] is used for the post-process of the initial transmission $\hat{t}$. We set the radius of guided filter window to $r = 20$, and regularization parameter to $\mu = 10^{-3}$ (the same values as the haze removal experiments of [16]). The results in a final transmission map $t(x)$.

**Selection of Dust Storm Color Map:** As Fig. 2 shows, we further compare various possible dust storm colors and select the final hypothetical dust storm color map $C$ for our dusty image synthetic method. According to our observation on a large number of real dusty images, the dust storm color is generally pale orange, and the red in RGB color space has the main impact on pale orange. Therefore, we respectively fixed the green and blue color values at 175 and 90, and changed the red color value within the possible color range of dust storm to obtain different dust storm color maps. The second row of the Fig. 2 gives the corresponding synthetic results of different hypothetical dust storm color maps. Considering the color of most real dusty scenes, we selects the RGB value of (230, 175, 90) as the input dust storm color map.

Considering the impact of the proposed method on synthetic image depth, a parameter $\lambda$ is introduced based on Eq. (1). Therefore, the equation transformed into the following expression:

$$I(x) = J(x)(1 - \lambda(1 - t(x))) + \lambda C(1 - t(x)), \tag{8}$$

where $I(x)$ is the target synthetic dusty image, $J(x)$ is the original clear image, and $C$ denotes the hypothetical dust storm color of the proposed method, $\lambda$ is a parameter. As shown in Fig. 3, it is a whole flow chart of the proposed dusty image synthetic method.

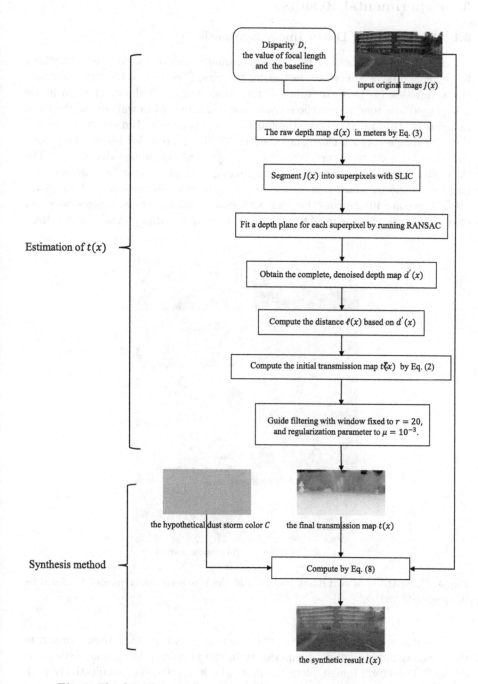

**Fig. 3.** The flow chart of the proposed dusty image synthetic method.

# 3   Experimental Results

## 3.1   Evaluation of Dusty Image Synthesis

To validate effectiveness of the proposed image dust storm synthetic method, five typical images are chosen as testing samples. Five images have the following characteristics: (1) they are from different image scenes. In detail, the first image is captured in a road scene, the second image is captured in wall scene, the third image is capture in city scene, the fourth image is captured in street scene, and the fifth image contain close shot details. (2) They are color image. Therefore, they can be used to observe dusty image synthesis on image details. (3) The third and fourth images have not sky region, and the former two images have large sky regions, which can be used to observe color distortion in sky regions. (4) The second image contains sharp shadows, and the other images have not sharp shadows, which can obtain more realistic image dusty synthetic results.

(a) real dusty image    (b) synthetic dusty image

**Fig. 4.** Comparison of real dusty images and the synthetic dusty images obtained by the proposed method.

As far as we know, there are few or even no existing image dust storm synthesis algorithms to validate superiority of the proposed dusty image synthetic method. Therefore, image dusty storm synthetic results were qualitatively evaluated. As Fig. 4 shows, we compare dusty images obtained by the proposed method with real dusty images downloaded randomly from the internet. From Fig. 4, we can observe that image dust storm synthetic results obtained by the proposed method eliminate color distortion in sky regions, and make image detail visible. As compared with the real-world dusty scenes, the synthetic dusty images obtained by the proposed method can be a substitute for real dusty images in many dedusting related post-processing tasks.

## 3.2 Parameter Discussion

We further discuss the influence of parameter $\lambda$ on the results of dust storm synthesis in this section, and find an ideal value of $\lambda$ for the proposed dusty image synthetic method to improve the visual results of synthetic dusty images.

During the specific synthetic experiment, we first use different values of the parameter $\lambda$ (0.4, 0.6, 0.8 and 1.0) for obtaining dusty images with different qualities. Figure 5 shows dust storm synthetic results from clear images, the columns from left to right are the original images and the synthetic dusty images obtained by the proposed method with $\lambda = 0.4$, $\lambda = 0.6$, $\lambda = 0.8$, and $\lambda = 1.0$.

According to Eq. (8) and Fig. 5, we can intuitively find that the higher the value of $\lambda$ is, the larger the weight of the hand-right term of Eq. (8), thus resulting in thicker dust storm on the synthetic image. On the contrary, higher weight of the hand-left term of Eq. (8) indicates lower dusty thickness in synthetic dusty image. Therefore, we draw a hypothetical conclusion that the damage to the original image depth by the synthetic dusty image increases with the increase of $\lambda$ value.

(a) original          (b) $\lambda$=0.4          (c) $\lambda$=0.6          (d) $\lambda$=0.8          (e) $\lambda$=1.0

**Fig. 5.** The visual synthetic dusty images with different $\lambda$ values obtained by the proposed method.

And then, in order to obtain more reliable value of $\lambda$ for higher quality of simulation results, we use an existing depth estimation method [17] to obtain the depth maps of original image and its different synthetic results of different values of $\lambda$. An example of the corresponding depth maps is illustrated in Fig. 6.

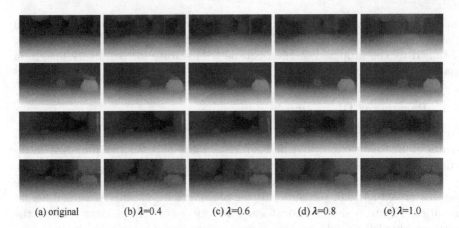

(a) original      (b) $\lambda$=0.4      (c) $\lambda$=0.6      (d) $\lambda$=0.8      (e) $\lambda$=1.0

**Fig. 6.** The scene depth estimation maps of original images and corresponding synthetic dusty images with different $\lambda$ values.

Image depths were quantitatively evaluated. Four commonly used indexes (including the Root Mean Squared Error (RMSE), Log Root Mean Squared Error (RMSE_log), Square Relative Error (Sq_Rel), and Absolute Relative Error (Abs_Rel)) are adopted as the evaluation criteria for comparison of depth map, where the smaller the values of all these indexes are, the less is the depth map of the original image damaged by the proposed dusty image synthetic method.

**Table 1.** Average absolute difference of scene depth between original image and synthetic dusty images with different $\lambda$ values.

| $\lambda$ | RMSE | RMSE_log | Abs_Rel | Sq_Rel |
|---|---|---|---|---|
| 0.4 | 4.731 | 0.147 | 0.023 | 0.077 |
| 0.6 | 6.509 | 0.188 | 0.056 | 0.121 |
| 0.8 | 9.591 | 0.228 | 0.120 | 0.168 |
| 1.0 | 11.881 | 0.255 | 0.224 | 0.236 |

Finally, by using the four evaluation indexes mentioned above, we quantitatively compare original clear image with four different depth maps of dusty images obtained by the proposed method, and the results are shown in Table 1. It can be found from the data in Table 1, as we set the parameter value higher, the four indexes become larger as well, which means that the synthetic dusty image obtained with the high-value parameter will cause more serious damage to the original image depth map. To sum up, we validate the previous assumption that the dusty images with higher quality and similarity to the dusty scenes in real world can be obtained when we set the input parameter $\lambda = 0.4$.

# 4 Conclusion

In this paper, we proposed a novel method for generating synthetic dusty images to alleviate the difficulty of collecting dusty images in image dedusting task needing a lot of dusty images. Specifically, the proposed method constructed a mathematical model of dusty image generation according to the classical optical model, and estimated the transmission map of original clear image for the generation of final synthetic dusty images. We qualitatively compared the synthetic dusty images obtained by the proposed method and real dusty images. Experimental results on a series of original images captured in clear weather demonstrate that the synthetic dusty images obtained by the proposed method can be used as a good substitute for real dusty images. In the future, we will try to improve the proposed synthetic method, and use deep learning based image dedusting methods to verify the effectiveness of synthetic data for improving image dedusting effect.

**Acknowledgment.** This work is partially supported by National Natural Science Foundation of China (61972187, 61772254), Fujian Provincial Leading Project (2017H0030, 2019H0025), Government Guiding Regional Science and Technology Development (2019L3009), and Natural Science Foundation of Fujian Province (2017J01-768 and 2019J01756).

# References

1. Jeong, K.M., Song, B.C.: Fog detection and fog synthesis for effective quantitative evaluation of fog-detection-and-removal algorithms. IEIE Trans. Smart Process. Comput. **7**(5), 350–360 (2018)
2. Sakaridis, C., Dai, D., Van Gool, L.: Semantic foggy scene understanding with synthetic data. Int. J. Comput. Vis. **126**(9), 973–992 (2018)
3. Jeong, K.M., Song, B.C.: Image synthesis algorithm for road object detection in rainy weather. IEIE Trans. Smart Process. Comput. **7**(5), 342–349 (2018)
4. Iizuka, S., Endo, Y., Hirose, M., Kanamori, Y., Mitani, J., Fukui, Y.: Object repositioning based on the perspective in a single image. Comput. Graph. Forum **33**, 157–166 (2014)
5. Harald, K.: Theorieder horizontalen sichtweite: Kontrast und sichtweite, vol. 12. Keim & Nemnich, Munich (1924)
6. McCartney, E.J.: Optics of the Atmosphere: Scattering by Molecules and Particles, p. 421. Wiley, New York (1976)
7. Nayar, S.K., Narasimhan, S.G.: Vision in bad weather. In: Proceedings of the Seventh IEEE International Conference on Computer Vision, vol. 2, pp. 820–827. IEEE (1999)
8. Ning, X., Li, W., Tang, B., He, H.: BULDP: biomimetic uncorrelated locality discriminant projection for feature extraction in face recognition. IEEE Trans. Image Process. **27**(5), 2575–2586 (2018)
9. Ren, W., Liu, S., Zhang, H., Pan, J., Cao, X., Yang, M.-H.: Single image dehazing via multi-scale convolutional neural networks. In: Leibe, B., Matas, J., Sebe, N., Welling, M. (eds.) ECCV 2016. LNCS, vol. 9906, pp. 154–169. Springer, Cham (2016). https://doi.org/10.1007/978-3-319-46475-6_10

10. Koschmieder, H.: Theorie der horizontalen sichtweite, beitrage zur physik der freien atmosphare. Meteorologische Zeitschrift **12**, 3353 (1924)

11. He, K., Sun, J., Tang, X.: Single image haze removal using dark channel prior. IEEE Trans. Pattern Anal. Mach. Intell. **33**(12), 2341–2353 (2010)

12. Fattal, R.: Single image dehazing. ACM Trans. Graph. (TOG) **27**(3), 1–9 (2008)

13. Hirschmuller, H.: Stereo processing by semiglobal matching and mutual information. IEEE Trans. Pattern Anal. Mach. Intell. **30**(2), 328–341 (2007)

14. Wang, L., Jin, H., Yang, R., Gong, M.: Stereoscopic inpainting: joint color and depth completion from stereo images. In: Proceedings of the IEEE Conference on Computer Vision and Pattern Recognition, pp. 1–8. IEEE (2008)

15. Achanta, R., Shaji, A., Smith, K., Lucchi, A., Fua, P., Süsstrunk, S.: SLIC superpixels compared to state-of-the-art superpixel methods. IEEE Trans. Pattern Anal. Mach. Intell. **34**(11), 2274–2282 (2012)

16. He, K., Sun, J., Tang, X.: Guided image filtering. IEEE Trans. Pattern Anal. Mach. Intell. **35**(6), 1397–1409 (2012)

17. Godard, C., Mac Aodha, O., Firman, M., Brostow, G.J.: Digging into self-supervised monocular depth estimation. In: Proceedings of the IEEE International Conference on Computer Vision, pp. 3828–3838 (2019)

# Research on Classification Model of Government Big Data Based on Deep Learning

Qiong Yuan[1], Duqiang Mai[2(✉)] (iD), Ying Sun[3], and Peiyu Liang[2]

[1] Institute of Information Security, Beijing Electronic Science and Technology Institute, Beijing, China

[2] Department of Electronics and Communication Engineering, Beijing Electronic Science and Technology Institute, Beijing, China
davidmai@yeah.net

[3] Department of Cryptography and Technology, Beijing Electronic Science and Technology Institute, Beijing, China

**Abstract.** Based on analyzing the textual features of Government Big Data, this paper proposes the Attention-BiLSTM-CNN hybrid classification model. Firstly, the model uses the BiLSTM model to extract the sentence vector of the text, and then introduces an attention mechanism to automatically learn the weight distribution of keywords, and finally uses the CNN model to calculate the probability of the category. And design experiments to verify the effect of the new model. The experimental results show that the classification model proposed in this paper has a good classification effect in the government big data corpus, which can improve the efficiency of government big data classification.

**Keywords:** Government big data · Attention-BiLSTM-CNN · Classification model

## 1 Introduction

With the rapid development of information technology, data has become an important strategic resource. The application of big data has penetrated into various fields such as finance, medical care, and government management, and has brought many conveniences to people's lives and production. The so-called Government Big Data is data owned and managed by the government, including public government notices, geographic information, and traffic data. Government big data has the characteristics of high value, wide coverage and great sensitivity. Therefore, the issue of security and confidentiality of government big data has become the focus of attention.

Classification and grading is an important means to ensure the security of Government big data. But classification and grading generally adopts the manual method, which has the problems of high labor cost and low efficiency. As deep learning technology gradually matures, deep learning technology can be used to realize automatic classification and grading of government big data. This paper studies the application of deep learning technology to the classification task of government big data, designs a classification model of government big data based on deep learning, and verifies the validity of the model through experiments.

© Springer Nature Switzerland AG 2020
X. Chen et al. (Eds.): ML4CS 2020, LNCS 12488, pp. 227–236, 2020.
https://doi.org/10.1007/978-3-030-62463-7_21

## 2 Related Work

Deep learning was initially applied in the fields of images and speech, and made major breakthroughs. With the development of deep learning technology, scholars try to apply deep learning to text classification tasks and get good results. Yoon Kim proposed the TextCNN model in 2014 and used Convolutional Neural Networks (CNN) to process text classification tasks [1]. Tang proposed a neural network model for emotion classification using GRU to model documents [2]. Lai proposed a cyclic convolutional neural network classification model [3]. This model first uses Bi-RNN to capture the contextual features of words, and then extracts the feature vector of the document through the convolution pooling operation, and finally the feature vector Input to the SoftMax layer to get the document category information. Yang introduced an attention mechanism for extracting important words in sentences, and proposed a hierarchical attention mechanism network for document classification [4]. To solve the problem that RNN cannot memorize information for a long time and CNN has insufficient ability to capture the contextual semantics of text, Yang Yunlong proposed a sentiment analysis model of GRU and capsule feature fusion [5].

This paper combines the textual features of government big data and makes full use of the advantages of multiple models, and proposes the Attention-BiLSTM-CNN hybrid classification model. The research of this model is introduced below.

## 3 Research on Classification Model of Government Big Data

The research of the model in this paper mainly includes three aspects of work: one is to collect the government big data corpus and analyze the textual features; the second is to combine the textual features of the government big data to design the government big data classification model; the third is to study the realization of the classification model.

### 3.1 Government Big Data Text Features

Government big data is different from other news texts. It is authoritative data issued by government departments, and its content is normative. This paper summarizes and analyzes the textual features of government big data by collecting and sorting a large number of open corpus of government big data. Its text mainly has the following characteristics:

(1) The total length of the text of government big data is relatively long. In the collected government big data corpus, the length of each text is not exactly the same, but most of the text content is longer, and some texts even contain hundreds of thousands of characters. It is found through statistics that the texts of government big data are mostly long texts.

(2) The text of government big data has multi-category label information. This paper collects government notices issued by the Chinese government website and government websites of several provinces. All notices are divided into 22 categories according to the subject. Although each government notice has a corresponding

attribution category, it is found during analysis that the notice content may contain other types of label information, which undoubtedly increases the difficulty of classification.

(3) The text redundancy of government big data is greater. In order to be able to objectively reflect the scientific and accurate content of the text, government departments will use column numbers and other methods to increase the persuasion of the content when writing the text, which will increase the text redundancy of government big data and thus affect the classification effect.

## 3.2 Design of Classification Model

The design idea of the classification model is as follows: First, by analyzing the textual features of government big data, the model requirements are obtained. Then, improve and combine the existing classification models to meet the model requirements of government big data. Finally, a government big data classification model with better results is designed. Figure 1 shows the design idea of the government big data classification model.

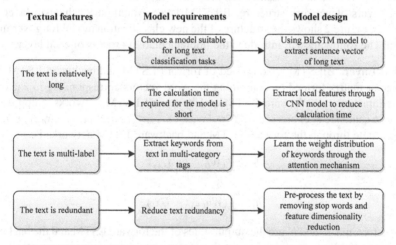

**Fig. 1.** The design idea of the classification model of government big data.

In view of the relatively long length of government big data texts, the new classification model needs to meet two needs. On the one hand, the model designed should be suitable for the classification task of long text. BiLSTM model is not only suitable for the classification task of long texts, but also able to capture the semantic dependency of the words before and after. Therefore, the designed classification model uses BiLSTM model for extracting sentence vectors of long text. On the other hand, considering the short calculation time required by the designed model, the advantage of the CNN model over other classification models is that it can extract local features of the text, and the time spent is relatively short. Therefore, the CNN model is used to extract local features in the designed model, which is used to calculate the probability distribution of text categories.

In view of the multi-label feature of government big data texts, the new classification model needs to be able to extract keywords of texts under different types of tags. The attention mechanism can automatically extract the keywords of the sentence according to the label information of the text, thereby obtaining the importance of each word for the text label. Therefore, an attention mechanism is integrated into the designed model to automatically learn the weight distribution of keywords.

In view of the large redundancy of government big data texts, the new classification model needs to be able to reduce the text redundancy. Methods such as removing stop words and feature dimensionality reduction are commonly used effective methods to reduce text redundancy. Therefore, in the designed model, the text is pre-processed by removing stop words and feature dimensionality reduction, so as to reduce the influence of extraneous words in the text on the classification model.

### 3.3 Classification Model of Government Big Data

This paper presents a new classification model for government big data, which consists of three layers of network structure: BiLSTM layer, Attention mechanism layer and CNN layer. Figure 2 shows the structure of the new classification model for government big data. The following will introduce the implementation process of each layer.

**BiLSTM Layer.** BiLSTM is composed of forward LSTM and backward LSTM. The word vector matrix formed by combining the word vectors corresponding to each sentence is input into the forward LSTM and backward LSTM networks. Suppose the vector of a sentence is expressed as $X = (x_1, x_2, \ldots, x_{t-1}, x_t, x_{t+1}, \ldots, x_{n-1}, x_n)$, and its output state through the forward LSTM and backward LSTM networks is:

$$h_t = f(Wh_{t-1} + Ux_t) \tag{1}$$

$$h'_t = f(Wh_{t+1\prime} + Ux_t) \tag{2}$$

In the formula, $h_t$ and $h'_t$ are the output states of the forward LSTM and the backward LSTM respectively, $h_{t-1}$ and $h_{t+1\prime}$ are the output states of the previous sentence and the latter sentence, $W$ and $U$ are the weight vectors of the output layer and input layer respectively. It can be seen from the formula that the forward LSTM output state is related to the corresponding word vector above, and the backward LSTM output state is related to the corresponding word vector below.

Finally, the output state $h_t$ of the forward LSTM and the output state $h'_t$ of the backward LSTM are longitudinally spliced to obtain the output state of BiLSTM:

$$H_t = h_t \oplus h'_t \tag{3}$$

**Attention Mechanism Layer.** The function of the attention mechanism layer is to increase the weight of keywords related to the category in the sentence to form a new

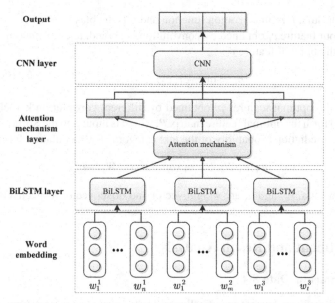

**Fig. 2.** Network structure of Attention-BiLSTM-CNN hybrid classification model.

sentence vector. Suppose a sentence passes through the BiLSTM network and the output is $H_1, H_2, \ldots, H_n$, and then the output result is input into the attention mechanism network to obtain the output vector $u_t$:

$$u_t = \tanh(W_d \cdot H_t + b_d) \tag{4}$$

$W_d$ and $b_d$ are the weight vector and bias vector of the attention mechanism. By comparing the similarity between the output vector $u_t$ and the original input vector $u_d$, the weight coefficient $\alpha_t$ is determined:

$$\alpha_t = \frac{\exp\left(u_t^T u_d\right)}{\sum \exp\left(u_t^T u_d\right)} \tag{5}$$

The larger the value of $\alpha_t$, the more important the word is in the sentence. The weight coefficient $\alpha_t$ is multiplied by the output result $H_1, H_2, \ldots, H_n$ of the BiLSTM network, and the result is used as the output $d$ of the attention mechanism network. The calculation formula is:

$$d = \sum_{i=1}^{n} \alpha_i H_i \tag{6}$$

**CNN Layer.** The output $d$ of the attention mechanism network is input into the CNN layer, and use different convolution kernel $W_k$ to convolve with the sentence vector matrix $d$ to obtain the output feature $c_i$. The calculation formula is:

$$c_i = f(W_k \cdot X_{i:i+k-1} + b) \tag{7}$$

In the formula, $f$ is an activation function and $b$ is the bias vector.

The output feature $c_i$ obtained by convolution is pooled, and the maximum value is selected as the optimal feature $p_i$:

$$p_i = \max(c_1, c_2, \ldots c_{i-k+1}) \qquad (8)$$

Finally, the optimal features $p_i$ obtained by different convolution kernels are fully connected to form the optimal feature matrix $P$, which is input into the SoftMax classifier to obtain the prediction probability of the text category. The calculation formula is:

$$p(y|P, W_s, b_s) = soft\max(P \cdot W_s + b_s) \qquad (9)$$

In the formula, $W_s$ and $b_s$ are weight vector and offset vector respectively.

## 4 Experiment and Result Analysis

### 4.1 Experimental Data Set

At present, there is no data set specifically used to train and test the government big data classification model. In order to verify the effect of the new classification model, this paper needs to build an experimental data set of the government big data classification model. Literature [6] collected information on policy announcements issued by the government through web crawler technology, and designed and implemented a government notice text classification system. This provides a new idea for the collation of the corpus of government big data.

In order to be able to collect a large number of government notices, this paper has surveyed the China Government website, provincial and autonomous region government websites. Figure 3 shows the government information disclosure page of the China Government website. From the figure, it can be seen that the China Government website has two different classification methods according to the type and subject of official documents, and each classification method is divided into many subcategories. Most government websites of provinces and autonomous regions also categorize government notices according to this classification method. However, government notices all contain information about the types of official documents, the classification method can be completed by extracting the keywords in the text. Therefore, the classification method according to the subject has become the classification target of this paper. The classification by subject includes 22 subcategories such as organizational structure, comprehensive government affairs, and national economic management. This classification method requires accurate classification by extracting semantic information from government big data.

This paper counts the number of government notices published on the Chinese government website and local government websites, and selects the websites with the large amount of total texts as the objects of collection. Among them, the number of government notice texts on the China Government Network and its local websites in Beijing, Hebei, Jiangsu, Qinghai and Guizhou is relatively large. The above website counts the total number of texts in each category, and selects the top ten data sets from the total

| Classified by type of official document (5485) | | | | | |
|---|---|---|---|---|---|
| **By subject (5485)** | | | | | |

Organization of the State Council (337)
Comprehensive Government Affairs (380)
National Economic Management, State-owned Assets Supervision (336)
Finance, Banking, and Auditing (358)
Land Resources, Energy (353)
Agriculture, forestry, water conservancy (398)
Industry, Transportation (403)
Commerce, customs, tourism (651)
Market supervision, production safety supervision (214)
Urban and Rural Construction, Environmental Protection (506)
Technology, Education (333)
Culture, Radio, Television, Press and Publication (130)
Health and Sports (220)
Population and family planning, work for women and children (23)
Labor, personnel, inspection (236)
Public Security, Security, and Justice (193)
Civil affairs, poverty

Search location ● title ○ full text    suitability ● blurry ○ Precise
Sort ● Relevance ○ Release date

Current position: Policy

| Serial number | title | Send text number | Date of writing | Release date |
|---|---|---|---|---|
| 1 | Notice of the General Office of the State Council on Printing and Distributing the Main Points of Government Affairs Opening Work in 2020 | Guobanfa [2020] No. 17 | June 21, 2020 | July 03, 2020 |
| 2 | Regulations on Cosmetics Supervision and Administration | National Order No. 727 | June 16, 2020 | June 29, 2020 |
| 3 | Notice of the General Office of the State Council on Carrying out the National Census of Poverty Alleviation | Guobanfa [2020] No. 7 | April 08, 2020 | June 29, 2020 |
| 4 | Notice of the State Council on the Provisional Adjustment and Implementation of the Administrative Regulations in the China (Hainan) Free Trade Pilot Zone | Guo Han [2020] No. 88 | June 18, 2020 | June 28, 2020 |
| 5 | Notice of the General Office of the State Council on Printing and Distributing the Reform Plan for the Division of Central and Local Financial Responsibilities and Expenditure Responsibilities in the Public Cultural Field | Guobanfa [2020] No. 14 | June 04, 2020 | June 23, 2020 |

**Fig. 3.** Government information disclosure page of China Government website.

number of texts. Table 1 shows the data distribution of the top ten texts in the data set. It can be seen from the table that there is a significant gap in the total number of texts in different categories, and the experimental data set must be balanced. In this paper, 3000 texts are extracted from each category in turn, as the experimental data set for that category.

### 4.2 Performance Evaluation Index

In this paper, Precision (P), Recall (R) and F1 values are selected as the evaluation indexes of the model. The calculation formulas for each index are as follows:

$$P = \frac{TP}{TP + FP} \tag{10}$$

$$R = \frac{TP}{TP + FN} \tag{11}$$

$$F1 = \frac{2PR}{P + R} \tag{12}$$

**Table 1.** The data distribution of the top ten in the total number of texts in the dataset.

| Num | Category | Original dataset | Select the amount of data |
|---|---|---|---|
| 1 | Urban and rural construction, environmental protection | 6694 | 3000 |
| 2 | Finance, banking, and auditing | 4009 | 3000 |
| 3 | Culture, radio, television, press and publication | 3332 | 3000 |
| 4 | Labor, personnel, inspection | 3220 | 3000 |
| 5 | Comprehensive government affairs | 3119 | 3000 |
| 6 | Technology, education | 2618 | 3000 |
| 7 | Agriculture, forestry, water conservancy | 2513 | 3000 |
| 8 | Market supervision, production safety supervision | 2371 | 3000 |
| 9 | Industry, transportation | 2203 | 3000 |
| 10 | National economic management, state-owned assets supervision | 1982 | 3000 |
| | Total | 32061 | 30000 |

In the formula, TP (True Positives) indicates the number of samples whose actual category is positive and the predicted category is positive; FP (False Positives) indicates the number of samples whose actual category is negative but the predicted category is positive; FN (False Negatives) indicates the number of samples in which the actual category is positive but the predicted category is negative; TN (True Negatives) indicates the number of samples in which the actual category is negative and the predicted category is also negative.

In general, the higher the precision and recall rate, the better the classification model. However, the precision rate and the recall rate are in a contradictory relationship. If the precision rate of the classification model is increased, the model recall rate may be reduced. The F1 value is the harmonic average of the precision rate and the recall rate, and it can reflect the relationship between the precision rate and the recall rate. The larger the F1 value, the greater the precision and recall rate of the model. Therefore, by analyzing the three evaluation indicators of precision rate, recall rate and F1 value, the larger the evaluation value, the better the performance of the classification model.

## 4.3 Experimental Comparison and Analysis

Based on the construction of the government big data corpus, the government big data corpus was used for the training and testing of six classification models, and finally the evaluation indexes of the six classification models were obtained. Table 2 shows the evaluation index results of six classification model experiments.

**Table 2.** Evaluation index results of six classification model experiments.

| Classification model | Precision | Recall | F1 values |
|---|---|---|---|
| NB | 0.667 | 0.652 | 0.633 |
| KNN | 0.652 | 0.659 | 0.650 |
| SVM | 0.713 | 0.709 | 0.701 |
| TextCNN | 0.739 | 0.740 | 0.737 |
| TextRNN | 0.579 | 0.575 | 0.569 |
| Attention-BiLSTM-CNN | **0.752** | **0.760** | **0.751** |

It can be seen from Table 2 that the Attention-BiLSTM-CNN hybrid classification model proposed in this paper performs best compared to other classification models. The results of the evaluation indicators are: Precision 0.752, Recall 0.760, and F1 value 0.751. All three parameters are above 75%, which basically meets the classification requirements of government big data. Compared with the TextCNN model, the Attention-BiLSTM-CNN hybrid classification model has a higher accuracy rate, indicating that after adding BiLSTM and attention mechanism to the TextCNN model, the contextual relationship of words can be used to effectively extract the features of the sentence, thereby improving the accuracy of classification. The TextRNN model is the worst among the six classification models, and the running time is too long. The reason for the low accuracy of the TextRNN model may be that the number of training samples is too small, and the feature information that can be extracted from the text is relatively limited.

In summary, the Attention-BiLSTM-CNN hybrid classification model proposed in this paper has a good classification effect on the government big data corpus, and the classification accuracy rate is more than 75%. Therefore, it is proved that this model has a better adaptability in the classification of government big data.

## 5  Conclusion

Based on the textual features of government big data, this paper proposes a the Attention-BiLSTM-CNN hybrid classification model. First, use BiLSTM to extract the sentence vector representation of words; then, introduce an attention mechanism to obtain the weight coefficients of the words in the sentence, and update to the sentence vectors extracted by BiLSTM to form a new sentence vector; finally, input the new sentence vector to CNN network. Calculate the probability distribution of the text category. In order to verify the effectiveness of the model in this paper, the corpus of government big data is used for the training and testing of six classification models. Experimental results show that the model in this paper is better than other models in the classification of government big data corpus, and can be applied to the text classification task of government big data.

# References

1. Kim, Y.: Convolutional neural networks for sentence classification. arXiv preprint arXiv:1408. 5882 (2014)
2. Tang, D., Qin, B., Liu, T.: Document modeling with gated recurrent neural network for sentiment classification. In: Proceedings of the 2015 Conference on Empirical Methods in Natural Language Processing, pp. 1422–1432 (2015)
3. Lai, S., Xu, L., Liu, K., et al.: Recurrent convolutional neural networks for text classification. In: Twenty-ninth AAAI Conference on Artificial Intelligence (2015)
4. Yang, Z., Yang, D., Dyer, C., et al.: Hierarchical attention networks for document classification. In: Proceedings of the 2016 Conference of the North American Chapter of the Association for Computational Linguistics: Human Language Technologies, pp. 1480–1489 (2016)
5. Yang, Y., Sun, J., Song, G.: Text sentiment analysis based on feature fusion of gated recurrent unit and capsule [J/OL]. J. Comput. Appl. 1–6 (2020)
6. Zhu, Z.: Design and implementation of government policy text classification system, Southeast University (2018)

# Research of Paper Recommendation System Based on Citation Network Model

Sun Jing and Sun Yu[✉]

School of Information Science, Yunnan Normal University, Kunming 650500,
Yunnan Province, China
sunyu_km@hotmail.com

**Abstract.** In view of the increasing number of existing papers, this paper is a study of paper recommendation system. The data set used in this paper is the DBLP citation network in AMiner. First of all, we build a three layers citation network graph model. In this model, we integrate the citation relationship, paper's feature information, co-authorship relationship and research field information into this model. Secondly, we proposed the algorithm PAFRWR. This algorithm combines three layers citation network graph mode with RWR. And, the search vector is constructed by word2vec model. Finally, in this experiment, using Recall@N and NDCG@N as evaluation metric. Then the restart probability of PAFRWR is determined by experiments. And the most effective search vector is determined by comparison. The Recall@N and NDCG@N of PAFRWR are higher than PageRank, LDA and Link-PLSA-LDA through the experiment. So the recommendation model and algorithm in this paper are more accurate and effective.

**Keywords:** Three layers citation network · RWR · Paper recommendation system

## 1 Introduction

With the increase of scientific researchers, the number of published papers is also increasing. Reading papers is one of the most important and time-consuming parts of scientific research. When researchers want to get the papers about the field for study, the traditional search engine can only search by keywords and phrases, then get the search results. But the results have a wide range and lack of pertinence. Therefore, the paper recommendation system is produced. This paper is the research about the paper recommendation system based on Citation Network.

In the first part, we conduct a certain amount of research on the recommendation system and the recommendation algorithm. Understanding the development status of researches. In the second part, we constructed a three layers citation network model. We integrate the citation relationship, paper's feature information, co-authorship relationship and research field information into this model. And adopts the DBLP-Citation-network data set of AMiner to build the three layers citation network model. In the third part, we put forward the algorithm of a paper recommendation system based on the three layers

© Springer Nature Switzerland AG 2020
X. Chen et al. (Eds.): ML4CS 2020, LNCS 12488, pp. 237–247, 2020.
https://doi.org/10.1007/978-3-030-62463-7_22

citation network model. We combined the three layers citation network model and RWR algorithm to form the paper recommendation algorithm (PAFRWR). Finally the experimental results show that, the evaluation index value of PAFRWR is better than the other three methods.

## 1.1 Research Status of Paper Recommendation System

In 1997, Resnick and Varian gave the definition of recommender system for the first time [1]. Then, Bollacker, Lawrence, Giles build the first paper recommendation system in 1998 [2]. This system uses web search engine and heuristic search to find the papers, and through the reference relationships between papers to find the recommendation of related papers.

In 2012, Wang et al. Proposed a method of paper recommendation based on the historical behavior of users [3]. Choochaiwattana [4] studies the application of tags in paper recommendation and proposes a paper recommendation mechanism based tag. Li Ran et al. [5] solved the cold start problem in the paper recommendation system. According to the preferences for research of users, a collaborative topic regression model based on frequent topic set preferences was proposed.

## 1.2 Research Status of Citation Network

At first, citation network was only used to Library and Information Science [6]. Through the continuous research of citation network, its application scope is more comprehensive. The citation network of the papers is a kind of network information body, which is formed by the reference relationships between the papers [7].

Strohman et al. [8] put forward a global citation recommendation for the first time, and they get the papers of citation recommendation by searching the whole papers. Tang et al. [9] studies a new problem in his paper, that is, citation recommendation based topic, and explores the topic distribution and citation relationship of the papers. In 2011, Shi Jie et al. [10] proposed to form a relationship collection with multiple attribute information in the citations, and then cluster more related citations according to the clustering algorithm, and finally recommend the results to users. Xiao Shibo et al. [11] use the graph model to analyze the relationship between the papers in the citation network and recommend for users. Chen Zhitao et al. [12] proposed a citation recommendation algorithm based on multi feature factor fusion. Based on the traditional citation recommendation model, integrated to the author related factors, overall influence factors and query related factors.

## 2 Related Work

### 2.1 Citation Recommendation

A paper needs a large number of references to support its point of view, the citation recommendation can provide appropriate references for researchers. According to the different citation method, citation recommendation can be divided into local citation recommendation and global citation recommendation. The purpose of local citation recommendation is to recommend relevant papers in the process of writing papers where the citation needs to be added. However the global citation recommendation refers to the recommendation for the whole paper, and provides a reference list for the target paper.

## 2.2  Random Walk with Restart

Random Walk with Restart (RWR) is a model proposed by Grady [13] in 2006. It mainly measures the similarity between network nodes through the relationship between the topology structure. The main ideas of RWR are as follows: (1) Starting from any vertex or any set of vertices in the graph, walk randomly along the edge of the graph to the next vertex. (2) In the random walk process, any vertex randomly selects the next adjacent node with a certain probability to move or selects the starting point to return to for random walk again. (3) After repeating the random walk process for a finite times and iterating for many times, the probability value of vertices in each graph tends to be stable, and the iteration ends. (4) Finally, the probability value of each vertex can be regarded as the similarity between the current vertex and the selected starting vertex. The formula for RWR is as follows:

$$\overrightarrow{r_i} = c\,\tilde{W}\,\overrightarrow{r_i} + (1-c)\overrightarrow{e_i} \tag{1}$$

Here, $\overrightarrow{r_i} = [r_{i,j}]$ is the relevance score, $c$ is the restart probability, $W = [w_{i,j}]$ is the weighted graph adjacency matrix, $\tilde{W}$ is matrix of W by standardizing, $\overrightarrow{e_i}$ is the identity matrix.

# 3  Analysis and Construction of Three-Layer Citation Network Graph Model

## 3.1  Analysis of Three Layers Citation Network Model

This part will analyze and build a three layers citation network model.

1. Citation network of papers
   A paper contains multiple references, and each reference, as an independent paper, also has its own references, which constitutes a citation network. The Fig. 1 is the citation network structure of the papers. As shown in the figure, paper $P_1$ quoted $P_3$, $P_4$ and $P_5$. If a paper $P_i$ quoted a paper $P_j$, there are directed edges to connect them. So $P_iP_j = 1$ when we building the paper citation network.
   In this paper, we get the word vector of feature information of the paper by word2vec. Then we calculate the mean value of the word vector, and the mean reflects the characteristic information of a paper. The formula for calculation is as follow: N is the total vocabulary of title and abstract of paper $i$, $w_j$ is the word vector of a word $j$ in paper $i$.

$$R_i = \frac{1}{N}\sum_{j=1}^{n} w_j \tag{2}$$

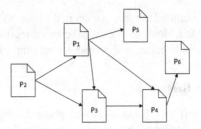

**Fig. 1.** Citation network of the papers

2. Co-authorship network

For an academic paper, it can have many authors, so there is a co-authorship relationship in the paper. The Fig. 2 is the co-authorship network diagram. As shown
in the figure, author $A_1$ and authors $A_2$, $A_4$, $A_5$ write the same paper. If there is a
co-authorship between author $A_i$ and author $Aj$, there is an undirected edge between
them. So $A_iA_j = 1$ when we building the co-authorship network.

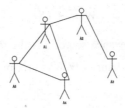

**Fig. 2.** Co-authorship network diagram

3. Relationship between papers and research fields

Research fields can help researchers directly locate the topic and research direction
of the paper when they study the papers. A paper can correspond to one or more
research fields. The Fig. 3 is the diagram of relationship between papers and research
fields. As shown in the figure, if the paper $P_i$ corresponds to a research field $F_j$, there
is an edge to connect them. So $P_iF_j = 1$ when building relationships between papers
and research fields.

4. Three layers citation network model

In this paper, we selects the feature information, the author and the research field
of the paper, and builds a three layers citation network model with the relationship
among them. The Fig. 4 is the three citation network model diagram.

### 3.2  Construction of Three Layers Citation Network Model

Before building the model, explain the symbols used. As follows in this Table 1, it is the
definition of related symbols.

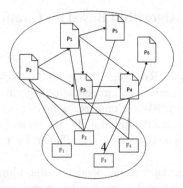

**Fig. 3.** Research field diagram

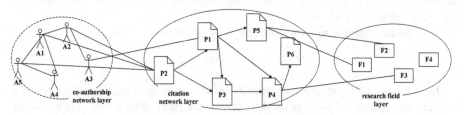

**Fig. 4.** Three layers citation network model

**Table 1.** Definition of related symbols

| Symbol | Symbol definition |
|--------|-------------------|
| A | Author sets |
| P | Paper sets |
| F | Field sets |
| $M_{AA}$ | Co-authorship matrix |
| $M_{PP}$ | Citation matrix |
| $M_{AP}$ | Matrix between papers and authors |
| $M_{PF}$ | Matrix between papers and fields |

When building this model, the matrix is used to express the relationship among the three layers networks. The following matrix is the matrix of three layers citation network graph model.

$$
M = \begin{matrix} & \begin{matrix} P & A & F \end{matrix} \\ \begin{matrix} P \\ A \\ F \end{matrix} & \left| \begin{matrix} M_{PP} & M_{PA} & M_{PF} \\ M_{AP} & M_{AA} & M_{AF} \\ M_{FP} & M_{FA} & M_{FF} \end{matrix} \right| \end{matrix} \tag{3}
$$

# 4  Recommendation Algorithm Based on Citation Network Model

## 4.1  Algorithm Design

In this paper, we combines the RWR algorithm with the proposed three layers citation network graph model, it formed the recommendation algorithm based on Citation Network Model, and we call it PAFRWR. Compared with the random walk algorithm, RWR can make the node walk around the initial node, not aimless walk in the random walk model. So RWR algorithm is more accurate and efficient for the determination of similar nodes.

In order to implement the paper recommendation algorithm based on the three layers citation network model, search and recommendation are combined. In this paper, S is used to represent the set of search vectors. Training the search information by word2vec and structure search vector. We normalize the transition probability matrix for assign a reasonable weight to each node.

The recommended algorithm in this paper is as follows:

$$C^{(t+1)} = (1 - \beta)MC^{(t)} + \beta s \tag{4}$$

Here $C$ is stationary distribution probability of each node. It represent the relevance between user search and nodes in the three layers citation network model. $\beta$ is the restart probability. $M$ is a transition probability matrix. $s$ is the initial search vector.

The pseudo code of the algorithm is shown below:

---

Algorithm: PAFRWR algorithm pseudocode.

---

Input: The matrix of three layer citation network model matrix $M$, Search vector set
      $S$, restart probability $\beta$.
Output: The N recommendation papers
1. Normalization the transition probability matrix
2. Normalization the search vector
3. Initial $C^0 = s$;
4. times=0;
5. d=0;
6. **while** (1):
7.     $C^{(t+1)} = (1 - \beta)MC^{(t)} + \beta s$;
8.     $d = C^{(t+1)} - C^{(t)}$;
9.     **if**  $d <$ Minimum convergence threshold:
10.         **break**
11.     **if**  times $>$ Maximum number of iterations:
12.         **break**
13.     times = times+1;
14. **break**
15. **return** Top n Papers with maximum C value;

---

## 4.2  Experiment and Analysis

1. Data

   In this paper, The data set used in this paper is DBLP-Citation-network downloaded from AMiner. Then, data of 63469 non repetitive papers from 2013 to 2019 are selected, and it include 152586 authors.

2. Evaluation index

   We use Recall and NDCG as evaluation indexes.

   Recall is an important index to evaluate recommendation results. In the field of information retrieval. The following is the calculation formula of recall. Here, N is the total number of recommended results. Q is the total number of search. $R(p)$ is the set of recommended results produced. $T(p)$ is the set of references of the current tested papers. $R(p) \cap T(p)$ is the set of papers recommended correctly.

$$\text{Recall@N} = \frac{1}{Q} \sum_{i=1}^{Q} \frac{R(p) \cap T(p)}{T(p)} \tag{5}$$

   In the paper recommendation system, it is considered that the recommendation results are in the current references of the tested papers, and the higher the recommendation results are in the list, indicating that the performance of the recommendation algorithm is better. NDCG is the value normalized by IDCG.

$$\text{NDCG@N} = \frac{1}{Q} \sum_{j=1}^{Q} \frac{DCG@N}{IDCG@N}$$

$$DCG@N = \sum_{i=1}^{N} \frac{2^{r_i} - 1}{log_2(i+1)} \tag{6}$$

$$IDCG@N = \sum_{i=1}^{Rel} \frac{2^{rel_i} - 1}{log_2(i+1)}$$

3. Restart probability experiment

   In this section, the restart probability parameter $\beta$ is determined by algorithm experiment. The following figure shows the change trend of Recall@N and NDCG@N under different restart probability $\beta$. In this experiment, N is selected as 50, 75 and 100. It can be seen that Recall@N and NDCG@N are the highest when $\beta = 0.3$. And the value of Recall@100 and NDCG@100 are the highest under different . So we choose the $\beta = 0.3$ as the restart probability (Fig. 5).

4. Search vector contrast experiment

   In PAFRWR algorithm, there are author search vectors, paper search vectors and field search vectors. Different search vectors have different effects on the recommendation results of this paper. Four search vectors are defined as follows:

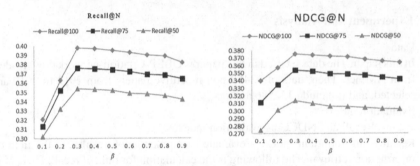

**Fig. 5.** Recall@N and NDCG@N under different β values

(1)  $s_1 = [s_p, s_a, s_f]$ include paper search vectors, author search vectors and field search vectors.

(2)  $s_2 = [0, s_a, s_f]$ include author search vectors and field search vectors.

(3)  $s_3 = [s_p, s_a, 0]$ include paper search vectors, author search vectors.

(4)  $s_4 = [0, s_a, 0]$ include only author search vectors.

The following table shows the results of the search vector comparison experiment (Table 2):

In this experiment, the N is 25, 50, 75, 100. Among them, the Recall@N and NDCG@N of search vector $S_1$ reach the maximum, which containing all three kinds of information. The results of $S_2$ and $S_3$ are very similar, but $S_3$ has higher Recall@N and NDCG@N than $S_2$. This shows that the search vector containing the content information of the paper can give the user better recommendations. So the search vector $S_1$ will be selected as the search vector of the PAFRWR in this paper.

(5)  PAFRWR algorithm comparison

We compare PAFRWR with PageRank, LDA and Link-PLSA-LDA. The following figure is a comparison of four algorithms. The Recall@N and NDCG@N of PAFRWR are higher than PageRank, LDA and Link-PLSA-LDA.

PageRank only considers the citation relationship between papers, but does not consider the author, content subject, research field and other specific information of the paper. Therefore, the Recall@N and NDCG@N of PageRank are the lowest among the four algorithms. LDA and Link-PLSA-LDA both build the theme model of the paper. The Link-PLSA-LDA combines the reference relationship between papers based on the topic model. So the result of Link-PLSA-LDA is slightly higher than that of LDA. Overall, the algorithm of this paper (PAFRWR) has better recommendation results (Fig. 6).

**Table 2.** Recall@N and NDCG@N under different search vectors

|  | Recall@25 | Recall@50 | Recall@75 | Recall@100 | NDCG@25 | NDCG@50 | NDCG@75 | NDCG@100 |
|---|---|---|---|---|---|---|---|---|
| $s_1$ | **0.297** | **0.354** | **0.376** | **0.398** | **0.2875** | **0.3136** | **0.3538** | **0.3753** |
| $s_2$ | 0.252 | 0.325 | 0.349 | 0.387 | 0.2527 | 0.3008 | 0.3425 | 0.3587 |
| $s_3$ | 0.263 | 0.338 | 0.357 | 0.392 | 0.2633 | 0.3027 | 0.3496 | 0.3631 |
| $s_4$ | 0.208 | 0.297 | 0.314 | 0.335 | 0.2118 | 0.2715 | 0.3038 | 0.3223 |

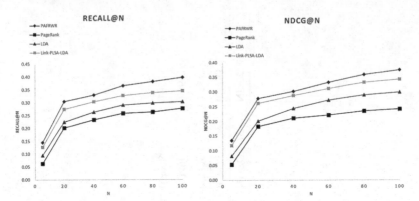

**Fig. 6.** Comparison of Recall@N and NDCG@N of different algorithms

## 5   Conclusions

In this paper, we constructs a three layers citation network graph model based on the DBLP citation network data set of AMiner. Which combines the content information, author information and research field information of the papers. Secondly, we proposed the algorithm PAFRWR. This algorithm combine three layers citation network graph mode with RWR. The restart probability $\beta = 0.3$ is determined by experiments, and the most effective search vector is determined. Finally, the Recall@N and NDCG@N of PAFRWR are higher than PageRank, LDA and Link-PLSA-LDA through the experiment.

For the paper recommendation method, it can also be improved from the following aspects. First, we can refine the structure and content of the network model, for example, you can add the same journal relationship of the papers, the same organization relationship of authors, and the relationship between research fields. Secondly, it can combine the user's historical behavior in the paper recommendation. According to the shortcomings, we can build more effective paper recommendation model and algorithm in the future work.

**Acknowledgments.** This work was supported by the project is the Yunnan Provincial Smart Education Key Laboratory Project,Key Laboratory of Education Informalization for Nationalities of Ministry of Education and the Yunnan University Innovation Research Team Project.

## References

1. Resnick, P., Varian, H.R.: Recommender systems. Commun. ACM **40**(3), 56–58 (1997)
2. Bollacker, K.D., Lawrence, S., Giles, C.L.: CiteSeer: an autonomous Web agent for automatic retrieval and identification of interesting publications. In: Proceedings of the 2nd International Conference on Autonomous Agents, pp. 116–123 (1998)
3. Wang, Y., Liu, J., Dong, X., Liu, T., Huang, Y.: Personalized paper recommendation based on user historical behavior. In: Zhou, M., Zhou, G., Zhao, D., Liu, Q., Zou, L. (eds.) NLPCC 2012. CCIS, vol. 333, pp. 1–12. Springer, Heidelberg (2012). https://doi.org/10.1007/978-3-642-34456-5_1

4. Choochaiwattana, W.: Usage of tagging for research paper recommendation. In: International Conference on Advanced Computer Theory and Engineering (2010)
5. Ran, L., Hong, L.: Academic paper recommendation algorithm based on frequent topic set preference. Appl. Res. Comput. (9) (2019)
6. Haifeng, W., Yiming, S.: On status QUO OF citation network research and the overview on its development. Comput. Appl. Softw. **29**(2), 164–168 (2012)
7. Yaru, D.: Structural modeling of citation network systems. Libr. Inform. Serv. **4**, 58–61 (1996)
8. Strohman, T., Croft, W.B., Jensen, D.: Recommending citations for academic papers. In: International ACM SIGIR Conference on Research and Development in Information Retrieval. ACM (2007)
9. Tang, J., Zhang, J.: A discriminative approach to topic-based citation recommendation. In: Theeramunkong, T., Kijsirikul, B., Cercone, N., Ho, T.-B. (eds.) PAKDD 2009. LNCS (LNAI), vol. 5476, pp. 572–579. Springer, Heidelberg (2009). https://doi.org/10.1007/978-3-642-01307-2_55
10. Jie, S., Derong, S., Tiezheng, N., et al.: A citation recommendation method based on multiple factors. J. Comput. Res. Dev. (s2) (2011)
11. Shibo, X., Sheng., F.: Research on intelligent recommendation algorithm of research papers based on citation graph model. Comput. Knowl. Technol. **15**(03), 196–198 (2019)
12. Zhitao, C., Shuqin, L., Bin, L., et al.: Citation recommendation algorithm based on multi-feature factor fusion. Comput. Eng. Des. **39**(7), 103–111 (2018)
13. Grady, L.: Random walks for image segmentation. Pattern Anal. Mach. Intell. **28**(11), 1768–1783 (2006)

# CellIdNet: Automatic Cell Instance Segmentation and Classification in Bone Marrow Examination with Double Backbone Networks

Shiyin He[1(✉)], Zheng Huang[1,2], Kai Chen[1], Jie Guo[1], and Weidong Qiu[1]

[1] Shanghai Jiao Tong University, Shanghai, China
{serenayin,huang-zheng,kchen,guojie,qiuwd}@sjtu.edu.cn
[2] Westone Cryptologic Research Center, Beijing 100070, China

**Abstract.** Bone marrow examination plays an important role in the diagnosis of a number of diseases, including leukemia, multiple myeloma, etc. However, the examination is slowed down by complicated manual work which requires professional knowledge. In this paper, deep learning methods are applied in bio-medical field to detect and count different types of cells in an automatic way instead of manual work in bone marrow examination. The process is mainly composed of two steps, segmentation and classification. Double backbone networks structure is employed to show better result. A two-step training strategy is also proposed. Mask-Region-Convolutional Neural Networks(Mask-RCNN) was used for detection and image segmentation to extract cells and then common architectures in convolutional neural networks pre-trained on ImageNet, such as Dense Convolutional Network and Deep Residual Network(ResNet), were fully experimented and compared in classification. Result of the first stage shows high efficiency to meet application requirements. For cell classification, InceptionV3 shew the best precision of 93% in five cell lines and ResNet-152 achieved 82% in twelve classes. This structure improves nearly three times of accuracy of a single Mask-RCNN.

**Keywords:** Medical image · Bone marrow examination · Cell detection and segmentation · Cell classification · Deep learning · Mask-RCNN

## 1 Introduction

Since deep learning has brought great productivity improvement, application in biomedical field has arisen great interest. One applicable problem is bone marrow examination in which cells that are differentiated from hematopoiesis

Supported by The National Key Research and Development Program of China under grant 2017YFB0802704 and 2017YFB0802202.

stem cells are counted in up to 30 types and stages. Numbers or proportions of different cells are then used in the diagnosis of a number of diseases, including leukemia, multiple myeloma, and anemia [15]. However, the examination is inefficient because pathologists have to manually identify at least 200 cells per patient under a light microscope. In many cases, adjacent stages may cause confusion even to pathologist. Therefore, applying automatic recognition to manual work is in great need. Thus, CellIdNet is developed in this paper to recognize different types of cells required in examination to help with blood components calculation and analysis.

Common algorithms for cells segmentation make use of color and contour features, known as watershed algorithm [11], circle detection [13], et al. Threshold is always selected as a filter, as in [6,12], which is the weak point to practical application. The Hough Transform was used in [13] to detect haematopoietic cells based on their round shape. Machine learning algorithms are also applied to learn about image features. [2] investigates the use of support vector machine classification based on Laplace edge features. In [14], Iterated Conditional Model was trained with contextual color and texture information to classify pixels. However, this kind of methods depend a lot on feature selection and image process. Sometimes, only one method is not enough to deal with complex bone marrow smear images, so [9] adopted all of above to segment nucleus and cytoplasm. Later, many variants of FCN, like [1], produce neuron segmentation with improvement. There are other nucleus center location research like [20], using multi-label prediction by CNN to locate center of nucleus in compress sensing form. These previous studies did achieve excellent segmentation quality of more than 90%, but they required complex pre-processing, careful feature selection and even multiple combination of algorithms. When illumination and white-balance settings are changed by different operators in reality, outcomes of these studies would be strongly affected, so they are hard to re-implement. However, deep learning algorithm makes easier segmentation possible with excellent performance and no one yet implemented that according to our knowledge.

Traditional method for classification is to extract features first and then apply machine learning algorithm to classify based on feature vectors. For example, [10] utilized Local Binary Patterns and Histogram of Oriented Gradients to generate features and a support vector machine to classify cellular signaling status. The traditional approach is effective but feature selection makes it unsensitive to other similar datasets. Later, Convolutional Neural Networks (CNNs) is introduced to recognition, as was used in [8] and [19]. However, both of them implemented detection based on traditional morphological process, along with low accuracy under 90% and no more than 6 classes. Recently, [21] studied deep neural network over 11 categories but results was far from satisfactory with mean average precision of only 0.659 and accuracy of 0.801 for cell classification. Therefore, further study on the performance of different neural network structures are needed, as well as a combined study of both cell instance segmentation and classification in bone marrow cells.

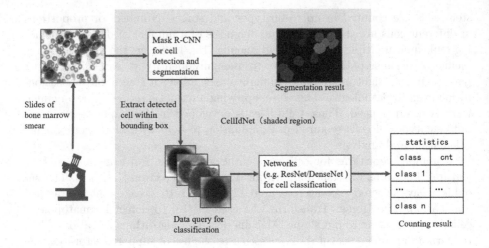

**Fig. 1.** Overall process of bone marrow cells detection and classification in a two-stage manner with double backbones

In this paper, a CellIdNet is proposed and tested as a two-stage work, similar to [6]. Bone marrow smear images are first sent into a Mask-Region Convolutional Neural Networks(Mask R-CNN) [3] for detection and segmentation to extract each cell. It shows an excellent performance with a mean average precision of 96.2%. And then each extracted cell is identified in five cell lines and twelve classes through various architectures, like Dense Convolutional Network(DenseNet) or Deep Residual Network(ResNet), achieving accuracy of up to 93% and 82% respectively. These networks form an integrated system that raises efficiency. With classification result, we can count each type and get bone marrow examination report. The main work is summarized in Figs. 1, 2.

The main contributions of this paper are summarized as follows:

1. A novel system with only deep learning methods is addressed for cell detection, segmentation and classification.
2. Try to recognize cells in bone marrow with all cell systems together.
3. Double backbone structure is employed to overcome low accuracy.
4. Deep learning models are fully compared and selected as the solution. The performance is also evaluated in different situation.

## 2    Methodology

Our network consists of two main stages. First is to detect and segment cells in a image with Mask R-CNN. Second, each detected cell is sampled from input image with bounding box and sent into cell classification networks, including all networks mentioned in the last subsection for comparison. The procedure is shown in Fig. 1 and a more detailed illustration of our network is shown in Fig. 2.

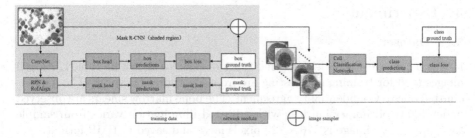

**Fig. 2.** Detailed illustration of CellIdNet

Images, along with their ground truth masks, are sent into Mask R-CNN model of training mode. As can be seen in the Fig. 2, we focus on bounding box and mask brunches of Mask R-CNN, so the class brunch is removed and it is achieved by the other backbone and head instead.

During cell classification, cell images sampled from input images of Mask R-CNN are sent into classification models, including AlexNet [7], PolyNet [22], VGGNets [16], ResNets [4], DenseNets [5] and Inception modules [17,18]. Transfer learning and fine-tunig were used during training. Weights are initialized with pre-trained models on ImageNet dataset [7]. In this way, the total loss remains the same form as defined as:

$$L = L_b + L_m + L_c \tag{1}$$

where $L_b$, $L_m$ and $L_c$ represent box loss, mask loss and class loss respectively. However, the definition of $L_c$ itself can vary based on the backbone we choose during experiment. Moreover, $L_b$ and $L_m$ do not exist at the same time with $L_c$ for the training strategy we employed.

### 2.1 Training

A complete Mask R-CNN is first trained in an end-to-end joint training manner to see how a single backbone works. For our CellIdNet, we explore a stage-wise training procedure. To cover the shortage of single backbone, a cell classification network is extended as described above. From Fig. 2 and loss definition, we assume that the two stages of CellIdNet do not affect each other. Therefore, we separate the procedure into cell detection and segmentation training (first stage) and cell classification training (second stage). In the first stage, Mask R-CNN is trained with bounding box annotations and segmentation masks in an all-in-one class. In the second stage, the extended cell classification networks and class head are trained while keeping the mask head and the bounding box head fixed. This two-step training strategy allows us to put more emphasis on the defective part–the classification brunch, and make more efficient improvement.

# 3   Experiment

## 3.1   Dataset

In this paper, we study the following types of cells, as shown in Table 1. All cell images used for training and testing are from a corporate hospital. The name is omitted due to Confidentiality agreement. The bone marrow smears are observed under 1000x oil immersion lens with adjusted illumination varies from sample to sample. Each image is $576 \times 720$ pixels large and saved in BMP format.

Data-set for cell detection and segmentation has 206 images with 148 pieces for training, 45 for validation and 13 for test. Though the amount looks small, they actually have 1800, 492 and 156 cells respectively. Figure 3(a) is an example of input data for Mask R-CNN with bounding boxes and segmentation masks. Cell classes are labeled in detailed classification, but in our double backbone structure, Mask-RCNN does not care about classes, so their classes are all labeled as "al" as placed on the top-left corner of dotted boxes in Fig. 3(c) . Nevertheless, detailed labels are remained as they will be used in second stage training.

In the data-set for cell classification, training set, validation set and test set were randomly selected under the proportion of 8:1:1.

Image augmentation is also adopted, including rotation, zooming, flipping and etc., to enlarge data-set.

**Table 1.** Types of cells studied in this paper and data-set for cell classification

| Five cell lines | Twelve classes | Trn set | Val set | Test set |
|---|---|---|---|---|
| Normoblast | Normoblast | 1020 | 135 | 128 |
| Plasma cell | Plasma cell | 170 | 17 | 18 |
| Monocyte | Monocyte | 209 | 23 | 22 |
| Lymphocyte | Lymphocyte | 260 | 32 | 36 |
| Granulocyte | Myeloblast | 230 | 29 | 32 |
| | Promyelocyte | 143 | 25 | 17 |
| | Myelocyte | 414 | 51 | 43 |
| | Metamyelocyte | 128 | 17 | 19 |
| | Band cell | 224 | 30 | 34 |
| | Segmented cell | 239 | 23 | 27 |
| | Platelet | 20 | 5 | 4 |
| | Smear cell | 309 | 33 | 41 |

## 3.2   Experiment

To understand the performance, standard detection metrics are used to evaluate Mask R-CNN, like average precision(AP). Mean AP(mAP) and mean Inter-section over Union(mIoU) are also adopted for Mask R-CNN in double back-bone structure, along with AP, the mean of average precision from 0.5 to 0.95

IoU threshold of all classes, $AP_{50}$ and $AP_{75}$(AP evaluated at 0.5 and 0.75 IoU threshold respectively) and AP over small($AP_s$), medium($AP_m$), and large($AP_l$) objects. Presicion, Recall and weighted F-1 score are used for cell classification stage.

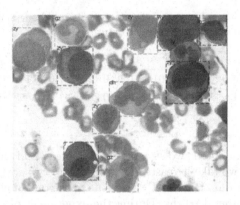

(a) ground truth mask with twelve classes labeled

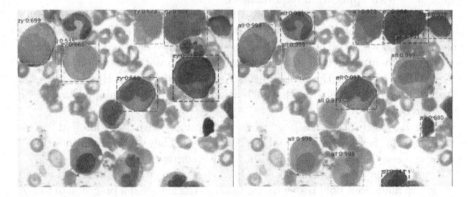

(b) prediction with a single Mask R-CNN under min detection confidence of 0.5

(c) prediction by Mask R-CNN in double backbone structure

**Fig. 3.** Mask ground truths and predictions from Mask-RCNN of single backbone and double backbone structure

**Single Backbone with Mask R-CNN.** We first conduct the experiment of Mask R-CNN only which should have meet the need. It can be seen from Fig. 3(b) that the coverage of mask is extremely low. Though we set a low detection confidence, it still shows no more than half of the objects detected. Moreover, it is also clear in Table 2 that single backbone achieved extremely low precision. As a comparison, first stage of CellIdNet is evaluated, corresponding to Fig. 3(c).

However, their annotation classes are different, which are twelve and one respectively, so conclusion about classification ability should wait for the second stage. But the high AP of first stage still guarantees us a better performance with double back bone, as long as cell classification achieves high accuracy.

**Table 2.** Comparison of end-to-end training of single backbone and two-step training of Mask R-CNN based double backbone structure

| Method | AP | $AP_{50}$ | $AP_{75}$ | $AP_s$ | $AP_m$ | $AP_l$ |
|---|---|---|---|---|---|---|
| Single backbone | 0.101 | 0.160 | 0.128 | −1.0 | 0.141 | 0.071 |
| Double backbone | 0.631 | 0.940 | 0.786 | −1.0 | 0.559 | 0.711 |

**CellIdNet.** As we adopted two-step training strategy and the function of two stages has no cross, we take it reasonable to evaluate two stage seperately.

As there was only one class, mAP represents to the rate of number of detected cells to number of ground truths. During the experiment of cell detection and segmentation, a mAP of 96.2% and a mIoU overlap rate of 84.7% were obtained, which meet with application requirements.

The performance is vividly revealed in Fig. 3(c). To compare the network's performance with input ground truth, output in Fig. 3(c) is also generated by image in Fig. 3(a). A cell on the edge of the image is also detected and segmented out and segmentation is nearly perfect. Numbers on the top-right corner of dotted boxes represent the probability score of cell detection. High scores even 1.0 are achieved. It is believed that increasing the number of training epochs can make further improvement.

For cell classification, each network was trained with two different classification methods–five systems and twelve classes. They were all trained in 50 epoches to see their learning ability and better compare the result.

Result can be seen in Table 3. The best performances on validation set and test set are made font bold respectively. ResNet-152 achieved a precision of 91.5% and InceptionV3 achieved 93.19% on test set and validation set in the classification of five major cell systems; when it comes to twelve classes, ResNet-152 still obtained the highest 81.9% on validation set and DenseNet-169 obtained even higher recall of 81.95% on test set. The performances are weakened with more complicated classification problem. However, this result first shows prospective improvement and possibility for classification among different cell systems all together. It is sure that cell classification can achieve equal or better precision in twelve classes than five with further studies.

In classification of five cell lines, variance of results is smaller with 15 out of 17 obtaining over 90% on validation set and half of them over 90% on test set. It is reasonable for that task was not so difficult and networks couldn't show big difference. When it came to twelve classes, improvement in architecture shew difference in results. Networks with shortcuts connections of layers, ResNets and

**Table 3.** Experiment result of cell classification

| Network | Evaluation set | Five cell lines (%) | | | Twelve classes (%) | | |
|---|---|---|---|---|---|---|---|
| | | Precision | Recall | F1-score | Precision | Recall | F1-score |
| AlexNet | Validation set | 0.36 | 6.02 | 0.68 | 0.30 | 5.48 | 0.57 |
| | Test set | 0.34 | 5.85 | 0.65 | 0.27 | 5.23 | 0.52 |
| PolyNet | Validation set | 91.24 | 91.10 | 91.11 | 77.66 | 78.10 | 77.67 |
| | Test set | 87.68 | 87.50 | 87.58 | 76.89 | 77.20 | 76.48 |
| VGG11_bn | Validation set | 91.70 | 91.88 | 91.77 | 77.35 | 77.38 | 76.95 |
| | Test set | 90.92 | 91.22 | 90.83 | 77.58 | 77.67 | 77.11 |
| VGG13_bn | Validation set | 92.38 | 92.41 | 92.38 | 80.83 | 80.95 | 80.42 |
| | Test set | 91.07 | 91.22 | 91.10 | 78.21 | 78.38 | 77.84 |
| VGG16_bn | Validation set | 92.20 | 92.41 | 92.27 | 79.92 | 79.05 | 79.11 |
| | Test set | 91.36 | 91.76 | 91.39 | 77.19 | 76.72 | 76.56 |
| VGG19_bn | Validation set | 92.10 | 92.15 | 92.10 | 79.15 | 78.57 | 78.36 |
| | Test set | 91.51 | 91.49 | 91.46 | 77.83 | 78.15 | 77.49 |
| ResNet18 | Validation set | 92.90 | 92.93 | 92.90 | 80.92 | 80.71 | 80.60 |
| | Test set | 91.12 | 91.22 | 91.14 | 77.12 | 76.96 | 76.81 |
| ResNet34 | Validation set | 92.52 | 92.41 | 92.45 | 80.91 | 80.71 | 80.47 |
| | Test set | 89.67 | 89.89 | 89.70 | 79.58 | 79.81 | 79.46 |
| ResNet50 | Validation set | 92.75 | 92.67 | 92.70 | 81.52 | 81.19 | 81.19 |
| | Test set | 89.80 | 89.63 | 89.69 | 79.14 | 79.33 | 78.80 |
| ResNet101 | Validation set | 93.08 | 92.93 | 92.95 | 80.14 | 80.71 | 80.08 |
| | Test set | 90.84 | 90.96 | 90.88 | 77.34 | 77.43 | 77.20 |
| ResNet152 | Validation set | 92.90 | 92.93 | 92.90 | **82.21** | **81.90** | **81.48** |
| | Test set | **91.50** | **91.76** | **91.55** | 79.79 | 78.86 | 77.88 |
| DenseNet121 | Validation set | 92.07 | 91.88 | 91.95 | 79.41 | 79.29 | 79.05 |
| | Test set | 89.10 | 88.56 | 88.78 | 77.54 | 77.43 | 77.23 |
| DenseNet161 | Validation set | 93.04 | 92.41 | 92.64 | 78.54 | 78.57 | 78.13 |
| | Test set | 90.56 | 90.16 | 90.24 | 80.51 | 80.29 | 79.77 |
| DenseNet169 | Validation set | 91.90 | 91.88 | 91.89 | 80.75 | 80.71 | 80.45 |
| | Test set | 88.76 | 88.83 | 88.79 | **81.98** | **81.95** | **81.39** |
| DenseNet201 | Validation set | 92.12 | 92.41 | 92.21 | 80.52 | 79.76 | 79.61 |
| | Test set | 89.25 | 89.63 | 89.34 | 78.04 | 78.15 | 77.81 |
| InceptionV3 | Validation set | **93.15** | **93.19** | **93.16** | 79.80 | 79.76 | 79.42 |
| | Test set | 90.14 | 89.89 | 89.79 | 78.31 | 78.86 | 78.28 |

DenseNets, performed better. Among them, deeper networks performed better. Also, ResNet-152 achieved the top in both classification task.

There are also others worth our attention. The past pattern shows that deeper the network, better it performs. However, in our case, shallow architecture can achieve even better than deep ones. For example, ResNet-18 achieved the same as ResNet-101 and ResNet-152 and was better than the most of the others, like

DenseNet-201. There are many same examples in Table 3. Surprisingly, AlexNet failed to learn anything. The same happened to VGGNets before batch normalization(bn). This apparently shows that deeper networks are not always better because backpropagation can be weakened and output of each layer can have big variance which makes it can't get to train. In our case, bn can solve the problem, as well as shortcuts in ResNets and DenseNets.

To understand the precision of CellIdNet, we have to combine two stages together. Learn from the training data flow in Fig. 2, we can see that the input of second stage depends on the detection result from first stage. Therefore, to get the final AP, we can simply multiply precision of two stages as an approximation, which can be defined as:

$$AP_{CellIdNet} = mAP \times Precision \tag{2}$$

where mAP and Precision represent to precision of first and second stage respectively.

In this way, we can get the highest performance of CellIdNet is 0.90 in five-class recognition and 0.79 in twelve-class recognition, which is three times higher than a single backbone. Our method significantly outperforms the base model.

## 4 Conclusion

This paper studied cell segmentation and classification system under high magnification cell images. According to the experiment and result, the following conclusions are obtained:

1. Deep learning method is competent to segmentation tasks in medical images, as it is in this study, and it has potential application value.
2. The double backbone structure achieves great effect, which outperforms a single Mask R-CNN by three times.
3. ResNet-152 is the most suitable network in our study.
4. In general, deeper networks perform better in classification, especially when the task is more challenging.
5. Twelve categories classification implies perspective possibility that recognition of bone marrow cells can be carried out among all types with one stage instead of seperating different cell systems apart.

The current results of this project have a good effect, but there are still room for improvement. In this study, we assume that stage-wise training strategy is suitable, but further study is worthy to be done in end-to-end joint training. Analysis of myeloid fluids, normoblasts and other cell lines also have more detailed classification requirements. Therefore, it is also necessary to expand the classification categories and network retraining in the later period.

# References

1. Chen, J., Banerjee, S., Grama, A., Scheirer, W.J., Chen, D.Z.: Neuron segmentation using deep complete bipartite networks. In: Descoteaux, M., Maier-Hein, L., Franz, A., Jannin, P., Collins, D.L., Duchesne, S. (eds.) MICCAI 2017. LNCS, vol. 10434, pp. 21–29. Springer, Cham (2017). https://doi.org/10.1007/978-3-319-66185-8_3
2. Han, J.W., Breckon, T.P., Randell, D.A., Landini, G.: The application of support vector machine classification to detect cell nuclei for automated microscopy. Mach. Vis. Appl. **23**(1), 15–24 (2012)
3. He, K., Gkioxari, G., Dollár, P., Girshick, R.: Mask R-CNN. In: Proceedings of the IEEE International Conference on Computer Vision, pp. 2961–2969 (2017)
4. He, K., Zhang, X., Ren, S., Sun, J.: Deep residual learning for image recognition. Corr abs/1512.03385 (2015)
5. Huang, G., Liu, Z., Van Der Maaten, L., Weinberger, K.Q.: Densely connected convolutional networks. In: Proceedings of the IEEE Conference on Computer Vision and Pattern Recognition, pp. 4700–4708 (2017)
6. Jia, H.: Classification and counting of white blood cell based on deep learning. Master's thesis, Shenzhen University (2017)
7. Krizhevsky, A., Sutskever, I., Hinton, G.E.: ImageNet classification with deep convolutional neural networks. In: Advances in Neural Information Processing Systems, pp. 1097–1105 (2012)
8. Lee, G.G.C., Haung, K.W., Sun, C.K., Liao, Y.H.: Stem cell detection based on convolutional neural network via third harmonic generation microscopy images. In: 2017 International Conference on Orange Technologies (ICOT), pp. 45–48. IEEE (2017)
9. Liu, H., Cao, H., Song, E.: Bone marrow cells detection: a technique for the microscopic image analysis. J. Med. Syst. **43**(4), 82 (2019)
10. Lohrer, M.F., et al.: Applying pattern recognition to high-resolution images to determine cellular signaling status. IEEE Trans. Nanobiosci. **16**(6), 438–446 (2017)
11. Ng, H., Ong, S., Foong, K., Goh, P., Nowinski, W.: Medical image segmentation using K-means clustering and improved watershed algorithm. In: 2006 IEEE Southwest Symposium on Image Analysis and Interpretation, pp. 61–65. IEEE (2006)
12. Prinyakupt, J., Pluempitiwiriyawej, C.: Segmentation of white blood cells and comparison of cell morphology by linear and naïve bayes classifiers. Biomed. Eng. Online **14**(1), 63 (2015)
13. Ramesh, N., Salama, M.E., Tasdizen, T.: Segmentation of Haematopoeitic cells in bone marrow using circle detection and splitting techniques. In: 2012 9th IEEE International Symposium on Biomedical Imaging (ISBI), pp. 206–209. IEEE (2012)
14. Reta, C., et al.: Segmentation and classification of bone marrow cells images using contextual information for medical diagnosis of acute leukemias. PLoS ONE **10**(6), e0130805 (2015)
15. Shujuan, W., Jianzhong, W., Zhenru, W.: Atlas of Current Hemocytology. People's Medical Publishing House, Beijing (2001)
16. Simonyan, K., Zisserman, A.: Very deep convolutional networks for large-scale image recognition. arXiv preprint arXiv:1409.1556 (2014)
17. Szegedy, C., et al.: Going deeper with convolutions. In: Proceedings of the IEEE Conference on Computer Vision and Pattern Recognition, pp. 1–9 (2015)
18. Szegedy, C., Vanhoucke, V., Ioffe, S., Shlens, J., Wojna, Z.: Rethinking the inception architecture for computer vision. In: Proceedings of the IEEE Conference on Computer Vision and Pattern Recognition, pp. 2818–2826 (2016)

19. Witmer, A., Bhanu, B.: Multi-label classification of stem cell microscopy images using deep learning. In: 2018 24th International Conference on Pattern Recognition (ICPR), pp. 1408–1413. IEEE (2018)
20. Xue, Y., Ray, N.: Cell detection with deep convolutional neural network and compressed sensing. arXiv preprint arXiv:1708.03307 (2017)
21. Yu, T.C., et al.: Automatic bone marrow cell identification and classification by deep neural network (2019)
22. Zhang, X., Li, Z., Change Loy, C., Lin, D.: PolyNet: a pursuit of structural diversity in very deep networks. In: Proceedings of the IEEE Conference on Computer Vision and Pattern Recognition, pp. 718–726 (2017)

# Mathematical Subject Information Entity Recognition Method Based on BiLSTM-CRF

Haoze Li[1,2], Tianwei Xu[2,3(✉)], and Juxiang Zhou[2,3]

[1] School of Information Science and Technology, Yunnan Normal University, Kunming, China
[2] Key Laboratory of Education Informalization for Nationalities,
Yunnan Normal University, Kunming, China
xutianwei@ynnu.edu.cn
[3] Yunnan Key Laboratory of Smart Education, Yunnan Normal University,
Kunming 650500, China

**Abstract.** Combining language conditional random field (CRF) and bidirectional long-term and short-term memory (BiLSTM) networks, a mathematical subject information entity recognition method based on BiLSTM-CRF is constructed to extract entity information in mathematical language. Experimental results show that compared with BiLSTM, BiLSTM-CRF improves the recall rate by nearly 5%, the accuracy rate by nearly 2%, and the F1 value by nearly 4%. The results of the BERT-CRF model are also significantly better than other models.

**Keywords:** Mathematical subject · Entity recognition · Deep learning

## 1 Introduction

Natural language processing is an interdisciplinary subject that combines computer science, artificial intelligence, and linguistics. In natural language processing, deep neural network methods are used in tasks such as named entity recognition [1], text classification [2], information extraction [3], machine translation [4], sentiment analysis [5], and question answering system [6]. Very good results have been achieved.

Mathematics is a very important subject in subject education. Mathematical language contains many theorems, conclusions and methods. It hides important information, but it is often overlooked, which makes our study of theoretical concepts not deep enough or the understanding of the topic deviate. For example, the classic zero point theorem:$y = f(x)$ is a continuous function on the interval a, b and f(a) and f(b) are different signs, then there is at least one f(x) makes $f(c) = 0$ hold. Continuous functions, different signs, zero points and other information can make us quickly find a breakthrough in solving problems when doing problems about this theorem, and then complete the problem. Therefore, in the field of mathematics, the recognition of mathematical information entities has very important research value and significance.

The name entity of the mathematics subject makes its name abstract and logically rigorous due to the characteristics of its own subject, which also makes the name recognition of mathematical information unique and complex. Compared with traditional

X. Chen et al. (Eds.): ML4CS 2020, LNCS 12488, pp. 259–268, 2020.
https://doi.org/10.1007/978-3-030-62463-7_24

recognition methods, deep neural networks are driven by data, which can automatically extract effective feature parts from them, and have obvious advantages when applied to unstructured, variable and unknown data. This paper presents a method of named entity recognition based on deep neural networks to identify and extract logical concepts such as objectively existing entities in mathematics and methods and theorems in mathematical information.

## 2  Ralated Work

### 2.1  CRF Model

The CRF model is a conditional random field model proposed by Lafferty et al. [7] to solve sequence labeling problems. Conditional random fields are widely used in natural language processing, especially in named entity recognition tasks. Results. In general, the conditional random field model can solve the long-term dependence problem between sequences when dealing with sequence problems and can fully learn the context information in the text. This probabilistic graphical model can solve the problem of labeling bias. Effect, so the current natural language processing field generally uses linear chain conditional random field model to solve the sequence labeling problem.

The principle of linear chain conditional random field application in named entity recognition task is shown in Fig. 1.

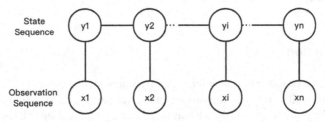

**Fig. 1.**  Linear chain conditional random field structure

It can be seen from the figure that the observation sequence $x = (x_1, x_2, \cdots x_n)$ and the state sequence $y = (y_1, y_2, \cdots, y_n)$, let $P(y|x)$ be a linear chain conditional random field, then $P(y|x)$ is a linear chain conditional random field, then the form of $(y|x)$ is defined as:

$$P(y|x) = \frac{1}{Z(x)} exp\left\{\sum_{i,k} \lambda_k f_k(y_{i-1}, y_i, x, i) + \sum_{i,l} \mu_j h_j(y_i, x, i)\right\} \qquad (1)$$

among them,

$$z(x) = exp\left\{\sum_{i,k} \lambda_k f_k(y_{i-1}, y_i, x, i) + \sum_{i,l} \mu_j h_j(y_i, x, i)\right\} \qquad (2)$$

where $\lambda_k$ and $\mu_j$ are corresponding weights, $f_k$ and $h_j$ are characteristic functions, and $z(x)$ is a normalization function.

Named entity recognition is actually to treat a sentence as a sequence to be observed, take each word in the sentence as a symbol, and set a state for each symbol, and finally maximize the $\lambda_k$ and $\mu_j$ parameters through the training set training to find To meet the conditional probability to complete the sequence labeling.

Given a sequence of input states, the optimal state sequence can be obtained as shown in formula (3):

$$y^* = \arg maxP(y|x) \tag{3}$$

## 2.2  LSTM Model

Long short-term memory network (LSTM) [8] is to solve the recurrent neural network (RNN) [9] In the actual training process, due to the problem of gradient disappearance, it is often impossible to use information that is too far away. Generally speaking, the memory ability of RNN after layer 7 tends to zero, and the long-term and short-term memory network emerged to solve this problem [10]. Hochreiter introduces long- and short-term memory cells. The main idea is to store information in a memory cell. Update, attenuation, input and output in the memory cell will be controlled by multiple gates. Parameters to decide whether to save or forget information in the memory unit. The LSTM cell diagram is shown in Fig. 2.

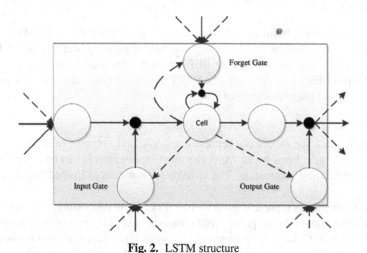

**Fig. 2.** LSTM structure

Let the input of LSTM at time t be $i_t$, the hidden layer and memory unit at time $t-1$ are $h_{t-1}$ and $c_{t-1}$, respectively, and output the hidden layer $h_t$ and memory unit $c_t$ at time t.

- Calculation gate information

$$\text{Input gate: } i_t = \sigma(w_{xi}x_t + w_{hi}h_{t-1} + w_{ci}c_{t-1} + b_i) \tag{4}$$

$$\text{Forgotten door}: f_t = \sigma(w_{xf}x_t + w_{hf}h_{t-1} + w_{cf}c_{t-1} + b_f) \tag{5}$$

$$\text{Output gate}: o_t = \sigma(w_{xo}x_t + w_{ho}h_{t-1} + w_{co}c_t + b_o) \tag{6}$$

- Calculate the value of the memory unit

$$c_t = f_t c_{t-1} + i_t \tanh(w_{xc}x_t + w_{hc}h_{t-1} + b_c) \tag{7}$$

- Calculate the value of the hidden layer at time $t$.

$$h_t = o_t \tanh(c_t) \tag{8}$$

where $w$ and $b$ both represent parameters, and $\sigma$ generally takes the sigmod function [11].

It can be seen from Fig. 2 and the above calculation formula that the input cell and the value of the memory cell without the gate are multiplied to input the input information to the memory cell. Forgetting the gate and multiplying the value at time $t - 1$ will get the attenuation of the memory unit. The output gate and the memory unit at time t are multiplied to output the information in the memory unit to the hidden layer, which affects the output of each gate at time $t + 1$.

## 2.3   BERT Model

BERT is a multi-layer bidirectional Transformer encoder based on fine-tuning. This model training requires massive data and powerful computing power to achieve. Google has open sourced two versions of the BERT model. This article uses Google to train Chinese corpus BERT Base version. The BERT model has two main tasks, namely input representation and pre-training tasks.

### Input Indication

For different tasks, the model can represent a single text sentence or a pair of texts in a sequence of words. For a given word, the input representation can be composed by a three-part Embedding summation. The visual representation of Embedding is shown in Fig. 3.

Token Embeddings represent word vectors. In Chinese processing, they can be word vectors or word vectors. In this paper, word vectors are used in experiments, because word vectors conform to Chinese characteristics. Segment Embeddings is used to distinguish between two sentences when doing a classification task that takes two sentences as input. Position Embeddings is the position information obtained through model learning.

### Pre-training Tasks

The BERT model uses two new unsupervised prediction tasks to preprocess BERT, which are Masked LM and next sentence prediction. The goal of pre-training is to build a language model. The BERT model uses a two-way Transformer. In order to train the deep two-way Transformer representation, a simple method is adopted: randomly cover some input words, and then predict those masked words. During the training process, the

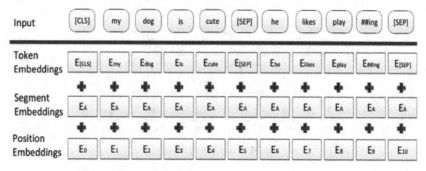

**Fig. 3.** BERT model input representation

original vocabulary of the word is randomly covered. It is different from the left-to-right language model pre-training. Masked LM randomly hides some words from the input. Its goal is to predict the original vocabulary of the masked word based on the context. It is different from the left-to-right language model pre-training. The representation learned by Masked LM can merge the left and right contexts. The bidirectional Transformer in the model does not know which words it will be required to predict, or which have been replaced by random words, so it must maintain a distributed contextual representation of each input word. In addition, since random replacement occurs only 1.5% of all words, it will not affect the model's understanding of the language.

## 2.4 BiLSTM-CRF Model

It can be seen that the long-short-term memory network basically solves the problem of the disappearance of gradients in the recurrent neural network, but when it is actually applied to natural language processing tasks, it will still be found that the long-short-term memory network model can only use the historical information of the foregoing, without considering The following text is also very important for the impact of the previous text, and for sequence labeling tasks, if the context information is not fully utilized, the prediction results will also have an impact. In response to this problem, scholars have proposed a bidirectional long-short-term memory network (Bidirectional LSTM, Bi LSTM) model. BiLSTM makes it possible to use contextual information at the same time, that is, information of the entire sequence.

The structure of BiLSTM is shown in Fig. 4. BiLSTM is composed of a forward LSTM network and a reverse LSTM network. Calculating the input sequence in two directions can make full use of the context information of the input sequence. The results of the calculation are simultaneously passed to the output layer for output.

## 3  The Proposed Model

In mathematics, the description of logical concepts such as methods and theorems usually has obvious hints, such as "XXXX method" and "XXXX theorem". Therefore, when judging whether a text sequence contains annotated entities, the start word can play

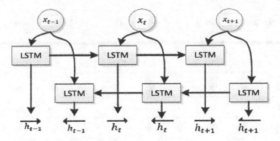

**Fig. 4** BiLSTM structure

a very important role and can obtain the strong dependency relationship between the before and after text. The BiLSTM model can not only capture the dynamic information of the time series, but also use the context information of the current word, and finally obtain a better dependency relationship.

Combining BiLSTM with CRF, so that BiLSTM can be used to extract the context information in the text sequence, and the accuracy of annotation can also be improved by CRF annotation information at the entire sentence level. The structure of the BiLSTM-CRF model is shown in Fig. 5.

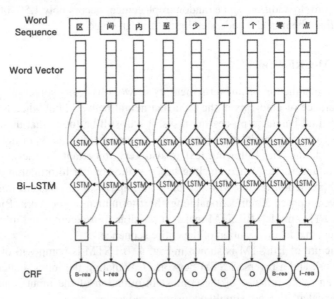

**Fig. 5.** BiLSTM-CRF model structure

The structure of BiLSTM-CRF model mainly includes word vector layer, BiLSTM network layer, and neural network language model CRF layer. The input of the model is sequence text, input according to each character, and the output is the label of each character, which represents whether it is part of the required entity. Each character in the input sequence is input into BiLSTM once through the word vector expression, and then the bidirectional expression of the text sequence containing the context information is established through the BiLSTM network. After the bidirectional expression of the BiLSTM neural network is obtained, it is merged and the combined expression After a layer of implicit exposure, input it into the CRF, and then calculate the label of each character in the sequence text through the CRF, compare it with the label to obtain the log likelihood of the input sequence, and then define it as the loss of the overall model. And in order to prevent overfitting, this experiment added a dropout layer to the model.

The important and complex information in mathematical information is mainly a large number of logical concepts such as the entities that actually exist in the subject and theorems and methods. In this paper, the BiLSTM-CRF model is applied to the information extraction of mathematics, the process is shown in Fig. 6. First, crawl relevant information from websites such as People's Education Textbooks and Baidu Encyclopedia, normalize each text and remove abnormal characters. Then, the entire text is cut through punctuation and paragraph control to form many small texts. The BiLSTM-CRF model is used to calculate the label sequence corresponding to each small text sentence. Finally, the final entity is found by combining the far text according to the label sequence. The algorithm of using BiLSTM-CRF model to extract information in mathematics is shown in Algorithm 1.

**Fig. 6.** Mathematics subject information extraction process

---

**Algorithm 1 : Mathematics subject information extraction algorithm**
Input: Enter the content of each mathematics document source
Output: Objectively existing set of reality, set of logical concepts in mathematical discipline
1. reality = {}, theoretical = {}; / * initialize the collection * /
2. source = normalize (source); / * Normalize mathematics discipline files * /
3. clips = token (source); / * Cut punctuation marks and paragraphs of the processed mathematics files to form text fragments and save them in the list clips * /
4. for clip in clips do; / * For each segment in clips * /
5. temp_reality, temp_theoretical = get_entity (clip); / * Enter each clips into the BiLSTM-CRF model to get the objectively existing entities and logical conceptual entities in the mathematical information * /
6. reality temp_reality; theoretical temp_theoretical; / * Put the objective knowledge points and logical concepts obtained in step 5 into reality, theoretical * /
7. end for;
8.return reality， theoretical；

---

## 4   Experimental Results and Analysis

### 4.1   Experimental Data

This experiment obtained a total of 10232 effective documents from the knowledge data shared by the People's Education Press electronic textbooks published by the People's Education Press, Mathematics Subject Network, Mathematics Resource Network, Mathematics China Network, Baidu Encyclopedia, and Wang Jianing, which basically covered primary and junior high schools, High school and advanced mathematics knowledge points, of which entity 4139, for these data, this article randomly selected according to the ratio of 7: 3 to form a training set and a test set.

### 4.2   Experimental Environment and Parameter Settings

In this experiments, the CPU is Intel 9750H, the memory is 16G, and the GPU is RTX1660Ti. The models used in this article are all built using Tensorflow, which is a deep learning framework developed by Google and widely used in the implementation of various machine learning algorithms. Some parameters are set as follows: dropout is 0.5, seq_length is 128, training learning rate is 3e−5, and Epochs is 30.

### 4.3   Experimental Results and Analysis

In the mathematics data set, HMM, CRF, BiLSTM, BiLSTM-CRF, BERT-CRF, BERT-softmax models were used for performance analysis. The experimental results are shown in Table 1 and Table 2.

**Table 1.** Comparison of the recognition results of the mathematical objective existence entities

|  | Recall | Precision | F1-score |
|---|---|---|---|
| HMM | 82.15% | 83.63% | 83.38% |
| CRF | 86.85% | 89.70% | 88.24% |
| BiLSTM | 95.11% | 96.44% | 95.32% |
| BERT-softmax | 38.37% | 59.26% | 46.58% |
| BERT-CRF | 96.42% | 95.73% | 96.07% |
| BiLSTM-CRF | 96.52% | 94.79% | 95.56% |

**Table 2.** Comparison of the recognition results of the mathematical logic concept entities

|  | Recall | Precision | F1-score |
|---|---|---|---|
| HMM | 82.38% | 74.82% | 78.41% |
| CRF | 81.69% | 88.05% | 84.65% |
| BiLSTM | 74.13% | 88.39% | 79.29% |
| BERT-softmax | 27.16% | 66.67% | 38.60% |
| BERT-CRF | 83.95% | 89.47% | 86.62% |
| BiLSTM-CRF | 82.20% | 91.11% | 86.32% |

It can be seen from Table 1 and Table 2 that:

- The method based on recurrent neural network (such as BiLSTM) is generally better than the method based on HMM, because the model based on RNN can effectively extract sequence features.
- Comparing the combination of CRF and deep neural network, it can be seen that the recall rate of BiLSTM-CRF is increased by nearly 5% compared to BiLSTM, the accuracy is increased by nearly 2%, and the F1 value is increased by nearly 4%. Compared with BERT-softmax, BERT-CRF has also been significantly improved. It can be seen that the addition of deep neural networks can provide better sequence characteristics for the model, thereby improving the model's effect.

## 5   Conclusion

In this paper, deep neural networks are applied to the recognition of information entities in mathematics, and a BiLSTM-CRF model is constructed to identify the objectively existing entities and logical concept entities in mathematical information. The Chinese text is vectorized word by word, and the bi-directional semantic features of the pretext and posttext are obtained using the BiLSTM network. Experimental results show that the BiLSTM-CRF model has a better recognition effect than HMM, CRF, and BiL-STM. In the future, we will try to construct an entity relationship extraction method for mathematics information.

**Acknowledgments.** We thank the anonymous reviewers. This work is supported by Doctoral Research launch project of Yunnan Normal University (No. 2019XJLK21), and Program for innovative research team (in Science and Technology) in University of Yunnan Province.

## References

1. Cho, M., Ha, J., Park, C., Park, S.: Combinatorial feature embedding based on CNN and LSTM for biomedical named entity recognitio. J. Biomed. Inform. **103**, 103381 (2020)

2. Celardo, L., Everett, M.G.: Network text analysis: a two-way classification approach. Int. J. Inf. Manage. **51**, 102009 (2020)
3. Haihong, E., Xiao, S., Song, M.: A text-generated method to joint extraction of entities and relations. Appl. Sci. **9**(18), 3795 (2019)
4. Farhan, W.: Unsupervised dialectal neural machine translation. Inf. Process. Manage. **57**(3), 102181 (2020)
5. Huang, M., Xie, H., Rao, Y., Feng, J., Wang, F.L.: Sentiment strength detection with a context-dependent lexicon-based convolutional neural network. Inf. Sci. **520**, 389–399 (2020)
6. Kodra, L., Meçe, E.K.: Question answering systems: a review on present developments, challenges and trends. Int. J. Adv. Comput. Sci. Appl. (IJACSA) **8**, 217–224 (2017)
7. Graves, A.: Long short-term memory. In: Supervised Sequence Labelling with Recurrent Neural Networks, pp. 1735–1780. Springer, Berlin (2012). https://doi.org/10.1007/978-3-642-24797-2_2
8. Mirza, A.H., Kerpicci, M., Kozat, S.S.: Efficient online learning with improved LSTM neural networks. Digital Sig. Process. **102**, 102742 (2020)
9. Lafferty, J.D., Mccallum, A., Pereira, F.C.N.: Conditional random fields: probabilistic models for segmenting and labeling sequence data. In: Eighteenth International Conference on Machine Learning. Morgan Kaufmann Publishers Inc., pp. 282–289 (2001)
10. Hocreiters, S.J.: Long short-termmemory. Neural Comput. **9**(8), 1735–1780 (1997)
11. Graves, A., Jaitly, N., Mohamed, A.R.: Hybrid speech recognition with deep bidirectional LSTM. In: Proceedings of 2013 IEEE Workshop on Automatic Speech Recognition and Understanding, pp. 273–278. IEEE Press, Washington, D. C. (2013)

# A Facial Fatigue Expression Recognition Method Based on Sparse Representation on the Low-Resolution Image

Ling Zhang[1], Xiao-lu Tian[1], Wenchao Jiang[1]([✉]), and Dongjun Ning[2]

[1] Faculty of Computer, Guangdong University of Technology, Guang Zhou, China
85711585@qq.com
[2] Taotall Technology Co., Ltd., Guangzhou, China

**Abstract.** In order to effectively improve the performance of facial fatigue expression recognition on the low-resolution image, a method of fatigue facial expression recognition based on Sparse Representation is proposed. At present, study on facial fatigue expression recognition is almost based on high-resolution, high-quality images. In the network environment, especially in the Internet of Things environment, the images of facial fatigue expression mostly are low-resolution images, in which the performance of facial fatigue expression recognition will degrade with image quality. In order to improve the recognition rate on the low-resolution image, a facial fatigue expression recognition method based on sparse representation on the low-resolution image is proposed. At the same time, in order to improve the credibility of the results of the experiment, Kendall's coefficient of concordance method was used to construct the low-resolution facial fatigue expression database—TIREDFACE, whose words mean human's tired-face. The TIREDFACE database consists of about 240 Image sequences of 10 university students. Size of all images is normalized to $92 \times 112$ pixels with 8-bit precision for grayscale values. Firstly, TIREDFACE database is set up, and then we exploit the discriminative nature of sparse representation to perform fatigue expression detection. Instead of using the generic dictionaries, we represent the test sample in an overcomplete dictionary whose base elements are the training samples themselves. Compressed sensing theory was used to solve its sparsest representation. With sufficient training samples for each fatigue status, including awake, slight fatigue and severe fatigue, it is done to represent the test samples as a linear combination of just those training samples from the same classification, and the sparse representations of the low resolution facial fatigue expression image of the identified test sample in the database TIREDFACE are given. After that, according to the sparsest representation solution, the low-resolution facial fatigue expression status classification is performed. In order to further investigate the advantages of this method, we compare this method with the linear classifier, the nearest neighbor (NN), support vector machine (SVM) and the nearest subspace (NS) on our database TIREDFACE. Experimental results show that the recognizing accuracy on low resolution facial fatigue expression images by the method of this paper is much higher than the linear classifier, the nearest neighbor (NN), support vector machine (SVM) and the nearest subspace (NS), etc. Therefore, the proposed method in this paper on the low-resolution facial fatigue expression recognition tasks is feasible.

© Springer Nature Switzerland AG 2020
X. Chen et al. (Eds.): ML4CS 2020, LNCS 12488, pp. 269–278, 2020.
https://doi.org/10.1007/978-3-030-62463-7_25

**Keywords:** Compressed sensing · Sparse representation · Fatigue expression recognition · Kendall's coefficient of concordance

## 1 Introduction

Fatigue is a symptom of an underlying disease and can be described in many ways, from feeling weak to being constantly tired or lacking energy. The main performance of fatigue is that people find they are difficult to concentrate, think, and easy to make mistakes, etc. [1].

According to China's Ministry of public security statistics, in 2017 China's fatigue driving caused by a total of 2568 traffic accidents, resulting in 1353 deaths, 3129 people were injured [2]. While, U.S. Highway Traffic Safety Administration (NHTSA) reported that 71000 people were injured [3]. With many years of development, the facial fatigue expression recognition yielded some results, however, most of their researches [4–12] were based on high-resolution, high-quality images. The compressive sampling (CS) (also called compressive sensing) theory [13, 14], which originally aims to address signal sensing and coding problems, has shown huge potential for other problems such as pattern recognition [15–17]. In this paper, a new method based on the theory of Sparse Representation-based Classification (SRC) is proposed to solve the problem of facial fatigue expression recognition on the low-resolution images. Based on a sparse representation computed by $\ell^1$ -minimization, sparsity in the facial fatigue recognition problem is properly harnessed, and the choice of features is no longer critical.

The remainder of this paper is organized as follows: Sect. 2 briefly review the CS theory. In Sect. 3, the method of Kendall's coefficient of concordance is used to construct the low-resolution facial fatigue expression database ——TIREDFACE. Section 4 shows the algorithm of Sparse Representation-based Classification (SRC) on the problem of facial fatigue expression recognition on the low-resolution images. The experiment verifications are presented in detail in Sect. 5, and Sect. 6 gives the conclusions and prospects.

## 2 Compressive Sensing (CS)

In this section, we briefly review the CS theory. Given a system of under-determined equation.

$$x_{m\times 1} = D_{m\times n}\partial_{n\times 1}, m < n \tag{1}$$

It is known that there is no unique solution in the Eq. (1), since the number of equations is less than the number of variables. In signal processing terms, the number of samples ($m$) is less than the length of the signal ($n$). However, according to the theory of compressive sampling (CS), if the signal is sparse, it is necessarily unique, and the signal can be reconstructed by practical algorithms. Suppose that the signal is $k$-sparse. That is, there are only $k$ non-zero values in $x$, and the remainder are all zeroes, which means that it is feasible to find the solution to Eq. (1) by using a brute force enumeration of all the possible $k$-sparse vectors of length $n$.

Mathematically speaking, this problem can be represented as,

$$\min ||\partial||_0, \text{ subject to } x = D\partial \tag{2}$$

Where $|| \ ||_0$ is the $l_0$-norm, and it denotes the number of non-zero elements in the vector. Equation (2) is known as an NP (non-deterministic polynomial) hard problem, and thus, there is not a practical solution to Eq. (1). The CS literatures indicates that under a certain situation on the projection matrix D, i.e., restricted isometric property (RIP), the sparsest solution to Eq. (1) can be gained by replacing the $l_0$-norm in Eq. (2) by its closest convex surrogate, the $l_1$-norm. Therefore, the solution to Eq. (2) is equivalent to the following $l_1$-norm. Minimization problem.

$$\min ||\partial||_1, \text{ subject to } x = D\partial \tag{3}$$

Where the means the minimization of the sum of absolute values of all elements in the vector, and serves as an approximation of the $l_0$-norm. In fact, the equality $x = D\partial$ is relaxed to take into account the existence of measurement error in the sensing process due to some small amount of noise. Suppose that the measurements are inaccurate and the noisy model is taken into consideration.

$$x = D\partial + e \tag{4}$$

Where $e$ is a stochastic or deterministic error term. Particularly, if the error term $e$ is assumed to be white noise, such as $||e||_2 < \varepsilon$, where $\varepsilon$ is a very small constant, a noise robust version of Eq. (3) can be defined as follows.

$$\min ||\partial||_1, \text{ subject to } ||x - D\partial||_2 < \varepsilon \tag{5}$$

In order to solve the $l_1$-minimization of Eqs. (3) and (5), many efficient algorithms have been developed. $l_1$-magic [16] and $l_1$-ls [17] are two typical algorithms based on the interior-point idea. The $l_1$-magic algorithm [16] recasts the $l_1$-minimization problem as a second-order cone program and then applies the primal log-barrier approach. The $l_1$-ls algorithm [17] is a specialized interior-point method for solving the large-scale $l_1$-regularized least-squares programs that uses the preconditioned conjugate gradients algorithm to compute the search direction.

## 3 Database—TIREDFACE

### 3.1 Kendall's Coefficient of Concordance

Because there was little research about the fatigue expression on the low-resolution images, there was also lacking of fatigue expression databases issued publicly by now. To evaluate the proposed algorithm in this paper, a facial fatigue expression database on the low-resolution images (TIREDFACE) is constructed by using the method of Kendall's coefficient of concordance [31–33].

Reliability of a test directly depends on its performance on consistency, stability and reliability of its final results. The coefficients of reliability refer to the quantitative indicators of reliability. The higher of the reliability coefficients are the more stable and reliable the test results will be. On the other hand, scorer reliability refers to the consistency of different raters scoring on a same object. When there are only two raters, we can adopt the methods of Spearman's rank-order correlation [32, 33] and Pearson correlation coefficient [20] to analyze reliability. However, when there are more than two raters $m$ ($m > 2$), Kendall's coefficient of concordance [31–33], multiple-series relevant coefficient and Cronbach's $\alpha$ can be used [33]. During the process of constructing our facial fatigue expression database (TIREDFACE), there are 20 raters, so the methods of Kendall's coefficient of concordance, multiple-series relevant coefficient and Cronbach's $\alpha$ can be used.

Suppose that $n$ subjects are evaluated by $m$ raters, $m > 2$, giving rise to $m$ variables. The Kendall's coefficient of concordance $W$ is based on the ratio between the variance of the sum of ranks to subjects, expressed by $s$, and the variance's maximum possible value of the sum of ranks related to the values of n and m, as formula (6).

$$W = \frac{s}{\max(s)} \tag{6}$$

$$\max(s) = m^2 n(n^2 - 1)/12 \tag{7}$$

Where the $max(s)$ in Eq. (7) denotes the value of $s$. There is a perfect agreement among all the ranks. Using $R_{ij}$ to represent the rank of the $i^{th}$ subject in the $j^{th}$ rater and then there is.

$$R_i = \sum_{j=1}^{m} R_{ij} \tag{8}$$

Where the $R_i$ in Eq. (8) represents the sum of ranks awarded to the $i^{th}$ subject, for $i = 1,..., n$ and $j = 1,..., m$. Accordingly, Kendall's coefficient $W$ can be presented as the following,

$$W = \frac{\left(n \sum_{i=1}^{n} R_i^2 - \left(\sum_{i=1}^{n} R_i\right)^2\right)/n}{m^2 n(n^2 - 1)/12} \tag{9}$$

When there are tied observations for a given variable, each tied observation is assigned a same rank, which is the arithmetic mean of the ranks of those tied observations. Then a correction factor should be added into the $W$ statistical formula (9), as shown in formula (10).

$$W = \frac{\left(n \sum_{i=1}^{n} R_i^2 - \left(\sum_{i=1}^{n} R_i\right)^2\right)/n}{m^2 n(n^2 - 1) - m \sum_{j=1}^{m} \sum_{k=1}^{g_j} (t_{kj}^3 - t_{kj})/12} \tag{10}$$

Where $g_i$ is the number of groups of ties for the $j^{th}$ variable and $t_{kj}$ is the number of observations in the $k^{th}$ group of ties, $k = 1,..., g_i$ and $j = 1,..., m$.

After we have obtained a value $W$ of Kendall's coefficient of concordance, we must test the significance of $W$, and the following formula (11) can be applied for testing the significance of $W$.

$$\chi^2 = k(N-1)W \sim \chi^2(N-1) \tag{11}$$

There is a strong relationship between $W$ and the Friedman $\chi^2$ statistic, $\chi^2 = k(N-1)W$ which has an approximate chi-squared distribution with $(N-1)$ degrees of freedom. Given a level $\alpha$, there is a formula (12).

$$P(\chi^2 > \chi^2_{1-\alpha}) = \alpha \tag{12}$$

Checking the distribution tables with $df = N\text{-}1$, a quintile threshold $\chi^2_{1-\alpha}$ can be obtained, Bring the value $W$ of Kendall's coefficient of concordance that we got by formula (10) into the formula (12), if $\chi^2 > \chi^2_{1-\alpha}(N-1)$, the score was significantly consistent, Otherwise, the score was significantly different.

### 3.2 TIREDFACE Database

We transform the score results of facial fatigue image in the database TIREDFACE into rank data, the results can be seen from Table 1.

**Table 1.** Rank data of our database TIREDFACE

| Statistical Magnitude | Value |
|---|---|
| $k$ | 20 |
| $W$ | 0.884 |
| $\chi^2$ | 1467.129 |
| $df$ | 83 |
| $p$ | <0.001 |

Where $k$ is the observation, $W$ is Kendall's coefficient of concordance, $\chi^2$ is the Friedman statistic, $df$ is the difference, $p$ is the significance. From Table 1 we can see that the value of Kendall's coefficient of concordance $W$ is 0.884, and the value of the significance of that we can use the facial fatigue database TIREDFACE to identify fatigue rank in this study.

## 4    Fatigue Recognition Algorithm

A sparse representation classifier (SRC) has been developed based on the CS theory [13, 14]. To the training samples in TIREDFACE, it can be represented as:

$$x_{k,test} = \partial_{k,1}x_{k,1} + \partial_{k,2}x_{k,2} + \cdots + \partial_{k,n}x_{k,n} = \sum_{i=1}^{n} \partial_{k,i}x_{k,i} + \varepsilon_k \tag{13}$$

Where $x_{k,test}$ is the test sample of the $k^{th}$ class, $k$ is 3 in this study (namely waking expression, slight fatigue expression, severe fatigue expression), $x_{k,i}$ is the $i^{th}$ training sample of the $k^{th}$ class, $\partial_{k,i}$ is the weight corresponding weight and $\varepsilon_k$ is the approximation error. For the all $c$ classes, it can be described instead as.

$$x_{k,test} = \partial_{1,1}x_{1,1} + \cdots + \partial_{k,1}x_{k,1} + \cdots + \partial_{k,n_k}x_{k,n_k} + \cdots + \partial_{k,n_c}x_{k,n_c} + \varepsilon$$

$$= \sum_{i=1}^{n_1} \partial_{1,i}x_{1,i} + \cdots + \sum_{i=1}^{n_k} \partial_{k,i}x_{k,i} + \cdots + \sum_{i=1}^{n_c} \partial_{c,i}x_{c,i} + \varepsilon \qquad (14)$$

In matrix-vector notation, Eq. (14) can be rewritten as,

$$x_{k,test} = D\partial + \varepsilon \qquad (15)$$

$l_0$-*norm* minimization problem should be solved for Eq. (14):

$$\min ||\partial||_0, \textit{subject to } ||x_{k,test} - D\partial||_2 < \varepsilon \qquad (16)$$

The solution of Eq. (16) is equivalent to $l_1$-*norm* minimization problem.

$$\hat{\partial} = \min||\partial||_1, \textit{subject to } ||x_{k,test} - D\partial||_2 < \varepsilon \qquad (17)$$

We then classify $x_{k\_test}$ based on these approximations by assigning it to the object class that minimizes the residual between $x_{k\_test}$ and $\hat{x}_{k\_test}$.

$$\min r_i(x_{k\_test}) = \left\| x_{k\_test} - D\delta_i(\hat{\partial}) \right\| \qquad (18)$$

To construct a sparse dictionary of fatigue expressions in super resolution images, we randomly select half of the 240 images from our database TIREDFACE as the training set and the rest for testing. In this case, we subsample the images from the original $92 \times 168$ to size $10 \times 12$. The pixel values of the down sampled image are used as 120-dimension features-stacked as columns of the matrix $D$ in the dictionary constructing process. Therefore, matrix $D$ has size $120 \times 120$, and the system $x_{k,test} = D\partial$ is underdetermined. With different training samples, the residuals with respect to the dozens of projected coefficients $\delta_i(\hat{\partial})(i = 1, 2, \ldots, n)$ are calculated, and different features corresponding to the heading largest coefficients are defined, namely, the few largest coefficients are associated with training samples from our database TIREDFACE.

## 5  Experiment Verification

### 5.1  Experiment Results and Analysis

Figure 1, 2 and 3 show some sample images from the TIREDFACE database.

The average accuracy rate obtained by our algorithm on 3-classes facial fatigue expression in 583 times on the TIREDFACE database is 91.7% ± 8%. From Table 2, we can know that the average accuracy rate of each facial fatigue expression recognition in the experiments, and from Table 3 we can see the results of classification of each fatigue expression in the experiments.

**Fig. 1.** Examples of waking expression from our TIREDFACE database

**Fig. 2.** Examples of slight fatigue expression from our TIREDFACE database

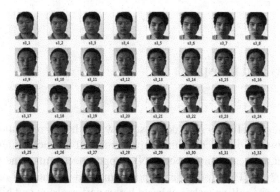

**Fig. 3.** Examples of severe fatigue expression from our TIREDFACE database

**Table 2.** Average accuracy on 3-class facial fatigue expression recognition

| Facial Fatigue Expression | Awaking | Slight Fatigue | Severe Fatigue |
|---|---|---|---|
| Accuracy | 93.3% | 90.1% | 92.4% |

**Table 3.** Results of classification on 3-class facial fatigue expression

| Fatigue Status | Awaking | Slight Fatigue | Severe Fatigue |
|---|---|---|---|
| Awaking Status | 56 | 2 | 0 |
| Slight Fatigue | 4 | 54 | 5 |
| Severe Fatigue Status | 0 | 4 | 55 |

## 5.2  Comparative Analysis

We compare this method with the linear classifier,the nearest neighbor (NN), support vector machine (SVM) and the nearest subspace (NS) on the database TIREDFACE, and the results are shown in Table 4.

**Table 4.** Comparison of Recognition rate (%) with different methods on our TIREDFACE

| Method | Accuracy |
|---|---|
| The Linear Classifier | 67% |
| The Nearest Neighbor (NN) | 84% |
| The Nearest Subspace (NS) | 87.3% |
| Support Vector Machine (SVM) | 80.3% |
| Sparse Representation Classifier (SRC) | 92.5% |

# 6  Conclusions and Prospects

## 6.1  Conclusion

Experimental results on TIREDFACE database show that the low resolution facial fatigue expression performance obtained by this paper's method is much better than those by the linear classifier,the nearest neighbor (NN), support vector machine (SVM) and the nearest subspace (NS). Therefore, the proposed method on the low-resolution facial fatigue expression recognition tasks achieved better performance and high accuracy.

## 6.2  Prospects

Here are some suggestions for further inquiry:

(1) In the actual environment, there are some problems of the face image such as occlusion, how to improve recognition rate on facial fatigue expression with occlusion, fuzzy and side face need to be further explored and researched.

(2) 3D facial fatigue expression recognition has been considered as a effective solution to 2D facial images, and it's interesting to investigate the performance of the presented method for 3D facial expression recognition.

**Acknowledgements.** This paper is funded by Scientific Project of Guangdong Provincial Transport Department (No. Sci & Tec-2016-02-30), Surface Project of Natural Science Foundation of Guangdong Province (No. 2016A030313703 and 2016A030313713).

# References

1. Chen, Y., Liu, W., Zhang, L., Yan, M., Zeng, Y.: Hybrid facial image feature extraction and recognition for non-invasive chronic fatigue syndrome diagnosis. Comput. Biol. Med. **64**, 30–39 (2015)
2. China Transportation Yearbook. China Transportation Yearbook GuangZhou (2012)
3. Dawn, R.: National survey on distracted and driving attitudes and behaviors. Washington, D.C. Technical report DOT HS 809 566 (2003)
4. Zhao, C., Zhang, X., Zhang, B.: Driver's fatigue expressions recognition by combined features from pyramid histogram of oriented gradient and contourlet transform with random subspace ensembles. IET Intell. Transp. Syst. **7**(1), 36–45 (2013)
5. Song, F., Tan, X., Liu, X.: Eyes closeness detection from still images with multi-scale histograms of principal oriented gradients. Pattern Recogn. **47**(9), 2825–2838 (2014)
6. Yang, Q., Li, C., Li, Z.: Application of FTGSVM algorithm in expression recognition of fatigue driving. J. Multimedia **9**(4), 527–533 (2014)
7. Wang, X.M., Zhang, J.J.: Recognition system for fatigue facial expression of driver. In: Applied Mechanics and Materials, **644–650**, 4174–4177 (2014)
8. Guo, Y.H., Liu, J.: A novel approach of eye detection based on Haar-Like feature and SVM. In: Applied Mechanics and Materials, **121–126**, 1863-1867 (2011)
9. Ahmed, R., Emon, K.E.K., Hossain, M.F.: Robust driver fatigue recognition using image processing. In: 2014 International Conference on Informatics, Electronics & Vision (ICIEV) (2014)
10. Wei, H., Wei, Z.: Driver fatigue recognition based on supervised LPP and MKSVM. In: 3rd International Conference on Digital Image Processing (ICDIP) (2011)
11. Shiwu, L., Linhong, W., Zhifa, Y.: An active driver fatigue identification technique using multiple physiological features. In: 2011 International Conference on Mechatronic Science, Electric Engineering and Computer (MEC) (2011)
12. Agrawal, U., Giripunje, S., Bajaj, P.: Emotion and gesture recognition with soft computing tool for drivers assistance system in human centered transportation. In: 2013 IEEE International Conference on Systems, Man, and Cybernetics (SMC) (2013)
13. Donoho, D.L.: Compressed sensing. IEEE Trans. Inform. Theor. **52**, 1289–1306 (2006)
14. Baraniuk, R.G.: Compressive sensing [lecture notes]. IEEE Signal Process. Mag. **24**, 118–121 (2007)
15. Candes, E.J., Wakin, M.B.: An introduction to compressive sampling. IEEE Signal Process. Mag. **25**, 21–30 (2008)
16. Wright, J., Yang, A.Y., Ganesh, A., Sastry, S.S., Ma, Y.: Robust face recognition via sparse representation. IEEE Trans. Pattern Anal. Mach. Intell. **31**, 210–227 (2009)
17. Wright, J., Ma, Y., Mairal, J., Sapiro, G., Huang, T.S., Yan, S.: Sparse representation for computer vision and pattern recognition. Proc. IEEE **98**, 1031–1044 (2010)

18. Zhu, S., Zeng, B., Gabbouj, M.: Adaptive sampling for compressed sensing based image compression. J. Vis. Commun. Image Representation **30**, 94–105 (2015)
19. Zhang, J., Yang, K.: Informational analysis for compressive sampling in radar imaging. Sensors **15**(4), 7136–7155 (2015)
20. Xu, Y., Zhang, B., Zhong, Z.: Multiple representations and sparse representation for image classification. Pattern Recogn. Lett. **68**, 9–14 (2015)
21. Huang, S., Yang, Y., Yang, D., Huangfu, L.: Class specific sparse representation for classification. Sig. Process. **116**, 38–42 (2015)
22. Babu, R.V., Parate, P., Acharya, Ä.K.: Robust tracking with interest points: a sparse representation approach. Image Vis. Comput. **33**, 44–56 (2015)
23. Li, K., Yang, J., Jiang, J.: Nonrigid structure from motion via sparse representation. IEEE Trans. Cybern. **45**(8), 1401–1413 (2015)
24. Yin, H.: Sparse representation with learned multiscale dictionary for image fusion. Neurocomputing **148**, 600–610 (2015)
25. Dong, W., Chang, F., Zhao, Z.: Visual tracking with multifeature joint sparse representation. J. Electron. Imaging **24**(1), 013006 (2015)
26. Li, W., Du, Q., Zhang, B.: Combined sparse and collaborative representation for hyperspectral target detection. Pattern Recogn. **48**(12), 3904–3916 (2015)
27. Du, P., Xue, Z., Li, J., Plaza, A.: Learning discriminative sparse representations for hyperspectral image classification. IEEE J. Sel. Top. Sign. Process. **9**(6), 1089–1104 (2015)
28. Nagesh, P., Li, B.: A compressive sensing approach for expression-invariant face recognition. In: Proceedings of IEEE Conference on Computer Vision and Pattern Recognition (CVPR), Miami, FL, USA, 20–25, pp. 1518–1525, June 2009
29. Candes, E., Romberg, J.: L1-magic: Recovery of sparse signals via convex programming. User Guide. L1-Magic Software (2005). http://users.ece.gatech.edu/~justin/l1magic/. Accessed 12 December 2005
30. Kim, S.J., Koh, K., Lustig, M., Boyd, S., Gorinevsky, D.: An interior-point method for large-scale-regularized least squares. L1-ls software. IEEE J. Sel. Top. Signal Process. 1, 606–617 (2007). http://www.stanford.edu/~boyd/l1_ls/. Accessed 26 February 2007
31. Kendall, M.G., Babington-Smith, B.: The problem of m rankings. Ann. Math. Stat. **10**, 275–287 (1939)
32. Sheskin, D.J.: Handbook of Parametric and Nonparametric Statistical Procedures, 4th edn. Chapman and Hall/CRC, Boca Raton (2007)
33. Banerjee, M., Capozzoli, M., McSweeney, L., Sinha, D.: Beyond kappa: a review of interrater agreement measures. Can. J. Stat. **27**, 3–23 (1999)

# Spontaneous Expression Recognition Based on Visual Attention Mechanism and Co-salient Features

Ling Zhang[1], Qiumin Ji[1], Wenchao Jiang[1(✉)], and Dongjun Ning[2]

[1] Faculty of Computer, Guangdong University of Technology, Guangzhou, China
85711585@qq.com
[2] Taotall Technology Co., Ltd., Guangzhou, China

**Abstract.** Spontaneous facial expression recognition has gained much attention from researchers in recent years, however most of the existing algorithms still encounter bottlenecks in performance due to too big redundant images data in the video. In this paper, we propose a novel co-salient facial feature extraction algorithm, combined with human visual attention mechanism and group data co-processing technology, which would largely reduce the redundant information in the original images and effectively improve the recognizing accuracy of facial expressions. Firstly, based on human visual mechanism, key frames of expression are dynamically derived from the original videos to capture the temporal dynamics of facial expressions. Secondly, using key sequence frames, salient regions are obtained by multiplicative fusion algorithm and in multi-images co-operative manner. Thirdly, we get rid of these salient regions due to their little deformation and low-correlation to facial expressions, and reduce the number of facial features data. At last, we extract Local Binary Pattern (LBP) features from the remainder of facial features and use Support Vector Machine (SVM) classifier to classify them respectively. Experimental results on dataset Cohn-Kanade plus and MMI showed that our proposed method can effectively improve the recognizing accuracy of spontaneous expression sequence.

**Keywords:** Dynamic continuous spontaneous · Dynamic sampling · Co-saliency · Reverse salient

## 1 Introduction

It is well known that emotion recognition has a broad application prospect in the fields of human-computer interaction, medical treatment, advertising, video games etc. [1]. In recent years, the continuous spontaneous expression has gained much more attention from researchers due to its spontaneity and imperceptibility. In spontaneous expression recognition, first of all, we need to extract expression frames by down-sampling the video sequence, and then those frames will be used to recognize the expression. Because of the huge amount of video and diversity of facial expression, there are two major challenges for spontaneous expression recognition: one is how to extract expression

© Springer Nature Switzerland AG 2020
X. Chen et al. (Eds.): ML4CS 2020, LNCS 12488, pp. 279–289, 2020.
https://doi.org/10.1007/978-3-030-62463-7_26

frames automatically and the other is how to extract the common features of the same type of expression. In terms of down-sampling the video sequence, the heuristic methods achieve better performance, but they need annotation of peak frames, thus they cannot be applied to continuous facial expressions without peak frames annotation. References [2, 3] proposed methods for dynamic down-sampling, which identify pruned samples sorted later in the pipeline. Glodek et al. [4] proposed a stochastic video down-sampling method to solve the problem of low accuracy for uniform down-sampling in extraction facial expression frames, however the accuracy of this method is still not high. In [5], an adaptive down-sampling method was proposed, in which the sampling granularity varies depending on whether a prediction tag is detected, the limitation of the system is whether it can correctly predict the tag or not. In [6], the training data is down-sampled so that frames with greater emotion strength than the average emotion intensity are left. Dahmane et al. [7] determines whether the time granularity of the lower sampling will be adjusted according to the change in the category labels predicted. This method requires the system to accurately predict the category label. Zhu et al. [8] reduced the number of video frames through a booting process. Jiang et al. [9] conducted a comparative study of random, guided and heuristic down-sampling schemes and found that the heuristic method obtains the highest accuracy. However, both the guided and the heuristic methods require annotation of peak expression frame of the video sequences, thus they cannot be applied to continuous spontaneous expression video sequences without peak expression frame annotation. In the aspect of co-salient, the Maximally Stable External Region, edge- based region and other salient region extracting algorithms were proposed in [10–13]. According to experimental results, the Maximally Stable External Region method has the highest recognition accuracy among the above methods, but its computational complexity is also the highest in most cases. Cai et al. [14] gave a general region extracted method based on affine invariant feature regions, but it is not robust to light changes.

We propose a method automatically down-sample the video sequence according to the changing frequency of face expression. There are both diversity and generality between the frames of spontaneous expression, therefore, we adopt the co-saliency region extraction algorithm to extract salient regions from the down-sampled image data. The contribution of this article includes:

(1) We use the visual attention mechanism [15, 16] to determine the appropriate down-sampling rate. The time granularity for adaptive down-sampling is inversely proportional to the changing frequency of facial expression.
(2) Using the algorithm of co-saliency region, we extract the region with small changes between facial expressions in the sampled image, and eliminate these regions.
(3) Improve the recognition rate of the dynamic sequential emotions effectively by using the SVM classifier based on posterior probability to perform feature extraction and classification of the image.

## 2   Key Frames Extraction Process

Firstly, Haar-like model is used to extract the region of interest of human face from facial expression images, so as to remove the unrelated parts of the expression. Secondly, the dynamic sampling algorithm is used to down-sample spontaneous facial expression sequences. Then, the collaborative saliency algorithm is used to detect and extract the area with small facial deformity. The area with less deformation intensity is removed and LBP features are extracted from the remaining image frames.

## 2.1  Down Sampled Facial Expression Video

Video is equally divided into smaller video segments, each segment is dynamically down-sampled, and each segment has its own down-sampling factor. Suppose video $I$ is divided into $N$ non-overlapping segments, and video $I_\psi$ contains $N$ frames, denoted as $\psi = \{m_0, m_0 + 1, m_0 + 2, \cdots, m_0 + N - 1\}$. The video sequence after down-sampling is denoted as $I\varphi$. The system processes $N$ frames of video clips at a time. We start with $m_0 = 0$, and the first $N$ frames form a video clip. Then another video clip is formed from the $N^{\text{th}}$ frame to the 2 $N$-1 frame, and so on, until the video ends. If the video sequence contains less than $N$ frames, a video clip is formed. The parameter $N$ we choose makes the duration of each segment to be 1 s, since 1 Hz is the maximum HVS limit according to the theory of vision and attention [17].

$I_\varphi$ is obtained by down-sampling $I_\psi$. The first step is to quantify the facial expression as a signal that changes over time. The frequency of the signal must correspond to changes in facial expressions. Due to the high frame rate, light flow can be used to quantify facial expressions. $\Delta I_n$ is the optical flow between two frames $I_n$ and $I_n$-1, the output of which is a motion vector, which sums all the pixels in the image to form a one-dimensional signal.

$$f(n) = \sum_x \|\Delta I_n(x)\|_2 \tag{1}$$

Where $f(n)$ is the temporal characteristic of a single-frame, $X$ is a pixel vector, and the second norm represents the magnitude. For the entire video segment $I_\psi$, the temporal feature can be represented by,

$$f(n) = [f(m_0), f(m_0 + 1), \cdots, f(m_0 + N - 1)] \tag{2}$$

Figure 1 demonstrates the optical flow and the generated temporal features of a video clip from frame $m_0$ to frame $m_0 + N$. In Fig. 1, sub figure $(a)$ is a video clip $(b)$ and $(c)$ are the optical flow features formed between the first two frames and between the last two frames in this video sequence, respectively. Figure 1$(d)$ is the temporal characteristics of the optical flow generated by this video clip.

In order to calculate the main frequency, DC offset must be removed firstly.

$$\tilde{f}_\phi = f_\phi - E(f_\phi) \tag{3}$$

Where $E(\cdot)$ is the expectation operator. The important reason for removing the DC offset is that for the actual data, $D_\varphi{}'(0)$ corresponds to a factor of 0 Hz, the DC offset will be greater than the other values of $D_\varphi$, so the DC offset may be chosen as the dominant frequency. $D_\varphi{}'$ is the discrete cosine transform $D_\phi = DCT(\tilde{f}_\phi)$. Figure 1(d) is the

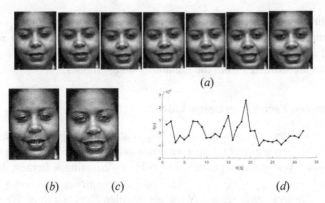

(a)

(b)          (c)                              (d)

**Fig. 1.** Time domain feature extraction diagram

curve of the frequency domain features transformed from the time domain features. The maximum energy corresponding to the frequency is calculated in the following Eq. (4).

$$\beta = \arg \max_{k} \left\| D_{\phi}(k) \right\| \tag{4}$$

Where $k$ is the frequency and $\left\| D_{\varphi}'(k) \right\|$ is the magnitude of $D_{\varphi'}(k)$. We down sample discrete signals by removing that do not change much in the signal. Therefore, we sample at the main frequency. The down-sampling factor $M$ is given by: (Maximum frequency/Main frequency). The index of the frequency $\beta$ can be converted to the main frequency $2\pi\beta/N$, and the maximum frequency index $N$ corresponds to the frequency $2\pi$, that is $M = N/\beta$. Accordingly, a sample of a video clip can be expressed as $I\varphi = I\psi \downarrow M$. When the temporal features have high frequency characteristics as $\beta \rightarrow N$, the down-sampling factor is close to 1 and all frames are retained. When temporal features have low frequency characteristics, the down-sampling factor increases, and most of the frames are removed.

We can also use vertex to sample the video into equal segments, each containing $N$ frames, with the central frame of the segment as the vertex, and frames far from the vertex will be deleted. After obtaining the main frequency $\beta$ of each video segment, the frames affected by the main frequency are selected for the duration of time. If the dominant frequency is high, we will select more frames at the vertices, and if the dominant frequency is low, we will select only a few frames near the vertices. The formula is in the following Eq. (5).

$$\varphi^{*}_{Apex} = \{n : n_0 - \beta/5 < n < n_0 + \beta/5\} \tag{5}$$

## 3   Co-salient Region Extraction Algorithm

Co-saliency detection is divided into two parts: the saliency detection and the collaborative filtering, which are based on contrast cue and spatial cue on cluster-level, respectively [18]. Then, with multiplicative fusion algorithm [19], each facial expression co-saliency map is generated.

Contrast cue represents uniqueness of the visual features in a single or multiple image and it is the most widely used measuring means of saliency. Suppose there are $C^k$ clusters, and contrast cue $w^c(k)$ of cluster $C^k$ is defined using its feature contrast to all other clusters, as shown in Eq. (6).

$$w^c(k) = \sum_{i=1, i \neq k}^{k} \left( \frac{n^i}{N} \left\| \mu^k - \mu^i \right\|_2 \right) \tag{6}$$

Where $n^i$ represents the pixel number of class $C^i$, and $N$ denotes the pixel number of all images, while $\mu^k$ and $\mu^i$ denote centers of the $k^{th}$ and the $i^{th}$ cluster, respectively.

In human visual system, the center region of the picture is more attractive than other regions [20]. With the increase of distance between the object and the image center, the attention gain depreciates. This is known as 'central bias criterion' in single image saliency detection. We extend this concept to the cluster-based methods, the spatial cue $w^s(k)$ of cluster $C^k$ is defined in Eq. (7).

$$w^s(k) = \frac{1}{n^k} \sum_{j=1}^{M} \sum_{i=1}^{N_j} \left[ N \left( \left\| z_i^j - o^j \right\|^2 \Big| 0, \sigma^2 \right) \cdot \delta \left[ \left( p_i^j \right) - C^k \right] \right] \tag{7}$$

Where $\delta(\cdot)$ is the Kronecker delta function, and $o^j$ is the center of the image $I^j$. The normalization coefficient $n^k$ is the pixel number of cluster $C^k$, and Gaussian kernel $N(\cdot)$ computes the Euclidean distance between pixel $z_j^i$ and the image center $o^j$. The variance $\sigma^2$ is the normalized radius of images.

In this paper, we use multiplicative fusion algorithm to calculate the co-salient probability $p(k)$ of cluster $C^k$ on cluster-level, as shown in Eq. (8).

$$p\left( C^k \right) = \prod_i w_i(k) \tag{8}$$

Where $w_i(k)$ represents contrast cue $w^c(k)$ and spatial cue $w^s(k)$. Based on cluster-level, salient value of pixel $x$ in cluster $C^k$ subjects to normal distribution $N$ significantly, and it is calculated as Eq. (9).

$$p\left( x \middle| C^k \right) = N \left( \left\| v_x, \mu^k \right\|_2 \middle| 0, \sigma_k^2 \right) \tag{9}$$

Where $v_x$ is the eigenvector of pixel $x$ and $\sigma_k$ is its variance in cluster $C^k$. Therefore, marginal probability $p(k)$ of saliency is obtained by summing the joint salient values $p(C^k) \cdot p(x| C^k)$ on all clusters, as shown in Eq. (10).

$$p(x) = \sum_{k=1}^{K} p\left( x, C^k \right) = \sum_{k=1}^{K} p\left( x \middle| C^k \right) \cdot p\left( C^k \right) \tag{10}$$

Finally, a co-salient map of multi-images in pixel-level is obtained.

We use the saliency region extraction algorithm mentioned above on a sequence of 6 key frames obtained in Sect. 2, and results are shown in Fig. 2. In Fig. 2, figure

(a) shows sequential key frames numbered $s1$ to $s6$ according to their sequences, and sub-figure (b) shows that corresponding co-salient map of each image. We can see that, in the sub-figure (b), most of the salient regions are distributed in the forehead and cheek area, and it shows that these regions vary less when the expression is going on.

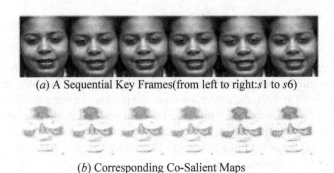

(*a*) A Sequential Key Frames(from left to right:$s1$ to $s6$)

(*b*) Corresponding Co-Salient Maps

**Fig. 2.** Collaborative saliency map extraction

Accordingly, we could remove these redundant portions and take the remainders as the facial expression features, as shown in Fig. 3. To more clearly display extracted LBP features after removing facial co-salient regions, In Fig. 3, we use only two images (*a*) and (*b*) corresponding to $s1$ frame (the first one from the left of Fig. 2(*a*)) and $s6$ frame (the last one from the left of Fig. 2(*a*)), as examples to show their feature maps after eliminating the co-salient regions in Fig. 2(*b*). Results of other frames are about similar with those of $s1$ frame and $s6$ frame.

(*a*) $s1$ Frame                    (*b*) $s6$ Frame

**Fig. 3.** Co-salient LBP maps corresponding to $s1$, $s6$ frame

In Fig. 3, only one person's sequential images are used to extract co-salient features, but there is commonality among different persons with the same expression. Accordingly, we perform some experiments on different people with the same expression, where each people's peak frame image is selected out from the expression sequential images, and then salient regions are cooperatively derived from these peak frame images as co-salient regions of expression, as shown in Fig. 4.

In Fig. 4, co-salient regions are extracted from different people with the same expression'happy', whose original images are as shown in Fig. 4(*a*). From Fig. 4(*b*), we can

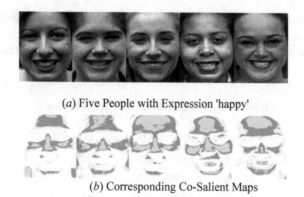

(*a*) Five People with Expression 'happy'

(*b*) Corresponding Co-Salient Maps

**Fig. 4.** Co-saliency maps of different individual's expression

see that eyes and mouth have large deformation, and forehead, cheeks and nose have low deformation strength, which are considered less significant compared with the former while representing expression 'happy'. If we exclude the latter regions from the expression features, amount of facial expression feature data would greatly reduce.

## 4   Experimental Results and Discussion

### 4.1   Video Down-Sampling Experiments

In order to illustrate the effectiveness and robustness of our proposed algorithm, we conduct down-sampling experiments on MMI database. MMI database is a positive database similar to CK+, whose video samples cover the whole process of facial expression development, starting with a neutral frame, going through a speeding up process, reaching its peak frame, and finally returning back to its neutral expression.

We used 196 video samples of six expressions from CK+ and MMI databases, where angry samples accounted for 16.3%, disgusted samples accounted for 14.7%, fear samples accounted for 14.2%, happy samples accounted for 17.8%, sad samples accounted for 16.3%, and astonished samples accounted for 20.4%, respectively.

### 4.2   Facial Expression Recognition Experiments

Firstly, co-salient regions in the facial expression image are derived from the sampled sequence frames by using our proposed algorithm. Secondly, these co-salient regions are got rid of from the original images, and then LBP texture features are extracted on the rest and taken as facial expression features. Let $y_j$ be the LBP eigenvector of the $j^{th}$ frame of video segmentation $\varphi^*$, and all $y_j$ ($j = 1,2,...$) consist of a set of feature vectors $\psi$ for all $\varphi^*$ segmentations. Finally, based on feature matrix $\psi$, SVM algorithm is used to train for facial expression classification according to their category label $C_i \in \{C_1,......,C_K\}$.

For comparison, the improved Adaptive DCT(Discrete Cosine Transform)-sample algorithm (AD-sample) and Active Appearance Model (AAM)+ Lucas Kanade (LK) method ALK mentioned in reference [21] are added. Experimental results are shown in Table 1.

**Table 1.** Experimental results on CK+ database

| Expression | ALK | AD-sample | Ours |
|---|---|---|---|
| Angry | 85.0% | 95% | 95% |
| Disgust | 82.7% | 85% | 90% |
| Fear | 87.8% | 83% | 85% |
| Happy | 83.7% | 90% | 100% |
| Sad | 79.3% | 85% | 91% |
| Surprised | 82.7% | 100% | 100% |
| Average | 83.5% | 89.8% | 93.5% |

Table 2 shows the experimental results of our algorithm, and the experimental results are as shown in Table 3 for MMI database.

**Table 2.** Experimental results of confuse percentage matrix for our algorithm

|  | Angry | Disgust | Fear | Happy | Sad | Surprise |
|---|---|---|---|---|---|---|
| Angry | 83.8% | 3.30% | 3.30% | 0% | 4.20% | 5.30% |
| Disgust | 5.50% | 80.7% | 5.40% | 0% | 2.10% | 6.30% |
| Fear | 7.50% | 3.30% | 83.3% | 0% | 0% | 5.90% |
| Happy | 2.30% | 2.50% | 1.50% | 85.6% | 0% | 8.10% |
| Sad | 6.40% | 4.40% | 0% | 0% | 81.1% | 8.10% |
| Surprise | 7.40% | 0% | 0% | 8.10% | 0% | 84.5% |

In Table 2 and Table 3, elements in the diagonal line of confuse percentage matrix represent recognizing rate of each type of expression. It can be seen from the above experimental results that the recognition rate of CK+ database is higher than the recognition rate of MMI database. The CK+ database, emoticons appear only once in each video, whereas the MMI database, expressions appear three times or more, so the expression data in CK+ database is simpler compared with that in MMI database. It can also be seen that the recognition rate of AD-sample is lower than that of our algorithm because our algorithm extracts salient regions and eliminates redundant regions, thereby reducing the number of features and ultimately improving the recognition accuracy.

**Table 3.** Experimental results on MMI database

| Expression | ALK | AD-sampl | Ous |
|---|---|---|---|
| Angry | 85.0% | 80.4% | 83.8% |
| Disgust | 82.7% | 77.2% | 84.7% |
| Fear | 87.8% | 77.1% | 87.3% |
| Happy | 83.7% | 82.5% | 93.6% |
| Sad | 79.3% | 79.2% | 85.1% |
| Surprised | 82.7% | 80.5% | 91.5% |
| Average | 83.5% | 79.5% | 87.7% |

In run time test, we completed two sets of experiments on each of the two databases separately. Finally, in the stages of data training and testing, our algorithm takes only 70% running times of the AD-sample algorithm.

## 5  Conclusion and Discussion

Spontaneous expression images conceal the emotion's imperceptible change in a mass of redundant information, and this paper presents an efficient co-salient regions segmentation algorithm, to get rid of subtly changing data from a sequence of facial images, and derives closely-related features for expression recognition. Accordingly, an improved performance of expression classification can be obtained. Images of different people with a same expression can also remove redundant information in this way, and it keeps the further study we plan to continue.

In this paper, based on the dynamic down-sampling model of human visual attention mechanism, and a down-sampling method based on the amount of information for facial expression is proposed. Then, the saliency region between facial expressions is extracted that using the saliency method and then removing the redundant area of facial expression. Experiments results on a standard dataset show that the proposed algorithm improves the facial recognition rate compared with the algorithm without sampling.

**Acknowledgements.** This paper is funded by Scientific Project of Guangdong Provincial Transport Department (No. Sci & Tec-2016-02-30), Surface Project of Natural Science Foundation of Guangdong Province (No. 2016A030313703 and 2016A030313713).

## References

1. He, M.H.: Spontaneous and artificial expression analysis of recognition studies. ANhui: University of Science and Technology of China, pp. 15–16 (2014)
2. Dahmane, M., Meunier, J.: Continuous emotion recognition using gabor energy filters. In: D'Mello, S., Graesser, A., Schuller, B., Martin, J.C. (eds.) ACII 2011. LNCS, vol. 6975, pp. 351–358. Springer, Heidelberg (2011). https://doi.org/10.1007/978-3-642-24571-8_46

3. Zhu, Y., Torre, F., Cohn, J.F., Zhang, Y.: Dynamic cascades with bidirectional boost strapping for action unit detection in spontaneous facial behavior. IEEE Trans. Affect. Comput. **2**(2), 79–91 (2011)

4. Glodek, M., et al.: Multiple classifier systems for the classification of audio-visual emotional states. In: D'Mello, S., Graesser, A., Schuller, B., Martin, J.C. (eds.) ACII 2011. LNCS, vol. 6975, pp. 359–368. Springer, Heidelberg (2011). https://doi.org/10.1007/978-3-642-24571-8_47

5. Dahmane, M., Meunier, J.: Continuous emotion recognition using gabor energy filters. In: D'Mello, S., Graesser, A., Schuller, B., Martin, J.C. (eds.) ACII 2011. LNCS, vol. 6975, pp. 351–358. Springer, Heidelberg (2011). https://doi.org/10.1007/978-3-642-24571-8_46

6. Savran, A., Cao, H., Shah, M., Nenkova, A., Verma, R.: Combining video, audio and lexical indicators of affect in spontaneous conversation via paricle filtering. In: Proceedings of the 14th ACM International Conference on Multimodal Interaction Workshops, pp. 485–492 (2012)

7. Dahmane, M., Meunier, J.: Continuous emotion recognition using gabor energy filters. In: D'Mello, S., Graesser, A., Schuller, Björn, Martin, J.C. (eds.) ACII 2011. LNCS, vol. 6975, pp. 351–358. Springer, Heidelberg (2011). https://doi.org/10.1007/978-3-642-24571-8_46

8. Zhu, Y., De la Torre, F., Cohn, J.F., et al.: Dynamic cascades with bidirectional bootstrapping for action unit detection in spontaneous facial behavior. IEEE Trans. Affect. Comput. **2**(2), 79–91 (2011)

9. Jiang, B., Valstar, M., Martinez, B., et al.: A dynamic appearance descriptor approach to facial actions temporal modeling. IEEE Trans. Cybern. **44**(2), 161–174 (2014)

10. Matas, J., Chum, O., Urban, M., et al.: Robust wide baseline stereo from maximally stable external regions. In: Proceedings of the British Machine Vision Conference, pp. 384–393 (2002)

11. Tuytelars, T., Van Gool, L.: Matching widely separated views based on affine invariant region. Int. J. Comput. Vis. **59**(1), 61–85 (2004). https://doi.org/10.1023/B:VISI.0000020671.28016.e8

12. Kadir, T., Zisserman, A., Brady, M.: An affine invariant salient region detector. In: Pajdla, T., Matas, J. (eds.) ECCV 2004. LNCS, vol. 3021, pp. 228–241. Springer, Heidelberg (2004). https://doi.org/10.1007/978-3-540-24670-1_18

13. Mikolajczyk, K., Tuytelars, T., Schmid, C., et al.: A comparison of affine region detectors. Int. J. Comput. Vis. **65**(1-2), 43–72 (2005). https://doi.org/10.1007/s11263-005-3848-x

14. Cai, H.P., Lei, L., Chen, T., et al.: A general approach for extracting affine invariant regions. J. Acta Electrinica Sin. **36**(4), 672–678 (2008). in Chinese

15. Ramirez, G.A., Baltrušaitis, T., Morency, L.P.: Modeling latent discriminative dynamic of multi-dimensional affective signals. In: D'Mello, S., Graesser, A., Schuller, B., Martin, J.C. (eds.) ACII 2011. LNCS, vol. 6975, pp. 396–406. Springer, Heidelberg (2011). https://doi.org/10.1007/978-3-642-24571-8_51

16. Nicolle, J., Rapp, V., Bailly, K., Prevost, L., Chetouani, M.: Robust continuous prediction of human emotions using multi-scale dynamic cues. In: Proceedings of the 14th ACM International Conference on Multimodal Interaction Work-shops, pp. 501–508 (2012)

17. Haber, R., Hershenson, M.: The Psychology of Visual Perception. Holt, Rinehart and Winston, New York (1973)

18. Huazhu, F., Xiao, C., Zhuowen, T.: Cluster-based co-saliency detection. IEEE Trans. Image Process. **22**(10), 3766–3778 (2013)

19. Lang, C., Nguyen, T.V., Katti, H., Yadati, K., Kankanhalli, M., Yan, S.: Depth matters: influence of depth cues on visual saliency. In: Fitzgibbon, A., Lazebnik, S., Perona, P., Sato, Y., Schmid, C. (eds.) ECCV 2012. LNCS, vol. 7573, pp. 101–115. Springer, Heidelberg (2012). https://doi.org/10.1007/978-3-642-33709-3_8

20. Hu, Y.X., Wan, L.: Heterogeneous image fusion based on visual attention mechanism. Comput. Eng. **41**(3), 247–252 (2015)
21. Shao, H., Wang, Y., et al.: Dynamic sequence emotional recognition based on AAM and optical flow method. Comput. Eng. Des. **38**(6), 1642–1656 (2017)

# Single Image Super-Resolution Reconstruction Using Nonlocal Low-Rank Prior

Zhongxing Zhang[1,2], Hui Liu[1,2(✉)], Qiang Guo[1,2], and Yuxiu Lin[1,2]

[1] School of Computer Science and Technology,
Shandong University of Finance and Economics, Jinan 250014, China
liuh_lh@sdufe.edu.cn
[2] Digital Media Technology Key Laboratory of Shandong Province,
Jinan 250014, China

**Abstract.** In most practical imaging applications, the images with high resolution is desired, but most of imaging images are generally low resolution in practice, which brings many problems. In this paper, we propose an effective image super-resolution reconstruction model using nonlocal low-rank prior. Firstly, this model uses the single image as data input, and the self-similarity inside the single image is used as prior knowledges to improve the matching degree of similar image patches. Then, the reconstruction progress is modeled with maximum a posterior probability framework. Finally, a nonlocal low-rank regularization is adopted to regularize the reconstruction process, which exploits the local and global information of image to improve the reconstruction effect. Experimental results show that the proposed method has achieved better results than the existing methods.

**Keywords:** Image super-resolution reconstruction · Maximum a posterior probability · Nonlocal low-rank regularization · Alternative minimization

## 1 Introduction

In vision-based practical applications, high resolution (HR) images are of great significance. Although image imaging technology has matured, the imaging image is generally low resolution (LR) due to the mutual constraints of imaging equipment, imaging environment and other factors. Super-resolution technology plays an essential role in improving image resolution, among which super-resolution reconstruction are widely used.

Tsai and Huang [1] proposed the method based on the frequency domain in 1984. This method is simple in theory and has certain advantages in the aspect of computation, but it is difficult to deal with noise and add priori knowledges. Due to the complex transformation relationship between the frequency domain and the spatial domain, it is hard to handle the situation with local motion. The

© Springer Nature Switzerland AG 2020
X. Chen et al. (Eds.): ML4CS 2020, LNCS 12488, pp. 290–296, 2020.
https://doi.org/10.1007/978-3-030-62463-7_27

iterative back projection method is proposed by Irani and Peleg [2]. Although the IBP method is intuitive and comprehensible, IBP is an inverse problem, which will cause non-unique solutions. The convex projection method is an iterative super-resolution reconstruction method. The POCS method is more flexible to introduce priori knowledges. However, the computation complexity of the method is high, which requires multiple iterations and projections. The maximum a posterior probability method (MAP) is an algorithm framework based on statistical probability, and the basic idea is derived from conditional probability. The LR image sequence is known as an observation result to estimate the unknown HR image. MAP has complete theoretical framework, flexible spatial domain model and the powerful capacity of containing prior knowledges, which shows that the MAP method has strong adaptability, flexibility and robustness. Thus this method can generate excellent reconstruction results, and it is an effective super-resolution reconstruction method.

In order to improve the quality of the reconstructed image, this paper applies the self-similarity of the image and the non-local low-rank prior (NLR) to the image super-resolution reconstruction technology, and combines MAP framework to establish a super-resolution reconstruction model (MAP-NLR) that fully utilizes the properties of the image. The experimental results show that the MAP-NLR has excellent performance than some advanced methods.

## 2  MAP-NLR Model

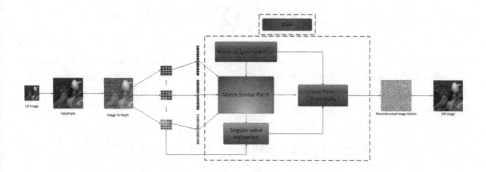

**Fig. 1.** MAP-NLR architecture

The MAP-NLR model can handle the low-resolution images. To exploit the non-local self-similarity of natural images, the method adopts non-local low-rank regularization terms to regularize the image reconstruction process. The reconstruction process of the model is shown in Fig. 1. Mathematically, image observation model is $y = Dx + n$, where $x$ is the high-resolution image to be reconstructed, $y$ is the low-resolution image, and $D$ is the degradation operator. $n$ represents the additive white Gaussian noise with the variance $\sigma_n^2$. The observation model can be formulated as the following MAP problem.

$$\hat{x} = \arg \max \log P(y|x)P(x) \tag{1}$$

$P(y|x)$ is the likelihood term characterized by the Gaussian distribution with variance $\sigma_n^2$, and $P(x) \propto \frac{1}{c} \exp(-\eta J(x))$ is set as Gibbs distribution. The MAP estimator of Eq. (1) leads to the following objective function.

$$\hat{x} = \arg \min \frac{1}{2\sigma_n^2} \|y - x\|_2^2 + \eta J(x) \tag{2}$$

In this paper, we adopt the nonlocal low-rank regularization model proposed in [3], which can also be formulated as a MAP estimator. Let $\tilde{\mathbf{R}}_j x \doteq [\mathbf{R}_{j_0}x, \mathbf{R}_{j_1}x, ..., \mathbf{R}_{j_{p-1}}x] \in \mathbb{R}^{n \times m}$ denote the matrix formed by the series of similar patches for exemplar image patch $x_j$ of size $\sqrt{n} \times \sqrt{m}$ centered at position $j$, where $j$ is the image patch index and $\tilde{R}_j$ denote the operator extracting the similar patches. For each exemplar patch $x_j$, the sets of similar patches can be formed by block matching, i.e., selecting the first $p$ closest patches to $x_j$ (including itself). Assuming that each group of similar patches are independent, the low-rank image prior can be expressed as

$$P(x) \propto \prod_j P(\tilde{R}_j x), \, P(\tilde{R}_j x) \propto \frac{1}{c} \exp(-\eta rank(\tilde{R}_j x)) \tag{3}$$

As minimizing the $rank(\cdot)$ is a NP-hard problem, the nuclear norm is often used as a convex surrogate of $rank$. By substituting Eq. (3) into Eq. (2), we have

$$\hat{x} = \arg \min \frac{1}{2\sigma_n^2} \|y - x\|_2^2 + \eta \sum_j \|L_j\|_* \, s.t. L_j = \tilde{R}_j x \tag{4}$$

where $L_j$ is the low-rank patch of $\tilde{R}_j x$, $\|L_j\|_*$ is the kernel norm of $L_j$ which represents the sum of singular values. To solve the objective function Eq. (4), the iterative direction multiplier is used to construct the augmented Lagrangian equation.

$$\hat{x} = \arg \min \frac{1}{2\sigma_n^2} \|y - x\|_2^2 + \mu \left\| \tilde{R}_j x - L_j + \frac{U_j}{2\mu} \right\|_F^2 + \eta \sum_j \|L_j\|_* \tag{5}$$

where $U_j$ is the Lagrangian multiplier, and $\mu$ is a constant value. Then, solving Eq. (5) is divided into two subproblems.

$$\hat{x} = \arg \min_x \frac{1}{2\sigma_n^2} \|y - x\|_2^2 + \mu \|\tilde{\mathbf{R}}_j x - L_j + \frac{U_j}{2\mu}\|_F^2,$$
$$L_j = \arg \min_{L_j} \mu \|\tilde{\mathbf{R}}_j x - L_j + \frac{U_j}{2\mu}\|_F^2 + \eta \|L_j\|_*$$

In this paper, the MAP estimation method is used to solve $L_j$, the singular value of $L_j$ is estimated using the singular value of $\tilde{R}_j x$.

$$\hat{\sigma}_i = \arg \max \log P(\sigma_i(L_j)|\sigma_i(\tilde{R}_j x)) \tag{6}$$

According to the Bayes theory, the Eq. (6) can be rewritten as

$$P(\sigma_i(L_j)|\sigma_i(\tilde{R}_jx)) = \frac{P(\sigma_i(L_j),\sigma_i(\tilde{R}_jx))}{P(\sigma_i(\tilde{R}_jx))} \propto P(\sigma_i(\tilde{R}_jx)|\sigma_i(L_j))P(\sigma_i(L_j)) \quad (7)$$

Supposing the distortion degree $f$ represents the degree of distortion of the singular value of the HR image patch and the LR image patch. In the above formula, the first part is fitted with Gaussian function , where the mean value is 0 and standard deviation is $f$,

$$P(\sigma_i(\mathbf{\tilde{R}}_jx)|\sigma_i(L_j)) = \frac{1}{\sqrt{2\pi}f_i} \exp(-\frac{\left(\sigma_i(\mathbf{\tilde{R}}_jx) - \sigma_i(L_j)\right)^2}{2f_i^2}) \quad (8)$$

$P(\sigma_i(L_j))$ is calculated by kernel density estimation and its probability density function is defined as the sum of a series of kernel functions, which the kernel function is determined by the neighborhood $\Omega_i$ size of 1×3 centered $\sigma_i(L_j)$. In the paper, the mean value of the kernel function is set to the Gaussian distribution with the mean value $\sigma_t(\tilde{R}_jx)$ and the standard deviation $h_i$. Then the probability density function of the singular value is defined as

$$P(\sigma_i(L_j)) \propto \sum_{\sigma_i(\tilde{R}_jx)\in\sigma(\tilde{R}_jx)} \frac{1}{\sqrt{2\pi}h_i} \exp(-\frac{(\sigma_i(L_j) - \sigma_t(\tilde{R}_jx))^2}{2h_i^2}) \quad (9)$$

By substituting the $i$ index of Eq. (8) and Eq. (9) into Eq. (7), we have

$$P(\sigma_i(L_j)|\sigma_i(\mathbf{\tilde{R}}_jx)) \propto \frac{1}{2\pi f_ih_i} \exp\left(-\left(\frac{\left(\sigma_i(\mathbf{\tilde{R}}_jx) - \sigma_i(L_j)\right)^2}{2f_i^2} + \frac{\left(\sigma_i(L_j) - \sigma_t(\mathbf{\tilde{R}}_jx)\right)^2}{2h_i^2}\right)\right)$$
$$(10)$$

Let the derivative of Eq. (10) be zero, the solution is $\hat{\sigma}_i(L_j) = \frac{f_j^2\sigma_t(\tilde{R}_jx)+h_i^2\sigma_i(\tilde{R}_jx)}{f_j^2+h_i^2}$.

Then, averaging the whole MAP estimator, i.e.,

$$\hat{\sigma}_i(L_j) = \frac{1}{|\Omega_i|} \sum_{\sigma_i(\tilde{R}_jx)\in\sigma(\tilde{R}_jx)} \frac{f_j^2\sigma_t(\tilde{R}_jx) + h_i^2\sigma_i(\tilde{R}_jx)}{f_j^2 + h_i^2} \quad (11)$$

Finally, the similar patch $\hat{L}_j$ of the low rank image is obtained,

$$\hat{L}_j = U(diag(\hat{\sigma}(L_j)))V^T \quad (12)$$

## 3   Experiment

In this section, we conduct experiments to test the model proposed in this paper. The standard optical flow experimental datasets are used as the test datasets, and the peak signal-to-noise ratio (PSNR) and structural similarity (SSIM) are adopted as evaluation indicators.

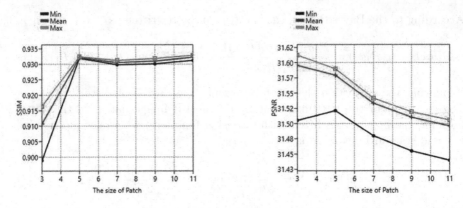

**Fig. 2.** The size of Patch comparison chart

## 3.1 Parameter Analysis

The complexity of the model is mainly determined by the size of image patch. In theory, the size of patch will not only influence the running speed of the experiment, but also the image registration accuracy and the quality of the reconstructed image.

Figure 2 shows that the indicators curve of the size of patch. It can be seen from figure that the patch of size 5×5, 7×7, 9×9, 11×11 have superior performance than the patch of size 3×3 in SSIM, and the patch of size 5×5 is the maximum point, comparatively. Except for 5×5, The patch of size 3×3 is exceed the patch size of 7×7, 9×9, 11×11, and the size 5×5 is only 0.0228 , 0.0161 lower than 3×3 in the maximum and mean value of PSNR, respectively, but the size 5×5 surpasses 3×3 in the minimum of PSNR. From the above, the MAP-NLR model sets image patch to the size 5×5.

## 3.2 Reconstruction Model Comparison

To test the proposed model, we compare MAP-NLR with other existing methods, such as: Nearest [4], Bicubic [5], LRTV [6], L2-L2 [7]. Figure 3 exhibits the result images of the MAP-NLR model and other reconstruction models, and Table 1 shows the values of indicators. It can be seen from the figure that all five models can reconstruct the HR image, but the quality of the reconstructed image is quite different. The HR image of MAP-NLR has the highest quality compared to other algorithms. The most important point is that the reconstructed images of the model have adequate high frequency information and retain the image details effectively. Meanwhile, Table 1 highlights the excellent performance of the MAP-NLR model for reconstructing the HR image in PSNR and SSIM.

**Table 1.** The values of PSNR(dB) and SSIM(%)

|          | Scale | Army          | Basketball    |
|----------|-------|---------------|---------------|
| Nearest  | ×2    | 29.05(0.85)   | 31.59(0.94)   |
| Bicubic  | ×2    | 31.38(0.90)   | 34.08(0.96)   |
| LRTV     | ×2    | 27.59(0.83)   | 31.81(0.95)   |
| L2-L2    | ×2    | 31.25(0.82)   | 34.74(0.88)   |
| MAP-NLR  | ×2    | **32.33(0.92)** | **35.23(0.97)** |
|          | Scale | Mequon        | *Mean*        |
| Nearest  | ×2    | 29.85(0.91)   | 30.16(0.90)   |
| Bicubic  | ×2    | 32.06(0.94)   | 32.51(0.93)   |
| LRTV     | ×2    | 28.33(0.91)   | 29.24(0.90)   |
| L2-L2    | ×2    | **32.94(0.86)** | 32.98(0.86)   |
| MAP-NLR  | ×2    | **32.94(0.95)** | **33.50(0.95)** |

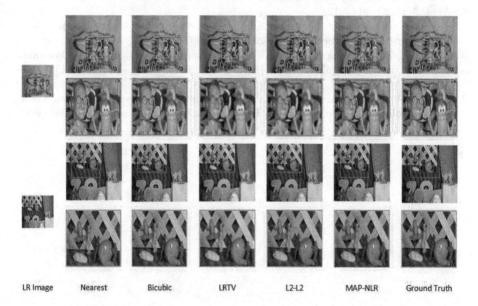

LR Image    Nearest    Bicubic    LRTV    L2-L2    MAP-NLR    Ground Truth

**Fig. 3.** Reconstructed results of different methods on test images

## 4    Conclusion and Future Work

In this paper, the MAP-NLR model is proposed for image super-resolution reconstruction. MAP-NLR takes into account the advantages of MAP framework, nonlocal self-similarity, and low rank in natural images. The experimental results show that the super-resolution model in this paper has excellent performance in image super-resolution reconstruction compared with other methods. However, some small details are lost in the low-rank truncation process. It is the focus of the next step to add the residual error to the model for maintain the image details further.

**Acknowledgement.** This study was supported in part by National Natural Science Foundation of China (Grant Nos. 62072274 and 61873145).

# References

1. Tsai, R.Y., Huang, T.S.: Multiframe image restoration and registration. Adv. Comput. Vis. Image Process. **8**(1), 317–339 (1984)
2. Irani, M., Peleg, S.: Improving resolution by image registration. Graph. Models Image Process. **53**(3), 231–239 (1991)
3. Huang, T., Dong, W., Xie, X., et al.: Mixed noise removal via laplacian scale mixture modeling and nonlocal low-rank approximation. IEEE Trans. Image Process. **26**(7), 3171–3186 (2017)
4. Olivier, R., Cao, H.: Nearest neighbor value interpolation. Int. J. Adv. Comput. Sci. Appl. **3**(4), 25–30 (2012)
5. Carey, W., Chuang, D., Hemami, S.: Regularity-preserving image interpolation. IEEE Trans. Image Process. **8**(9), 1293–1297 (1997)
6. Shi, F., Cheng, J., Wang, L., et al.: LRTV: MR image super-resolution with low-rank and total variation regularizations. IEEE Trans. Med. Imaging **34**(12), 2459–2466 (2015)
7. Zhao, N., Wei, Q., Basarab, A., et al.: Fast single image super-resolution using a new analytic solution for $\ell_2$-$\ell_2$ problems. IEEE Trans. Image Process. **25**(8), 3683–3697 (2016)

# Product Consistency Joint Detection Algorithm Based on Deep Learning

Bo Liu[1], Jianhou Gan[2(✉)], Jun Wang[2], and Bin Wen[1]

[1] School of Information Science and Technology, Yunnan Normal University,
Kunming 650500, China
[2] Key Laboratory of Education Informatization for Nationalities, Ministry of Education,
Yunnan Normal University, Kunming 650500, China
10800662@qq.com

**Abstract.** At present, the consistency detection of products mainly relies on manual method which has defects such as difficulty in identifying similar categories and low efficiency. Recently, the methods which are based on image classification have been widely used in it. However, image classification is based merely on simple visual features without making full use of other categrizable attributes. The approach this paper adopt to solve the problem is a joint detection algorithm which is based on deep learning. This algorithm locates product areas and implements image segmentation according to the default information code. Then it takes the segmented image as input of convolutional neural network model for classification. A reflection-assisted decision-making algorithm is used to solve the defect that the approximate class output probability is unstable. Finally, the consistency judgment is made according to the extracted default class information and product category. The experimental results obtained in this research show that the proposed algorithm can segment the products image accurately and quickly. The joint decision-making algorithm has better accuracy and stability than the single classification network.

**Keywords:** Product consistency · Image segmentation · CNN · Transfer learning · Joint decision

## 1 Introduction

China has now become the world's largest manufacturing country. Product consistency testing is an important step during the processes of production, transportation and distribution in the fields of both industrial manufacturing and logistics. For example, a reminder technology for product shelf life based on graphic recognition technology is proposed. By extracting the warranty date of the image containing the shelf life information, the inconsistent expired products are screened out [1].

The traditional product consistency test method uses manual visual inspection to observe whether there is product confusion or not through human eyes. This method is inefficient and highly dependent on personal experience. With the development of the

© Springer Nature Switzerland AG 2020
X. Chen et al. (Eds.): ML4CS 2020, LNCS 12488, pp. 297–311, 2020.
https://doi.org/10.1007/978-3-030-62463-7_28

Internet of Things technology, more automated methods of product consistency detection have been presented. L. Cui et al. proposed a new matching method in theory [2]. Matching the properties of the part to ensure consistent product assembly. C. H. Zhao et al. introduced the method of target anomaly detection by hyperspectral imaging technology [3, 4]. Hyperspectral imaging technology combines image with spectral analysis to express the external and internal features of the product. C. Xie et al. proposed a classification method to attach wireless sensors (Radio Frequency Identification, RFID) in products, then used it to carry or gather product information for classification purposes [5, 6]. K. Wamg et al. proposed a classification method that the product information is recorded by barcode, and the product information is decoded by the terminal to classify the product [7]. With the successful application of deep learning in the field of image processing and item classification, T. J. Li et al. uses a pre-trained convolutional neural network to identify the target object and obtain the location and category information of the target [8, 9].

In this paper, we mainly focus on solving the defects of similar items in the traditional product consistency detection method, such as difficulty in identifying similar items, cumbersome manual operation and slow processing speed. In this paper, a product consistency detection algorithm based on deep learning is proposed, which automatically extracts preset information and product images during the detection process, identifies product categories and makes product batch consistency judgments. Firstly, the method constructs an image classification model based on MobileNet network, and uses the transfer learning to train it. The last hidden layer is trained through fixing model parameters except the last hidden layer and the image classification model uses the collected predefined class data as input. Secondly, image segmentation is performed on the detected product image to extract a single product image. In order to reduce the operation complexity, a method of locating the product area by preset information code and implementing image segmentation is designed. At the same time, the preset standard class information extraction is completed. The classification model is used to classify the segmented product images. Aiming at the defects of the classification probability value which is too close and unstable when classifying similar material class by image, an auxiliary decision algorithm based on illumination reflection is proposed. Finally, the determination of product consistency is completed according to the preset standard class information and product classification results. The experimental results show that this algorithm can accurately segment the products image which is to be inspected and the processing time has a large-scale adaptability. Using the auxiliary decision algorithm will effectively change the distribution of the similar class output probability values and increase the variance. Meanwhile, it has a classification output function that corrects the original judgment error.

## 2   Related Work

### 2.1   CNN Introduction

Image object classification and detection are two important basic problems in computer vision research, and also the basis of other high-level visual tasks such as image segmentation, object tracking and behavior analysis [10]. In recent years, the deep learning has been successfully applied in technical areas including handwriting recognition, speech recognition, image recognition and natural language processing. Alex-Net has won an overwhelming victory in the ImageNet image classification competition in 2012. Subsequently, more powerful network structures such as VGGNet (Visual Geometry Group), Inception, ResNet (Deep Residual Network), etc. have been proposed. The classification accuracy rate can be comparable to or even exceed the human accuracy rate in the standard test set. But with the improvement of model accuracy is the huge cost of computation, storage space and energy consumption, they are difficult to accept for mobile applications. Google's team proposed a lightweight identification network MobileNet [11], which greatly reduces convolution computation, providing a near real-time processing speed. The emergence of these technologies has made it possible to detect product consistency based on image recognition.

Convolutional neural Network (CNN) consists of a feature extraction convolutional layer and a feature sampling layer. The convolution kernel with shared weights is used to extract the spatial features of the image [12]. As a feature filter, Convolution kernels of different sizes extract the feature of each scale field of view domain, and transform the input original image into a feature map. The traditional convolution kernel is manually designed to extract the established features such as edge detection. In CNN, the parameters in the convolution kernel are trained by a large number of labelled data. The convolution kernel is obtained after the network best fit with the training data set. CNN can effectively avoid the limitations of manual design features and greatly improve the accuracy of recognition. However, in the image classification task, the approximate class image has high similarity, which causes the output probability value of the CNN classification network to be too close or classified incorrectly.

## 2.2  MobileNet Model and Transfer Learning

MobileNet is an efficient, lightweight neural network structure proposed by Google. The goal is for visual applications used in mobile and embedded devices. In the model miniaturization method, the MobileNet model is based on the depth-wise separable convolutions, which can decompose the standard convolution into a deep convolution and a dot convolution. This approach is firstly calculated by using a two-dimensional convolution kernel with a channel number of one. After the process of channel-by-channel convolution is finished, the three-dimensional 1 * 1 convolution kernel is used to process the previously output feature map. After using the decomposition convolution, the amount of calculation is greatly reduced, while the accuracy decreases very little. In the network structure, deep convolution and point convolution are treated as two independent modules, and a large number of point convolution operations save considerable computation time. MobileNet proposes two hyperparameters: width multiple and resolution multiple. The width multiple is mainly to reduce the number of channels proportionally, and the resolution multiple is mainly to reduce the size of the feature map proportionally. The final experimental results show that the model parameters can be reduced to a large extent while ensuring the performance.

Transfer Learning [13] is a new machine learning method that uses the prior knowledge to solve problems in different but related domains. The purpose is to transfer the prior knowledge to solve target domain problems which only have a small number of training data sets or even have no labelled data [14]. People can reuse and inherit the knowledge they have learned in the current learning. Thus, their abilities are enhanced. In the field of machine learning, the same ability is represented by multiplexing the parameters obtained from other training sets which are used for new learning tasks. The advantage of transfer learning is that it is suitable for small datasets, avoiding the over-fitting of small data on large models and maintaining the superior feature extraction capability of large neural networks. Transfer learning is mainly implemented by freezing part of the convolutional layer of the pre-training model, which trains the remaining convolutional layers and fully connection layers.

## 3   Consistency Detection Based on Combination of CNN and Illumination Reflection

At present, the object image classification technology based on deep convolutional network is very mature and the classification accuracy is close to the human level in the ImageNet Challenge. However, the difficulties of applying it to product consistency detection include: 1. The confusion of product state makes it impossible to extract a single image effectively for further analysis. 2. The classification accuracy of products with high similarity is unacceptable if the image classification is the only one to be used. In order to solve the problems above, this paper proposes a product consistency detection algorithm which is shown in Fig. 1. The algorithm consists of data collection, preprocessing and joint decision making.

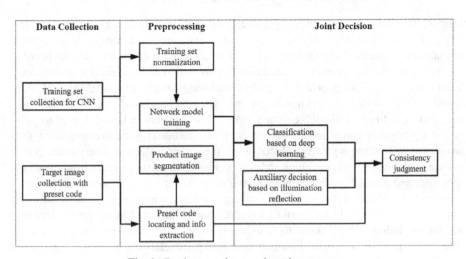

**Fig. 1.**  Product consistency detecting process

Data Collection: The specific data required to be collected at this stage include: a training set of the convolutional neural network and an image to be detected with a preset information code. The training data set could be collected from standard class and its historical confusion class. In the paper, they are product images which will be used for the transfer training of MobileNet. The acquired image which contains the product to be tested and the preset information code is used for image segmentation, preset information extraction and input data for consistency judgment.

Preprocessing: This stage consists of preset information code locating and information extraction, product image segmentation, training set normalization and model training. The preset information code positioning means to locate code position in the image based on the preset information code characteristics. The information extraction is parsing the standard class information contained in the code. The product image segmentation calculates the product area according to the positioning result before obtaining images of single product. The training set normalization is to sort the training set images according to categories, and then to remove the image which is blurred or has unsuitable size, even is obvious distorted. The image is transformed to meet the model input requirements. The model training uses a normalized training set to perform transfer learning training in the MobieNet. It is a method that includes three steps: 1. freezing hidden layer parameters except the last layer, 2. using the training set to train the last hidden layer, 3. changing the number of neurons in the output layer to meet the actual demand. Eventually, a custom training classification network model is obtained.

Joint Decision: This stage consists of object classification based on deep learning, auxiliary decision based on illumination reflection and consistency decision. The object classification based on deep learning obtains classification probability values by inputting the single-sample product image which is obtained from pre-processing into the classification model. If the probability distribution is consistent with the definition of similar classes, then the auxiliary system of illumination reflection is used for joint decision making. This auxiliary system converts the material reflectivity into the class probability value. The final probability is a weighted summation from probability values of network model and illumination reflection system. It is used to predict the single product class. The consistency judgment is made on the basis of each product class and the extracted standard class information, representing whether the batch products are consistent or not.

The main idea of the algorithm is to pre-define the position of the object which will be tested and segment the individual product images by the information code. Then deep neural network and auxiliary decision system is used in the step of classification. Finally, the product consistency is judged. The key points in the whole algorithm contain: 1. Product image segmentation. 2. Joint decision mechanism. 3. Consistency detection.

## 3.1 Product Image Segmentation

Image segmentation is an important preprocessing method in image processing and pattern recognition. The purpose of image segmentation is to divide the image into several non-overlapping sub-regions, so that the features in the same sub-region have certain similarities and the features among different sub-regions show obvious differences [15]. The macroscopic meaning is to extract semantic related regions. Since the training data for deep learning is mostly a single instance of this class, the parameters of model learned are fitting for a single object. However, there is a very low probability of occurrence in reality in the ideal single object image. The objects often cover each other. The input image of the classification network is distorted, missing or oversaturated. Thereby, the detection accuracy is reduced.

The traditional image segmentation method relies on the color, brightness and texture of the pixel to determine the similarity. It is easy to generate incorrect segmentation. Segmentation methods that is frequently used are threshold, boundary detection, regional method, etc. [16]. In recent years, the deep learning has been applied to the field of image segmentation. The idea is to transform the segmentation problem into a single pixel classification problem. Typical methods are Faster R-CNN (Regions with CNN) [17] based on the candidate region and the end-to-end method FCN (Fully Convolutional Networks) [18]. The accuracy is sufficient in these methods, but they also fail to solve the missing features caused by the mutual coverage of objects. If the product location is pre-standardized, the segmentation method of deep learning will result in significant computational waste and processing delay.

In this paper, a method to estimate the products area based on the preset product information code position and then segment the original image which a single product image will be obtained.

---

**Algorithm 1** Target image segmentation using preset information code

**Input:** the products to be tested and class information code

**Output:** the images of single product and class information Sinfo

**Step1.** Attaching the preset information code at the check station, ruling the product placement position according to the rules. The specific steps are as follows:

  **Step1.1.** Selecting a certain graphic coding method to contain information (such as barcode, QR (Quick Response) code, this paper experiment uses the QR code), and adding the product class information $S_{info}$ into the code.

  **Step1.2.** According to the shape characteristics of the preset code, the product placement area is designed. The length of the QR code is required to be $L_{qr}$. The product placement area is $a*L_{qr}: b*L_{qr}$, where $a$ and $b$ are the scaling parameters. In this area where the products are placed, it can be divided into $n$ blocks according to actual needs.

  **Step1.3.** The product to be tested is accurately placed in the specified single product area.

**Step2.** After confirmation, collecting the image which contains the product and preset information code. The specific steps are as follows:

  **Step2.1.** The test station puts the product in the specified position according to the configuration in step 1.3, after all the products are placed in the n areas. Send confirmation.

  **Step2.2.** After receiving the confirmation information, the system acquires the image to be detected. When the system receives the confirmation information sent in step 2.1, the image acquisition device is enabled to acquire the image to be detected. This image should include both the reserved information code and the whole product area.

  **Step2.3.** The image to be detected is transmitted. Transfer the image obtained in step 2.2 to the processing module.

**Step3.** Locate the preset code and segment test image, obtain images of single product and preset information. The specific steps are as follows:

**Step3.1** Binarizing the image obtained in step 3 and performing contour detection to obtain a contour set tree: $T_C = \{c_1: \{c_{11}, c_{12}... ...\}, c_2: \{c_{21}... ...\}... ...c_n\}$

**Step3.2.** The anchor point of preset code is filtered according to the nested relationship of outer frame and ratio of length $R_l$. The upper left and lower right positioning point $\{PC_{lu}, PC_{rd}\}$ of the preset code frame can be obtained. Thereby the preset code area $RG_C$ and its length $lqr$ in the image also can be obtained.

**Step3.3.** Using the known parameters $a$ and $b$ in step 1.2 and the length $l_{qr}$ in step 3.2, the pixel positioning points $\{PM_{lu}, PM_{rd}\}$ of product area to be detected are calculated. The product area $RG_M$ and the image tilt angle $\theta$ in the image are obtained.

**Step3.4.** If $\theta \neq 0$, rotate and correct image.

**Step3.5.** Using the predefined n-region assignment configuration in step 1.2, the single product region $RG_O$ is obtained.

**Step3.6.** The image is segmented according to $RG_M$ and $RG_O$. The image of preset code and the images of single product are obtained. The information code is decoded to obtain a reserved message $S_{info}$. The images of single product are saved.

end

---

## 3.2 Joint Decision Mechanism

The CNN-based classification network has a significant effect on classification among the more differentiated categories. The output class probability has a large variance and is highly practical. However, the classification results between approximate subclasses are not sufficiently robust and stable. In the experiment, similar classes occupy the main probability distribution. The probability values between similar classes are too close. A slight change in the detected image can result in a large change in the output value of the classification network, which ultimately leads to incorrect classification results. In product detection application, the approximate subclass is the main cause of confusion.

In addition to image, classification methods based on material detection are also widely used. For example, near infrared spectroscopy (Near Infrared, NIR) chemical imaging is a new analytical tool developed rapidly in recent years [19]. Its advantage lies in the analysis of material composition. At present, attempts have been made to combine spectral analysis with machine learning methods to detect object. J. Ning et al. use the method of near infrared spectroscopy combining with neural network to determine the degree of fermentation of Pu'er tea [20]. H. Jiang et al. propose a method that uses the combination of chemical composition data obtained by infrared spectroscopy and clustering method for testing plastic beverage bottles [21].

The classification probability values obtained by the neural network classification model are extremely close and fallible when the shape and color of objects are highly similar. In this paper, the material reflectivity is used as the auxiliary decision parameter to solve the confusion of approximate subclass. This method improves accuracy and stability of the consistency detection in a low cost manner.

---

**Algorithm 2** Product classification using joint decision mechanism

**Input:** The image of single product to be tested

**Output:** The set including class quantity information $S_{class}=\{class_0=a, class_1=b..., class_k=x\}$

**Step1.** Resizing the image of the single product obtained in product image segmentation step 3.6 to match the input size requirement of classification model in the preprocessing.

**Step2.** Converting image adjusted into a tensor as the input of classification model. Get the classification probability vector $V_{class}=[P_{class0}, P_{class1}, ... P_{classk}, P_{Blank}]$.

**Step3.** Sorting the components in $V_{class}$ from big to small and obtaining the maximum probability value $P_m$. If the difference between the first two probability values ($P_m$, $P_{m-1}$) is greater than 30% of the maximum probability value $P_m$, the product belong to the class of $P_m$, otherwise the illumination reflection assisted decision system is activated.

**Step4.** The light reflection system is composed of a light source, a receiving system and a data processing module. The selection of the light source requires a high luminous intensity in the wavelength range from 0.4 μm to 2.5 μm, and the divergence angle of the light source incident on the object is less than 10 mrad. The receiving system receives the light reflected back by the measured object. A test beam is emitted to the position where the product is placed, and the intensity of the reflected light is received by the photosensitive sensor. According to the definition of reflectivity the radiant energy reflected by the object accounts for the percentage of the total radiant energy. The data processing module converts the light intensity into an electrical signal by photoelectric converter and calculates the reflectivity $f_0$ according to the definition of reflectivity. The product auxiliary probability value $P_s$ is defined as shown in Equation (1):

$$P_s = \frac{\left|f_0 - f_{m-1_{std}}\right|}{\left|f_0 - f_{m_{std}}\right| + \left|f_0 - f_{m-1_{std}}\right|} \tag{1}$$

where $f_{(m)std}$ denotes the typical reflectivity of the material with the maximum probability of classification.

$f_{(m-1)std}$ denotes the typical reflectivity of the material with the second maximum probability of classification.

**Step 5.** The final classification probability value $P_f$ is defined as shown in Equation (2):

$$P_f = w_m * p_m + w_s * p_s \tag{2}$$

where $w_m$ and $w_s$ denote predefined weight parameters. If $P_f$ is greater than the classification threshold $T$, the product belongs to the class corresponding to $P_m$. Otherwise, the product belongs to the class corresponding to $P_{m-1}$.

**Step 6.** The images which are divided and saved in the segmentation step 3.6 are sequentially disposed by steps 1.1 to 1.5. Finnally, a set including class quantity information $S_{class}=\{class_0=a, class_1=b..., class_k=x\}$ is obtained.

**end**

---

## 3.3   Consistency Detection

Product consistency refers to the logo and structure of a batch product should be consistent with the type of qualified products. The macroscopic consistency is expressed in the same category as the product, and there is no other confused class. After the key points of the algorithm are proposed in this paper: product image segmentation and joint decision mechanisms are implemented. Accordingly, the class of each product in the batch can be confirmed and a consistency determination will be worked out conclusively. The output of classification can be obtained from historical confused data. When the new class of confused product appears, the existing algorithm fails. Then, the new confused class should be added to the classification model output, and the classification network can be reset. The adaptive consistency detection model can be obtained again after that.

The detailed steps for the consistency detection based on convolutional neural network combined with illumination reflection are:

**Step 1.** Collecting images of expected products, including standard and confused classes. Clearning and organizing the data: (1) removing image such as blur, unsuitable size, obvious distortion and incomplete items. (2) Adjusting the image with excessive tilt angle to improve classification accuracy. (3) Resizing to the input image.
**Step 2.** Training a classification neural network model using the training set in step 1.

> **Step 2.1.** Defining the input and output of the neural network. In the detected image, there may be cases that n detection areas are not filled. So a class named Blank must be added. It can be the background image of detection station. The final output is the probability value of k class, adding the probability value of class of Blank.
> **Step 2.2.** Constructing a convolutional neural network. In order to improve the accuracy of this model, a training method based on transfer learning is adopted. The output is probability values of $k + 1$ classes;
> **Step 2.3.** Dividing collected images into training sets, validation sets and test sets.
> **Step 2.4.** Defining the loss function and the accuracy calculation method.
> **Step 2.5.** Iterating training and updating parameters.

**Step 3.** Performing segmentation for the image of product to be tested
**Step 4.** The product image obtained in step 3 is processed through the classification model from step 2. If the classification probability values are too close, a joint decision mechanism is adopted.
**Step 5.** According to the preset information obtained in step 3 and the product class information obtained in step 4, it is determined whether the batch products are consistent or not. The standard class $Class_{std}$ is parsed from the preset information $S_{info}$ before traversing the product class list $S_{class}$ obtained in step 4. If there is only $Class_{std}$ in $S_{class}$ except $Class_{Blank}$, then the true of the consistency of the batch product can be determined, otherwise it is false.

# 4 Experiments

## 4.1 Experimental Setup

The experimental environment in this paper utilizes the deep learning framework Tensorflow and computer vision library OpenCV to implement the proposed algorithm. TensorFlow is an open source software library for high performance numerical computation whose flexible architecture allows easy deployment of computation across a variety of platforms and from desktops to clusters of servers to mobile and edge devices. It originally developed by researchers and engineers from the Google Brain team within Google's AI organization. OpenCV is free for both academic and commercial use. OpenCV was designed for computational efficiency with a strong focus on real-time applications. It can takes advantages of the hardware acceleration of the underlying heterogeneous compute platform.

The experimental data are normal commercial including plastic bottles, cans and glass bottles. Parts of plastic bottles and glass ones are similar in shapes and colors in the experimental data. The preset information code uses a QR code to record class information string: "plastic", "can" and "glass". The training set includes 4 types of images (adding class of Blank) and are resized to 224 * 224 pixels. The images quantity of each class are about 1000 and the capability is about 400 M.

The model used for transfer learning is MobileNet. The accuracy of the test set after training was about 93%. The chosen suitable single detection area size is 30 * 12 cm. When the number of product area n is 3, the scaling factor $a$ is 4, $b$ is 6.4.

The experimental items include:

- Testing the accuracy and execution time of the image segmentation algorithm on different detection sets (including the time of saving images).
- Testing the classification probability value of the typical sample to be tested only processed by Mobilnet classification network.
- Testing the classification probability value of the similarity sample after being processed by the joint decision mechanism (where $w_m$ is 0.4, $w_s$ is 0.6, $f_{std}$ of glass is 0.8, $f_{std}$ of plastic is 0.4,).

## 4.2 Results and Discussion

Experiment one: To test the accuracy and execution time of the image segmentation algorithm on different detection sets. In this experiment, image segmentation was performed on the image to be tested containing different numbers of products. The execution time of this algorithm is recorded, and the accuracy is obtained according to the integrity and correctness of the single product image after the segmentation. The experimental results are shown in Table 1 and Fig. 2 below:

**Table 1.** The results of image segmentation

| The number of detected objects | Accuracy rate | Execution time |
| --- | --- | --- |
| 3 | 100 | 0.505 s |
| 5 | 100 | 0.512 s |
| 10 | 100 | 0.527 s |
| 20 | 100 | 0.552 s |

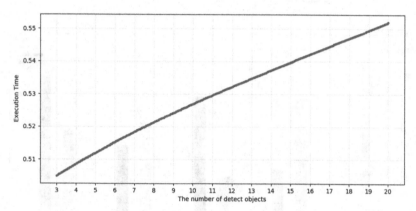

**Fig. 2.** The results of image segmentation

It can be seen from Table 1 and Fig. 2 obviously that this algorithm can accurately segment the area where the products to be tested are located. Observing the processing time, it can be inferred that the main processing time of this segmentation algorithm spend in is for locating QR code and then calculating its area. The linear increase of the amount of test set does not lead to a linear increase of the processing time. Thereby, this algorithm is advantageous for implementing large-scale detection.

Experiment two: To test the classification probability value of the typical sample to be tested only processed by Mobilnet classification network. In this experiment, three types of products (plastic bottles, glass bottles, and cans) in the test data were classified by merely using the MobileNet classification model. In order to test the classification accuracy of convolutional neural networks on appearance similar classes, two contrastive groups have been set. One group are higher in appearance similarity: plastic bottles (transparent) and glass bottles (transparent). While another group has lower appearance similarity: plastic bottles (opaque) and glass bottles (opaque).

Classification probability values are generated by MobileNet only. The experimental results are shown in Table 2 and Fig. 3 below:

**Table 2.** The classification results of MobileNet network

| Test sample | Plastic | Cans | Glass | Blank |
|---|---|---|---|---|
| Metal Cans | 0.06 | 99.91 | 0.03 | 4.4e-6 |
| Plastic Bottle (opaque) | 68.12 | 8.9e-6 | 31.88 | 7.1e-6 |
| Glass Bottle (opaque) | 28.26 | 2.13 | 68.55 | 1.06 |
| Plastic Bottle (transparent) | 55.23 | 0.51 | 44.15 | 0.11 |
| GlassBottle (transparent) | 51.21 | 0.03 | 48.73 | 0.03 |

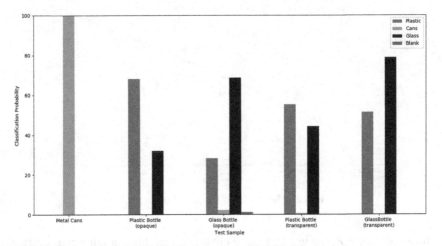

**Fig. 3.** The classification results of MobileNet network

It can be concluded from Table 2 and Fig. 3 that when MobileNet is used as the classification model only, the output probability values which come from the class with large difference in shape and color account for a large distribution space. This indicates that they can be availably judged by the classification model. But when the shape and color difference of the object to be tested are small, the output probability values between similar classes share little differences, even the classification result according to the maximum probability value is erroneous.

Experiment three: To test the classification probability value of the similarity sample after being processed by the joint decision mechanism. In this experiment, the products that are confused in experiment 2 are used as test data. They are plastic bottles (transparent) and glass bottles (transparent). The classification model is changed to the joint decision mechanism which is based on MobileNet network and illumination reflection assistance. The classification probability values are obtained with updated configuration. The typical parameters $w_m$ is 0.4, $w_s$ is 0.6, $f_{std}$ of glass is 0.8 and $f_{std}$ of plastic is 0.4. The experimental results are shown in Table 3 and Fig. 4:

It can be concluded from Table 3 and Fig. 4 that when the material-based joint decision system is added, the distribution of the similar class output probability values will

**Table 3.** The classification results of Joint Decision

| Test sample | Plastic | Cans | Glass | Blank |
|---|---|---|---|---|
| Plastic Bottle (transparent) | 37.09 | Null | 62.66 | Null |
| GlassBottle (transparent) | 70.48 | Null | 29.49 | Null |

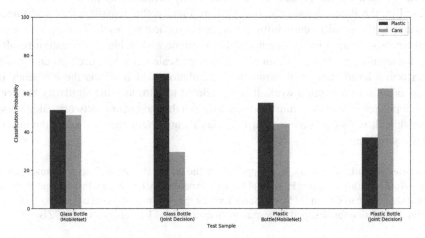

**Fig. 4.** The classification results of Joint Decision

be effectively changed and the variance will be increased. The auxiliary decision-making mechanism can correct the error classification output as well. The weight parameters $w_m$ and $w_s$ depend on the validity of the material detection method.

The essence of joint decision-making is to increase the data dimension and to use more categorizable attributes. The data distribution which is difficult to classify becomes easy to distinguish by adding new attributes. Therefore, the newly added features should be distributed independently from the image features. The weight $w$ represents the parameter of the optimal classification function $f(x)$. The parameter $w$ can be determined by the objective weighting method. This method means that the weight should reflect the influence degree of each feature on the final result. It can be generally obtained by normalizing the reliability of different features. Assume that the image classification accuracy is $p_i$ and the material classification accuracy is $p_m$. The weight vector $[w_m, w_s]$ can be obtained from normalizing the vector $[p_i, p_m]$.

Furthermore, the neural network is a nonlinear function of pixel features. The linear weighting method used in this paper still contains a large number of artificial designed factors such as weight selection. This leads to limited generalization of the model. In the future work, you can consider the material properties as input of the neural network and choose to add them in the appropriate layer. Then, the joint decision weight can be automatically learned by the machine to get stronger generalization ability.

## 5   Conclusion

This paper proposes a product consistency joint detection algorithm based on convolutional neural network and illumination reflection to solve the consistency problem in batch products. Firstly, the image to be detected is segmented to obtain a single product image to be tested and standard class information. Then, through the pre-trained MobileNet network, the product class is effectively identified by joint decision which is based on the illumination reflection system. Finally, the consistency is determined according to the standard class information and classification result. This algorithm are advantageous in rapid image segmentation, accurate and stable classification results. The shortcoming of this algorithm is that in large-scale detection, each product needs to perform a large number of parameters calculation, which affects the efficiency of this algorithm. In the future work, it is considered to introduce the similarity between product images for pre-determination and reduce redundant neural network calculation. Meanwhile, it is also possible to adjust the classification sequence and adopt the idea of cascade screening.

**Acknowledgement.** This work was supported by the National Natural Science Foundation of China (No. 61862068), Yunnan Provincial Smart Education Key Laboratory Project, and Program for innovative research team (in Science and Technology) in University of Yunnan Province, and Starting Foundation for Doctoral Research of Yunnan Normal University (No. 2017ZB013).

## References

1. Wang, B., Yunlu, W.: Research on product quality reminder technology based on graphic recognition technology. Technol. Innov. Appl. **29**, 6–7 (2016)
2. Lirong, C., Dejing, K., Haijun, L.: Matching via majorization for consistency of product quality. Q. Technol. Quant. Manag. **13**(4), 439–452 (2016)
3. Chunhui, Z., Xiaohui, L., et al.: Research advance on anomaly detection for hyperspectral imagery. J. Electron. Meas. Instrum. **28**(8), 803–811 (2014)
4. Ouyang Aiguo, O., Jian, W., et al.: Application of hyperspectral imaging in nondestructive detection of agricultural products. Guangdong Agric. Sci. **23**, 164–171 (2015)
5. Chuan, X., Ying, P.: Research of intelligent garbage collection collaboration based on RFID technology. J. Sic. Univ. **48**(3), 556–560 (2011)
6. Yuyan, S., Hong, Y., et al.: An intelligent logistics tracking system based on WSN. J. Comput. Res. Dev. **48**, 343–349 (2011)
7. Kun, W., Xiaoshan, L., et al.: Design of automatic sorting system for logistics package based on bar code recognition. J. Luoyang Institute Sci. Technol. **26**(2), 66–70 (2016)
8. Xiru, W., Guoming, H., et al.: Fast Visual identification and location algorithm for industrial sorting robots based on deep learning. Robot **38**(6), 711–719 (2016)
9. Tianjian, L., Bin, H., Jiangyu, L., et al.: Application of convolutional neural network object detection algorithm in logistics warehouse. Comput. Eng. **44**(6), 176–181 (2018)
10. KaiQi, H., Weiqiang, R., et al.: A review on image object classification and detection. Chin. J. Comput. **37**(6), 1225–1240 (2014)
11. Howard, A.G., Zhu, M., Chen, B., et al.: Mobilenets: efficient convolutional neural networks for mobile vision applications. arXiv:1704.04861v1 (2017)

12. Shun, Z., Yihong, G., et al.: The development of deep convolutional neural network and its applications on computer vision. Chin. J. Comput. **40**(144), 30 (2017)
13. Jialin, P.S., Qiang, Y.: A survey on transfer learning. IEEE Trans. Knowl. Data Eng. **22**(10), 1345–1359 (2010)
14. Zhuang, F.Z., Luo, P., He, Q., Shi, Z.Z.: Survey on transfer learning research. J. Softw. **26**(1), 26–39 (2015)
15. Lili, Z., Feng, J.: Survey on image segmentation methods. Appl. Res. Comput. **34**(7), 1921–1928 (2016)
16. Jiang, F., Gu, Q., Hao, H.Z., Li, N., Guo, Y.W., Chen, D.X.: Survey on content-based image segmentation methods. J. Softw. **28**(1), 160–183 (2017)
17. Ren, S., He, K., Girshick, R., Sun, J.: Faster R-CNN: towards real-time object detection with region proposal networks. IEEE Trans. Pattern Anal. Mach. Intell., 91–99 (2016)
18. Long, J., Shelhamer, E., Darrell, T.: Fully convolutional networks for semantic segmentation. In: Proceedings of the 28th IEEE Conference on CVPR, pp. 1337–1342. IEEE Computer Society, Washington (2015)
19. Chu, X., Lu, W.: Research and application progress of near infrared spectroscopy analytical technology in China in the past five years. Spectroscopy Spectral Anal. **34**(10), 2595–2605 (2014)
20. Jingming, N., Xiaochun, W., Zhengzhu, Z., et al.: Discriminating Fermentation Degree of Pu'er tea based on NIR spectroscopy and artificial neural network. Trans. Chin. Soc. Agric. Eng. **29**(11), 255–260 (2013)
21. Hong, J., Chenyang, J., et al.: Spectral analysis of plastic beverage bottles based on cluster analysis. Infrared Laser Eng. **47**(8), 358–363 (2018)

# Research on Image Recognition Method of Ethnic Costume Based on VGG

Qinchuan Lei[1], Bin Wen[1(✉)], Zhaoxiang Ouyang[2], Jianhou Gan[3], and Kangwei Wei[1]

[1] School of Information, Yunnan Normal University, Kunming 650500, China
315773170@qq.com
[2] School of Information, Dehong Teacher's College, Mangshi 678400, China
[3] Key Laboratory of National Education Informationization of Ministry of Education,
Yunnan Normal University, Kunming 650500, China

**Abstract.** This paper uses deep learning related algorithms to classify and recognize the ethnic costume images of the Wa and Yi people. In the deep learning framework Tensorflow, the deep convolution network VGG model was migrated to the ethnic costume recognition task, and the image recognition of ethnic costume based on VGG was realized. By adjusting the size of the convolution kernel and the number of convolution layers of the VGGNet network model, a neural network model of ethnic costume image recognition suitable for the classification and recognition of the Wa and Yi people is constructed. By training the images in the ethnic costume image library, iteratively adjusting the parameters of Batch_size, Epoch, Dropout and other parameters of the network model, the comparative experiments are on, and the classification recognition rate of ethnic costume images under different network model parameters is analyzed.

**Keywords:** Ethnic costume image · Convolution Neural Network · Image recognition

## 1 Introduction

Most of the ethnic minorities in various regions have their own unique costume totem culture. With the passage of time, these unique ethnic costumes still retain the style and cultural characteristics of their own ethnic groups, and can represent and reflect the history and cultural development process of their ethnic groups [1]. The current research on the digitalization of ethnic costumes mainly includes the digital protection of ethnic costumes, the construction of a digital image library for ethnic costumes, the construction of a library of ethnic costume elements and so on. Since ethnic costumes have more advantages than other ordinary natural images in terms of color, style, pattern and pattern, the use of digital images to protect the national costume culture has the more obvious technical advantages [2]. Zhao Haiying [3] mentioned that national museums and art galleries in France, Netherlands, United States, China and other places have used digital methods to protect, inherit and share traditional ethnic costume resources. Zhao Haiyan [4] established a relatively standardized education resource database of ethnic

© Springer Nature Switzerland AG 2020
X. Chen et al. (Eds.): ML4CS 2020, LNCS 12488, pp. 312–325, 2020.
https://doi.org/10.1007/978-3-030-62463-7_29

costume images. Gao Fei [5] took the field shots, then vectored and cutout the patterns in the images of ethnic costumes, so as to extract the pattern elements in the images of ethnic costumes and construct a library of ethnic costume elements. In recent years, image classification and recognition has been a hot research direction in image processing technology. As a branch of computer vision, image classification and recognition technology has achieved good applications [6]. With the emergence of convolution neural networks, the classification and recognition of images has been brought into a new research field, and some scholars have begun to build image recognition classification network models based on convolution neural networks. Convolution neural networks have also surpassed the traditional pattern recognition and machine learning algorithms in more and more fields, achieving top performance and accuracy. These achievements are mainly realized by increasing the number of neural network layers, increasing the number of training samples, and improving the training and learning algorithms [7, 8]. At present, the models for image classification and recognition based on convolution neural networks are LeNet [9], AlexNet [10], VGG [11], GoogLeNet [12], ResNet [13] and so on. In AlexNet, the GPU is used for network acceleration training for the first time, and a Dropout operation of randomly deactivated neurons is proposed to reduce the overfitting. Using the Relu activation function instead of the traditional Sigmoid and Tanh functions, the idea of LRN local response normalization is proposed. The VGG network model reduces the training parameters by stacking multiple $3 \times 3$ convolution kernels to replace the original large-scale convolution kernels. GoogLeNet was proposed by the Google team in 2014, and it won the first place in the classification task of the ImageNet competition in the same year. GoogLeNet introduces the Inception structure to fuse feature information of different scales, uses a $1 \times 1$ convolution kernel for latitude reduction and mapping processing, and adds two auxiliary classifiers to help training, discarding the fully connected layer and using the average pooling layer to greatly reduce the model parameter. ResNet was proposed by Microsoft Labs in 2015, and it also won the first place in the classification and target detection in the ImageNet competition in the same year, and the first place in target detection and image segmentation in the COCO dataset. ResNet uses an ultra-deep network structure (breaking through 1000 layers), proposes to use the residual module, discards Dropout and uses Batch Normalization to accelerate training. This article will focus on the costume images of the Wa and Yi people. Using deep learning technology, based on the VGG network model, a neural network model is built in the Tensorflow framework that can realize the classification recognition of the Wa and Yi ethnic costume images, and the problem of ethnic costume image classification recognition is converted into a process of objective function solution. By constantly iteratively optimizing and adjusting the parameters of the ethnic costume image classification and recognition network model, the accuracy of the network model classification and recognition in the prediction data set is improved, which provides reference for the construction of the ethnic costume image classification and recognition network model.

## 2 VGG-Based Image Recognition of Ethnic Costumes

### 2.1 Constructing a Network Model for Ethnic Costumes Recognition

The construction of the ethnic costumes classification and recognition model is mainly based on the VGG network, which is commonly used image classification model. Based on the VGG network model, the number of convolution kernels and the number of convolution layers are adjusted, and the deep convolution network VGG model is transferred to the task of ethnic costume recognition, and finally the classification and recognition of ethnic costume images are realized. VGGNet was proposed by the famous research group VGG (that is, Visual Geometry Group) of Oxford University in 2014, and it won the first place of the positioning task and the second place of the classification task in the ImageNet competition in the same year. The structure of the VGGNet model is shown in Fig. 1 [11]:

| ConvNet Configuration | | | | | |
|---|---|---|---|---|---|
| A | A-LRN | B | C | D | E |
| 11 weight layers | 11 weight layers | 13 weight layers | 16 weight layers | 16 weight layers | 19 weight layers |
| input (224 × 224 RGB image) | | | | | |
| conv3-64 | conv3-64 **LRN** | conv3-64 **conv3-64** | conv3-64 conv3-64 | conv3-64 conv3-64 | conv3-64 conv3-64 |
| maxpool | | | | | |
| conv3-128 | conv3-128 | conv3-128 **conv3-128** | conv3-128 conv3-128 | conv3-128 conv3-128 | conv3-128 conv3-128 |
| maxpool | | | | | |
| conv3-256 conv3-256 | conv3-256 conv3-256 | conv3-256 conv3-256 | conv3-256 conv3-256 **conv1-256** | conv3-256 conv3-256 **conv3-256** | conv3-256 conv3-256 **conv3-256** |
| maxpool | | | | | |
| conv3-512 conv3-512 | conv3-512 conv3-512 | conv3-512 conv3-512 | conv3-512 conv3-512 **conv1-512** | conv3-512 conv3-512 **conv3-512** | conv3-512 conv3-512 **conv3-512** |
| maxpool | | | | | |
| conv3-512 conv3-512 | conv3-512 conv3-512 | conv3-512 conv3-512 | conv3-512 conv3-512 **conv1-512** | conv3-512 conv3-512 **conv3-512** | conv3-512 conv3-512 **conv3-512** |
| maxpool | | | | | |
| FC-4096 | | | | | |
| FC-4096 | | | | | |
| FC-1000 | | | | | |
| soft-max | | | | | |

Fig. 1. VGGNet model structure diagram

There are mainly six different configurations in the VGGNet model, namely A, A-LRN, B, C, D, and E. In the six different configurations of VGGNet, there are different network depths that can be divided into VGG11, VGG13, VGG16 and VGG19. At the same time, the VGGNet model compares the effect of using LRN (local response normalization) on the final result under the same network depth and convolution kernel size. In the network models C and D of VGG16, we can see the comparison of the effect of the convolution kernel size of 1 and the convolution kernel size of 3. Because there are the small amount of data and the small scale in the classification and recognition of ethnic costume images, we is mainly based on the network model structure in the VGG16 network D and fine-tune the size and number of convolution kernels to complete the classification and recognition of ethnic costume images. In the process of constructing the neural network model of ethnic costume image classification and recognition, it is different from the classic VGG convolution neural network model. The depth of the convolution neural network and the choice of the size of the convolution kernel are very important. How to give full play to the advantages of the algorithm depends on the choice of the number of convolution layers and the choice of the size of the convolution kernel. The classic VGG16 neural network includes: convolution layer, ReLu layer, maxpool layer, fully connected layer and softmax layer. It is shown in Fig. 2 [11]:

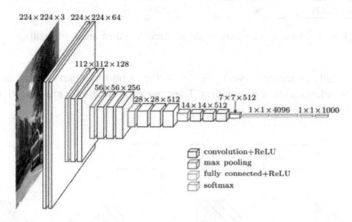

**Fig. 2.** Classic VGG16 network model structure

It can be seen from Fig. 2 that the classic VGG16 network totally contains 16 layers. The step size of the convolution kernel is 1, the padding value is 1, the size of the maxpool is 2, and the step size is 2. In order to reduce the required parameter training in the VGG16 network, the large-scale convolution kernel is replaced by stacking multiple 3 × 3 convolution kernels. In order to give full play to the advantages of the algorithm in the image recognition model of ethnic costumes, we replaced 5 × 5 convolution kernels by stacking two 3 × 3 convolution kernels and replaced 7 × 7 convolution kernel by stacking three 3 × 3 convolution kernels and other operations. A smaller convolution kernel structure will be introduced and modified to make this structure simple and reusable. The neural network structure used in this paper is shown in Fig. 3:

It can be seen from Fig. 3 that the image recognition model of ethnic costumes used in this paper is similar to the classic VGG16 network. Stacking multiple 3 × 3 convolution kernels, using the Relu activation function behind each convolution layer, and using Maxpooling for sub-sampling to complete feature extraction. The use of Dropuot random inactivation neuron operation in front of FC (fully connected layer) reduces over-fitting, and finally completes the recognition of ethnic costume images

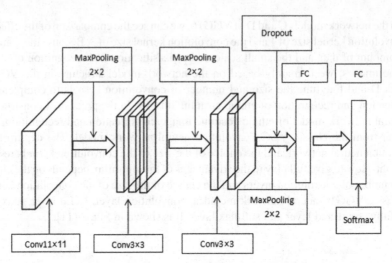

**Fig. 3.** The neural network model of ethnic costume image recognition

through FC (fully connected layer). We divide Fig. 3 into two parts, which are feature extraction and classification recognition. The network structure we introduced in feature extraction is shown in Fig. 4:

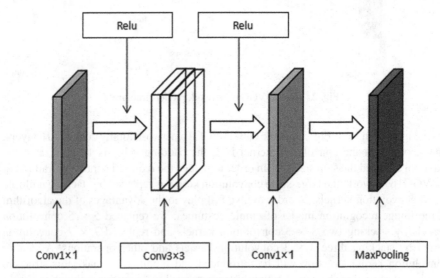

**Fig. 4.** Network structure introduced by feature extraction

In Fig. 4, $1 \times 1$ convolution kernel is used at the beginning and the end, which facilitates the connection of the upper and lower layers so that the neural layers of the convolution kernel can be directly connected. Using three $3 \times 3$ convolution kernels, firstly the features of the image are effective extracted, and secondly, it ensures that too many parameters are not introduced to like the larger convolution kernels. Each convolution layer in Figs. 3 and 4 contains a ReLu layer. Although the number of layers of the network is increased, the training time is reduced and the performance of the algorithm is improved. The neural network activation function used in the classic VGG neural network structure is the ReLu function. The ReLu function can indeed reduce training time, improve algorithm performance, and converge better than the sigmoid and tanh functions, so this paper prefers ReLu Activation function. In actual experiments, we found that increasing the depth of the network did not cause an explosive increase in training time. On the contrary, it is a slow growth process. More importantly is that the results obtained can be significantly improved. However, after only the convolution neural layer and ReLu processing, each pixel in the image contains only a small area of information around it, resulting in information redundancy. If you continue to use images that contain redundant information, not only the performance of the algorithm will be reduced, but also the translation invariance of the algorithm will be destroyed. In order to improve the performance of the algorithm, we used the MaxPooling layer to subsample the image. The MaxPooling operation divides the image into some small areas, calculates a value for each area, and then takes the calculated value to the maximum value. In order to make the image more contrastive, we also added a normalization layer in the convolution neural network structure. The normalization layer requires an algorithm for calculating the average value in order to increase the contrast of the image, and then adjusts each pixel of the image according to certain rules to make the main part and the back part of the image more distinguishable. Finally, using the Softmax regression layer outputs the costume image category with the highest probability, and it is more convenient to measure the accuracy of the algorithm.

After determining the Relu activation function and the construction of the network model, next we complete the optimization algorithm and solve the objective function. At present, the optimization algorithms commonly used in machine learning include gradient descent algorithm, second-order optimization algorithm, adjacent gradient descent algorithm, coordinate descent algorithm, alternating direction multiplier algorithm and so on, which are respectively applicable to different types of optimization problems. In this paper, the gradient descent method is used to optimization. The gradient descent method is a method of continuously iteratively solving the optimal value of the objective function along the direction of gradient descent, that is, using the negative gradient direction as the search direction. The calculation process of the gradient descent method is shown in Algorithm 1:

| | **Algorithm 1** Gradient descent algorithm |
|---|---|
| **Inupt:** | Input objective function $J(\theta)$; |
| **Output:** | Optimal solution of objective function $\theta_{k+1}$; |

(1)　Assign the initial value $\theta_k$ to the parameter $\theta$ to calculate the gradient function of the objective function, namely the reciprocal of the objective function $J(\theta)$ at $\theta_k$:

$$\nabla J(\theta_k) \tag{1}$$

(2)　Change the value of $\theta$ so that $J(\theta)$ to decrease in the direction of gradient descent:

$$\theta_{k+1} = \theta_k - \alpha \nabla J(\theta_k) \tag{2}$$

Where $\alpha$ represents the search step size in the negative direction of the gradient, and the step size is determined by the linear search method;

(3)　$\theta_{k+1}$ assign the value to $\theta_k$ in formula (1), and then iterate iteratively according to steps 1 to 3 until convergence, so that after multiple iterations, the optimal value $\theta_{k+1}$ is got.

The algorithm is implemented for different objective function attributes, and the improvement strategies used for gradient descent is also different. This paper also selects specific optimization algorithms based on specific objective functions.

## 2.2 Build a Network Model Based on TensorFlow Framework

Based on the basic structure of the constructed network, we will build an ethnic costume recognition network based on the TensorFlow framework. First, we must calculate the parameters of each layer in the ethnic costumes recognition network, which are as shown in Table 1:

**Table 1.** Tensorflow build neural network parameters

| Layer_name | Kernel_size | Kernel_num | Padding | Stride |
|---|---|---|---|---|
| Conv1 | 11 | 48 | [1, 2] | 4 |
| Maxpool1 | 3 | Null | 0 | 2 |
| Conv2 | 3 | 128 | [2] | 1 |
| Conv3 | 3 | 128 | [2] | 1 |
| Maxpool2 | 3 | Null | 0 | 2 |
| Conv4 | 3 | 192 | [1] | 1 |
| Conv5 | 3 | 192 | [1] | 1 |
| Conv6 | 3 | 128 | [1] | 2 |
| Maxpool3 | 3 | Null | 0 | 2 |
| FC1 | 2048 | Null | Null | Null |
| FC2 | 1024 | Null | Null | Null |
| FC3 | 2 | Nul | Null | Null |

After the network parameters are set, a neural network model is built based on the Tensorflow framework. The main steps are as follows:

Step 1: Create a Model to define the parameters of the network, including the height and width of the image, and the categories of classification.

Step 2: Create a convolution layer, Maxpool layer, etc. in the ethnic costumes recognition network based on the Tensorflow framework. The parameters for creating the convolution layer and Maxpool layer are shown in Table 1. The convolution layer network parameters of the second and third layers have replaced the $5 \times 5$ convolution kernels in the original structure with two $3 \times 3$ convolution kernels, and Relu activation function is used behind each convolution layer.

Step 3: Flatten the feature matrix. The feature matrix after the previous convolution and pooling is flattened to obtain a $6 \times 6 \times 128$ one-dimensional feature vector.

Step 4: Use the Dropout function in Tensorflow to randomly inactivate neurons by a certain percentage to prevent over-fitting. The parameter used in this paper is 0.2, which means that the rate of inactivation is 20%.

Step 5: Define the fully connected layer. The Relu activation function is also used when defining the fully connected layer. It is noted that the last layer of fully connected does not use the Relu activation function because the softmax processing is finally required to convert the output into a probability distribution. The number of output nodes in the last layer is the number of classifications. This paper is set to 2 to realize the recognition of two types of costume of the Wa and Yi nationalities.

After the above steps, the construction of a network model for ethnic costume recognition based on the Tensorflow framework was completed. Next, we will train the defined model and optimize the model parameters.

## 2.3  Ethnic Costumes Recognition Network Model Training

After the ethnic costumes recognition network model is built, the model needs to be called for training. At the same time, by continuously training the model to optimize the training parameters, the recognition efficiency is optimized. There are eight main steps in this course:

Step 1: Define the file locations of the training set and verification set.

Step 2: Save the weights of the training model.

Step 3: Input the parameters such as the width, height, batch_size, and number of training iterations of the training image. We define the batch_size as 32 and the number of iterations as 10 times.

Step 4: Define the image generator for the training set.

Step 5: Get the value of category and index.

Step 6: Define the image generator for the validation set.

Step 7: Instantiate the ethnic costumes recognition network.

Step 8: Train ethnic costumes recognition models. In the training of ethnic costume recognition models, we chose the Adam optimizer and set the initial learning rate to

0.0005, and then we printed the accuracy rate. At the same time, we monitor the loss rate of the validation set and save the weight value of the model that has the best recognition effect in training to complete the optimization of the training parameters.

After the above 8 steps, the training of ethnic costume recognition model was completed and the optimization of training parameters was completed.

### 2.4  Ethnic Costume Recognition Based on Tensorflow Framework

After completing the construction of the ethnic costume recognition network model, the ethnic costume images will be transferred to the network and the costume recognition will be carried out. There are three main steps:

Step 1: Input the parameters of the images to be recognized, including parameters such as the size of the image.
Step 2: Expand the dimension of the image to read its batch dimension.
Step 3: Instantiate the network structure and load it into the network model to complete the recognition of ethnic costumes.

After the above three steps, the reading and the classification and recognition of ethnic costume images are completed. The value type returned here is the most probable. The returned value determines the ethnic classification of the read image to complete the recognition.

## 3  Experiment and Analysis

### 3.1  Experiment Operating Environment

In order to complete the experiment of classification and recognition of ethnic costume images, we used the current mainstream deep learning development framework Tensor-Flow. TensorFlow is a deep learning framework released by Google in 2015. The initial version only supported symbolic programming. In 2019, Google launched the official version of TensorFlow 2, which improved many of the defects of the original TensorFlow 1.0 version and simplified the writing of code and the use of API. The official version of TensorFlow 2 is selected in the ethnic costume classification and recognition experiment. The configuration environment of the experiment operation is shown in Tables 2 and 3:

### 3.2  Experiment Results and Discussion

In order to analyze the effect of different parameter settings in the neural network model of ethnic costume recognition on the final classification and recognition rate results, we compared the classification recognition effects of the model under different parameters Batch_size, Dropout, Step_per_epoch. The structure of the two network models is the same, only some of the parameter settings are different, and the parameter settings are shown in Table 4 and Table 5 when training the network.

**Table 2.** Hardware configuration table

| Hardware | Configuration |
|---|---|
| CPU | Inter Xeon E5 V4 |
| GPU | Nvidia GTX 1080Ti*2 |
| RAM | 16 GB DDR4*2 |
| Hard disk | Samsung 500 GB SSD*2 |

**Table 3.** Software configuration environment

| Software | Configuration |
|---|---|
| Language | Python |
| Dependency library | OpenCV, Cudnn, CUDA |
| Deep learning framework | Tensorflow 2.0 |

**Table 4.** VGGNet_v1 training parameter settings

| Parameter | Parameter value |
|---|---|
| Train_set number | 6400 |
| Test_set number | 1600 |
| Batch_size | 32 |
| Dropout | 0.2 |
| Learning_rate | 0.0005 |
| Epoch | 10 |
| Step_per_epoch | 200 |

In the parameter setting table, Train_set number is the amount of training set data, Test_set number is the amount of test set data, Batch_size is the number of images trained for each batch of images, Dropout is the inactivation rate of randomly inactivated neurons in the fully connected layer, Learning_rate is Learning rate, Epoch is the number of training iterations, and Step_per_epoch is the number of iterations per round. After we set different parameters to the same train data set, we plotted the training loss and verification loss of the $v_1$ and $v_2$ models in the same chart, and at the same time plotted the training accuracy and the test accuracy on the same chart to compare.

As we can be seen from the experimental results in Fig. 5 and Fig. 6, the loss value decreases continuously with the increase of epoch during training, and the accuracy of image classification and recognition gradually increases with the increase of epoch. After 10 iterations, the accuracy rate of ethnic costume image classification and recognition reached 71.59%.

**Table 5.** VGGNet_v2 training parameter settings

| Parameter | Parameter value |
|---|---|
| Train_set number | 6400 |
| Test_set number | 1600 |
| Batch_size | 20 |
| Dropout | 0.25 |
| Learning_rate | 0.0005 |
| Epoch | 20 |
| Step_per_epoch | 320 |

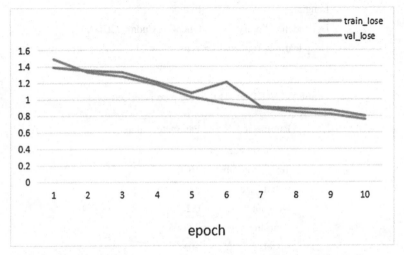

**Fig. 5.** Training loss and verification loss of VGGNet_v1 model

By analyzing the accuracy of model recognition in Fig. 7 and Fig. 8, we find that the accuracy of the network increasing the number of iteration steps to 20 times does not increase significantly with the increase of iteration steps but decreases. According to the neural network model comparing these two network parameters in Table 6, the accuracy of the classification and recognition of the Wa and Yi people's costume images was found to be set to 32, the Dropout value was set to 20%, and the Epoch value was set to 10, our network model has a better recognition rate.

### 3.3 Experimental Discussion

In summary, the results show that when we train the constructed network model, we do not set the training iteration number as large as possible. As the number of Epoch increases, the number of weight updates in the neural network model will also increase, easily causing over-fitting. At the same time, from the results in Table 6 and the change

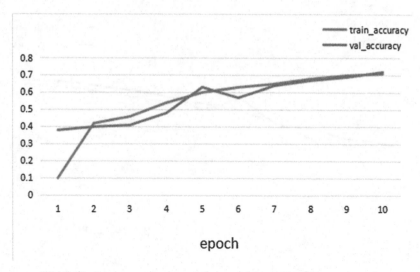

**Fig. 6.** Training set accuracy and test set accuracy of VGGNet_v1 model

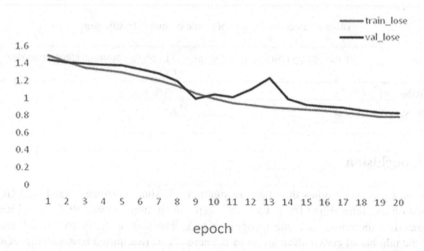

**Fig. 7.** Training loss and verification loss of VGGNet_v2 model

chart of training test accuracy rate, it can be seen that our ethnic costume classification and recognition network model based on VGGNet has a good classification and recognition effect on the input ethnic costume image data. Among them, in the VGGNet_v1 network model, the average recognition accuracy of Wa group costume images reached 71.78%, and the average recognition accuracy of Yi family costume images reached 70.08%.

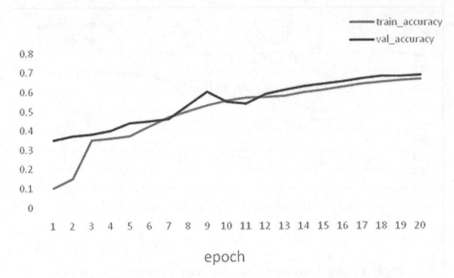

**Fig. 8.** Training set accuracy and test set accuracy of VGGNet_v2 model

**Table 6.** Recognition rate of ethnic costume classification

| Model | Wa nationality costume recognition rate (%) | Yi people's costume recognition rate (%) |
|---|---|---|
| VGGNet_v1 | 71.78 | 70.08 |
| VGGNet_v2 | 68.84 | 68.37 |

## 4 Conclusion

This paper mainly studies the image recognition of ethnic costumes based on VGG. Based on the TensorFlow framework, the deep convolution network VGG model was migrated to the ethnic costume recognition task. The size of the convolution kernel and the number of convolution layers of the network were adjusted based on the VGG neural network model, then a neural network model for classification and recognition the costume image is constructed. At the same time, adjusting the parameters of the model through the training model improves the accuracy of the model for the recognition of ethnic costume images. In the study of the classification and recognition of ethnic costume images, the recognition efficiency needs to be further improved and currently only needs training of the Wa and Yi nationalities. In the next step, we will try to improve some other convolution neural network models to realize more kinds of ethnic costume recognition.

**Acknowledgment.** The research is supported by a National Nature Science Fund Project (61862068), Yunnan Key Laboratory of Smart Education, Program for innovative research team (in Science and Technology) in University of Yunnan Province, Kunming Key Laboratory of Education Information, and Starting Foundation for Doctoral Research of Yunnan Normal University (2017ZB013).

# References

1. Ch, Y.: Feature Extraction and Recognition of Ethnic Costumes. Guizhou University for Nationalities, Guiyang (2018)
2. Sh, X.: Research and Implementation of Content-based Image Retrieval Technology of Ethnic Costume. Yunnan Normal University, Kunming (2016)
3. Zh, H.: Research on Digital Collection Standard of Traditional National Costume. Beijing University of Posts and Telecommunications, Beijing (2019). (Chinese)
4. Zh, H.: Research on Educational Resources Retrieval of National Costume Images Based on Convolutional Neural Network. Yunnan Normal University, Kunming (2018)
5. Fei, G.: Research on the Construction of a Digital Learning Platform for Ethnic Costume Patterns. Yunnan Normal University, Kunming (2018)
6. Lanying, L.: Construction of a Learning System for National Costumes Based on Image Recognition Technology. Yunnan Normal University, Kunming (2017)
7. Yunxing, G.: Research on Medical Image Super-Resolution Reconstruction Algorithm Based on Convolutional Neural Network. Jinan University, Guangzhou (2018)
8. Yuanyuan, W.: Research on Image Depth Estimation Based on Convolution Neural Network. Xi'an University of Technology, Xi'an (2018)
9. LeCun, Y., Bottou, L., Bengio, Y., Haffner, P.: Gradient-based learning applied to document recognition. Proc. IEEE **86**(11), 2278–2324 (1998)
10. Krizhevsky, A., Sutskever, I.I., Hinton, G.: ImageNet classification with deep convolutional neural networks. Commun. ACM **60**(6), 84–90 (2017)
11. Simonyan, K., Zisserman, A.: Very Deep Convolutional Networks for Large-Scale Image Recognition. https://arxiv.org/abs/1409.1556 (2014)
12. Christian, S., Wei, L., Yangqing, J., et al.: Going deeper with convolutions. In: 2015 IEEE Conference on Computer Vision and Pattern Recognition (CVPR). IEEE Computer Society, Boston (2015)
13. He, K., Zhang, X., Ren, S., et al.: Deep residual learning for image recognition. In: IEEE Conference on Computer Vision & Pattern Recognition, pp. 770–778. IEEE Computer Society, Las Vegas (2016)

# National Defense Education Resource Recommender of High Education Institutions Based on Knowledge-Aware Generative Adversarial Network

Lin Tang[1,2] and Lin Liu[3(✉)]

[1] Key Laboratory of Educational Informatization for Nationalities Ministry of Education, Yunnan Normal University, Kunming 650500, China
[2] Yunnan Key Laboratory of Smart Education, Kunming 650500, China
[3] School of Information, Yunnan Normal University, Kunming 650500, China
maitanweng2@163.com

**Abstract.** As there are issues of cold start and data sparsity in existing recommendation algorithms, this paper integrates the representation learning of knowledge graph into the recommendation process of generative adversarial model, and proposes a national defense education resource recommender of high education institutions based on knowledge-aware generative adversarial (KA-IRGAN). This model extracts the related entities and relations from the comment information of resource, and constructs the feature related knowledge sub graph of national defense education resource. Then the knowledge graph embedding method is utilized to transfer the entities in the knowledge sub graph as low-dimensional vectors, so as to obtain the low-dimensional feature vector of items and students for constructing the scoring function of item preference. The generative and discriminative modules are designed in the generative adversarial framework based on knowledge-aware. The generator and discriminator achieve Nash equilibrium in the process of iterative calculation, so that the final distribution of generative model is close to the real distribution of recommended resources. The experimental results on two benchmark datasets show that KA-IRGAN is superior to the existing IGRAN model in terms of recommendation accuracy and ranking quality, and is able to provide more efficient recommendation for national defense education system of high education institutions.

**Keywords:** National defense education · Recommendation system · Knowledge graph · Generative adversarial network

## 1 Introduction

Strengthening national defense education building, especially the national defense education for reserve talents in high education institutions, which has important significance to ensure the security of ethnic minority areas, promote national unity and consolidate national defense [1]. Many research institutions have made in-depth discussion about national defense education ideas in view of the "Internet +" in recent years

© Springer Nature Switzerland AG 2020
X. Chen et al. (Eds.): ML4CS 2020, LNCS 12488, pp. 326–335, 2020.
https://doi.org/10.1007/978-3-030-62463-7_30

[2]. Although current national defense education website integrates a large number of high quality resources of national defense education courses, experts and media, and so on, there are few studies and applications on personalized recommendation of national defense education resources.

At present, there are two difficulties in the research of national defense education resources recommendation for high education institutions. First of all, the resources of national defense education in high education institutions cover a wide range, not only including the curriculum resources of military theory formulated by the ministry of education, but also the resources of military technological trends, national policies, current affairs and news of border, which will contain a wealth of knowledge from national defense education fields. Secondly, the problems of cold start and data sparsity exist in the defense education resources recommendation. The consciousness of students' independent learning for defense education is still weak, so it is difficult to obtain the feedback data of students' rating on resources. A few of labeled feedback data can be obtained only by means of manual annotation, which cause the problem of data sparsity and "cold start" between new users and recommended resources.

With the introduction of generative adversarial networks (GAN) and knowledge graph (KG) into recommendation system, the above key problems of current studies have been solved in a certain degree in recent years. On the one hand, the application of GAN on the recommendation algorithm provides a new researching idea to the problem of data sparsity and cold start. The generator and discriminator can work collaboratively, so that the recommendation model is trained based on labeled and unlabeled data. This way of semi-supervised learning can not only make full use of the labeled data to learn the sample distribution, but also utilize the generator to generate supplementary samples for the discriminator. On the other hand, the rise of knowledge graph also provides a new idea for the improvement of personalized recommendation system. Knowledge graph represents the information in a structured form as a semantic network, which is understandable to people, and is easier to be organized, managed and utilized. Therefore, this paper propose an knowledge-aware information retrieval network (KA-IRGAN), which combine the knowledge graph and the theory of generative adversarial. This model makes full use of knowledge entity associations in the field of national defense education to obtain more accurate and efficient recommendation with the lack of labeled data.

## 2   Related Studies

Current researches of recommendation system based on deep learning can be roughly divided into three categories according to the idea of traditional recommendation algorithm: content-based recommendation algorithm, collaborative filtering and hybrid recommendation algorithm [3, 4].

For content-based recommendation, deep learning is mainly used to extract hidden features of items and users from item content, user preference and feedback information (such as users' browsing records), and generate recommendations by calculating the matching degree of hidden features of users and items. When there are rich auxiliary information for users and items, deep learning model is an effective tool of feature extraction. Researchers have carried out several researches of this field since 2013, which include multi-layer perceptron [5], convolutional neural network (CNN) [6, 7], recurrent neural network (RNN) [8] and deep belief network [9].

For the collaborative filtering algorithm based on deep learning, the feedback data between users and items is regarded as the training data of deep learning model. Through the model training, the hidden representation of users or items can be obtained, so as to provide the recommendation results. Researchers have successively applied multi-layer perceptron [10], restricted Boltzmann machine [11], autoencoder [12] and RNN [13] to collaborative filter since 2007. However, the insufficient feedback data will cause the problem of data sparsity in this kind of model. Especially, generative adversarial network has been introduced into the field of recommendation algorithm in 2017 [14]. This study integrated the classical generative retrieval model and the modern discriminative retrieval model into a unified "minimax game" framework. Hereafter, further studies have been carried out by integrating matrix factorization and RNN into the generative adversarial framework [15]. Nonetheless, these studies are still based on collaborative filtering algorithm, where a large amount of user-item feedback data is required.

## 3   The Problem Description and Model Framework

### 3.1   The Vector Representation of Resource Items and Students Based on Knowledge Graph

**Step1:** The "relevant entities" are extracted from the user's comments on the resource items, and the "feature related knowledge sub graph" of the defense education resource is constructed. Students' comments on national defense education resources are mainly based on text, pictures, videos and so on. There are not only a large number of attribute information related to these resources, but also student preferences for these resources. The entity in the comment is the basic language unit that carries the feature information. Among them, the most intuitive and convenient feedback information is the high-quality explicit feedback called "rating" (such as the user rating of a military video); when there is no explicit feedback, the student's online behavior (including browsing history, collection history, purchase history, etc.) can be used as the implicit feedback indirectly that reflect students' opinions.

Through named entity recognition (NER) and entity linking (EL) technology, this paper analyzes and extracts the related entities and their link relations corresponding to the knowledge map from the comment information of national defense education resources. In this paper, we extract the related entity set from the comment information by NER based on conditional random fields (CRF), to make a traversal for entity name in named entity database of KG. Then, the entities are linked to the corresponding entity in KG by EL, and the ambiguity problem between the entities is solved.

Given the students set $U = \{u_n\}_{n=1}^{N}$ and the defense education resource set $I = \{i_m\}_{m=1}^{M}$, the total number of students and resources are denoted as $N$ and $M$ respectively. Each item of $I$ has its related description information and students' comment information (such as name, type, function and so on). Then the resource items set selected by a specific student $u$ is $I_u = \{i_m^u\}_{m=1}^{P}$, and $P$ is the total number of resource items selected by student $u$. In the comment information of $p$ th ($p = 1, 2, \ldots, P$) choosed by student $u$, we can extract $k_p$ related entities $e_p^u$ from KG $G$, the related entity set of item $i_p^u$ was denoted as $R_p^u = \{e_{p,j}^u\}_{j=1}^{k_p}$. Therefore, according to the analysis of all items in $I^u$, we obtain the related entities set $R^u = \{R_p^u\}_{p=1}^{p_u}$ of student $u$ (based on history behaviors). Each entity of $R^u$ is connected by different correlation, so as to construct a "feature related knowledge sub graph".

**Step2**: Through the link relationship of KG, the feature related knowledge subgraph of student-item is expanded. In the process of extracting related entity in step 1, the number of entities is limited by the number of comments on the resource. If there is less comment information, there will be less related entities extracted from resource, that causes the problem of sparse data and single feature. In order to enrich the related entity set $R^u$ of student $u$, $R^u$ is regarded as the seed entities set in KG $G$, and is expanded to $k$ order related entity set of through the link relationship of KG. Given the KG $G$ and the set of $n$ seed entities $E(n < |E|)$, in the KG entity-relation-entity triple $(h, r, t) \in G$, if the head $h$ is the set of $k - 1$ order related entities $h \in E^{k-1}$, then the tail $t$ is the $k$ order related entities set: $E^k = \{t|(h, r, t) \in G, \ \& \ h \in E^{k-1}\}$, $k = 1, 2, \ldots, H$, where $H$ is the maximum order, so as to expand the related knowledge sub graph of defense education resource.

**Step3:** Using KG embedding method, the low-dimensional vector representation of entities in knowledge sub graph is obtained, and the low-dimensional vector representation of items and students in defense education resource is further obtained.

In this paper, TransR is used to represent head and tail entities in the KG entity-relation-entity triple $(h, r, t)$ as $k$ dimension vector $\mathbf{v}_h^k$ and $\mathbf{v}_t^k$, and the relationship $r$ is represented as $d$ dimension vector $\mathbf{v}_r^d$. Based on the $k \times d$ dimension transformation matrix $\mathbf{M}_r^{k \times d}$, the $k$ dimension head and tail entity vector is mapped to the $d$ dimension relation space: $\mathbf{v}_h^d = \mathbf{v}_h^k \mathbf{M}_r^{k \times d}$ and $\mathbf{v}_t^d = \mathbf{v}_t^k \mathbf{M}_r^{k \times d}$. The corresponding score function is defined as L2 norm $f = \|\mathbf{v}_h^d + \mathbf{v}_r^d - \mathbf{v}_t^d\|$.

The set of related entities extracted from the feedback information of item $i$ selected by student $u$ is expressed as $R_i^u = \{e_{i,j}^u\}_{j \in J}$, then each entity $e_{i,j}^u$ in $R_i^u$ represents the feature of items $i$ in different degree. Therefore, we give weight to different entitiy $e_{i,j}^u$ as $\omega_j$ on the basis of low-dimensional vector representation. The greater the weight $\omega_j$, the stronger the entity's $e_{i,j}^u$ ability to express the characteristics of item $i$; if the entity $e_{i,j}^u$ is not able to reflect the feature of item $i$, or only represent a few portion of feature, then $\omega_j \approx 0$. At this time, the vector of the items selected by student $u$ is expressed as $\mathbf{v}_i^u = \sum_{j \in J} \omega_j \mathbf{v}_{e_{i,j}^u}$. If the total number of items selected by student $u$ is $P$, then the vector of student $u$ is expressed as $\mathbf{u}^u = \mathbf{B}^u + \sum_{i=1}^{P} \mathbf{v}_i^u$, where $\mathbf{B}^u$ is the deviation of student $u$.

**Step4**: The implicit feedback matrix of student-item is constructed, and the scoring function of student-item pair is calculated based on knowledge-aware.

The implicit feedback is collected through students' online operation records, which is a useful complement to item feature information when there is little or no students' comments (explicit feedback). Given the set of students $U = \{u_n\}_{n=1}^{N}$ and the set of defense education resource items $I = \{i_m\}_{m=1}^{M}$, the implicit feedback of student $u$ to item $i$ is denoted by $Q_i^u$, then $Q_i^u = 1$ means that student $u$ has given feedback to item $i$, otherwise $Q_i^u = 0$. By this way, an implicit feedback matrix $\mathbf{Q}$ with $N \times M$ dimension can be constructed. The score of student $u$ to item $i$ is expressed as:

$$p(u, \ i) = \sigma\left(\mathbf{u}^u \odot \mathbf{v}_i^u\right), \quad p(u, \ i) > 0 \tag{1}$$

Among the above equation, $\mathbf{u}^u$ and $\mathbf{v}_i^u$ are vector representations of student $u$ and item $i$ obtained by step 1 to step 3. $\odot$ denotes the inner product of vectors, and $\sigma$ is sigmoid function used to calculate score of $(u, i)$.

## 3.2  Recommendation Process of IRGAN Framework

In IRGAN framework, the student's information set $Q = \{q_u\}_{u \in U}$ (either preference information or students' historical behavior data), the items set $I = \{i_m\}_{m=1}^M$, and an item recommendation rating matrix $\mathbf{R} = \{r_{ui}\}_{u \in U, i \in I}$ with $N \times M$ dimension are given. $r_{ui}$ represents the rating of student $u$ to item $i$, then the recommendation process can be described as obtaining a list of items $I^u$ related with $q_u$ based on student's information $q_u$. Assuming that there is a real distribution $p_{true}(i|q_u, r)$ of items related to user information $q_u$, and the relevance degree of $q_u$ and item $i$ is $r$. The generator try to fit the real distribution $p_{true}$ as much as possible according to the input user information $q_u$, and generates "pseudo items" $i'$ with a probability of $G_\theta(i'|q_u, r)$. Meanwhile, the discriminator $D$ attempts to distinguish whether the current item $i$ is real or "pseudo items" $i'$ from the generator $G$, and provides the relevance degree $r$ of current item $i$ and $q_u$, so as to make the generator $G$ approaching to the real distribution. Iterative learning is carried out by maximizing and minimizing the objective function shown in Eq. (2) respectively. Finally, the generator $G$ and the discriminator $D$ are optimized together in the strategic game, and the Top-N items are recommended by the optimized rating $r$ for student $u$.

$$J^{\theta^*, \phi^*} = \min_\theta \max_\phi \sum_{u \in U} (\mathbb{E}_{p_{true}(i|q_u, r)}[\log D_\phi(i|q_u, r)]$$
$$+ \mathbb{E}_{G_\theta(i'|q_u, r)}[\log(1 - D_\phi(G_\theta(i'|q_u, r)))]) \tag{2}$$

Among the above Eq. (2), $f_\phi(q_u, i)$ is a discriminative recommendation model (or discriminative scoring function). $f_\phi$ is used to estimate the probability $D_\phi$ of student information $q_u$ associated item $i$ by the sigmoid function $\sigma$ of discriminative scoring.

$$D_\phi(i|q_u, r) = \sigma(f_\phi(q_u, i)) = \frac{\exp(f_\phi(q_u, i))}{1 + \exp(f_\phi(q_u, i))} \tag{3}$$

The parameter $\phi$ of $D_\phi$ is solved by Eq. (4) using the stochastic gradient descent method ($f_\phi$ is differentiable for $\phi$):

$$\phi^* = \arg\max_\phi \sum_{u \in U} (\mathbb{E}_{p_{true}(i|q_u, r)}[\log \sigma(f_\phi(q_u, i))]$$
$$+ \mathbb{E}_{G_{\theta^*}(i'|q_u, r)}[\log(1 - \sigma(f_\phi(q_u, i')))]) \tag{4}$$

If the generative recommendation model is $g_\theta(q_u, i')$, then the probability $G_\theta$ of selecting a specific item $i'$ from $I = \{i_m\}_{m=1}^M$ can be provided by the softmax function $\delta$:

$$G_\theta(i'|q_u, r) = \delta(g_\theta(q_u, i')) = \frac{\exp(g_\theta(q_u, i'))}{\sum_{i \in I} \exp(g_\theta(q_u, i))} \tag{5}$$

The parameter $\theta$ of $G_\theta$ is updating by Eq. (6)

$$\theta^* = \arg\min_\theta \sum_{u\in U} \mathbb{E}_{G_\theta(i'|q_u,r)}\left[\log\left(1 - \sigma\left(f_{\phi^*}(q_u, i')\right)\right)\right]$$
$$= \arg\max_\theta \sum_{u\in U} \mathbb{E}_{G_\theta(i'|q_u,r)}\left[\log\left(1 + \exp\left(f_{\phi^*}(q_u, i')\right)\right)\right] \tag{6}$$

In the field of recommender, the generator $G_\theta(i'|q_u, r)$ is a discrete distribution of item $i' \in I$, which is updated by the strategic gradient descent method of reinforcement learning. It can be seen from Eq. (6) that the generative process can be regarded as a simple single step Markov decision process (MDP). $G_\theta$ is the probability of current state $(q_u, i)$ to the next state $(q_u, i')$, and $r$ is the reward of environmental feedback. $\log(1 + \exp(f_\phi(q_u, i')))$ severs as a strategy and an item is selected from candidate, that is the feedback reward to the next state. In order to reduce the variance caused by the strategic gradient, we can take the strategic gradient as the baseline by setting it to 1, which is replaced by Eq. (7):

$$V(q_u, i) = 2\sigma f_\phi(q_u, i) - 1 \tag{7}$$

### 3.3 IRGAN Recommendation Algorithm Based on Knowledge-Aware

To integrate the KG embedding learning process into the recommendation process of IRGAN, this paper proposes a knowledge-aware IRGAN (KA-IRGAN) framework.

In IRGAN, the hidden feature vectors $(\mathbf{u}^u, \mathbf{v}_i^u)$ are obtained by matrix factorization of the item recommendation score matrix $\mathbf{R} = \{r_{ui}\}_{u\in U, i\in I}$, and then define the preference score function $s(u, i) = \mathbf{u}^u \odot \mathbf{v}_i^u + b_i$. The difference of our approach with IRGAN is that, we first represent users and resource items as low-dimensional vector $\mathbf{u}^u$ and $\mathbf{v}_i^u$ based on KG through steps 1–3 in Sect. 3.1. Then, the generative recommendation model is defined as follows:

$$g_\theta(q_u, i) = \mathbf{u}^u \odot \mathbf{v}_i^u + b_i \tag{8}$$

The probability of selecting a specific item $i'$ is $G_\theta(i'|q_u, r)$ that the probability of state-transition is obtained from Eq. (5). Then generated (selected) "pseudo" student-item pair that $(u, i) \xrightarrow{G} (u, i')$ is input to the discriminator $D_\phi$. $D_\phi$ identifies student-item sample pair that input, and take the student-item preference score as the discriminative score or feedback reward:

$$f_\phi(q_u, i) = p(u, i) = \sigma\left(\mathbf{u}^u \odot \mathbf{v}_i^u\right) \tag{9}$$

It is passed to the generator $G_\theta$. As the influence of global bias and user bias will be gradually eliminated in top-N recommendation, so they are ignored in Eqs. (8) and (9). The model parameter $\phi$ and $\theta$ are updated by Eq. (4) and (6). With the continuous iteration, an optimal distribution is finally obtained, that is the recommended items list of student.

# 4 Experimental Result

## 4.1 Experiment Environment

For the hardware environment, the Intel Core i7-4720HQ@2.60 GHz of quad-core processors and the internal storage with 8.00 GB are deployed, which also supports GTX960 graphics card of CUDA7.5 framework. For software environment, there are Windows 10 x64, pcharm compiling environment, Anaconda 5.1 with Python 3.5 interpreter and CPU version of tensorflow1.12.1.

## 4.2 Datasets and Experimental Setting

First of all, we have carried out a wide range of data collection and organization various defense education websites, and obtained a certain amount of data. There are 1047 of text resource data and 2304 of military theory course text data obtained from websites. The statistics of datasets is shown in Table 1:

**Table 1.** Statistics of national defense education resources.

| Resource Website | Military theory | Defense mobilization | Patriotism education | Military technology | Defense news | Policies & regulations | Others |
|---|---|---|---|---|---|---|---|
| http://www.gfjyzx.com/ | 19 | 13 | 18 | 31 | 38 | 0 | 15 |
| http://www.guofangjiaoyu.net/ | 33 | 3 | 4 | 0 | 74 | 38 | 21 |
| http://www.mod.gov.cn/edu | 10 | 7 | 412 | 12 | 21 | 15 | 13 |
| https://military.china.com/gfjy/ | 15 | 12 | 60 | 18 | 22 | 0 | 14 |
| http://www.gf81.com.cn/ | 20 | 15 | 15 | 12 | 24 | 0 | 23 |

Secondly, to compare with other collaborative filtering recommendation algorithms, especially IRGAN model, two datasets that Movielens (100k) and FilmTrust are adopted in our experiment. To do experiments. These two datasets are randomly divided into training and testing datasets based on the ratio of 4:1. MovieLens dataset is about movie scoring proposed by GroupLens team of the University of Minnesota, which was divided into many sub datasets by size, such as 100k, 1M, 10M, 20M and the latest dataset. FilmTrust is a small dataset crawled from movie website in 2011. The Movielens (100k) and FilmTrust dataset are used to compare KA-IRGAN with the original IRGAN model. The evaluation criteria used in experiment includes: the "accuracy" that reflect the proportion of students' interested items in all items (Precision@k), the normalized discounted cumulative gain(NDCG@k) usually used to measure ranking quality, and mean average precision (MAP).

### 4.3 Experimental Results and Analysis

The comparison results of KA-IRGAN and original IRGAN on Movielens (100k) and FilmTrust datasets are shown in Table 2 and Table 3. The experimental results are illustrated in Fig. 1, which shows the change of Precision@k with the number of recommended items k on Movielens (100k) dataset.

**Table 2.** Item recommendation results and performance (Movielens-100 k).

|          | Pre@5  | Pre@10 | NDCG@5 | NDCG@10 | MAP    |
|----------|--------|--------|--------|---------|--------|
| IRGAN    | 0.3750 | 0.3140 | 0.4009 | 0.3723  | 0.2418 |
| KA-IRGAN | 0.4789 | 0.4465 | 0.4734 | 0.4115  | 0.2632 |

**Table 3.** Item recommendation results and performance (FilmTrust).

|          | Pre@5  | Pre@10 | NDCG@5 | NDCG@10 | MAP    |
|----------|--------|--------|--------|---------|--------|
| IRGAN    | 0.2928 | 0.2671 | 0.3155 | 0.2913  | 0.2120 |
| KA-IRGAN | 0.3854 | 0.3247 | 0.3334 | 0.3480  | 0.2632 |

As we can see from the experimental results, KA-IRGAN has obtained significant improvement on Precision@k and NDCG@k, when the number of recommended items $k = 5$ and $k = 10$, and get better results than IGRAN on three evaluation criteria. Meanwhile, the values of Precision@k gradually decreases with the increase of the recommended items number k, which tends to be stable when $k = 15$. These experimental results indicate that KA-IRGAN has a reliable training process compared with IRGAN, and has the distinct advantage to complete resource recommendation owing to the knowledge representing abilities of KG.

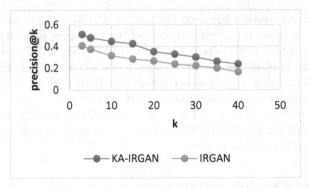

**Fig. 1.** The comparison of accuracy between KA-IRGAN and IRGAN with different k

## 5 Conclusion

For the personalized recommendation of national defense education resource of high education institutions, there are issues of cold start and data sparsity due to insufficient analysis of resource and student feature in existing recommender studies. This paper proposes a national defense education resource recommender of high education institutions based on knowledge-aware generative adversarial named KA-IRGAN. Based on the comment information of student on resource items and historical behavior in the network, this model introduces knowledge graph embedding method to learn the low dimension feature of national defense education resources and students, and integrate these features into generative adversarial model of information retrieval. The parameters of the generator and discriminator are optimized through minimax game iteration, where the distribution of generator is approaching to the real item distribution, so as to achieve more accurate recommendation.

**Acknowledge.** We would like to thank the supports of Program for innovative research team (in Science and Technology) in University of Yunnan Province and Kunming Key Laboratory of Education Informatization.

This research was supported by the National Natural Science Foundation of China (no. 61862067), and the Doctor Science Foundation of Yunnan normal university (No. 01000205020503090).

## References

1. Yu, M.: Thoughts on strengthening national defense education in border areas under the new situation. Nat. Defence **04**, 48–50 (2016)
2. Cong, D., Li, J., Bo, S., et al.: Research on the main problems and countermeasures of the development of national defense education in colleges and universities. Beijing Educ. **12**, 30–32 (2018)
3. Huang, L., Jiang, B., Lu, S., et al.: A survey of recommendation system based on deep learning. J. Comput. Sci. **41**(07), 1619–1647 (2018)

4. Zhang, S., Yao, L., Sun, A., et al.: Deep learning based recommender system: a survey and new perspectives. ACM Comput. Surv. (CSUR) **52**(1), 5 (2019)
5. Yu, W., Zhang, H., He, X., et al.: Aesthetic-based clothing recommendation. In: Proceedings of the 2018 World Wide Web Conference, Lyon, France, pp. 649–658 (2018)
6. Gong, Y., Zhang, Q.: Hashtag recommendation using attention-based convolutional neural network. In: International Joint Conference on Artificial Intelligence (IJCAI), New York, USA, pp. 2782–2788 (2016)
7. Zhou, J.X., Liu, X.D., Gan, J.H.: Exploiting best practice of deep convolutional features for national costume image retrieval. Int. J. Performability Eng. **14**(4), 621–630 (2018)
8. Cui, Q., Wu, S., Liu, Q., et al.: MV-RNN: a multi-view recurrent neural network for sequential recommendation. IEEE Trans. Knowl. Data Eng. **14**(8), 317–331 (2018)
9. Wang, D., Liang, Y., Xu, D., et al.: A content-based recommender system for computer science publications. Knowl. Based Syst. **157**, 1–9 (2018)
10. Deldjoo, Y., Elahi, M., Quadrana, M., et al.: Using visual features based on MPEG-7 and deep learning for movie recommendation. Int. J. Multimedia Inf. Retrieval **7**(4), 207–219 (2018)
11. Wu, X., Yuan, X., Duan, C., Wu, J.: A novel collaborative filtering algorithm of machine learning by integrating restricted Boltzmann machine and trust information. Neural Comput. Appl. **31**(9), 1–8 (2018). https://doi.org/10.1007/s00521-018-3509-y
12. Zhuang, F., Luo, D., Yuan, N. J., et al. Representation learning with pair-wise constraints for collaborative ranking. In: Proceedings of the 10th ACM International Conference on Web Search and Data Mining, Cambridge, UK, pp. 567–575 (2017)
13. Liu, Q., Wu, S., Wang, L.: Multi-behavioral sequential prediction with recurrent log-bilinear model. IEEE Trans. Knowl. Data Eng. **29**(6), 1254–1267 (2017)
14. Wang, J., Yu, L., Zhang, W., et al.: IRGAN: a minimax game for unifying generative and discriminative information retrieval models. In: Proceedings of the 40th International ACM SIGIR conference on Research and Development in Information Retrieval, pp. 515–524. ACM (2017)
15. Zhao, W., Chai, H., Wang, B., et al.: Leveraging long and short-term information in content-aware movie recommendation. arXiv preprint arXiv:1712.09059 (2017)

# Dictionary Learning Based on Structural Self-similarity and Convolution Neural Network

Ling Zhang, Wenchao Jiang$^{(\boxtimes)}$, and Wen Xiang

Faculty of Computer, Guangdong University of Technology, Guangzhou, China
85711585@qq.com

**Abstract.** Aiming at insufficient detailed description problem caused by the loss of edges during a single low-resolution (LR) image's reconstruction process, a novel algorithm for super resolution image reconstruction is proposed in this paper, which is based on fusion of internal structural self-similarity dictionary and external convolution neural network parameters learning model. Firstly, for solving training samples too scattered problem, besides external database, an internal database is constructed to learn a dictionary of the single image's structural self-similarity by multi-scale decomposition approach. Secondly, nonlocal regularization constraint is calculated on the priori knowledge, which is obtained from the internal database of the single LR image. Thirdly, similar block pairs of high and low-resolution samples in the external database are input into a convolution neural network for learning the parameters of reconstructing model. After all, combined parameters learned and the internal dictionary, the single LR image is reconstructed, and by iterative back-projection algorithm its result is improved. Experimental results show that, compared with state-of-the-art algorithms, such as Bicubic, K-SVD algorithm and SRCNN algorithm, our method is more effective and efficient.

**Keywords:** Super resolution · Structural self-similarity · Convolution natural network · Nonlocal regularization · Block matching

## 1 Introduction

Super-resolution image reconstruction refers to the technology of improving low-quality and low-resolution images to recover high-resolution images, and has important applications in military, medical, public safety and computer vision. The general way of super-resolution image reconstruction is to learn from a large number of high-resolution images to reconstruct high-frequency details of the low-resolution image [1–4]. Performances of these algorithms are not so satisfied, for they are likely to be affected by training data that are too scattering to effectively represent a given image. On the other hand, structural information of this given image is valuable for itself reconstruction but little attention has been paid so far [5–7]. Freedman et al. [8] pointed out that there were many structural self-similar blocks distributed within one single image region, so several related studies on local structural self-similarity extraction are also reported [9, 10].

© Springer Nature Switzerland AG 2020
X. Chen et al. (Eds.): ML4CS 2020, LNCS 12488, pp. 336–346, 2020.
https://doi.org/10.1007/978-3-030-62463-7_31

However, they cannot effectively deal with irregular texture blocks, which are sparsely or infrequently appeared in a single image. Error matching between image blocks will bring a lot of fake textures and make it difficult to guarantee the good effect of reconstruction. In order to solve the problems, this paper proposes a super-resolution image reconstruction method based on the structural self-similarity of the single image. Similar structural blocks of the same scale and different scales are extracted from the single image and used to set up an internal dictionary model. Then the weights of the dictionary are learned by external sample images trained with a convolution neural network. With these data, a reconstructive model adaptive to the given single image is obtained and information of the single image is made best use of. Experimental results verify the effectiveness of our proposed algorithm when compared with other state-of-the-art approaches.

## 2    Parameters Learning Model Based on Convolution Neural Network

In this paper, reconstructing parameters of the single LR image are learned by a Super-Resolution Convolution Neural Network (SRCNN) [11] framework, which has been proved to have a great capability in extracting the essential features of data sets. Firstly, block pairs of Low-Resolution (LR) and High-Resolution (HR) images from the external database are matched to each other to achieve matching pairs of image blocks. Then these blocks are regarded as samples and input into SRCNN, which consists of three layers of convolution layers, including feature extracting, non-linear mapping and high-resolution image reconstructing parameters achieving, respectively. The framework of our SRCNN is shown in Fig. 1, and three convolution layers of SRCNN deep learning algorithms are expressed as the following equations:

$$Y_1 = \max\{0, W_1 \cdot X + B_1\} \tag{1}$$

$$Y_2 = \max\{0, W_2 \cdot X + B_2\} \tag{2}$$

$$Y_3 = \max\{0, W_3 \cdot X + B_3\} \tag{3}$$

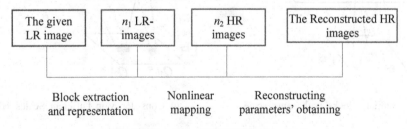

**Fig. 1.**  Structure of SRCNN framework

In Eq. (1)–(3), matrix X represents the original single LR image, $Y_i$ ($i = 1, 2, 3$) represents output of each convolution layer, $W_i$ ($i = 1, 2, 3$) and $B_i$ ($i = 1, 2, 3$) represent the neuron convolution kernel and neuron bias vector, respectively. Symbol '·' represents a convolution operation, whose result is then processed by the ReLu activation function max $\{0, x\}$. With matching pairs of image blocks, this neural network frame needs to learn parameters set $\Phi = \{W_1, W_2, W_3, B_1, B_2, B_3\}$, which are estimated by minimizing the error loss between the last output of neural network and HR image. Given a HR image $Y$ and its corresponding LR image $X$, its loss function could be described by using its mean square error $L(\Phi)$, as shown in Eq. (4).

$$L(\Phi) = \frac{1}{n} \sum_{i=1}^{n} \|F(X_i, \Phi)\|^2 \tag{4}$$

Equation (4) can be solved by stochastic gradient descent and back-propagation algorithm together.

## 3   Extraction of Self-similarity Feature

### 3.1   Self-similarity on the Same and Multi-scale Images

Image self-similarity refers to as similar features available among the various regions of the entire image. Researches [12] had shown that, for a $5 \times 5$ image block in a natural image, there were a large number of image blocks in the same scale and different scales can be found in the image. A statistic shows that more than 90% of image blocks can find at least 9 similar image blocks in the same scale of itself; more than 80% of image blocks can find at least 9 similar image blocks of different scales. Based on this image similarity mechanism, we can extract a lot of redundant information of the image itself on the same scale and different scales. Frank et al. [13] pointed out that there were two characteristics in general images: one is that a large number of similar structural regions appear in the whole image; and the second is that these structural similarities can keep consistent on multiple scales of the image, as shown in Fig. 2a).

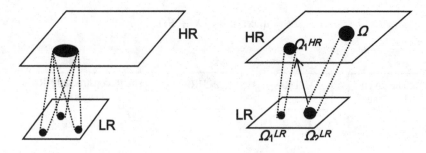

a) Similar Blocks with same scale image

b) Consistency in Different Scales HR

**Fig. 2.** Similar image blocks with same scale and different scales in a single image

Since there are so many structure-similar images blocks in the same scale and different scales of a single LR image, we will benefit if we make the best use of these structural similarities in its reconstruction. The basic scheme of our algorithm is shown in Fig. 2b), where HR represents a high-resolution image, and LR represents a corresponding low-resolution image thereto. Size of HR image is $s$ times that of LR image. Suppose $\Omega_1^{HR}$ and $\Omega_2^{HR}$ represent two similar blocks with different scales in HR image, and size of $\Omega_2^{HR}$ is $s$ times that of $\Omega_1^{HR}$. The corresponding image blocks of $\Omega_1^{HR}$ and $\Omega_2^{HR}$ in LR image are $\Omega_1^{LR}$ and $\Omega_2^{LR}$. In this case, $\Omega_1^{LR}$ and $\Omega_2^{LR}$ in the LR image form a pair of similar image blocks with different scales. Suppose scaling factor between the HR and LR images is the same with that between $\Omega_1^{HR}$ and $\Omega_2^{HR}$, and then size of $\Omega_1^{HR}$ in HR image is exactly the same as $\Omega_2^{LR}$ in LR image. Accordingly, in recovering block $\Omega_1^{LR}$ to form block $\Omega_1^{HR}$ in HR image, $\Omega_2^{LR}$ could provide helpful additional information for it.

In this paper, LR and HR similarity block pairs in same scale and different scales are derived from images of the external and internal database by using a non-local block matching method. Then, these block pairs are treated as training samples for dictionary learning to reconstruct the image to be restored.

### 3.2  Non-local Self-similar Block Matching

Researchers have found that [15, 16], natural images have abundant similarities in regions of texture, edge and so on. A low resolution (LR) image can restore its missing details based on this structural high-frequent similarity. It seems that exploiting the similarities between nonlocal patches distributed in different regions of the image can achieve higher image reconstruction resolution [14].

This paper presents a regularization constraint item based on non-local block similarity. Suppose $X_i$ represents the $i^{\text{th}}$ block of a single LR image $X$, its similar blocks $X_i^l$ ($l = 1, 2, ..., L$) are firstly searched within $X$ itself, which are then used to estimate $X_i$ by their linear combination. The main idea of non-local constraint is that central point $P_i$ of block $X_i$ can be represented by the weighted average of central point $P_i^l$ of block $X_i^l$ ($l = 1, 2, ..., L$), which could be described in Eq. (5).

$$P_i = \frac{1}{L} \sum_{l=1}^{L} w_i^l P_i^l \tag{5}$$

Suppose that each weight vector $w_i$ is a matrix of vectors $w_i^l$, which consists of weight matrix $B$. Each $P_i$ is made up of by $P_i^l$, which consists of dictionary matrix $\Psi$. The nonlocal regularization constraint can be expressed as Eq. (6).

$$\alpha = \|(I - B)\psi\alpha\|_2^2 \tag{6}$$

In Eq. (6), $I$ is an identity matrix and $\alpha$ is the 2-norm constraint parameter of nonlocal regularization.

## 4 Structure Self-similarity Extraction

### 4.1 Low Resolution Degraded Model

LR images are caused by blurring, down-sampling or noise pollution of HR images [17]. The whole degraded process could be approximated as a linear one, as shown in Eq. (7).

$$X = HSY + n \tag{7}$$

Where $Y$ and $X$ are reconstructed HR image and the original LR image, respectively. $H$ represents the down-sampling operation, $S$ is the fuzzy operator, and $n$ is the noise pollution matrix. In order to accurately estimate the HR image matrix $Y$, some priori knowledge or regular constraint items of the image need to be introduced, as shown in Eq. (8).

$$\hat{Y} = arg \min_{Y} \|X - HSY\|_F^2 + \lambda\alpha \tag{8}$$

In Eq. (8), $Y$ is the reconstructed HR image, and $\|X - HSY\|_F^2$ represents error term in observation, $\alpha$ is the regular constraint item in Eq. (6), $\lambda$ is the weighted balance parameter of regular item.

We deformalize the degraded model constrained by the Eq. (6), and on behalf of the formula (8), the final algorithm is got in Eq. (9).

$$\hat{Y} = arg \underbrace{\min}_{y} X - HSY_F^2 + \lambda\alpha_1 + \mu(I - B)\psi\alpha_2^2 \tag{9}$$

Parameter $\mu$ represents the regularization parameters. After all, we use the iterative back-projection algorithm to further enhance the image reconstruction performance.

### 4.2 Algorithm in Detail

Firstly, with prior knowledge of the non-local self-similarity in the original LR image, search the best match blocks of the initial super-resolution image with multi-scale method, and then take them as an internal dictionary to learn non-local regularization constraints. Depth learning is generally trained with a large amount of data, but in this case, a relatively small training set consisting of 91 images [3] is used for training. Best match blocks in LR and HR images of these training samples are found out and made up to be a lot of pairs, which would be used to compose of an external dictionary. Both these two kinds of samples are input to convolution neural network for modeling self-structure similarity of LR image. After all, with the non-local regularization constraints learned from internal dictionary, the original LR image is reconstructed. There are four steps: initial interpolation, non-local blocks matching, neural network model learning and non-local regularization constraints.

- Initial Interpolation. In this paper, cubic bilinear interpolation algorithm, the commonly used algorithm for LR image reconstruction, is selected to build its original HR image, which is later used for LR image's self-similarity extraction.

- Non-local blocks matching. The original HR image is partitioned point-by-point into blocks, which are then matched with each other to obtain structure similar blocks. There are two categories of similar image blocks, which are the ones with the same scale and the other ones with different scales. In the case of the same scale blocks set, suppose represents the $i^{th}$ one, we match it across the whole blocks set with the same size to search its closest similar couple block. The difference between the searched block and the current block $\hat{x}_i$ is calculated as Eq. (10).

$$e_i^l = \left\| \left( \hat{x}_i^l - \hat{x}_i \right) \right\|_2^2 \tag{10}$$

- Neural network model training. Convolution neural network model has a strong feature learning ability; therefore, we use SRCNN algorithm with a three-layer structure for dictionary training. Finally, corresponding network model parameters $\Phi$ are obtained.
- Non-local Regularization Constraints. Based on the obtained parameters of convolution neural network model, combined with nonlocal regularization and dictionary data, this section builds a reconstructed image according to Eq. (9).

### 4.3 Algorithm Enhancement

We use the iterative back-projection algorithm to enhance our reconstructed image, which is based on a down-sampling image degradation model with sub-pixel displacement [18] Firstly, multi-frames of LR image are sampled in sequence and registered, and then errors between the LR image simulation and its observation results are iteratively back-projected to the HR image. Suppose that there are $K$ sequential observation LR images, described as $f_k$ $(m_1, m_2)$ with resolution $M_1 \times M_2$. Size of the estimated HR image $f$ $(n_1, n_2)$ is enlarged by $s$ times, which means resolution of the estimated HR image $N_1 \times N_2 = (sM_1) \times sM_2$). Using the Iterative Back-Project (IBP) method to estimate the HR image can be described as Eq. (11).

$$\hat{f}^{n+1}(n_1, n_2) = \hat{f}^n(n_1, n_2) + \sum_{m_1, m_2} \left( g_k(m_1, m_2) - \hat{g}_k^n((m_1, m_2)) \right) \times h^{BP}(m_1, m_2; n_1, n_2)$$

$$\tag{11}$$

In Eq. (11), $\hat{g}_k^n$ represents the $k^{th}$ simulation result of LR image in the $n^{th}$ iteration, generated by the actual displacement information of LR images. $h^{BP}(m_1, m_2; n_1, n_2)$ is the back-projection kernel, which determines how error affects the HR image construction during each iteration. We use a down-sampling rate of $s = 3$, and the displacement of sub-pixel $(x, y)$ are $(0, 0)$, ..., $(1, 3)$, respectively. We get 8 LR observation images and corresponding simulation images, and calculate errors between these two kinds of images. At last, we obtain the HR image according to the Eq. (11).

### 4.4 Algorithm Implementation

The proposed self-similar similarity convolution neural network algorithm is divided into two processes, training and reconstruction, respectively.

In order to clarify the algorithm in this paper more clearly, the algorithm flow chart is shown in Fig. 3.

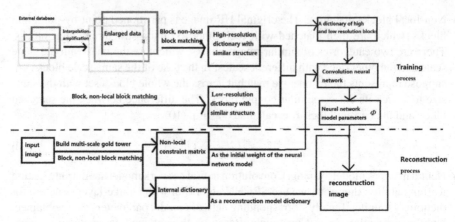

**Fig. 3.** Algorithm flow chart

### 4.5 Non-local Regularization Constraints Example

In this process, non-local regularization constraints of the single LR image are obtained by its structure self-similarity blocks' representation. According to Eq. (6), non-constraints are iteratively calculated by weighted average matrix made up by each block's similar representation, and a dictionary, which is consisted of by similar blocks' central points. A simple example of this process is shown in Fig. 4*a*). Four clustering results of structure blocks are also shown in Fig. 4*b*), which represent non-local regularization constraints of the example image.

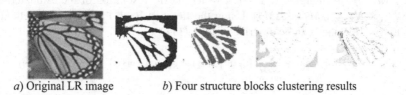

*a*) Original LR image                    *b*) Four structure blocks clustering results

**Fig. 4.** An example of non-local regularization constraints

## 5   Experimental Results and Analysis

### 5.1   Experimental Setup

In order to verify the validity of our proposed method, three international public SR databases are used, which are Set5, Set14 and Urban100, and three-layer convolution neural network is used to for model learning. The first layer has $9 \times 9$ size and 64 convolution kernels and neurons, the second layer has $1 \times 1$ and 32, and the third layer has $5 \times 5$ and 1. In the experiment, Bicubic interpolation, K-SVD and convolution neural networks are selected as contrast analysis approaches to compare with the performance of our proposed method, based on indicator of Peak Signal to Noise Ratio (PSNR).

## 5.2   Experimental Setup

In order to evaluate the quality of image reconstruction, we compare the performance of these methods on PSNR. Taking three images in database Set14 as an example, the reconstructed results with four approaches are shown in Fig. 5 under the condition of magnification of 3 times. From them, local information restored by ours is clearer and more delicate, and global reconstructed images are more approaching to the original images.

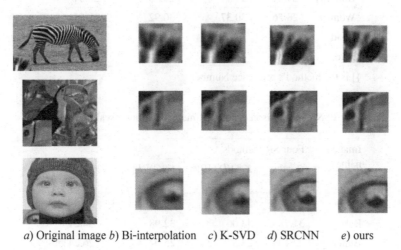

*a*) Original image *b*) Bi-interpolation    *c*) K-SVD    *d*) SRCNN      *e*) ours

**Fig. 5.**   Three reconstructed results from Set14 with up scaling factor 3

PSNR of each approach is shown in Table 1, 2 and 3. Bicubic interpolation method takes the lowest place, only 22.101 db and the best contrast algorithm can reach 40.642 db, while our proposed method can reach the highest 42.204 db.

**Table 1.**  PSNR comparison of four sr methods with up-scaling factor as 2

| Image name | Four SR methods | | | |
|---|---|---|---|---|
| | *Bicubic* | *K-SVD*[19] | *SRCNN* [11] | *Our method* |
| Zebra | 30.63 | 31.73 | 33.29 | 34.94 |
| Baby | 37.07 | 38.25 | 38.30 | 39.42 |
| Butterfly | 27.43 | 30.65 | 32.20 | 33.33 |
| Woman | 32.14 | 34.49 | 34.94 | 36.05 |
| Head | 34.86 | 35.59 | 35.64 | 36.86 |
| Bird | 36.81 | 39.93 | 40.64 | 42.20 |

[] is the method's reference Number

**Table 2.** PSNR comparison of four SR methods with up-scaling factor as 3

| Image name | Four SR methods | | | |
|---|---|---|---|---|
| | Bicubic | K-SVD [19] | SRCNN [11] | Our method |
| Zebra | 26.63 | 26.92 | 28.87 | 29.94 |
| Baby | 33.91 | 35.08 | 35.01 | 36.22 |
| Butterfly | 24.04 | 25.94 | 27.58 | 28.84 |
| Woman | 28.56 | 30.37 | 30.92 | 32.55 |
| Head | 32.88 | 33.56 | 33.55 | 35.86 |
| Bird | 32.58 | 34.57 | 34.91 | 36.20 |

[] is the method's reference Number

**Table 3.** PSNR comparison of four SR methods with up-scaling factor as 4

| Image name | Four SR methods | | | |
|---|---|---|---|---|
| | Bicubic | K-SVD [19]* | SRCNN [11]* | Our method |
| Zebra | 24.08 | 24.92 | 25.97 | 27.03 |
| Baby | 31.78 | 33.06 | 32.98 | 34.21 |
| Butterfly | 22.10 | 23.57 | 25.07 | 26.24 |
| Woman | 26.46 | 27.89 | 28.21 | 30.42 |
| Head | 31.59 | 32.21 | 32.19 | 34.06 |
| bird | 30.18 | 31.71 | 31.98 | 33.27 |

[] is the method's reference Number

Our method has an average declining rate 9.17% in PSNR, higher than other methods.

## 6 Conclusions

The proposed algorithm considers the reconstruction of a single super-resolution of image based on self-structure similarity within the image. The algorithm derives self-similarity of the training samples through the scale decomposition of the image, and makes full use of the structural self-similarity of the input image to solve the problem that training samples are too scattered for representing the LR image. The intrinsic structure self-similarity of the image is obtained through the nonlocal regularization constraint. Finally, the iterative back-projection algorithm is used to further optimize the reconstructive effect. Compared with state-of-the-art algorithms such as Bicubic, KSVD and SRCNN, the proposed algorithm can achieve better reconstructive performance.

**Acknowledgement.** This paper is funded by Scientific Project of Guangdong Provincial Transport Department (No. Sci & Tec-2016-02-30), Natural Science Foundation of Guangdong Province under Grant 2018A030313061, in part by the Guangdong Science and Technology Plan under Grant 2017B010124001, Grant 201902020016, and Grant 2019B010139001.

# References

1. Li, Z.H.X., He, H., Wang, R., et al.: Single image super-resolution bidirectional group sparsity and directional features. Image Process. **9**(24), 2874–2888 (2015)
2. Dong, C., Loy, C.C., He, K., et al.: Image super-resolution using deep convolutional networks. IEEE Trans. Pattern Anal. Mach. Intell. **2**(38), 295–307 (2016)
3. Yang, J., Wright, J., Huang, T., et al.: Image super-resolution via spare representation. IEEE Trans. Image Process. **19**(11), 2861–2873 (2010)
4. Timofte, R., Smet, V., Gool, L.: Anchored neighborhood regression for fast example-based super-resolution. In: IEEE International Conference on Computer Vision, Sydney, pp. 1920–1927. IEEE (2013)
5. Timofte, R., De Smet, V., Van Gool, L.: A + : adjusted anchored neighborhood regression for fast super-resolution. In: Cremers, D., Reid, I., Saito, H., Yang, M.-H. (eds.) ACCV 2014. LNCS, vol. 9006, pp. 111–126. Springer, Cham (2015). https://doi.org/10.1007/978-3-319-16817-3_8
6. Yang, C.Y., Yang, M.H.: Fast direct super-resolution by simple functions. In: IEEE Interaction Conference on Computer Vision, Sydney, pp. 561–568. IEEE (2013)
7. Dai, D., Timoft, R., Vangool, L.: Jointly optimized regressors for image super-resolution. Comput. Graph. Forum. **34**(2), 95–104 (2015)
8. Freedman, G., Fattal, R.: Image and video upscaling from local self-examples. ACM Trans. Graph. **30**(2), 12 (2011)
9. Protter, M., Elad, M., Takeda, H., Milanfar, P.: Generalizing the nonlocal-means to super-resolution reconstruction. IEEE Trans. Image Process. **18**(1), 36–51 (2009)
10. Mairal, J., Bach, F., Ponce, J., Sapiro, G., Zisserman, A.: Non-local sparse models for image restoration. In: IEEE International Conference on Computer Vision. Kyoto, Japan, pp. 2272–2279. IEEE (2009)
11. Dong, C., Loy, C.C., He, K., Tang, X.: Learning a deep convolutional network for image super-resolution. In: Fleet, D., Pajdla, T., Schiele, B., Tuytelaars, T. (eds.) ECCV 2014. LNCS, vol. 8692, pp. 184–199. Springer, Cham (2014). https://doi.org/10.1007/978-3-319-10593-2_13
12. Gkasner, D., Bagon, S., Irani, M.: Super-resolution from a single image. In: International Conference on Computer Vision, Kyoto, pp. 349–356. IEEE (2009)
13. Candocia, F.M., Principe, J.C.: Super-resolution of images based on local correlations. IEEE Interact. Neural Netw. **2**(10), 372–380 (1999)
14. Dong, W.S., Zhang, L., Shi, G.M., et al.: Nonlocally centralized sparse representation for image restoration. IEEE Trans. Image Process. **4**(22), 1620–1630 (2013)
15. You, X., Xue, W., et al.: Single image super-resolution with non-local balanced low-rank matrix restoration. In: International Conference on Pattern Recognition, Cancun, vol. 10, no. 23, pp. 1255–1260. IEEE (2016)
16. Xu, J., Zhang, L., Zuo, W., et al.: Patch group based nonlocal self-similarity prior learning for image denoising. In: Proceedings of IEEE Conference on Computer Vision, Santiago, pp. 244–252. IEEE (2015)

17. Tekalp, A.M.K., Sezan, M.I.: Hight-resolution image reconstruction from lower-resolution image sequences and space varying image restoration. In: Proceedings of the IEEE International Conference on Acoustics. Speech and Signal Processing, San Francisco, pp. 169–172 (1992)
18. Lu, Y., Imanura, M.: Pyramid-based super-resolution of the under sampled and subpixel shifted image sequence. Int. J. Syst. Technol. **12**, 254–263 (2002)

# Short-Text Feature Expansion and Classification Based on Non-negative Matrix Factorization

Ling Zhang[1], Wenchao Jiang[1(✉)], and Zhiming Zhao[2]

[1] School of Computers, Guangdong University of Technology, Guangzhou 510006, China
jiangwenchao@gdut.edu.cn
[2] Multiscale Networked System (MNS) Group, Informatics Institute, University of Amsterdam, Science Park 904, 1098XH Amsterdam, The Netherlands

**Abstract.** In this paper, a Non-negative Matrix Factorization Feature Expansion (NMFFE) approach was proposed to overcome the feature-sparsity issue when expanding features of short-text. Firstly, we took the internal relationships of short texts and words into account when segmenting words from texts and constructing their relationship matrix. Secondly, we utilized Dual regularization Non-negative Matrix Tri-Factorization algorithm (DNMTF) to obtain the words clustering indicator matrix, which was used to get the feature space by dimensionality reduction methods. Thirdly, words with close relationship were selected out from the feature space and added into the short-text in order to solve the sparsity issue. The experimental results showed that the accuracy of short text classification of our NMFFE algorithm increased 25.77%, 10.89% and 1.79% on three datasets: Web snippets, Twitter sports and AGnews respectively compared with Word2Vec algorithm and Char-CNN algorithm. It indicated that the NMFFE algorithm was better than BOW algorithm and the Char-CNN algorithm in terms of classification accuracy and algorithm robustness.

**Keywords:** Short text classification · Feature extension · Non-negative matrix factorization · Correlation

## 1 Introduction

Short texts are convenient in human communication, and have prevalent on the social networks nowadays. Short text classification is one of the challenges due to its natural sparsity, noise words, syntactical structure and colloquial terminologies [1]. Those topics attracted lots of research attention in the field of short text expansion and classification research.

Due to the imitation of words and low-frequency of terms in short text, the bag-of-words (BOW) representation has limits in analyzing short texts [2]. One possible solution for handling sparsity is to expand short text by appending new features based on semantic information extracted from Web searching, lexical databases or provided by machine translations [3], which are called an external resource-based approaches.

© Springer Nature Switzerland AG 2020
X. Chen et al. (Eds.): ML4CS 2020, LNCS 12488, pp. 347–362, 2020.
https://doi.org/10.1007/978-3-030-62463-7_32

Web searching [4] based feature extension technologies need to interact frequently with search engines, and result in high communication overhead and low efficiency for data analysis. Knowledge bases or lexical databases, such as Wikipedia and HowNet for concept taxonomies [5–7] or topic models [8, 9] are used to enrich short text representations. However, these feature extension methods have high dependencies on the integrity of external resources, and often time consuming. Moreover, these predefined topics and categories are domain-specialized or language-specific.

Using rules or statistical information hidden in the context of short texts is another kind of approaches to extend features, which are called the self-contained resource approaches [10, 11, 22–24, 27]. Mining hidden information in short texts plays a key role in feature extension. A self-aggregation-based topic model (SATM) [22] has been reported recently, which assumes short texts are sampled from long pseudo-documents, and then topic modeling is conducted by finding "document-ship" for each short text. U. K. Sikdar et al. [10] described a deep learning approach to recognize Amharic named entities from a large dataset annotated with six different classes, trained on various language independent features together with word vectors, which were the semantic information taken by an unsupervised learning algorithm, word2vec. The word vectors were merged with a set of specifically developed language independent features and together fed to the neural network model to predict the classes of the words. Zhang et al. [11] proposed a character-level convolutional network model for short text classification without any knowledge on the syntactic or semantic structures of a language. Nevertheless, these works ignore the relevance of the words in short texts. In the case of limited words, the association between words can be used as additional information to serve as an important basis for feature expansion and solve the problem of sparse features of the short text.

This paper considers two forms of information: inter-type and intra-type relationships between words and short texts. Based on these two kinds of data relations, the feature space is obtained by dimension reduction of word clustering indicator, which is obtained by non-negative matrix tri-factorization [12]. Then, according to the correlation between words, closely related features in the feature space are selected to expand the text feature vector, and this can effectively solve the problem of feature sparseness.

## 2   Related Works

Feature expansion is essential to classify short texts, and it has been mainly focusing on two kinds of approaches by now, Latent Dirichlet Allocation (LDA) topic model [40, 42, 43] and Word Embedding [29–31, 35–38, 42]. Y. Xu used LDA for clustering words or documents into "topics", and based on a "topic-word" probability distribution model, the closely-related words were found and selected out to expand feature space of words [42]. W. Xia, et al. chose the liveness of each user as a feature, and modelled it as the weighted value for the user. They improve the precision of topic detection and tracking, by including the user feature into LDA model to expand the feature of short texts [40]. Yu, et al. [43] used the Dirichlet Multinomial Mixture (DMM) model as the main framework and extended short texts with the potential feature vector representation of the words by combining the user-LDA topic model, and achieved a good performance as an external extension of short texts. The complexity of Probabilistic Graphical Model hampers the development of LDA, and the computational cost of LDA results in bigger penalty compare with the improvement of this algorithm.

On the other hand, word embedding presents another kind of words representation, converting per word into a continuous vector space with dimensionality reduction [32, 33]. Semantic expansion of words is then obtained by clustering of vectors. Recently, researches have widely employed deep learning-based approaches for word embedding model. Google developed a Word2Vec tool based on Bengio neural network for word embedding [24]. In fact, Word2Vec predicted words based on their context by using one of two distinct neural models: CBOW [33, 35, 38, 39] and Skip-Gram [10, 29, 31, 34, 36, 37, 40].

P. Wang et al. proposed a framework to expand short texts, based on skip-gram model to learn word embeddings from large-scale unstructured text data. By using additive composition over word embeddings from context with variable window width, the representations of multi-scale semantic units in short texts were computed [37]. In literature [36], distributed word embeddings were learned by skip-gram algorithm through a neural network architecture, and then they were combined into a sentence representation to predict the semantic relations between short texts. W. X. Liang et al. proposed a global and local word embedding-based topic model (GLTM) for short texts [34]. They trained global word embeddings from large external corpus and employed the continuous skip-gram model with negative sampling (SGNS) to obtain local word embeddings. Utilizing both the global and local word embeddings, their method could distill semantic related information between words which could be further leveraged by Gibbs sampler in the inference process to strengthen semantic coherence of topics.

G. X. Xun et al. used Continuous Bag of Words (CBOW) to provide additional semantics for short text corpus, and incorporated it into each short document's model to establish a Gaussian topic in the vector space [39]. In addition, a discrete background mode over word types was also added to complement the continuous Gaussian topics model. In literature [38], by using word embedding features, L. Sang et al. expanded and enriched the words density in the short texts, and semantic similarities of short texts were calculated for effective learning. This method combined external sources of word semantic information with the short text structure information. A. J. Pascual et al. presented a Contextual Specificity Similarity (CSS) algorithm [33] for document similarity measure, where documents were represented as arrays of their word vectors, and then Inverse Document Frequency (IDF) of the words were added into to define the closeness degree between documents.

Although Word2Vec has an outstanding performance in synonymous words analysis, it still relies on local context so much, lacking of global statistical information of short texts. Accordingly, in 2014, Jeffrey Pennington et al. presented a new model based on the words ice and steam to illustrate how to generate meaning from word occurrence, and how to result a global word vectors representing that meaning [23]. They defined it as GloVe, whose training was performed on aggregated global word-word co-occurrence statistics from a corpus, and the resulting representations showed interesting linear substructures of the word vector space [37]. Comparative study [41] showed that its effectives for the Arabic language processing, and pointed õut that the appropriate starting point for word vector learning might be indeed with ratios of co-occurrence probabilities rather than the probabilities themselves. The shortcoming of GloVe was also mentioned in literature [25], demanding a large-scale corpus and big enough storage resource.

Both approaches mentioned above cannot work without huge corpus data support. Opposite to the large-scale learning algorithms, this paper studies on feature expansion by short text itself. There are three aspects of relations taken into consideration, including word-to-word, word-to-text and text-to-text, to make use of more relatedness information from short text. We use this method as an alternative to the aforementioned relation features, in cases where only limited amounts of training data are available.

# 3  Algorithm Framework

Given a short text set $T = \{t_1,..., t_m\}$ and a word set $W = \{w_1, ..., w_n\}$. The goal is to group the texts $\{t_1, ..., t_m\}$ into $k$ clusters, in the meantime also grouping the words $\{w_1, ..., w_n\}$ into $k$ clusters. The relationship matrix $R$ describes the inter-type relationships between texts and words. The correlation matrix $A_t$ and $A_w$ represent the intra-type relationships of texts and words, respectively. The clustering indicator matrix $F$ represents the clustering result of words, whose element $F_{ij}$ represents the possibility that $w_i$ belongs to cluster $k_j$. Similarly, the clustering indicator matrix $G$ represents the clustering result of short texts. Since the short text category label of training set is known, the matrix $G$ can be obtained. In this way, the feature expansion for short texts is transformed into the clustering of texts and words jointly.

The overall framework of our algorithm is based on non-negative matrix factorization, including four steps: feature space establishment, feature expansion, feature space updating and short text classification, as shown in Fig. 1.

The feature space of the short text itself describes the possibility of the word belonging to the category. Based on training texts, we construct a relationship matrix to describe membership of word-to-text, and two correlation matrixes to describe intra-type relation of text-to-text and word-to-word respectively. Under the manifold regularization, the nonnegative matrix factorization algorithm is used to build words clustering indicator matrix. After removing some evenly distributed features in the indicator matrix, a dimension-reduced feature space is constructed. The feature of short text is to extend by the correlation between the features in the feature space and the text features. The updating of feature space is to predict the clustering indicator value of the unknown feature with the clustering indicator average value of the known feature in the same text, and then add the new feature into the feature space. The classifier is to divide the testing samples into different categories by using an SVM algorithm.

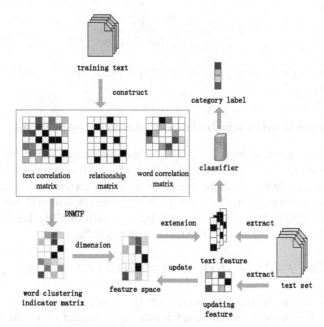

**Fig. 1.** Framework of the proposed algorithm

# 4 Feature Space Construction Based on DNMTF

## 4.1 Non-negative Matrix Tri-Factorization

The feature space is constructed by factorization of the relationship matrix. Firstly, according to the label data of the short text training set, the clustering indicator matrix $G$ can be directly obtained, which is part of the relationship matrix $R$ in the non-negative matrix tri-factorization [13]. Then, with manifold regularization constraint added, word clustering indicator matrix $F$ is obtained by decomposition.

The relation matrix $R$ is decomposed into three matrices, $F$, $S$ and $G$, noted as: $R \approx FSG^{T}$. Matrix $F$ and $G$ are clustering indicator matrix corresponding to two types of entities respectively, and matrix $S$ is an equilibrium matrix with multi-dimension, which would guarantee the accuracy of low-dimensional matrix representation.

## 4.2 Construction of Relationship and Correlation Matrix

The construction of the relationship matrix $R$ follows the natural relationship between text and word. If the word $w_i$ appears in the text $t_j$, then $R_{ij} = 1$, otherwise $R_{ij} = 0$.

The construction of the correlation matrix $A_t$ and $A_w$ is based on statistics information between text and words. The calculation of correlation strength between two samples $x_i$ and $x_j$ is shown in Eq. (1).

$$A_{ij} = \frac{B(x_i, x_j)}{\sum_{x_a, x_b \in T(W)} B(x_a, x_b)} \tag{1}$$

Where $B(x_i, x_j)$ is the number of words (text) co-occurrence by sample $x_i$ and $x_j$ in $T$ (word set $W$).

### 4.3   Relationship Matrix Factorization with Manifold Regularization

According to the manifold hypothesis [14], if two samples $x_i$ and $x_j$ are similar in geometric structure, then the practical significance of these two samples is also similar, which is reflected in clustering labels. Therefore, we propose a novel algorithm based on the dual regularization non-negative matrix tri-factorization algorithm (DNMTF) [15] to capture the intra-type and inter-type relationship among entities. The relationship matrix factorization based on manifold regularization is shown in Eq. (2).

$$J_1 = \left\| R - FSG^T \right\|^2 + \mu tr\left(F^T L_w F\right) + \phi tr\left(G^T L_t G\right) \quad s.t. F, S, G \geq 0 \tag{2}$$

Where $\mu, \phi > 0$ are the regularization parameters, used to balance the reconstruction error of DNMTF in the first item and graph regularizations in the second and third terms in Eq. (2). $L_w = D_w - A_w$ is the graph Laplacian of the data graph which reflects the label smoothness of the data points, and $L_t = D_t - A_t$ is the graph Laplacian of the feature graph which reflects the label smoothness of the feature $D_w$ and $D_t$ are diagonal matrix, whose entities are column sum of Aw and At, noted as $D_{ii}^w = \sum_j A_{ij}^w$, $D_{ii}^t = \sum_j A_{ij}^t$, respectively.

Since labels of training set are known already, the clustering indicator matrix $G$ can be directly obtained as part input of $J_1$. The objective function in Eq. (2) can be rewritten into Eq. (3).

$$J_1 = tr\left(\left(R - FSG^T\right)\left(R - FSG^T\right)^T\right) + \mu tr\left(F^T L_w F\right) + \phi tr\left(G^T L_t G\right)$$

$$= tr\left(RR^T\right) - 2tr\left(RGS^T F^T\right) + tr\left(FSG^T GS^T F^T\right) + \mu tr\left(F^T L_w F\right) + \phi\left(G^T L_t G\right) \tag{3}$$

Introduce Lawrencian multiplier $\alpha_n \times k$, $\beta_m \times k$ and $\gamma_k \times k$ for constraint $F \geq 0$, $G \geq 0$ and $S \geq 0$, respectively. Accordingly, the Lawrencian function is shown in Eq. (4).

$$L = tr\left(RR^T\right) - 2tr\left(RGS^T F^T\right) + tr\left(FSG^T GS^T F^T\right) + \mu tr\left(F^T L_w F\right)$$
$$+ \phi tr\left(G^T L_t G\right) + tr\left(\alpha F^T\right) + tr\left(\beta G^T\right) + tr\left(\gamma S^T\right) \tag{4}$$

In solving the matrix $S$, we take the matrix $F$ and $G$ as the given conditions, and then let the partial differential $\frac{\partial L}{\partial S} = 0$, then we derive Eq. (5).

$$\gamma = 2F^T RG - 2F^T FSG^T G \tag{5}$$

Using KKT condition [16] $\gamma_{ij}S_{ij} = 0$. Then we can get Eq. (6).

$$[F^T RG - F^T FSG^T G]_{ij}S_{ij} = 0 \tag{6}$$

According to Eq. (6), matrix $S$ follows the following updating, as shown in Eq. (7).

$$S_{ij} \leftarrow S_{ij}\frac{[F^T RG]_{ij}}{[F^T FSG^T G]_{ij}} \tag{7}$$

In solving the matrix $F$, we take the matrix $S$ and $G$ as the given conditions, and then let the partial differential $\frac{\partial L}{\partial F} = 0$. Then we get Eq. (8).

$$\alpha = 2RGS^T - 2FSG^T GS^T - 2\mu L_w F \tag{8}$$

Replace $L_w = D_w - A_w$ into Eq. (8) and use KKT condition [16] $\alpha_{ij}F_{ij} = 0$. Then we can get Eq. (9).

$$[RGS^T - FSG^T GS^T - \mu D_w F + \mu A_w F]_{ij}F_{ij} = 0 \tag{9}$$

According to Eq. (9), matrix $F$ follows the following updating, as shown in Eq. (10).

$$F_{ij} \leftarrow F_{ij}\frac{[RGS^T + \mu A_w F]_{ij}}{[FSG^T GS^T + \mu D_w F]_{ij}} \tag{10}$$

---

**Algorithm 1** Feature space construction

---

**Input**: the number of clusters $k$, regularization parameters $\mu$, $\phi$, and maximum number of iterations $I$, relationship matrix $R$, correlation matrices $A_t$, $A_w$, clustering indicator matrix $G$.
**Output**: feature space $H$.
Steps in Detail: $F$
while not convergent and number of *iterations* $< I$
  Update $S$ by Eq. (7)
  Update $F$ by Eq. (10)
end while
  Get $H$ by dimension reduction of $F$
return $H$

---

## 5  Feature Extension Based on Self-resources

### 5.1  Feature Expansion

Suppose there are $p$ feature words in the feature space $H_{p \times k}$, which is the output of Algorithm 1. Then, from space $H$, there are $q$ $(p >> q)$ features $f_i$ $(i = 1, ..., q)$ are chosen out to compose of a subset of the feature space $H$, denoted as $H^*_{q \times k}$, which contains and only contains those $q$ features. Then, multiply $H^*$ with feature space $H$ to get matrix $E_{q \times p}$, as shown in Eq. (11).

$$E = H^* \cdot H^T \tag{11}$$

Where the matrix $E$ describes $f_i$ $(i = 1, ..., q)$ correlation with all features in space $H$.

In order to select features for expansion conveniently, the matrix $E$ is compressed, and the values of each column are added and the mean is calculated to get the vector $e$ with dimensions $p$, as shown in Eq. (12).

$$e(j) = \frac{\sum_{i=1}^{q} E_{ij}}{q}, \quad j = 1 \cdots p \tag{12}$$

Vector $e$ describes the relevance between each feature word in the feature space $H$ and feature representation $f_i$ $(i = 1, ..., q)$ in the subspace $H^*$. In addition to the existing text features, the first $K$ features are selected to expand the short text according to the relevance in $e$.

## 5.2 Feature Space Update

In the process of extending the features of the short text, there is a possibility: some features extracted from the short text are not included in the feature space $H$. At this time, the feature space has an insufficient feature expansion. Therefore, before the feature expansion of the short text, the text features should be firstly detected to see whether update of space H to cover all new text features is needed. There are two kinds of new features needed to update:

(1)   the feature does not exist in the feature space $H$
(2)   the feature is not the one that had been deleted after dimension reduction on clustering indicator matrix.

Suppose there are features needed to be updated, and their corresponding clustering indicator matrix is $H^{**}$. Due to the correlation between input data, $H^{**}$ can be calculated based on $H^*$, as shown in Eq. (13).

$$H_i^{**}(j) = \frac{\sum_{g=1}^{q} \boldsymbol{H}_{gj}^*}{q}, j = 1 \cdots k, i = 1 \cdots a \tag{13}$$

Finally, $H^{**}$ is incorporated into $H$ to obtain an enlarged feature space, based on which feature expansion is carried out. Here, $H^*$ is a subset of the feature space $H$.

## 5.3 Algorithm Description

### Algorithm 2 Feature expansion

---

**Input**: short text set $T = \{t_1, \ldots, t_g\}$, the number of clusters $k$, feature space , the number of features to be expanded $K$

**Output**: feature vector $v(t_i)$ of $T$

Initialize the $v(t_i) = \{0\}$ based on $H$

for each $t_i = \{f_1, \ldots, f_{q+a}\}$ of $T$

    Get $H^*$ of $t_i$

    if $a \neq 0$

        for each $f_b$ $(b = 1, \ldots, a)$

            Get $H_i^{**}$ by Eq. (13)

            Update $H$

        end for

    end if

    Get $E$ by Eq. (11)

    Get $e$ by Eq. (12)

    for each $d = \{1, \ldots, K\}$

    Select the features fc with the maximal value in $e$

        if $f_c \notin t_i$

            Add $f_c$ to $t_i$

            $d{+}{+}$

        end if

    end for

    for each feature $f_d \in t_i$

        the $f_d$ position $v(t_i)$ is set to 1

    end for

end for

return $v(t_i)$

## 6 Experiments and Discussion

### 6.1 Experimental Datasets

This paper verifies the effectiveness of the proposed method using three datasets. In the experiment, the open source tool libsvm is used as the text classifier. The first dataset, Web snippets, obtained from Web search by Phan et al. [17], is a commonly used short text classification test set. The data set contains 8 categories, including 10060 training sets and 2280 test sets, with an average text length of 17.93. Specific information is listed in Table 1.

The second data set is the Twitter 100k, published by Hu et al. [18]. The text is written by users in an informal language and is subject to the number limitation of words. Without class label in this data set, only sports-related data are selected out, and used as experimental data for sport-item data classification after they are manually tagged and the final 6 items, including 3000 training sets and 630 test sets, are left with an average text length of 12.95. The specific information is listed in Table 2.

**Table 1.** Web snippets dataset

| Class | Training set | Testing set |
|---|---|---|
| Business | 1200 | 300 |
| Computers | 1200 | 300 |
| Culture-Arts-Entertainment | 1880 | 330 |
| Education-Science | 2360 | 300 |
| Engineering | 220 | 150 |
| Health | 880 | 300 |
| Politics-Society | 1200 | 300 |
| Sports | 1120 | 300 |

**Table 2.** Twitter sports dataset

| Class | Training set | Testing set |
|---|---|---|
| Baseball | 500 | 100 |
| basketball | 500 | 100 |
| Football | 400 | 80 |
| Golf | 400 | 50 |
| Rugby | 800 | 200 |
| Swimming | 400 | 100 |

The third data set is the AGnews data obtained by Zhang [19] et al., and the 4 classes with the largest amount of are selected to construct the data set, including 120,000 training sets and 7600 test sets, with an average text length of 38.82. The specific information is listed in Table 3.

**Table 3.** AGnews dataset

| Class | Training set | Testing set |
|---|---|---|
| World | 30000 | 1900 |
| Sports | 30000 | 1900 |
| Business | 30000 | 1900 |
| Sci/Tech | 30000 | 1900 |

## 6.2  Parameters Selection

In Eq. (2), the regularization parameters $\mu$ and $\phi$ are selected according to one of the three evaluation indexes, Purity [20], Normalized Mutual Information (NMI) [21] and Adjusted Rand Index (ARI) [26]. Purity calculates the proportion of correctly clustered documents in the total number of documents. NMI measures the degree of similarity between the two clustering results, and ARI measures the degree of coincidence between the clustering results and the real situation. In the process of relationship matrix factorization, the regularization parameter is set to $\mu = \phi$. Based on different value of $\mu$, the DNMTF method with random initialization is carried out for 50 times, and the comparison results are shown in Fig. 2.

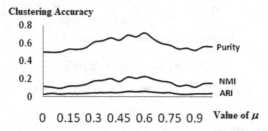

**Fig. 2.** Effect of different regularization parameter $\mu$

From Fig. 2, we can see that the clustering accuracy arrives the highest when $\mu = 0.6$, with any one of three evaluation indexes. Accordingly, in the following experiments of matrix factorization, we set up the regularization parameter to be $\mu = 0.6$.

The Web snippets data set has 4775 features, Twitter sports data set has 1248 features, and AGnews data set has 6582 features. The selection of feature extension number $K$ directly affects the classification results. Therefore, different parameters $K$ are selected on three data sets for comparative experiments, and the results are shown in Fig. 3(a)–(c), respectively. We can see that no matter which data set, even if there is only one feature is added, and the accuracy of classification results increase rapidly to be close to the optimal value 1. The reason for that is the feature with the strongest relevance to the short text is found in the feature space according to Eq. (12), which must be the most indicative feature in a certain category. Expansion by this feature will allow other short texts of the same category to enlarge their feature representation, in case they did not have it before. The similarity between the sparse feature vectors of the same category is greatly improved, which has a positive impact on the classification results.

When the number of extended features gradually increases, the accuracy of classification results increases comparatively constant until it reaches the peak point of each dataset, then it begins to decline slightly, as shown in Fig. 3 (a)–(c).

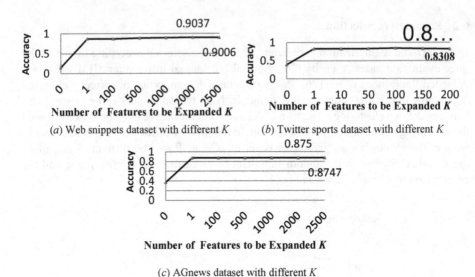

(a) Web snippets dataset with different $K$

(b) Twitter sports dataset with different $K$

(c) AGnews dataset with different $K$

**Fig. 3.** Results of parameter $K$ on three datasets

## 6.3   Compared Algorithms

In order to verify the effect of NMFFE algorithm, we compare NMFFE with BOW and Char-CNN, namely word bag method and character level convolutional neural network method without considering semantic information. The results are shown in Table 4. and the corresponding best results in the table are all in bold font. In the study [11], the accuracy of BOW algorithm and Char-CNN algorithm on AGnews data set was 88.81% and 87.18%, respectively. In our experimental environment and data processing operations, our experimental results shown in Table 4 are little different with those presented by study [11].

**Table 4.** Comparison results of classification accuracy on 3 datasets

| Datasets | BOW | Char-CNN | NMFFE |
|---|---|---|---|
| Web snippets | 0.646 | 0.625 | **0.9037** |
| Twitter100k | 0.7346 | 0.5234 | **0.8435** |
| AGnews | 0.8421 | 0.8571 | **0.875** |

From Table 4, we can find that in the respect of dataset size, the Char-CNN algorithm performs well in big datasets but perform less in small datasets, where the limited training data cannot cover the overall distribution of data, and lead to the over-fitting of convolutional neural network.

In the respect of data integrity, text length of the AGnews dataset is relatively long, and its sufficient corpus makes the three algorithms perform well in text classification. The accuracies of their classification results have small differences. The similarity between test dataset and training dataset of Web snippets (co-occurrence of keywords) is not as high as the other two datasets, making the BOW algorithm based on word frequency statistics on this dataset less effective.

The overall performance of the proposed NMFFE algorithm achieves better classification results than those of the other two algorithms, and the robustness on datasets with different sizes is better than the two latter. BOW algorithm and Char-CNN algorithm are more suitable for large-scale datasets.

The running time of the three algorithms is compared on three data sets, and the results are shown in Fig. 4. The execution time of BOW algorithm is shorter than the other two algorithms, and it is more obvious on large datasets, mainly because the model of BOW algorithm is relatively simple. NMFFE algorithm takes the longest time in the feature expansion process, because it involves a lot of matrix operations. When the number of feature extensions $K$ increases, the running time also increases. The Char-CNN algorithm model consists of 6 convolution layers and 3 full connection layers.

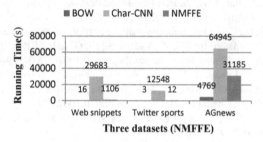

**Fig. 4.** Comparation of running time

# 7 Conclusions

Different from vector-form based feature expansion method of short texts, we proposed a method using K relevant features as a self-contained subset to extend feature space of short texts. Without relying on the external resources, words clustering indicator matrix was obtained from text dataset itself through graph dual regularization non-negative matrix tri-factorization (DNMTF). After dimension reduction, feature space was obtained as the basis for feature expansion, and then the most relevant features extracted within the dataset itself were selected to enlarge the feature space of short texts. Experimental results showed that NMFFE algorithm performed better than Word2Vec algorithm and Char-CNN algorithm in accuracy of classification. However, the datasets used in this paper were all open datasets which actually had been pre-processed. However, the main challenge of short-text feature expansion and classification is the online and real-time data processing. So, we will adjust our method to adapt the real-time online environments in the future.

**Acknowledgements.** This paper was funded by Scientific Project of Guangdong Provincial Transport Department (No. Tec-2016-02-30), Natural Science Foundation of Guangdong Province under Grant 2018A030313061, in part by the Guangdong Science and Technology Plan under Grant 2017B010124001, Grant 201902020016, and Grant 2019B010139001. The last author is also partially supported by the European Horizon 2020 research and innovation program by the ENVRI-FAIR project (824068), the BLUECIOUD project (862409), and the ARTICONF project (825134).

# References

1. Rafeeque, P.C., Sendhilkumar, S.: A survey on short text analysis in web. In: Third International Conference on Advanced Computing. https://doi.org/10.1109/icoac.2011.616 5203

2. Heap, B., Bain, M., Wobcke, W.: Word vector enrichment of low frequency words in the bag-of-words model for short text multi-class classification problems. e-print (2017). https://arxiv.org/abs/170905778H

3. Tommasel, A., Godoy, D.: Short-text feature construction and selection in social media data: a survey. Artif. Intell. Rev. 49(3), 301–338 (2016). https://doi.org/10.1007/s10462-016-9528-0

4. Kang, W., Qiu, H.Z., Jiao, D.D.: Search-based short-text classification. Appl. Electron. Tech. (2018). https://doi.org/10.16157/j.issn.0258-7998.181392

5. Li, X., Su, Y., Ma, H., Cao, L.: Combining statistical information and semantic similarity for short text feature extension. In: Shi, Z., Vadera, S., Li, G. (eds.) IIP 2016. IAICT, vol. 486, pp. 205–210. Springer, Cham (2016). https://doi.org/10.1007/978-3-319-48390-0_21

6. Li, J., Cai, Y., Cai, Z., Leung, H., Yang, K.: Wikipedia based short text classification method. In: Bao, Z., Trajcevski, G., Chang, L., Hua, W. (eds.) DASFAA 2017. LNCS, vol. 10179, pp. 275–286. Springer, Cham (2017). https://doi.org/10.1007/978-3-319-55705-2_22

7. Li, P.P., He, L., Wang, H.Y., et al.: Learning from short text streams with topic drifts. IEEE Trans. Cybern. 48(9), 2697–2710 (2018)

8. Vo, D.T., Ock, C.Y.: Learning to classify short text from scientific documents using topic models with various types of knowledge. Expert Syst. Appl. 42(3), 1684–1698 (2015)

9. Zhang, H., Zhong, G.Q.: Improving short text classification by learning vector representations of both words and hidden topics. Knowl. Based Syst. 102(C), 76–86 (2016)

10. Sikdar, U.K., Gambäck, B.: Named entity recognition for amharic using stack-based deep learning. In: Gelbukh, A. (ed.) CICLing 2017. LNCS, vol. 10761, pp. 276–287. Springer, Cham (2018). https://doi.org/10.1007/978-3-319-77113-7_22

11. Zhang, X., Zhao, J.B., Yann, L.C.: Character-level convolutional networks for text classification. In: 29th Annual Conference on Neural Information Processing Systems (NIPS), Montreal, Canada, pp. 649–657 (2015)

12. Wang, D.Q., Lu, C.W., Wu, J.J.: Softly associative transfer learning for cross-domain classification. IEEE Trans. Cybern. (1) (2019). https://doi.org/10.1109/tcyb.2019.2891577

13. Cheng, X., Guo, J., Liu, S.: Learning topics in short texts by non-negative matrix factorization on term correlation matrix. In: 13th SIAM International Conference on Data Mining, May 2013, Texas, USA, pp. 749–757 (2013)

14. Borg, I.: A note on the positive manifold hypothesis. Pers. Individ. Differ. 134(1), 13–15 (2018)

15. Shang, F.H., Jiao, L.C., Wang, F.: Graph dual regularization non-negative matrix factorization for co-clustering. Pattern Recogn. 45(6), 2237–2250 (2012)

16. Boyd, S., Vandenberghe, L.: Convex Optimization, vol. 3, pp. 63–107. Cambridge University Press, Cambridge (2004)

17. Phan, X.H., Nguyen, L.M., Horiguchi, S: Learning to classify short and sparse text & web with hidden topics from large-scale data collections. In: Proceeding of the 17th International Conference on World Wide Web, pp. 91–100 (2008)
18. Hu, Y.T., Zheng, L., Yang, Y.: Twitter 100k: a real-world dataset for weakly supervised cross-media retrieval. IEEE Trans. Multimedia **20**(4), 927–938 (2018)
19. Zhang, X.: AG's News Topic Classification Dataset Version 3, Updated, 09 September 2015
20. Hassani, M., Seidl, T.: Using internal evaluation measures to validate the quality of diverse stream clustering algorithms. Vietnam J. Comput. Sci. **4**(3), 171–183 (2016). https://doi.org/10.1007/s40595-016-0086-9
21. Yang, S., Zhang, L.: Non-redundant multiple clustering by nonnegative matrix factorization. Mach. Learn. **106**(5), 695–712 (2016). https://doi.org/10.1007/s10994-016-5601-9
22. Quan, X.J., Kit, C.Y., Ge, Y.: Short and sparse text topic modeling via self-aggregation. In: 1st International Workshop on Social Influence Analysis/24th International Joint Conference on Artificial Intelligence (IJCAI), Buenos Aires, Argentina, pp. 2270–2276 (2015)
23. Pennington, J., Socher, R., Manning, C.D.: GloVe: global vectors for word representation. In: Proceedings of the 2014 Conference on Empirical Methods in Natural Language Processing, Doha, Qatar, pp. 1532–1543 (2014)
24. Mikolov, T., Chen, K., Corrado, G.: Efficient estimation of word representations in vector space (2013). https://arxiv.org/abs/1301.3781v3
25. Jameel, S., Bouraoui, Z., Schockaert, S.: Unsupervised learning of distributional relation vectors. In: The 56th Annual Meeting of the Association for Computational Linguistics, Melbourne, Australia, pp. 1–11 (2018)
26. Robert, V., Vasseur, Y., Brault, V.: Comparing high dimensional partitions, with the coclustering adjusted rand index, 18 May 2017
27. Li, Z.H., Yang, Z.H., Shen, C.: Integrating shortest dependency path and sentence sequence into a deep learning framework for relation extraction in clinical text. BMC Med. Inform. Decis. Mak. **19**(1), 22 (2019)
28. Tsapatsoulis, N., Djouvas, C.: Opinion mining from social media short texts: does collective intelligence beat deep learning. Front. Rob. AI **5**, 138–146 (2019)
29. Hassan, A., Mahmood, A.: Deep learning approach for sentiment analysis of short texts. In: The 3rd IEEE International Conference on Control, Automation and Robotics, Nagoya, Japan, pp. 705–710 (2017)
30. Wang, Z.L., Li, S., Chen, G.: Deep and shallow features learning for short texts matching. In: The 5th IEEE International Conference on Progress in Informatics and Computing, Nanjing, Peoples R China, pp. 51–55 (2017)
31. Severyn, A., Moschitt, A.: Learning to rank short text pairs with convolutional deep neural networks. In: The 38th Annual International ACM SIGIR Conference on Research and Development in Information Retrieval, Santiago, Chile, pp. 373–382 (2015)
32. Jinarat, S., Manaskasemsak, B., Rungsawang, A.: Short text clustering based on word semantic graph with word embedding model. In: Joint 10th International Conference on Soft Computing and Intelligent Systems/19th International Symposium on Advanced Intelligent Systems, Toyama, Japan, pp. 1427–1432 (2018)
33. Jiménez Pascual, A., Fujita, S.: Text similarity function based on word embeddings for short text analysis. In: Gelbukh, A. (ed.) CICLing 2017. LNCS, vol. 10761, pp. 391–402. Springer, Cham (2018). https://doi.org/10.1007/978-3-319-77113-7_31
34. Liang, W.X., Feng, R., Liu, L.X.Y.: GLTM: a global and local word embedding-based topic model for short texts. IEEE Access **6**, 43612–43621 (2018)
35. Al-Azani, S., El-Alfy, E.S.M.: Using word embedding and ensemble learning for highly imbalanced data sentiment analysis in short Arabic text. In: The 8th International Conference on Ambient Systems, Networks and Technologies/7th International Conference on Sustainable Energy, Madeira, Portugal, vol. 109, pp. 359–366 (2017)

36. De Boom, C., Van Canneyt, S., Demeester, T.: Representation learning for very short texts using weighted word embedding aggregation. Pattern Recogn. Lett. **80**, 150–156 (2016)
37. Wang, P., Xu, B., Xu, J.M.: Semantic expansion using word embedding clustering and convolutional neural network for improving short text classification. Neurocomputing **174**, 806–814 (2016)
38. Sang, L., Xie, F., Liu, X.J.: WEFEST: word embedding feature extension for short text classification. In: The 16th IEEE International Conference on Data Mining, Barcelona, Spain, pp. 677–683 (2016)
39. Xun, G.X., Gopalakrishnan, V., Ma, F.L.: Topic discovery for short texts using word embeddings. In: The 16th IEEE International Conference on Data Mining, Barcelona, Spain, pp. 1299–1304 (2016)
40. Xia, W., He, Y.X., Tian, Y.: Feature expansion for microblogging text based on Latent Dirichlet Allocation with user feature. In: The 6th IEEE Joint International Information Technology and Artificial Intelligence Conference, Chongqing, China, pp. 228–232 (2011)
41. Naili, M., Chaibi, A.H., Ben, G.: Comparative study of word embedding methods in topic segmentation. Procedia Comput. Sci. **112**, 340–349 (2017)
42. Xu, Y.: Research on Short Text Classification Based on Word Vectors and Topics, vol. 5, pp. 5–14. Huazhong University of Technology, Wuhan (2018)
43. Yu, J., Qiu, L.R.: ULW-DMM: an effective topic modeling method for microblog short text. IEEE Access **7**, 884–893 (2019)

# Cross-Project Software Defect Prediction Based on Feature Selection and Transfer Learning

Tianwei Lei[1]([⊠]), Jingfeng Xue[1], and Weijie Han[1,2]

[1] School of Computer Science and Technology, Beijing Institute of Technology, Beijing 100081, China
absherry123@163.com
[2] School of Space Information, Space Engineering University, Beijing 101416, China

**Abstract.** Cross-project software defect prediction solves the problem that traditional defect prediction can't get enough data, but how to apply the model learned from the data of different mechanisms to the target data set is a new problem. At the same time, there is the problem that information redundancy in the training process leads to low accuracy. Based on the difference of projects, this paper uses MIC to filter features to solve the problem of information redundancy. At the same time, combined with the TrAdaboost algorithm, which is based on the idea of aggravating multiple classification error samples, this paper proposes a cross-project software prediction method based on feature selection and migration learning. Experimental results show that the algorithm proposed in this paper has better experimental results on AUC and F1.

**Keywords:** Transfer learning · TrAdaboost · MIC · Cross-project software defect prediction

## 1 Introduction

With the rapid development of information technology in recent decades, the scale of software is becoming larger and larger, and the software vulnerabilities are becoming hidden, which makes the assurance of the software's quality more difficult to achieve [1].

Software defect prediction can discover the defects in software before the application is put into production, to reduce the cost of subsequent manual testing and the development cycle [2]. It is difficult for many organizations to obtain enough historical versions to build data sets for defect prediction in practical, and it will lead the problem of cold start is encountered in the initial prediction. A possible solution to this problem is to use the data of other projects to build a model to predict the software defects of the target project, that is, cross-project software defect prediction [3]. But there are also some problems will also lead the low training accuracy in the cross-project software defect prediction, such as information redundancy, class imbalance [4].

© Springer Nature Switzerland AG 2020
X. Chen et al. (Eds.): ML4CS 2020, LNCS 12488, pp. 363–371, 2020.
https://doi.org/10.1007/978-3-030-62463-7_33

This paper propose a cross-project software defect prediction method based on feature selection and transfer learning. The main contributions of this paper are as follows:

1. This paper proposes a method of cross-project software defect prediction based on the combination of feature selection and transfer learning and the experiments show that the defect prediction is effective.
2. MICs method is used to solve the problems of data redundancy and feature dimension explosion;
3. After feature selection, the MuTrAdaboost algorithm is proposed to increase the TrAdaboost algorithm. MuTrAdaboost enhances the weight of instances which is like the target project data, and the final model is formed by multiple training.

The experimental results show that the MuTrAdaboost algorithm is better than the TrAdaboost algorithm in AUC and F1.

## 2   Related Works

Cross-project software defect prediction [5], as mentioned above, is mainly to build a defect prediction model for the target project by using the data sets already collected by other projects, so as to predict and analyze the data of the target project. And it is mainly based on traditional software defect prediction technology, aiming at finding similarities between different factors and using potential links to propose a reasonable and efficient prediction model.

In recent years, the transfer learning method in the field of machine learning provides a good idea for the prediction of cross project [6]. Transfer learning [7] is a machine learning method which uses the existing knowledge to solve the problems in the related fields.

Project variation is the first problem to be solved in the direction of cross project defect prediction [8]. He [9] have carried out the corresponding research from the perspective of the disaster problem of the feature dimension of data sets. Amasaki [10] try to use unsupervised learning to remove the irrelevant features and instances from the target project to improve the accuracy.

In the field of transfer learning, Dai et al. [11] proposed the TrAdaboost method, which applies the idea of AdaBoost to transfer learning to improve the instance weight of target classification task and reduce the instance weight of target classification task. TraAdaboost method is one of the classical research of transfer learning, and there are also some research works in the field of software defect prediction based on this algorithm. Chen [12] uses data preprocessing and TrAdaboost to predict defects from the perspective of reducing the weight of negative instances in the source project. Shen [13] considers the way of multiple projects migrating one target project, proposes two improved prediction algorithms based on TrAdaboost, and constructs the final prediction model by inheritance learning.

# 3  Framework of the Approach

The approach of this paper is illustrated as Fig. 1.

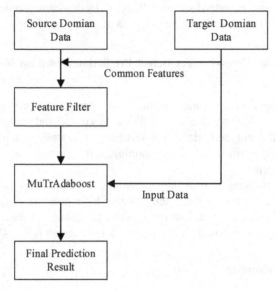

**Fig. 1.** The framework of the proposed approach

1. Firstly, the data from different project are divided into two domains, source domains and target domain. Our goal is to use the data in source domain to build prediction model to predict the fault in source data. The common features in both source data domain and target data domain are reserved for the next step.
2. The method MIC is used to filter the features. We do the MIC calculation between features of target dataset and source dataset and choose the features which have higher MIC results.
3. We build the prediction model with MuTradaboost which is an improved TrAdaboost algorithm. After the building process, the part of target data is inputted into model as test data to get final prediction result.

# 4  Key Technologies

## 4.1  Feature Selection Based on Maximum-Information-Coefficient MIC

Before the formal establishment of the model, it is necessary to filter the features in the training project, exclude the feature vectors that are too different from the features of the target project, and select the feature vectors that are highly correlated with the features in the training project as the basis for model construction.

In our experiment, we use the maximum information coefficient (MIC) to calculate the correlation between two vectors. It is a method proposed by David et al. [14] to express the dependence between two groups of variables.

The MIC calculation is done between the same features in source dataset and target dataset, and the features is sorted by their MIC result by descending order. We select the required number of features to build the model in training.

## 4.2 MuTrAdaboost, Cross-Project Defect Prediction Based on Improved TrAdaboost

After filtering the appropriate features of the source project, we proposed a improved TrAdaboost algorithm MuTrAdaboost to build the data model and predict the defects. In our method, we still adopt the basic idea of TrAdaboost: to strengthen the weight of the samples that are misclassified in a certain training, and hope that they can be correctly classified in the next iteration.

The core of our thinking is: while strengthening the weight of the samples that have been wrongly divided, we give those samples that have been wrongly divided many times higher weight to quickly adjust the model and update the data. Therefore, the defect prediction algorithm based on the improved TrAdaboost is as follows:

---

**MuTrAdaboost Algorithm**
*Input*
Dataset from source data field $T_s = \{T_{d_1}, T_{d_2}, \dots, T_{d_{N_s}}\}$

Sample target dataset $T_t$
The unlabeled data set $S$
Base learning algorithm Learner
Maximum number of iterations $N$
Error limits $\varepsilon$

---

**Algorithm**
1: Initialize the weight factor $\beta = 1$, err$= zeros(1 \dots n + m)$ and the weight vector $w^t = (\omega_1^1, \dots, \omega_{n+m}^1)$

$$\omega_i^1 = \begin{cases} \dfrac{1}{n}, & i = 1 \dots n \\ \dfrac{1}{m}, & i = n+1 \dots n+m \end{cases}$$

2: For each instance $t = 1 \dots N$ do

3:  Normalize the weight vector $p^t = \dfrac{w^t}{\sum_{i=1}^{n+m} \omega_i^t}$ ;

4:  Use the $p^t$ as the weight data to train the base learning algorithm learner , return the hypothesis $h_t$

5:  Calculate the error rate $\varepsilon_t$ of $h_t$ on $T_t$,

$$\varepsilon_t = \sum_{i=n+1}^{m} \dfrac{\omega_i^t}{\sum_{j=1}^{n+m} \omega_j^t} |h_t(x_i) - c(x_i)|$$

6:   Update the weight factor $\beta_t = \frac{\varepsilon_t}{1-\varepsilon_t}$, and the error rate $\varepsilon_t$ must be less than 0.5, otherwise set $\varepsilon_t = 0.5$

7:   For each instance $j = 1\ldots n+m$ do

8:      Calculate the err[j]:

$$err[j] = \begin{cases} err[j] + 1, & |h_t(x_i) - c(x_i)| \geq \varepsilon \\ err[j], & |h_t(x_i) - c(x_i)| < \varepsilon \end{cases}$$

9:   End for

10:  Update the weight vector:

$$\omega_i^{t+1} = \begin{cases} \omega_i^t \beta_t^{|h_t(x_i)-c(x_i)|} & i = 1\ldots n\&|h_t(x_i) - c(x_i)| < \varepsilon \\ \omega_i^t \beta_t^{|h_t(x_i)-c(x_i)|\cdot(-ln(err[i])+1)} & i = 1\ldots n\&|h_t(x_i) - c(x_i)| \geq \varepsilon \\ \omega_i^t \beta_t^{-|h_t(x_i)-c(x_i)|} & i = n+1\ldots n+m\&|h_t(x_i) - c(x_i)| < \varepsilon \\ \omega_i^t \beta_t^{-|h_t(x_i)-c(x_i)|\cdot errs[i]} & i = n+1\ldots n+m\&|h_t(x_i) - c(x_i)| \geq \varepsilon \end{cases}$$

11: End for

---

*Output*

$$h_f(x) = \begin{cases} 1, & \prod_{t=[N/2]}^{N} \beta_t^{-h_t(x)} \geq \prod_{t=[N/2]}^{N} \beta_t^{-\varepsilon} \\ 0, & otherwise \end{cases}$$

---

## 5   Experiments

### 5.1  Dataset

To show the performance of the software defect prediction of cross-project, we select the open NASA dataset and SOFTLAB dataset as the experimental set and compare the experimental results. We use NASA dataset as the source dataset and SOFTLAB data set as the target dataset for prediction. These datasets are available on the PRIMISE website.

We selected 10 items of NASA dataset, 3 items of SOFTLAB dataset, and their common features as the initial data feature set. Table 1 shows their details.

### 5.2  Experimental Indicators

In this paper, AUC (area under ROC) and F-measure are used as evaluation criteria. AUC refers to the area surrounded by ROC curve and axis A good prediction model should have a high precision at the same time of high recall. However, the high recall is often achieved at the cost of low precision, so we introduce the indicator F-measure which is used to measure the harmonic average of the two.

**Table 1.** Source data and target data

| Source data | | | |
|---|---|---|---|
| Project | Examples | %Defective | Description |
| CM1 | 327 | 12.84 | Space craft instrument |
| KC3 | 194 | 9.38 | Storage management |
| MC1 | 1988 | 2.31 | Video guidance system |
| MC2 | 161 | 32.30 | Video guidance system |
| MW1 | 403 | 7.69 | A zero gravity experiment |
| PC1 | 1109 | 6.94 | Flight software |
| PC2 | 745 | 2.15 | Flight software |
| PC3 | 1077 | 12.44 | Flight software |
| PC4 | 1287 | 13.75 | Flight software |
| PC5 | 1711 | 27.53 | Flight software |
| Target data | | | |
| ar3 | 63 | 12.70 | Embedded controller |
| ar4 | 107 | 18.69 | Embedded controller |
| ar5 | 36 | 22.22 | Embedded controller |

## 5.3 Experiment Comparison

In order to test the experimental effect of this paper, we designed three comparative experiment: separate TrAdaboost experiment; separate MuTrAdaboost experiment for the data; and use mic for feature selection first, and MuTrAdaboost experiment for the selected samples.

We select all NASA data sets as the data of the source data, and select one SOFTLAB data set for the target data at a time, and randomly select 10% of the target data as the test data for the experiment. Table 2 shows the comparison of experimental indicators of three algorithms on multi-source data sources. In TrAdaboost and MuTrAdaboost experiments, we used all the common features of each data set and the target data set in the source data domain. In feature filtering, we selected some features according to the order of MIC value. We can see that the use of MuTrAdaboost algorithm can significantly improve the AUC index of the experiment compared with TrAdaboost algorithm. In addition to dataset ar5, the F-measure index can also be improved to some extent. The third column of Table 2 shows that the index selected by adjusting the proportion of using features is the best after using the MIC algorithm for feature filtering and MuTrAdaboost algorithm. We can see that the use of MIC algorithm can improve the AUC index AUC and F-measure after the method have been improved to some extent. The defect rate of ar3 project is only 12.3%, which leads to the poor performance of the three algorithms.

**Table 2.** Comparison of multi-source data experiments

| Metric | TraAdaboost | MuTraAdaboost | Mic + MuTraAdaboost |
|---|---|---|---|
| ar3 | | | |
| AUC | 0.571 | 0.667 | 0.733 |
| F-measure | 0.275 | 0.333 | 0.375 |
| ar4 | | | |
| AUC | 0.545 | 0.745 | 0.812 |
| F-measure | 0.325 | 0.364 | 0.453 |
| ar5 | | | |
| AUC | 0.577 | 0.793 | 0.833 |
| F-measure | 0.375 | 0.372 | 0.449 |

Table 3 shows the data of AUC and F1 indexes with different proportion of characteristic number filtered by MIC value in each data set. It can be seen that ar3, ar4 and ar5 data sets have the best data results when selecting 70%, 80% and 70% of the number of features, respectively. Figures 2, 3 and 4 show the change of AUC and F1 indexes of target data sets ar3, ar4 and ar5 with the increase of feature proportion of filtering, where abscissa is the proportion of feature quantity selected after feature filtering to total feature quantity, while ordinate is the value range of AUC and F1.

From the experimental comparison of multi-source data, it can be seen that the introduction of MuTrAdaBoost can improve the accuracy of the experiment, while the use of MIC to filter features rather than select all features for the experiment can also improve the effect of the experiment to a certain extent, and at the same time, the speed

**Table 3.** Experimental results of feature selection scale of target domain dataset

| | 0.10 | 0.20 | 0.30 | 0.40 | 0.50 | 0.60 | 0.70 | 0.80 | 0.90 | 1.00 |
|---|---|---|---|---|---|---|---|---|---|---|
| ar3 | | | | | | | | | | |
| AUC | 0.21 | 0.28 | 0.34 | 0.40 | 0.50 | 0.53 | 0.57 | 0.56 | 0.57 | 0.53 |
| F1 | 0.07 | 0.10 | 0.15 | 0.18 | 0.24 | 0.26 | 0.28 | 0.27 | 0.27 | 0.27 |
| ar4 | | | | | | | | | | |
| AUC | 0.21 | 0.36 | 0.44 | 0.54 | 0.63 | 0.69 | 0.73 | 0.75 | 0.74 | 0.74 |
| F1 | 0.06 | 0.10 | 0.18 | 0.25 | 0.28 | 0.31 | 0.34 | 0.36 | 0.36 | 0.36 |
| ar5 | | | | | | | | | | |
| AUC | 0.27 | 0.30 | 0.42 | 0.52 | 0.66 | 0.77 | 0.83 | 0.82 | 0.83 | 0.81 |
| F1 | 0.05 | 0.13 | 0.20 | 0.27 | 0.34 | 0.38 | 0.41 | 0.41 | 0.41 | 0.41 |

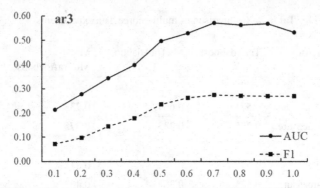

**Fig. 2.** AUC and F1 change chart of ar3 value with feature proportion

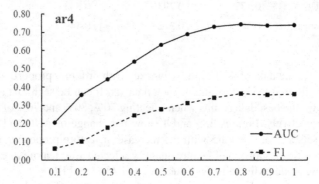

**Fig. 3.** AUC and F1 change chart of ar4 value with feature proportion

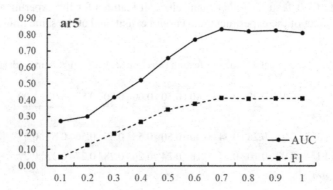

**Fig. 4.** AUC and F1 change chart of ar5 value with feature proportion

of the experiment has been improved due to the reduction of redundant features involved in the calculation.

In summary, from AUC and F1 experimental indicators, the experimental effect of MuTrAdaboost using all features is better than that of traditional TrAdaboost algorithm using all features, which shows that our idea of multiple weighting of wrong samples is more effective than that of TrAdaboost algorithm, which only weights the samples that are wrongly divided each time. The MuTrAdaboost algorithm filtered by MIC features is better than that using all features. This shows that the idea of feature selection is correct, and the use of feature vector MIC value as the basis of feature selection is also effective.

## 6 Conclusion

In this paper, we propose a cross-project software defect prediction method based on the combination of feature selection and transfer learning. And the experiments show that the method we supposed is better than that of traditional TrAdaboost algorithm.

The next research direction is to explore the experimental effect of the algorithm on larger datasets, and to carry out more comparative experiments for more indicators to explore the improvement direction of the updated algorithm.

**Acknowledgement.** This work was supported by the National Key Research and Development Program of China under Grant 2016QY06X1205.

## References

1. Pizzi, N.J.: A fuzzy classifier approach to estimating software quality. Inf. Sci. **241**, 1–11 (2013)
2. Nam, J., et al.: Heterogeneous defect prediction. IEEE Trans. Softw. Eng. **44**, 874–896 (2017)
3. Xia, X., et al.: HYDRA: massively compositional model for cross-project defect prediction. IEEE Trans. Softw. Eng. **42**, 977–998 (2016)
4. He, Z., et al.: An investigation on the feasibility of cross-project defect prediction. Autom. Softw. Eng. **19**(2), 167–199 (2012)
5. Hall, T., et al.: A systematic literature review on fault prediction performance in software engineering. IEEE Trans. Softw. Eng. **38**, 1276–1304 (2012)
6. Pan, S.J., Yang, Q.: A survey on transfer learning. IEEE Trans. Knowl. Data Eng. **22**(10), 1345–1359 (2010)
7. Zhuang, F., et al.: Survey on transfer learning. J. Softw. **26**(1), 26–39 (2015). (in Chinese)
8. Chen, X., et al.: A survey on cross-project software defect prediction methods. Chin. J. Comput. **041**(001), 254–274 (2018). (in Chinese)
9. He, P., et al.: An empirical study on software defect prediction with a simplified metric set. Info. Softw. Technol. **59**(mar), 170–190 (2015)
10. Amasaki, S., Kawata, K., Yokogawa, T.: Improving cross-project defect prediction methods with data simplification. In: Software Engineering Advanced Applications IEEE (2015)
11. Dai, W., Yang, Q., Xue, G., et al.: Boosting for transfer learning. In: Proceedings of the 24th International Conference on Machine Learning, Corvallis, USA, 20—24 June 2007, pp. 93–200. ACM, New York (2007)
12. Chen, L., et al.: Negative samples reduction in cross-company software defects prediction. Inf. Softw. Technol. **62**, 67–77 (2015)
13. Fagui, M., et al.: Cross-project software defect prediction based on instance transfer. J. Front. Comput. Sci. Technol. **10**, 43–55 (2016)
14. Reshef, D.N., et al.: Detecting novel associations in large data sets. Science **334**(6062), 1518–1524 (2011)

# A Swarm Enhanced Light Gradient Boosting Machine for Crowdfunding Project Outcome Prediction

Shuang Geng[1], Miaojia Huang[2(✉)], and Zhibo Wang[2]

[1] College of Management, and Greater Bay Area International Institute for Innovation, Shenzhen University, Shenzhen, People's Republic of China
[2] College of Management, Shenzhen University, Shenzhen, People's Republic of China
2016040301@email.szu.edu.cn

**Abstract.** Crowdfunding is an emerging financing method that project founders could obtain funding from vast investors through the online platform. Therefore, investigating the critical features of crowdfunding projects to forecast the project outcomes have become indispensable. This research draws upon some potential factors and introduces a swarm enhanced light gradient boosting machine (S-LightGBM) model to forecast the crowdfunding performance. Text mining and lexicon-based sentiment analysis methods were employed to derive the linguistic and sentiment features of project descriptions. This study compares the predictive power of logistic regression, support vector machine, light gradient boosting machine, and S-LightGBM on 5916 crowdfunding projects between 2017 and 2018. The result shows that the S-LightGBM approach achieves superior accuracy results than conventional methods. The usefulness of linguistic and sentiment features was also investigated and discussed. This research contributes to the existing research on machine learning methods and crowdfunding and provides fundraisers guidance for the presentation and illustration of innovative projects on crowdfunding platforms.

**Keywords:** Crowdfunding · Machine learning · Sentiment analysis · Particle swarm optimization

## 1 Introduction

Crowdfunding is an innovative Internet financing mechanism for entrepreneurs to secure financial resources from a large number of people online to support a specific venture [1, 2]. As a novel capital-raising model, it has offered a strong and supportive approach for small- and medium-sized enterprises (SMEs) that play the most vigorous roles in the market whereas have huge challenges in securing funds at the beginning stage. The project financing performance and the sustainable development of SMEs are closely related to the economic development of a country. Many studies have provided strong evidence towards the significance of crowdfunding [3].

However, the flourishing crowdfunding platforms are facing many practical problems such as the information asymmetry issues, low financing success rates, etc. [4].

© Springer Nature Switzerland AG 2020
X. Chen et al. (Eds.): ML4CS 2020, LNCS 12488, pp. 372–382, 2020.
https://doi.org/10.1007/978-3-030-62463-7_34

Although prior studies on the crowdfunding dynamics reveal that fundraisers' social capital as well as the project description can help alleviate the information asymmetry issue [5], many backers lack effective ways to identify the project quality and they are always plagued by the project authenticity, quality, feasibility. Some fundraisers show high interests and strong expertise in some projects via self-pledging or via posting reviews with multiple users to create the promising illusion of the projects [6]. Some projects fail to reach the desired goal and cannot offer the promised rewards to the investors, which arouse investors' mistrustful attitudes to the projects and consequently jeopardize the sustainable development of crowdfunding. Therefore, an appropriate prediction method would undoubtedly offer potential countermeasures that are urgently needed to the backers. From the perspective of methodology, the accessibility of crowdfunding data in recent years accelerates the application of data-driven methods in crowdfunding, for instance, predicting crowdfunding outcomes using logistic regression (LR), support vector machines (SVM), etc. Nevertheless, most of the extant research emphasize only on the predictive power of key project attributes.

In this study, we employ linguistic features and sentiment score based on project description, project information to predict project outcomes (success, fail, or live). We integrate particle swarm optimization (PSO) method with light gradient boosting machine to identify the most useful predictive features and estimation parameters, thus improve the prediction performance. Results demonstrate that incorporating linguistic and sentiment project features in most cases contribute to the prediction performance whereas in very few cases, noise may be introduced and thus undermines the performance. Parameter tuning through PSO effectively obtains better results than traditional parameter tuning methods, such as grid search and stepwise method.

The structure of this study goes as follows. Literature review based on the studies on project performance prediction with the analysis techniques of machine learning, text mining and sentiment analysis is presented in Sect. 2. Section 3 introduces methodology based on LGBM and PSO. Dataset description and experimental settings are presented in Sect. 4. Section 5 reports the results of multiple machine learning methods. Some practical implications based on the experimental results are also included in this part. We finally conclude this research with contributions, limitations, and future research.

## 2 Literature Review

### 2.1 Project Performance Prediction

**Data-Driven Techniques: Machine Learning in Crowdfunding.** There are different crowdfunding models, including reward-based, equity-based, and debt-based, among which reward-based crowdfunding is the most frequently discussed [2]. To date, many studies have focused on the crowdfunding performance and the influencing factors using various methods, including qualitative, quantitative, and big data analysis. Recent years have witnessed many researchers shift their focus from traditional qualitative or quantitative methods to data-driven techniques to forecast project outcomes and find continuous improvement on the prediction accuracy. For example, Greenberg et al. [7] adopted SVM and Decision Trees (DT) to predict project success at the beginning stage. Etter et al. [8] collected time-series features of the campaign state, including the amount of pledged money and social features such as the number of backers. K-nearest Neighbor (KNN), Markov Chain, and SVM were adopted to forecast the fundraising performance. Quitian et al. [9] collected data through the survey to seek patterns of participants' crowdfunding behaviors and analyzed the data based on two machine learning techniques (DT and clustering).

**Text Mining and Sentiment Analysis in Crowdfunding.** Prior research on the influencing factors focused more on numeric data, while some scholars seek to analyze the textual and unstructured data on crowdfunding page. For example, Desai et al. [10] used two different models: SVM with Radial Basis Function (RBF kernel) and l2-regularized LR to investigate the linguistic features of projects for binary project outcome classification. Sokolova and Perez [11] analyzed a dataset with multiple dimensions, including campaign page, narratives of Tweets and the information diffusion process based on text mining and graph analysis. Yuan et al. [12] employed Domain-Constraint Latent Dirichlet Allocation (DC-LDA) topic modeling to extract latent semantics from the project description and subsequently forecast the financing success. Lee and Park [13] used deep learning and topic modeling methods to investigate the correlation between project narratives (visual and textual) and the donors' funding motivation. The results reveal that charitable project initiators should use neither too miserable nor too happy pictures of recipients. Wang et al. [14] analyzed the textual contents from the sentiment perspective on the pledge results. The research indicates that positive sentiment in the detailed blurb promotes the likelihood of success.

To sum up, although many researchers have used qualitative research methodology and quantitative research approaches to identify key influencing factors to forecast the crowdfunding performance, current researches focusing on the crowdfunding dynamics based on machine learning are still lacking. Secondly, many extant researches consider the key factors from the perspective of crowd-funders whereas textual features of the projects are scarcely deeply mined and empirically investigated. Therefore, we seek to bridge the above two research gaps by integrating the usefulness of linguistic and sentiment features to predict the outcomes.

## 3   Methodology

### 3.1   Standard PSO

PSO is a random search combinatorial metaheuristic algorithm [15]. In a D-dimensional target space with the whole population of n particles, each particle's current position Xi represents a potential solution and its velocity is denoted as vi. c1 and c2 represent the learning factors and the adaptive value can be evaluated by the fitness function. The best position of an individual particle is expressed by pi whereas the best position of the whole community is denoted by Pbest. In each iteration, the updating rules of the particle position and velocity are given by the following two equations. PSO is a popular swarm intelligence method applied in a variety of areas because it has fewer parameters and it is simple, effective, and fast to converge. This is also the reason why we adopt PSO to improve the performance of conventional machine earning methods.

$$v_i^{t+1} = v_i^{t+1} + c_1{}^*U_t^t(pBest_i^t - p_i^t) + c_2{}^*U_2^t(gBest_i^t - p_i^t) \tag{1}$$

$$p_i^{t+1} = p_i^t + v_i^{t+1} \tag{2}$$

## 3.2  Light Gradient Boosting Machine

Light gradient boosting machine (LightGBM) is an improved version of gradient boosting framework, which supports parallel training in an efficient way. As an advanced boosting model in machine learning, it is developed by Microsoft in 2017 [16] and has so far surpassed other machine learning models in various aspects, including lower computational cost, higher efficiency, better prediction accuracy and stronger capability of minimizing over-fitting issues. It is also a good prediction method in dealing with high-dimensional and large-scale datasets. Distinct from level-wise growing method of other decision tree-based algorithms, LightGBM adopts leaf-wise growing method to save computation power and to effectively improve the branching gain (as shown in Fig. 1). Therefore, instead of controlling maximum tree depth, LightGBM limits the maximum number of leaves in the training process.

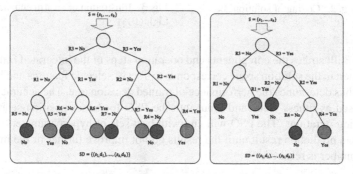

**Fig. 1.** Level-wise growing method and leaf-wise growing method of decision trees

We chose LightGBM as our target methods because of its capability of processing categorical features, good prediction power and computation efficiency. A further step is taken to improve the performance of LightGBM integrated with PSO to obtain the optimal set of parameters and features in the model training process. LightGBM contains multiple hyperparameters corresponding to its strategies controlling its training process. Instead of manually tuning these hyperparameters, researchers developed stepwise parameter tuning algorithm [17], Tree-structured Parzen Estimator [18] and some other algorithms to find the best combination of different hyperparameters. However, research on effective hyperparameter tuning is still very limited. We proposed an integrated framework called swarm enhanced LightGBM that utilizes PSO to tune the hyperparameters of LightGBM, as well as identifying the best features for prediction.

As illustrated in Fig. 2, the selection of set of features and hyperparameters are coded as particle positions in the constrained solution space for PSO. The selection of each feature is coded as binary values for each dimension. Seven hyperparameters are used for tuning, including maximum number of leaves for each tree, frequency for bagging, fraction of bagging, L1 regularization, L2 regularization, fraction of feature, and minimum samples for each leave. We selected these seven hyperparameters for comparison purpose with existing parameter tuning frameworks, such as Optuna [17].

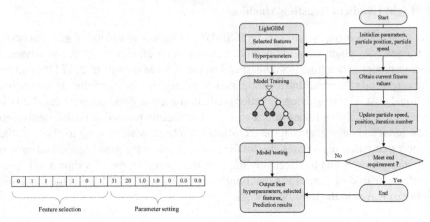

| Feature selection | Parameter setting |

**Fig. 2.** Coding of solution set

**Fig. 3.** Illustration of swarm enhanced LightGBM

Figure 3 illustrates the components and operation steps of the integrated framework. Each position in the solution space corresponds to a set of selected features and hyperparameters, thus determines the performance of trained decision tree. The accuracy results are then used as fitness value and input into the PSO for better position searching in the following iterations. The PSO will end with best sets of hyperparameters, selected features, and prediction result until the results do not improve further or the maximum iteration number is reached.

## 4 Experiments

### 4.1 Dataset

To empirically predict the outcome of crowdfunding projects, we collected 5916 projects created between June, 2017 and February, 2018 collected by Yang [19] from Kickstarter, which is globally famous for the biggest reward-based crowdfunding platform and many scholars have used dataset from Kickstarter for their researches [20]. After removing samples with missing and invalid values, 5516 projects are used in our experiment. The project feature information can be divided into three categories, namely project information (i.e. description, pledge, goal, comment), linguistic features (i.e. Flesch Kincaid Grade Level, Gunning Fog Score), and description sentiment. Details is provided by Yang [19].

### 4.2 Experiment Settings

Two popular algorithms in solving classification problems and crowdfunding project success prediction, namely, logistic regression and support vector machine, are implemented for performance evaluation and comparison. The description of multiple variations of each method and their key parameter settings are reported in Table 1. The parameter setting for PSO is reported in Table 2.

**Table 1.** Algorithm description and parameter settings

| Algorithm | Parameter setting |
|---|---|
| LR1 | Class_weight = default (equal); One v.s. Rest |
| LR2 | Class_weight = default (equal); One v.s. One |
| LR3 | Class_weight = balanced; One v.s. One |
| LR_pso | Class_weight = default (equal); Feature selection: PSO; One v.s. One |
| SVM1 | Kernel: Linear; One v.s. One |
| SVM2 | Kernel: Linear; One v.s. Rest |
| SVM3 | Kernel: Polynomial |
| SVM4 | Kernel: Gaussain |
| SVM_pso | Kernel: Polynomial; Feature selection: PSO |
| LGBM1 | Parameter = Default |
| LGBM 2 | Parameter tuning: Stepwise (Optuna) |
| LGBM3 | Parameter tuning: Feature importance ranking |
| LGBM_pso1 | Parameter tuning: PSO |
| LGBM_pso2 | Feature selection: PSO |
| LGBM_pso3 | Parameter tuning: PSO + Feature selection: PSO |

**Table 2.** PSO parameter setting

| Parameter | Value |
|---|---|
| W (inertia weight) | 0.8 |
| $c_1$ (learning speed) | 2 |
| $c_2$ (learning speed) | 2 |
| $r_1$ (random constant) | 0.6 |
| $r_2$ (random constant) | 0.3 |
| pN (number of particles) | 30 |
| Dim (dimension of solution space) | Feature number + parameter number |
| Max_iter (maximum iteration number) | [50, 100, 200, 500, 1000] |

## 5  Results and Discussion

### 5.1  Results

We adopt F1-score, which is a harmonic average of model accuracy and recall rate, as performance evaluation metric in this research. The F1-scores of each class (Success, Fail, Live) are averaged based on class sample size, thus provide weighted F1 score for the focal method. The classification accuracy ranges from 35.05% to 85.04% as shown in Table 3.

**Table 3.**  Classification accuracy of three data groups

| Method | Data group | | |
|---|---|---|---|
| | Group 1 | Group 2 | Group 3 |
| | Project information | Project information + linguistic features | Project information + linguistic features + description sentiment |
| LR1 | 0.5118 | 0.5743 | 0.6343 |
| LR2 | 0.7475 | 0.7694 | 0.7686 |
| LR3 | 0.7356 | 0.7559 | 0.7542 |
| LR_pso | 0.7736 | 0.7846 | 0.7846 |
| SVM1 | **0.3505** | 0.5245 | 0.5228 |
| SVM2 | 0.5422 | **0.4257** | **0.4299** |
| SVM3 | 0.7593 | 0.7551 | 0.7542 |
| SVM 4 | 0.7399 | 0.6174 | 0.5963 |
| SVM_pso | 0.7846 | 0.7880 | 0.7736 |
| LGBM1 | 0.8191 | 0.8267 | 0.8233 |
| LGBM2 | 0.8276 | 0.8343 | 0.8301 |
| LGBM_pso1 | **0.8301** | **0.8504** | 0.8350 |
| LGBM3 | 0.8183 | 0.8233 | 0.8233 |
| LGBM_pso2 | 0.8225 | 0.8301 | 0.8352 |
| LGBM_pso3 | 0.8267 | 0.8343 | **0.8394** |

Among the set of logistic regression methods (LR1, LR2, LR3, LR_pso), one-vs-one strategy proves to be more effective than one-vs-rest strategy in dealing with multi-class classification problem. Feature selection through PSO leads to an increase in the prediction accuracy of 1.52% to 2.61% based on the results of LR1, LR2, and LR3. The helpfulness of PSO based feature selection is also proved by the results of support vector machine methods (SVM1, SVM2, SVM3, SVM 4, SVM_pso). SVM_pso achieves an increase in prediction accuracy from 1.43% to 43.75%. It worths mentioning that SVMs that employ a nonlinear kernel perform better than SVMs using linear kernel in solving this multi-class classification problem. LightGBM methods achieve the best average results among the three group of methods. This confirms the effectiveness of LightGBM in solving multi-class classification problems. The PSO enhanced LightGBM by feature selection and parameter tuning helps to increase the prediction results to 85.04% with the second group of data features.

Horizontal comparison between the prediction results of three groups with different set of features provides two major findings: 1) on the basis of project information features, incorporating linguistic features of project description effectively enhances the project outcome prediction accuracy, 2) the inclusion of description sentiment score does not always improve the prediction performance. Overall, employing project information and linguistic features of project description provides the best sets of results among multiple algorithms.

## 5.2  Discussion

**Effective Features for Project Outcome Prediction.** This research employs two feature selection strategies for crowdfunding project outcome prediction. The results in 3 suggest that PSO based feature selection strategy outperforms the other two strategies. The feature importance ranking strategy obtains the worst results as it mainly relies on the manual deletion of least important features which may ignore the correlations among multiple features and do not explore other combined feature deletion choices.

Three group of features are examined in this project outcome prediction task. While key project information provides satisfactory prediction results of 83.01% by using LightBGM with PSO based parameter tuning, incorporating project description linguistic features further improves the prediction performance with 2.03%. The interpretation of these results provides two key implications: 1) The investors' decision is significantly connected with the characteristics of project itself and other crowd-funders. 2) The readability, complexity, and amount of efforts required for project description digesting, to a certain extent, adjust the investors' intention to invest projects. The second implication suggests crowd-funders should carefully phrase their project description for better readability. We also explored the helpfulness of incorporating sentiment features of project descriptions, which was consistent with Wang et al. [14]'s research that positive sentiment in the campaign description would contribute to successful campaigns. However, the contribution of sentiment feature in the prediction task only received limited improvement of the results (0.42%–0.51%) based on the combination of the first two groups of features. Therefore, we reserve our agreement on the previous findings about the prediction effectiveness of description sentiment.

**Comparison between Parameter Tuning Methods.** Parameter tuning has been an important issue in machine learning. We compared two stepwise methods and PSO based method in this research. Stepwise parameter tuning has proved effective in previous research [17], while employing swarm intelligence method for parameter tuning attracts less attention from researchers. The stepwise method performs well in searching for effective combination of features as proved in our results. However, it still may result in sub-optimization due to its sequential feature selection. Swarm intelligence method in its nature is a heuristic and stochastic algorithm which can effectively search through the solution space for optimal sets of parameters. Therefore, the converged results obtained by PSO based LightGBM (LGBM_pso1) outperforms stepwise based LightGBM (LGBM2) with each group of features.

# 6  Conclusion, Limitation, and Future Research

## 6.1  Conclusion

Swarm enhanced light gradient boosting machine (S-LightGBM) is used to forecast fundraising outcomes (success, live, failure), whose experimental results are compared to those of the other two commonly used models, namely, logistic regression and support vector machine. Three sets of project features, including project key information, linguistic features of project description, sentiment score of project description are employed in the fundraising outcome classification task. PSO is implemented for feature selection and parameter tuning of focal machine learning algorithms. Results reveal that the proposed PSO based LightGBM outperforms the other algorithms in project outcomes prediction with the accuracy rate ranging from 83.01% to 85.04% with different feature sets. The findings of this research contribute to researchers' knowledge of swarm enhanced machine learning algorithms, as well as crowd-funders' deeper understanding of how project reporting impacts investors' decisions and thus determines the project outcomes. Although employing swarm intelligence algorithms can reduce the sub-optimization risk of stepwise parameter tuning methods, it requires more computation resources to obtain superior results. For crowd-funders, besides the key project information, which is largely decided by the project nature, the textual features of project description offer space for improvement thus impacts the investors' reading experience and, more importantly, investment decision.

## 6.2  Limitation and Future Research

There still exists some limitations in this study. Firstly, this research focuses on the crowdfunding project outcome prediction domain, which results in constraints of the generalization of research findings. Secondly, we selected three popular and previous applied machine learning algorithms for experiment evaluation. More diverse categories of algorithms may help to confirm our findings. Thirdly, we employed project data on Kickstarter platform which is English language based. Expanding the experiment with dataset from platforms using a different language may provide more interesting findings. We leave these to our future research agenda.

**Acknowledgement.** This study is supported by National Natural Science Foundation of China (71901150), Guangdong Province Soft Science Project (2019A101002075), Guangdong Province Educational Science Plan 2019 (2019JKCY010), Guangdong Province Bachelor and Postgraduate Education Innovation Research Project (2019SFKC46).

# References

1. Howe, J.: Crowdsourcing: How the Power of the Crowd is Driving the Future of Business. Random House, York City (2008)
2. Light, A., Briggs, J.: Crowdfunding platforms and the design of paying publics. In: Proceedings of the 2017 CHI Conference on Human Factors in Computing Systems, pp. 797–809 (2017). https://doi.org/10.1145/3025453.3025979
3. Lam, P.T.I., Law, A.O.K.: Crowdfunding for renewable and sustainable energy projects: an exploratory case study approach. Renew. Sustain. Energy Rev. **60**, 11–20 (2016). https://doi.org/10.1016/j.rser.2016.01.046
4. Miglo, A., Miglo, V.: Market imperfections and crowdfunding. Small Bus. Econ. **53**(1), 51–79 (2018). https://doi.org/10.1007/s11187-018-0037-1
5. Davis, B.C., Hmieleski, K.M., Webb, J.W., et al.: Funders' positive affective reactions to entrepreneurs' crowdfunding pitches: the influence of perceived product creativity and entrepreneurial passion. J. Bus. Ventur. **32**(1), 90–106 (2017). https://doi.org/10.1016/j.jbusvent.2016.10.006
6. Kickstarter. Creator questions.https://help.kickstarter.com/hc/en-us/categories/115000492154-Creator-questions. Accessed 25 Sept 2015
7. Greenberg, D., Pardo, B., Hariharan, K., et al.: Crowdfunding support tools: predicting success & failure. In: Proceedings of the 31st ACM SIGCHI Conference on Human Factors in Computing Systems, pp. 1815–1820 (2013). https://doi.org/10.1145/2468356.2468682
8. Etter, V., Matthias, G., Patrick, T.: Launch hard or go home: predicting the success of Kickstarter campaigns. In: Proceedings of the 1st ACM Conference on Online Social Networks, pp. 177–182. ACM, Boston (2013). https://doi.org/10.1145/2512938.2512957
9. Quitian, O.I.T., Lis-Gutiérrez, J.P., Viloria, A.: Supervised and unsupervised learning applied to crowdfunding. In: Smys, S., Tavares, J.M.R.S., Balas, V.E., Iliyasu, A.M. (eds.) ICCVBIC 2019. AISC, vol. 1108, pp. 90–97. Springer, Cham (2020). https://doi.org/10.1007/978-3-030-37218-7_11
10. Desai, N., Gupta, R., Truong, K.: Plead or Pitch? The Role of Language in Kickstarter Project Success. Department of Computer Science Stanford University Stanford (2015)
11. Sokolova, K., Perez, C.: The digital ingredients of donation-based crowdfunding. A data-driven study of Leetchi projects and social campaigns. J. Decis. Syst. **27**(3), 146–186 (2018). https://doi.org/10.1080/12460125.2019.1587133
12. Yuan, H., Raymond, Y., Lau, K., et al.: The determinants of crowdfunding success: a semantic text analytics approach. Decis. Support Syst. **91**, 67–76 (2016). https://doi.org/10.1016/j.dss.2016.08.001
13. Lee, D., Park, J.H.: The relationship between a charity crowdfunding project's contents and donors' participation: an empirical study with deep learning methodologies. Comput. Hum. Behav. **106** (2020). https://doi.org/10.1016/j.chb.2020.106261
14. Wang, W., Zhu, K., Wang, H., et al.: The impact of sentiment orientations on successful crowdfunding campaigns through text analytics. IET Softw. **11**(5), 229–238 (2017). https://doi.org/10.1049/iet-sen.2016.0295
15. Kennedy, J.F., Kennedy, J., Eberhart, R.C., et al.: Swarm Intelligence. Morgan Kaufmann, Burlington (2001)
16. Ke, G., Meng, Q., Finely, T., et al.: LightGBM: a highly efficient gradient boosting decision tree. In: Advances in Neural Information Processing Systems, pp 3146–3154. Morgan Kaufmann Publishers, San Mateo (2017)
17. Akiba, T., Sano, S., Yanase, T., et al.: Optuna: A Next-generation Hyperparameter Optimization Framework.https://github.com/optuna/optuna. Assessed 2019

18. Bergstra, J., Bardenet, R., Bengio, Y., et al.: Algorithms for hyper-parameter optimization. In: Proceedings of the 24th International Conference on Advances in Neural Information Processing Systems, pp. 2546–2554 (2011)
19. Yang, K.L.: Kickstarter crowdfunding projects dataset. Peking University Open Research Data Platform, V1 (2018). https://doi.org/10.18170/DVN/LTF18B
20. Colombo, M.G., Franzoni, C., Rossi-Lamastra, C.: Internal social capital and the attraction of early contributions in crowdfunding. Entrepreneurship Theory Pract. 39(1), 75–100 (2015). https://doi.org/10.1111/etap.12118

# Satellite Signal Blind Detection Method Based on Time-Frequency Diagrams Semantic Segmentation

Zheyu Dong, Hao Luo, Zhigang Wu, and Shuzhang Zhang$^{(\boxtimes)}$

Beijing University of Posts and Telecommunications, Beijing, China
zhangshuzhuang@bupt.edu.cn

**Abstract.** Signal blind detection is an important signal processing function of software radio, especially in satellite signal detection. However, threshold selection, high density and low energy signals are still substantially difficult issues to resolve. This paper proposes a signal blind detection method based on time-frequency diagrams semantic segmentation. First draw the time-frequency diagram of the signal. On this basis, the generalization ability of deep learning is used to transform the radio signal recognition problem into a semantic segmentation problem in the field of image recognition, which improves the intelligence and anti-interference ability of blind signal detection. Finally, combined with spectrum analysis, the result graph of semantic segmentation is post-processed to complete the blind detection of the signal. The experimental results show that the accuracy of the validation set composed of two communication satellite samples reaches 97%, the errors of the center frequency and bandwidth are 0.005 MHZ and 0.012 MHZ, which can well satisfy production requirements.

**Keywords:** Signal blind detection · Time-frequency diagram · Spectrum analysis · Semantic segmentation · Deep learning

## 1 Introduction

In modern satellite communication systems, signal blind detection is an important function at the receiving end. Signal blind detection means that without any a priori information, only the existing signals are detected through observation data and the information such as the center frequency and bandwidth of the signal is obtained. Signal blind detection was initially applied to the government's spectrum supervision and interception of signal reconnaissance in the military field. In recent years, with the development of the Internet of Things and cognitive radio technology, signal blind detection has gradually become the focus of commercial attention.

Current methods commonly used for signal blind detection include traditional communication methods, detection methods based on spectrograms, and morphological detection methods based on time-frequency diagrams.

© Springer Nature Switzerland AG 2020
X. Chen et al. (Eds.): ML4CS 2020, LNCS 12488, pp. 383–393, 2020.
https://doi.org/10.1007/978-3-030-62463-7_35

The traditional communication methods include matched filter detection [1], energy detection [2], cyclic feature detection [3], etc. Cyclic feature detection determines the existence of a signal by detecting whether the signal has a cyclic feature. This method requires the signal to have a cyclic stationary feature, exhibit spectral correlation at a non-zero cyclic frequency, and the uncertainty of the noise in the channel will directly affect the detection result, so the detection effect in the actual channel is not good.

The detection method based on the spectrogram generally draws the spectrogram of the channel first and detects the signal by setting a threshold or looking for the rising and falling edges in the spectrum [4]. Most of this method requires professionals to analyze the signal characteristics and then make manual judgments, and it is difficult to choose a reasonable threshold. When the signal is dense and the workload is large, the accuracy of manual analysis will be greatly reduced.

The morphological detection method based on time-frequency diagrams introduces the time dimension in the spectral features, draws the time-frequency diagrams of the channel, and then uses image morphology opening and closing operation processing [5], edge detection [6] and other methods to obtain the optimized image. The disadvantage is that many signals in the real environment are close to each other and have a small bandwidth, so it cannot be directly applied to the actual channel environment.

Deep learning has better generalization ability and higher recognition accuracy in image positioning tasks [7], video target tracking [8], remote sensing image classification [9], and other fields. Literature [10] is the first to compare time-frequency diagrams with the combination of deep learning and the modulation method used to identify the signal provides an idea for signal blind detection. However, there is currently no generally accepted method for signal visualization. The most suitable machine learning model for signal blind detection needs further exploration. Due to unpredictable interference, uneven background noise, and base fluctuations caused by spectrum leakage in the signal of the real production environment, a deeper network model needs to be introduced to improve the generalization ability and anti-interference ability, so the deep learning model is suitable for time-frequency blind detection of signals in the image.

Based on this assumption, this paper proposes a technical idea for signal blind detection: first, draw a time-frequency diagrams as a featured image, and convert the blind detection of satellite signals into the semantic segmentation of the image; then use the latest achievements in image recognition in the field of deep learning to detect signals in time-frequency images.

This paper will focus on the above ideas. Section 2 discusses the work related to this paper. Section 3 presents the method design of the detection model. Section 4 verifies the effectiveness of the method in this paper through experiments. Section 5 presents the conclusions.

## 2  Related Work

### 2.1  Time-Frequency Diagrams

The time-frequency diagrams reflect the law that the frequency domain characteristics of the signal change with time, and can more intuitively show the energy distribution of the signal. Literature [11] used time-frequency diagrams for sound conversion. Literature [12] proposed to use the STFT (short-time Fourier transform) transform combined with the Gaussian window function to achieve time-frequency diagrams drawing. The first sidelobe attenuation of the Gaussian window is up to 55dB. It is commonly used to truncate non-periodic signals and has a low-frequency resolution, which is not suitable for the scene of this paper. Based on the STFT conversion, replace the Hanning which is more suitable for processing complex spectrum and has less spectrum leakage and fluctuation. Then directly draws the grayscale image, and significantly increases the FFT length in the calculation process.

### 2.2  Deep Learning

Using deep learning to do image target detection is the core problem in the field of computer vision. Well-known frameworks for target detection include R-CNN [13], SSD [14], etc. Literature [10] based on the idea of target detection was the first to use the YOLO framework to identify the signal modulation method, and achieved good results on self-built data sets. Inspired by this work, we use deep learning for the first time in the blind detection of signals. Image semantic segmentation is a pixel-level classification method, which can obtain the information that needs to be expressed according to the semantic features of the image itself [15]. Famous frameworks for image semantic segmentation include FCN [16], DenseNet [17], etc. In 2018, Google 's latest DeepLab-V3+ deep learning model [18] achieved an 89% success rate on the PASCAL VOC 2012 semantic image segmentation data set without adding any post-processing. The semantics represented by pixels can only be signals or noise, and the pixel value will change significantly when the signal occurs, so the image semantic segmentation is suitable for the detection of signals in time-frequency diagrams.

## 3  Proposed Method

This chapter first designs the drawing method of the time-frequency diagrams then designs the structure of the semantic segmentation network and finally outputs the segmentation result optimized by post-processing.

### 3.1  The Drawing of Time-Frequency Diagrams

The time-frequency diagram reflects the energy intensity of the signal at any time through the "brightness and darkness" of the pixel point and can reflect the characteristics of the frequency domain and the time domain at the same time. In this paper, a time-frequency image with an accuracy that can meet the requirements of quantitative calculation is drawn as the input of the subsequent detection model.

The first step is to perform short-time Fourier transform, STFT is a mathematical transform of FFT transform, the purpose is to determine the frequency and phase of the sine wave in the local area of the signal. Taking the signal $s(\tau)$ as an input parameter, the window function $h^*(t)$ selects Hanning window, where the FFT length N and the window length L are 65536, the calculation formula of STFT and the calculation formula of Hanning window are defined as:

$$STFT(n, k) = \sum_{m=-\infty}^{\infty} s(m)h^*(n - m)e^{\frac{-j2\Pi mk}{N}} \tag{1}$$

$$h^*(t) = \frac{1}{2[1 + cos(\frac{2\Pi t}{N-1})]} \tag{2}$$

The second step is the time-domain accumulation. The result of each STFT transformation is a slice of the signal at a certain instant t within the sampling time $t0\tilde{t}1$. The slice contains L values in the frequency domain and corresponds to the ordinate of the time-frequency diagram. For pixels, each slice is accumulated in the time domain as the abscissa of the time-frequency diagram and corresponds to the pixels of the abscissa of the time-frequency diagram. The purpose is to obtain the complete time-domain features within the sampling time $t0\tilde{t}1$. The third step is taking the logarithm and pixel value mapping. The time-frequency diagram uses the lightness and darkness of pixels to describe the energy value of the current position. To assign pixel values to a reasonable interval, it takes logarithm on f(k) to obtain a new pixel value dbf(k) in decibels (dB). To distinguish the signal from the noise floor and make the time-frequency diagram drawn more readable, all pixel values need to be mapped into the pixel space of $0\tilde{2}55$, F is the final time-frequency diagram results, where max and min are the maximum and minimum values of the signal energy, respectively:

$$dbf = 20log * \sum_{k=0}^{N-1} f_k \tag{3}$$

$$F = \frac{255}{(max - min)}dbf \tag{4}$$

The number of pixels in the frequency domain direction of the time-frequency map obtained through the above steps reaches 32768. For easy observation, the image aspect ratio is set to 1:1, so a single image can contain more than 1 billion pixels. It is known that the minimum signal bandwidth of the satellite signal is about 16 KHZ. The time-frequency diagram drawn in this paper can represent the bandwidth of 2.7 KHZ in the frequency domain. Therefore, even the smallest signal can guarantee 5 pixels in the frequency domain. The bandwidth can achieve the purpose of accurate measurement.

## 3.2  Signal Blind Detection Method Based on Time-Frequency Diagrams Semantic Segmentation

Semantic segmentation can not only abstract high-level semantic features but also accurately capture the local details of the image, which can meet the accuracy requirements of signal blind detection. To establish the mapping relationship between pixels and tags, the encoder-decoder structure is used to reconstruct the input image. The input of the encoder is a time-frequency diagram, and the output of the decoder is the semantics after segmentation. This paper uses Deeplab-V3+ deep learning framework as the main framework for signal blind detection. The principle of the network is shown in Fig. 1.

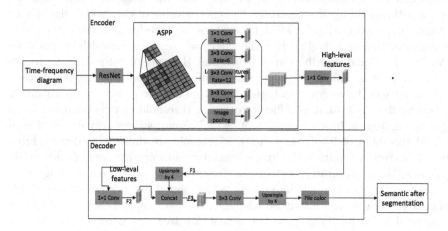

**Fig. 1.** The principle of the semantic segmentation network

Encoder structure: The original input image is abstracted into high-level semantics through deep convolution. The high-level semantics is good at capturing the most critical features. The network structure that introduces residual calculation can effectively extract feature information. The problem solved by the residual network is to allow the network layers to be deepened as much as possible while maintaining the accuracy rate.

The high-level semantics obtained by feature extraction have the problem that when the image to be predicted has high resolution, fine classification, and the receiving field is insufficient to receive all the features. The solution is to resample the convolution features of a single scale, insert dilates with different dilate rates into the original multi-filter and use dilated convolution as the filter for re-sampling. It avoids the loss of detail caused by the stretching and zooming of time-frequency images. In addition to effectively increasing the field of view of the filter, it can also capture the contextual features of the image from multiple angles. The module is mainly composed of a $1 \times 1$ convolution, three void convolutions with different dilate rates and an average pooling.

Decoder structure: The problem solved by the decoder is to gradually restore the high-level semantic information abstracted by the encoder structure to the original segmentation details. The high-level semantic information output from the encoder is unsampled by four times to obtain the eigenvalue F1, and the low-level semantic information in the residual network is passed through a $1 \times 1$ convolution down channel to obtain the eigenvalue F2. F2 and F1 are combined to obtain the eigenvalue F3. In this way, the mapping relationship between pixels and tags is established.

### 3.3    Post-processing Combined with Spectrum Analysis

The spectrum analysis searches for the start and end position of the signal by setting a threshold in the spectrum. The energy value of the "wide" signal with a bandwidth greater than 10 MHZ far exceeds the threshold, so the success rate of signal detection is high; while the "narrow" signal bandwidth is less than 1 MHZ. The signal distribution is dense and the energy value is close to the threshold, so it is difficult to detect successfully. The spectrum analysis method is relatively easy to determine whether the signal is present, but the disadvantage of this method is that it is difficult to find a threshold that is common to all signals. Figure 2(a, b) are examples of "wide" signals successfully detected with a fixed threshold of 24 dB, and "narrow" signals are difficult to detect. From the segmentation results of the image semantic segmentation network detection, it is possible to accurately detect the "narrow" signal without adding other processing, but because the individual "wide" signal exceeds the representable range of a single image, it may be misjudged as two signals. Besides, if there are large pixel value fluctuations in the non-signal area, the detection result may appear false positive. Figure 2(c, d) are examples of the "wide" signal misjudged and the "narrow" signal successfully detected.

Therefore, this paper adopts a post-processing method of fusion spectrum analysis, which integrates the spectrum analysis method into the resulting diagram after semantic segmentation, and performs a posterior filtering on the resulting diagram of the semantic segmentation network: combining the spectrum analysis to set the threshold, respectively complete the "wide" signal truncated due to exceeding the image size and filter the false positive signal due to the fluctuation of the pixel value of the non-signal part, optimize the output of the deep learning model. The experiment in the next chapter can prove that this method can achieve good results.

## 4    Experiments

### 4.1    Experimental Environment

The experiment uses Tensorflow open-source framework, CPU version Xeon E5, memory 64G, operating system RedHat version 7.4, Cuda version 10.2, Python version 3.6, Inter IPP version 2018, GPU version Geforce GTX 1080TI.

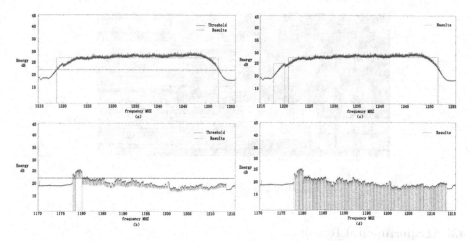

**Fig. 2.** (a) The "wide" signal successfully detected by the spectrum analysis. (b) Difficulties in detecting "narrow" signals. (c) Misjudged "wide" signal. (d) Successfully detected "narrow" signals.

## 4.2   Data and Data Set

In the experiment, we used data collected from 98.0°E satellites and 134.0°E satellites. Table 1 records the sampling parameters. The content of the data includes data networks, test transmission, broadcast programs, TV programs, etc. The signal types include continuous signals and burst signals, and the signal-to-noise ratio range covers 0–50 dB. Because the sampling data comes from the actual production environment, there is unpredictable noise interference.

**Table 1.** Sampling parameters

| Satellite name | Sampling time | Sampling bandwidth | Frequency range | Data size |
|---|---|---|---|---|
| 98.0°E | 5S | 180 MHZ | 950–1490 MHZ | 16.4G |
| 134.0°E | 5S | 200 MHZ | 950–1370 MHZ | 25.8G |

First, draw the collected data into a time-frequency diagram with a size of 32768 × 32768 as a data set using the method designed in this paper. Then, use random scaling, random interception, left-right up-down rotation, adding random noise and other methods to expand the data set and crop the image to two sizes of 1024 × 1024 and 512 × 512. Finally, use the pixel normalization method for image preprocessing. Examples of the data set are shown in Fig. 3(a–d), which uses a combination of machine labeling and manual labeling. The continuous signal is labeled green, the burst signal is labeled red, and the noise floor is labeled black. The data set contains a total of 500 images, randomly select 70% of them as training set and 30% as verification set.

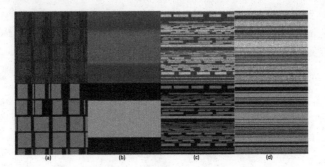

**Fig. 3.** Examples of data sets

## 4.3 Experimental Results

Use the above data set to train the semantic segmentation model of the image. The training parameters are shown in Table 2. When the final round is 30,000, the model reaches the training termination condition.

**Table 2.** Training parameters

| Parameter | Value |
|-----------|-------|
| Pre_trained_model | ResNet-101 |
| Start_learning_rate | 0.007 |
| End_learning_rate | 0.000001 |
| Learning_rate_policy | Poly |
| Weight_decay | 0.0002 |
| Output_stride | 16 |

To filter the impact of weak signals on the experiment, this paper ignores all signals with a signal-noise ratio of less than 1 dB. Common evaluation indicators for signal detection include recall, precision, and accuracy, as shown in formulas 5–7 where TP represents real cases, FP represents false-positive cases, and FN represents false-negative cases.

$$Recall = \frac{TP}{(TP + FN)} \tag{5}$$

$$Precision = \frac{TP}{(TP + FP)} \tag{6}$$

$$Accuracy = \frac{TP}{(TP + FP + FN)} \tag{7}$$

As shown in Table 3, the test results of unused post-processing methods and used post-processing methods are compared. Experiment 1 shows the detection results of 134.0°E satellites with 98.0°E satellite data as training samples; Experiment 2 shows the detection results of 134.0°E satellites with 98.0°E and 134.0°E satellite data as training samples. It can be seen from the table: without post-processing methods, the detection accuracy of the semantic segmentation network has reached more than 93%, and after using post-processing methods, the accuracy rate has reached more than 97%.

**Table 3.** Experimental results

| Methods | Experiment | TP | FP | FN | Recall | Precision | Accuracy |
|---|---|---|---|---|---|---|---|
| Unused post-processing methods | Experiment 1 | 209 | 13 | 1 | 99.52% | 94.14% | 93.72% |
| | Experiment 2 | 209 | 9 | 1 | 99.52% | 95.87% | 95.43% |
| Used post-processing methods | Experiment 1 | 210 | 5 | 0 | 100% | 97.67% | 97.67% |
| | Experiment 2 | 210 | 2 | 0 | 100% | 99.05% | 99.05% |

To further prove the effectiveness of the method, we compare literature [5] with the center frequency and bandwidth given in our work. As shown in Table 4, the literature [5] selected a TDMA stimulation signal with a sampling bandwidth 8 MHz and an FFT length of 4096, then used the average value of the 50th center frequency error and bandwidth error as the error estimate. In this paper, the average value of the center frequency and bandwidth of 50 real signals randomly measured is used as the error estimate. The estimated error of the bandwidth is 0.012 MHZ which is lower than the 0.028 MHZ in the literature [5]. It is worth noting that the frequency range occupied by a single-pixel under the sampling bandwidth of 200 MHZ in the literature [5] method reaches 48.8 KHZ, which is much larger than the smallest signal 16 KHZ that may exist in the satellite channel. Such a signal can only have a bandwidth of 1/3 pixel in the frequency domain of the image. However, successful detection of the signal requires that the signal has a bandwidth of at least one pixel in the frequency domain, so the method in literature [5] cannot be directly used in the actual production environment.

**Table 4.** Experimental comparison

| Method | Signal-noise ratio | Center frequency error | Bandwidth error |
|---|---|---|---|
| Literature [5] | 8 dB | 0.004 MHZ | 0.028 MHZ |
| This paper | 1–8 dB | 0.005 MHZ | 0.012 MHZ |

## 5    Conclusion

This paper proposes a signal blind detection method based on semantic segmentation of time-frequency diagrams, which characterizes signals as time-frequency diagrams as features and applies semantic segmentation to signal blind detection for the first time. The accuracy rate can reach more than 97%, and the errors of center frequency and bandwidth are 0.005 MHZ and 0.012 MHZ respectively, which meets the accuracy and precision requirements of blind detection of satellite signals. Therefore, it can be directly applied to the actual production environment.

## References

1. Wen-Jiang, F., Yu, G., Zhi-Yuan, H.U.: Spectrum sensing technologies in cognitive radio. J. Chongqing Univ. (Nat. Sci. Ed.) (2007)
2. Yang, L.I.U.: Signal detection and frequency estimation of high dynamic burst signals based on discrete Fourier transform. Telecommun. Eng. **56**(5), 557–561 (2016)
3. Derakhshani, M., Le-Ngoc, T., Nasiri-Kenari, M.: Efficient cooperative cyclostationary spectrum sensing in cognitive radios at low SNR regimes. IEEE Trans. Wirel. Commun. **10**(11), 3754–3764 (2011)
4. Yi, L.: A blind detection algorithm of burst signal based on cyclic spectrum. Sig. Process. (2011)
5. Wang, Q., Zhang, G., Hu, J., et al.: A signal detection algorithm based on time-frequency analysis. Radio Commun. Technol. **2016**(1), 38–42
6. Wenxiu, Z., Hao, M.: FSK signal recognition based on time-frequency map ant colony edge detection. J. Xian Univ. Posts Telecommun. **24**(01), 56–61 (2019)
7. Simonyan, K., Zisserman, A.: Very deep convolutional networks for large-scale image recognition. arXiv preprint arXiv:1409.1556 (2014)
8. Guan, H., Xue, X.-Y., An, Z.-Y.: Advances on application of deep learning for video object tracking. Acta Automatica Sinica **42**(6), 834–847 (2016)
9. Kun, T., Xue, W., Peijun, D.: Research progress of the remote sensing classification combining deep learning and semi-supervised learning. J. Image Graph. (2019)
10. Zhou, X., He, X., Zheng, C.: Radio signal recognition based on image deep learning. J. Commun. **040**(007), 114–125 (2019)
11. Zhang, J., Ling, Z., Liu, L.-J., Jiang, Y., Dai, L.-R.: Sequence-to-sequence acoustic modeling for voice conversion. IEEE/ACM Trans. Audio Speech Lang. Process. (2019)
12. Colominas, M.A., Meignen, S., Pham, D.H.: Time-frequency filtering based on model fitting in the time-frequency plane. IEEE Sig. Process. Lett. (2019)
13. Girshick, R., Donahue, J., Darrell, T., et al.: Rich feature hierarchies for accurate object detection and semantic segmentation. In: Proceedings of the IEEE Conference on Computer Vision and Pattern Recognition, pp. 580–587 (2014)
14. Liu, W., et al.: SSD: single shot MultiBox detector. In: Leibe, B., Matas, J., Sebe, N., Welling, M. (eds.) ECCV 2016. LNCS, vol. 9905, pp. 21–37. Springer, Cham (2016). https://doi.org/10.1007/978-3-319-46448-0_2
15. Xuan, T., Qi, D., Liang, W.: Review of image semantic segmentation based on deep learning. J. Softw. **30**(02), 250–278 (2019)

16. Long, J., Shelhamer, E., Darrell, T.: Fully convolutional networks for semantic segmentation. In: Proceedings of the IEEE Conference on Computer Vision and Pattern Recognition, pp. 3431–3440 (2015)
17. Huang, G., Liu, Z., Van Der Maaten, L., et al.: Densely connected convolutional networks. In: Proceedings of the IEEE Conference on Computer Vision and Pattern Recognition, pp. 4700–4708 (2017)
18. Chen, L.-C., Zhu, Y., Papandreou, G., Schroff, F., Adam, H.: Encoder-decoder with atrous separable convolution for semantic image segmentation. In: Ferrari, V., Hebert, M., Sminchisescu, C., Weiss, Y. (eds.) ECCV 2018. LNCS, vol. 11211, pp. 833–851. Springer, Cham (2018). https://doi.org/10.1007/978-3-030-01234-2_49

# A Hybrid Distributed Frequent Itemset Mining Method with Its Application in Medical Diagnosis

Mingxue Zhang and Fuyuan Xiao(✉)

School of Computer and Information Science, Southwest University, No. 2 Tiansheng Road, Beibei District, Chongqing 400715, China
doctorxiaofy@hotmail.com, xiaofuyuan@swu.edu.cn

**Abstract.** The world has entered an era of globalization, which represents the explosion of information. The application and development of big data technology make the scale of medical data geometric growth. People cannot intuitively see the correlation and the implicit relationship between complex medical data, which leads to the situation of more data and less knowledge. By using Spark and a variety of data pre-processing techniques and machine learning related algorithms, we implemented a platform which could help patients recommend more accurate treatment plans, help doctors analyze the relationship between diseases, and provide more natural results through a visual interface. Besides, we proposed a distributed frequent itemset mining algorithm (DSDFIM) based on the adjacency list and Spark. After evaluating, the proposed algorithm could reduce data transportation once between main memory and secondary storage, and improved the speed of data processing through distributed computing, compared with the classic algorithm. Meanwhile, it could solve the problem of merging frequent itemset of the same item under different independent paths.

**Keywords:** Data mining · Medical data · Frequent item mining

## 1 Introduction

With the development of big data, especially the popularity of wearable intelligent devices, making the scale of medical data increases geometrically. Due to the complexity of medical data, people cannot get the formal and professional medical services. Besides, through the search engine on the Internet, the search results of illness and symptoms often mislead patients, which would increase their worries. Applying data mining in the medical field to actual systems can serve both doctor and patients [1]. In order to find the relationship between data, many association algorithms, such as Apriori and FP-Growth [2], and analysis system have been built.

This research is supported by the National Natural Science Foundation of China (No. 62003280), Research Project of Education and Teaching Reform in Southwest University (No. 2019JY053), Fundamental Research Funds for the Central Universities (No. XDJK2019C085), and Chongqing Overseas Scholars Innovation Program (No. cx2018077).

© Springer Nature Switzerland AG 2020
X. Chen et al. (Eds.): ML4CS 2020, LNCS 12488, pp. 394–403, 2020.
https://doi.org/10.1007/978-3-030-62463-7_36

Qiu proposed the YAFIM algorithm, which accelerated the speed of mining through parallel computing [3]. Shi used Spark's engine and distributed mining further to expedite the processing speed of YAFIM [4]. Miao launched a distributed FP-Growth algorithm based on the Spark [5]. Zhang used unique weights and constrained subtrees to avoid scanning to obtain conditional pattern bases and establish FP-Trees [6]. Li calculated the weights for each item separately and constructs FP-Tree [7]. Yin used adjacency lists to reduce I/O once [8]. Li improved the shortcomings of weighted frequent itemset for mining and storage by using the weighted support of records in the two-dimensional table. This method saved the process of searching the first conditional pattern library by traversing the weighted FP tree [9]. Xiao integrated Dempster-Shafer theory with belief entropy, the new method could help to decrease the uncertainty caused by subjective human cognition to improve decision making [10]. Xiao proposed a generalized intelligent quality-based approach for fusing multi-source information, which could fuse the multi-complex-valued information while maintaining a high quality of the fused result [11]. Gao improved the accuracy of uncertainty measuring by optimizing Tsallis entropy [12]. Liu improved the result of detecting the number of unknown targets by proposing a new method based on Elbow method [13]. Li proposed a new method based on the similarity measure. The aggregation operator is presented according to Pythagorean fuzzy information, instead of being provided in advance by decision makers, which can reduce human subjectivity [14].

Our goal is to use Spark engine and machine learning algorithms to analyze medical data, which could provide accurate treatments and analysis results for patients and doctors. In this research process, the following issue will be studied and solved:

1) Medical data processing: Because medical records have characters of colloquialism and inconsistent, this study used word segmentation tools, synonyms tools, unique hot codes, and regularization methods to characterize the data set. Besides, we used Spark's productive machine learning library to complete the recommendation of the optimal treatment method based on the features provided by the user.

2) Frequent item mining: The traditional FP-Growth algorithm does not make corresponding optimization on the medical data set. And by using the adjacency list as the storage structure can reduce the I/O times between the main memory and the secondary storage. Still, it has an inaccurate problem in meagering frequent itemset with the same item stored in multiple independent paths. We synthesized several frequent itemset mining methods [4, 5, 8] and proposed a hybrid method called DSDFIM, which used adjacency list and matrix as structure and Spark for distributed computing and divided the data into disease and symptom by tags to solve the problem.

## 2  FP-Growth

### 2.1  Scan Database and Building FP-Tree

First, the algorithm scans the data set stored in the database on the hard disk (Fig. 1a) to obtain all frequent items and stores them in the header table in the descending order of minimum support. We assume the minsupport = 2.

The algorithm secondly scans each transaction in the database, deletes the items in the original item set that are less than the minimum support. The remaining items are arranged in descending order of their frequency (Fig. 1b). After obtaining the rear-ranged data set, it reads the transactions in sequence. First, T1 = {2, 1, 5} is read. The algorithm creates a path of $\emptyset \to I2 \to I1 \to I5$, and sets each degree counts as 1 because the path is independent. Secondly, the algorithm continues reading in sequence, increases item's degree by 1 when the path it had with the previous overlapping, and accessed from the empty set if there is no coincident path. The algorithm continues to read transactions according to this idea until the end, and gets a complete FP-Tree.

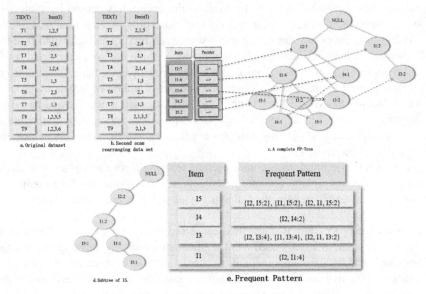

**Fig. 1.** The flow of FP-Growth

## 2.2 Generating Conditional Pattern Base and Frequent Itemset

After building, the algorithm digs up from the item with the minimum support degree in the item header table until the item with the second max support degree. The conditional pattern base is a subset consisting of the prefix path with the lowest node (suffix) in the FP tree. After obtaining the FP subtree. The algorithm assigns the value of the leaf node to each node above it, and deletes the node whose count is lower than the minimum support. From this conditional pattern base, frequent itemset can be obtained by recursive mining. Taking *I5* as an example (Fig. 1d). First, it uses *I5* as the leaf node to get the subtree. Because *I5* has two nodes, when the common ancestor nodes on the path of the two nodes are consistent, its value becomes the sum of the leaf nodes. The conditional pattern base is used to create conditional frequent item tree. The algorithm merges the same item on the same side and deletes the item that does not meet the minimum support. Finally, a frequent itemset is produced through a conditional frequent items tree. The items in the tree are merged to create the frequent items. Taking item *I3* as an example, although *I2* and *I3* both have a support level of 4, when the combination is {*I2, I1*}, the support for *I3* cannot be added together because they come from two different paths. When calculating this, we should use 2 as degree. Thus, the result was shown in Fig. 1e.

# 3   DSDFIM Algorithm

## 3.1   Principle of DSDFIM Algorithm

Due to the FP-Growth algorithm have to scan the database twice, the operating efficiency is not very ideal. We propose a Disease and Symptom Distributed Frequent Itemset Mining Algorithm Based on Adjacency List and Spark (DSDFIM). Because most medical records only recorded one disease, and the number of documented complications only accounted for 23% of the entire data set. The items in each transaction in the data set are labeled based on the medical dictionary. After that, the data is distributed and calculated to obtain the adjacency list and frequent itemset, and then combined to get the complete frequent item set. The entire process is shown in Fig. 2.

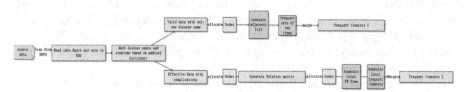

**Fig. 2.**  The flow of DSDFIM Algorithm

**For Valid Data with Only One Disease**

*Step 1.* The algorithm obtains data from HDFS, and reads it into Spark and store it in RDD. Then, it scans the data, marks the disease name, and replaces the first element in the thing and store it in *F_List1*. As shown in Fig. 3.

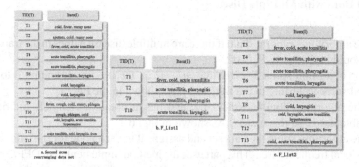

**Fig. 3.**  The result of splitting the dataset

*Step 2*. The algorithm divides *F_List1* into *G* groups, named *g1, g2 ... gk* (*k* < nodes.num ()) to the computing nodes. After the split, the Spark engine can be used to calculate each group, and the adjacency list of each group can be obtained very quickly. The items that do not meet the requirements are deleted according to the minimum support. As shown in Fig. 4.

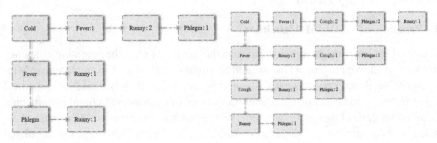

**Fig. 4.** The adjacency List of *g1* and *g2*

*Step 3*. The algorithm obtains binary frequent itemset from the two adjacency lists of *g1* and *g2*, which are {*cold, fever: 1*}, {*cold, runny: 2*}, {*cold, phlegm: 1*}, {*flow spit, fever: 1*}, {*runny, phlegm: 1*}; {*cold, fever: 1*}, {*cold, cough: 2*}, {*cold, phlegm: 2*}, {*cold, runny: 1*}, {*fever, runny: 1*}, {*fever, cough: 1*}, {*fever, phlegm: 1*}, {*cough, phlegm: 2*}, {*cough, runny: 1*}, {*runny, sputum: 1*}.

*Step 4*. The algorithm integrates binary frequent item sets in different nodes. According to the Apriori algorithm, it can be known that all its supersets are infrequent item sets when an item set is an infrequent item set. The minimum support for this combination can be combined with binary frequent item sets to obtain higher element frequent items sets as {*sputum, runny, cold: 2*}, {*cough, sputum, cold: 2*}, {*fever, runny, cold: 2*}. So the final result is: {*cold, fever: 2*}, {*cold, runny: 3*}, {*cold, sputum: 3*}, {*cold, cough: 2*}, {*fever, cough: 2*}, {*cough, Sputum: 2*}, {*sputum, runny: 2*}, {*sputum, runny, cold: 2*}, {*cough, sputum, cold: 2*}, {*fever, runny, cold: 2*}.

**For Valid Data with Multiple Diseases**

*Step 1*. The algorithm scans the data. If there are multiple disease names in the transaction, the data would be stored in *F_List2*.

*Step 2*. The algorithm splits *F_List2* into *m1, m2 ... mn* (*n* < nodes.num ()) to form a matrix, and sorts according to the frequency. As shown in the Fig. 5.

*Step 3*. The algorithm combines m1, m2, and m3 to obtain a complete matrix and a Header Table based on this matrix. As shown in Fig. 6.

*Step 4*. The algorithm splits the data set and assigns it to each node for the calculation to generate a local frequency tree. This part uses the method implemented in DFPS to ensure that each item could be computed accurately. By using the DFPS algorithm, frequent itemset can be mined in parallel without delivering messages. Finally, the algorithm generates local frequency itemset.

**Fig. 5.** The matrix of *m1*, *m2* and *m3*

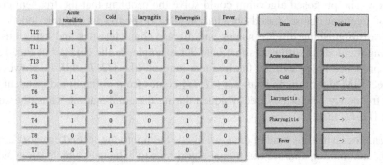

**Fig. 6.** Header table

*Algorithm1*: Generate frequent item for the data with only one disease
Input: dataset, minsupport=2
Output: Frequent itemset

1) Read file
2) Set label according to the Medical Dictionary as F_List1RDD
3) Distributed F_List1RDD into different Groups: Gk
4) Map<Node: String, Map<Node: String, edge: Int >>//Store the adjacency list
5) **foreach** transaction t in Gk
6)   **foreach** item Ii in t
7)     **if** Ii in Map
8)       **if** Ij in list where j≠i
9)         Ij.edge++
10)       **else**
11)         add(Ij, 1)
12)     **else**
13)       **add** Ii to a new list **in** Map
14)   **end**
15) **end**
16) **foreach** item Ii **in** list
17)   **foreach** Ij **in** item
18)     **if** Ii.edge <minSupport
19)       del(Ij)
20)     **else if** (Ii, Ij) ∉ frequent set
21)       **add**((Ii, Ij) **into** frequent set
22)       **if** (Ij, Ii) **in** Map
23)         del(Ij, Ii)
24)   **end**
25) **end**
26) aggregating local FP of each Node
27) **foreach** ItemSet (Ii,Ij) **in** frequent set //loop until generate all frequent set
28)   **if** (Ij,Ii) **in** frequent set
29)     (Ii,Ij).degree++
30)     del((Ij,Ii))
31)   **foreach** ItemSet(Ii, Ik) **in** frequent set
32)     **if** (Ij, Ik) **or** (Ik, Ij) ∉frequent set
33)       (Ii, Ij, Ik).weight=min((Ii,Ij).degree, (Ii,Ik).degree, (Ij, Ik).degree)
34)   **end**
35) **end**

*Algorithm2*: Generate frequent item for the data with only one disease
Input: dataset, minsupport=2
Output: Frequent itemset

1) Read file
2) Set label according to the Medical Dictionary as *F_List2RDD*
3) Distributed F_List2RDD into different Groups: Gk
4) matrix :Int // Storage matrix information
5) **foreach** transaction t in Gk
6)   **foreach** item Ii **in** t
7)     j=0;
8)     **do**{
9)       **if** map(I, j).name==li
10)         map(I, j)==1
11)       **else**
12)         j++
13)     }**while**(j==t.length)
14)   **end**
15) **end**
16) aggregating local matrix of each Node
17) Map<item: String, freq: Int> // Build FP-Tree
18) **foreach** j **in** item.value
19)   Set map< matrix (0,j).name, 0>
20)   **foreach** i **in** transaction.value
21)     **if** matrix (j,i)==0
22)       freq++
23)     **end**
24) **end**
25) Generate Header Table
26) Call for **DPFS** to generate frequent itemset

**Fig. 7.** DSDFIM algorithm

*Step 5.* The DSDFIM algorithm is shown in Fig. 7. The algorithm combines local frequent itemset to get the final result: {*acute tonsillitis, fever: 2*}, {*cold, fever: 2*}, {*acute tonsillitis, pharyngitis: 2*}, {*acute tonsillitis, Laryngitis: 4*}, {*cold, laryngitis: 4*}, {*acute tonsillitis, pharyngitis: 4*}, {*acute tonsillitis, cold, laryngitis: 2*}, {*Acute tonsillitis, Cold, fever: 2*}.

## 4   Experimental Results Analysis

### 4.1   Algorithm Performance

**Accuracy.** The proposed algorithm could solve the problem that the frequent combination of frequent itemset based on adjacency list or matrix in multiple independent paths by distinguishing disease names and disease characteristics and using different processing methods in different situations. For example, in the data set of Fig. 1e, the method proposed in [8] will get the frequent itemset, like {*I2, I1, I3: 4*}. However, in the correct FIM algorithm, the data set {*I1, I3*} comes from different independent path, and the frequency cannot be directly added during the combination. Thus, the correct result is {*I2, I1, I3: 2*}.

When processing this kind of data, we divide them into the second category according to the assumption and use traditional algorithms to calculate, which can avoid this problem. Through the DSDFIM algorithm, the results are shown in Fig. 8 and 9, which can see the relationship compared with the original dataset, it indicates that cold and acute tonsillitis have a very close relationship and often appear as complication. Thus, the DSDFIM algorithm could make the relationship between the data easier to explain, and could help doctors and patients find a better treatment, the association of complications and the invisible connection between the data.

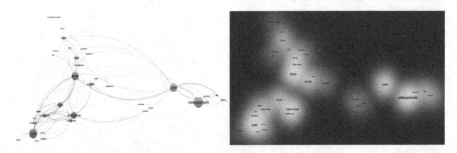

**Fig. 8.** Relationship of the data set before DSDFIM

**Running Time.** The adjacency list used in this article has a complexity of O (1) when adding nodes and edges. Although the complexity is O (N + E) at the time of construction compared to the storage structure of the tree and the time consumption is not significantly optimized, even under partial support compared with the PFP algorithm calculated in parallel, the calculation time is significantly improved compared to the traditional FP-Growth and Apriori algorithms due to the distributed computing and reducing I/O between main memory and secondary storage once.

**Fig. 9.** Relationship of the data set after DSDFIM

**Memory Usage.** The algorithm used adjacency lists and matrices as the storage structure. By using the characteristics of RDD in Spark. The algorithm scanned the data set only once. Besides, by using Spark to distributed compute on multiple nodes, it could significantly ease the pressure of main memory.

### 4.2  Medical Data Analysis

Through the DSDFIM algorithm proposed in this article, it can be intuitively found that the association of complications and the invisible connection between the data. Diseases such as osteoporosis have a very low probability of being related to the first two diseases in characteristics, and these are consistent with our cognition from life.

Besides, according to statistical analysis of the previously processed data, we used one-hot codes to encode data, and the platform could analyze and transform feature vectors from various dimensions to the same dimension. For example: {upper respiratory tract infection, low fever} The corresponding one-hot codes are:

{ *100000000000000000000000000000000000000000000000000000000000000000000000,*
*100000000000000000000000000000000000000000000000000000000* }

We converted two numbers into decimal base:

{ *295150156979166511104, 4503599627370496* }

At the same time, we complemented the features by using frequent itemset obtained by DSDFIM. It is used to shrink the feature dimensions to 4. For example, the original input feature of the user is {sneeze, runny nose, stuffy nose}. Then the available prediction from frequent itemset is {sore throat, cold}. Therefore, the first feature in the prediction is added to get {sneezing, runny nose, stuffy nose, sore throat}. Currently, the decimal value corresponding to the feature vector is listed below:

{ *6044629098073145873530888,   30223145490365729367654400,   151115727451*
*828646838272, 590295810358705651712* }

For the features in the dataset that are higher than four dimensions, we used the PCA algorithm for data dimensionality reduction. Finally, the value of the previously converted decimal feature matrix is too large, which is not conducive for data analysis. Therefore, the z-score standardized method is used to convert the data to obtain the characteristic effect. The result is shown below:

{ *1.52180576, 0.16850216, −0.50814964, −1.18215827* }

After the feature vector dimension is unified and normalized, the features are trained by K-Means algorithm. We specified the number of clusters is 3, and set 20 as the iteration round. According to the characteristics previously input by the user and the result of clustering, we could get a user group which is similar with the user. At the same time, because the user in the medical forum will get multiple answers, each data is weighted according to the order as a benchmark, and this was defined as the patient's preference for different treatments. Then we used the ALS algorithm to perform matrix decomposition and collaborative filtering and recommend their treatment plan to the user, as well as getting the user's potential for the treatment plan and score. As shown in the Fig. 10.

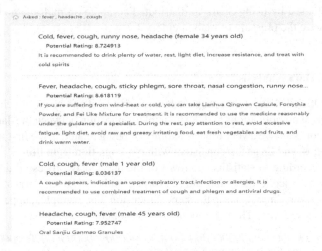

**Fig. 10.**  Recommend result

## 5   Conclusions

This research used the Spark engine to analyze medical data, and we proposed a hybrid algorithm, called DSDFIM algorithm, which split and calculated the data through a distributed engine. Meanwhile, the disease names and disease features in the data set were labeled and the data set was split according to different situations. The adjacency list and matrix were used to replace the original prefix tree storage structures. Through verification, this method optimized problem of meagering frequent item sets with the same item under multiple independent paths and reduced the I/O of disk and memory once and had a high processing efficiency compared with the traditional FP-growth comparison. Besides, the data platform integrated the methods in Spark's ML library and combined them to realize the function of data analysis. There is still room for the optimization of the proposed algorithm in this paper, such as reducing the number of iterations when mining frequent itemset, load balancing strategies in the distributed system, and scalability that means expanding the classic data set, such as mushroom and chess.

# References

1. Ding, Z., Liu, Y., Jing, S., Zhang, X.: A review of medical data mining. Smart Health **2**(10), 54–56 (2016)
2. Han, J., Pei, J., Yin, Y.: Mining frequent patterns without candidate generation. In: ACM SIGMOD International Conference on Management of Data, pp. 1–12. ACM Press, New York (2000)
3. Qiu, H., Gu, R., Yuan, C., Huang, Y.: YAFIM: a parallel frequent itemset mining algorithm with spark. In: 2014 IEEE International Parallel & Distributed Processing Symposium Workshops, Phoenix, AZ, USA, pp. 1664–1671. IEEE Computer Society (2014)
4. Shi, X., Chen, S., Yang, H.: DFPS: distributed FP-growth algorithm based on Spark. In: 2017 IEEE 2nd Advanced Information Technology, Electronic and Automation Control Conference (IAEAC), Chongqing, China, pp. 1725–1731. IEEE (2017)
5. Miao, Y., Lin, J., Xu, N.: An improved parallel FP-growth algorithm based on Spark and its application. In: 2019 Chinese Control Conference (CCC), Guangzhou, China, pp. 3793–3797. IEEE (2019)
6. Zhang, W., Luo, K.: A parallel FP-growth mining algorithm based on Spark framework. Comput. Eng. Sci. **39**(08), 1403–1409 (2017)
7. Li, Y., Yin, S.: Mining algorithm for weighted FP-growth frequent item sets based on ordered FP-tree. Int. J. Eng. Manag. Res. **9**(5), 154–158 (2019)
8. Yin, M., Wang, W., Liu, Y., Jiang, D.: An improvement of FP-Growth association rule mining algorithm based on adjacency table. MATEC Web Conf. **189**, 10012 (2018)
9. Li, Y., Yin, S.: Mining algorithm for weighted FP-tree frequent item sets based on two-dimensional table. J. Phys. Conf. Ser. **1453**, 012002 (2020)
10. Xiao, F.: EFMCDM: evidential fuzzy multicriteria decision making based on belief entropy. IEEE Trans. Fuzzy Syst. **28**, 1477–1491 (2020)
11. Xiao, F.: GIQ: a generalized intelligent quality-based approach for fusing multi-source information. IEEE Trans. Fuzzy Syst. (2020). https://doi.org/10.1109/TFUZZ.2020.2991296
12. Gao, X., Liu, F., Pan, L., Deng, Y., Tsai, S.-B.: Uncertainty measure based on Tsallis entropy in evidence theory. Int. J. Intell. Syst. **34**(11), 3105–3120 (2019)
13. Fan, L., Deng, Y.: Determine the number of unknown targets in Open World based on Elbow method. IEEE Trans. Fuzzy Syst. (2020). https://doi.org/10.1109/TFUZZ.2020.2966182
14. Fei, L., Deng, Y.: Multi-criteria decision making in Pythagorean fuzzy environment. Appl. Intell. **50**, 537–561 (2019). https://doi.org/10.1007/s10489-019-01532-2

# A DNS Server Classification Method Based on Long-Term Behavior Features

Rui Wang, Hao Luo, Zhigang Wu, and Shuzhang Zhang[✉]

Beijing University of Posts and Telecommunications, Beijing, China
zhangshuzhuang@bupt.edu.cn

**Abstract.** Obtaining the overall status of domain name system (DNS) by DNS measurement is of great significance to DNS protection and Internet traffic optimization. The accurate identification of the DNS server type is one of the challenging problems in the field of DNS measurement, given that the identification simply by protocol fields will cause inaccurate results owing to factors such as server misconfiguration and network attacks. This paper proposes a DNS server classification method based on long-term behavioral features, in which mass data of DNS records will be analyzed to extract long-term behavioral features, and machine learning (ML) will be applied to classify its types. Experimental results show that the method can effectively identify recursive and authoritative services, 25% higher in accuracy than that relying only on protocol fields.

**Keywords:** DNS measurement · Recursive DNS server · Authoritative DNS server · DNS server type identification · Machine learning

## 1 Introduction

DNS provides a service for mapping domain names to IP addresses, working as one of the basic services of the Internet [1,2]. Relevant statistics demonstrate that DDoS attacks on the DNS have mounted year by year, and the complexity of attack techniques has also increased significantly [3].

DNS servers can be roughly divided into two categories: authoritative DNS servers and recursive DNS servers. Getting type of servers rather essential in our comprehension of the overall architecture of the DNS. And when protecting DNS, it is necessary to take targeted measures according to the distinct characteristics of these two DNS servers. For authoritative servers, we need to protect those belonging to important domains of countries, operators, and some enterprises; while concerning recursive services, those with more visits are in higher priority. Therefore, we must distinguish between authoritative servers and recursive servers. Under most circumstances, DNS measurement is applied to obtain such DNS server information as type and traffic. Accurate judgment of the type of DNS server constitutes one of the major challenges in DNS measurement research. Common methods for DNS classification include active measurement

X. Chen et al. (Eds.): ML4CS 2020, LNCS 12488, pp. 404–412, 2020.
https://doi.org/10.1007/978-3-030-62463-7_37

and passive measurement [4,5]. Active measurement works to send packets to the server, and make judgments on the response to obtain server information. Passive measurement acquires information by deploying probes on the network to capture DNS traffic. Blind spots and identification deviations exist in active measurement due to DNS server access restrictions and other reasons. While passive measurement helps improve the accuracy since it is based on authentic DNS traffic, and obtain some information unavailable for active measurement by properly setting the probe. However, currently common passive classification methods determine the type of DNS server on the basis of the DNS protocol field and sometimes lead to inaccurate judgment on account of network fluctuations and the complexity of actual configuration of the DNS server.

In order to identify DNS servers more accurately, by studying the long-term behavior of DNS servers, identifying DNS servers in combination with protocol field information, and employing 4 classifiers of machine learning, a DNS server classification model based on long-term behavior features is proposed, and the effectiveness of the method is tested and evaluated. The structure of the paper is as follows: the second part reviews related work, the third part introduces the classification model, the fourth part analyzes the experimental results and evaluation, and the fifth part is the conclusion.

## 2  Related Researches

DNS server classification is a critical research direction in DNS measurement. According to the measurement method, it can be divided into passive classification and active classification.

Researches related to active classification abound. Callejo [6] proposed a DNS measurement classification method, designed a JavaScript-based DNS request script, injected the script into online advertisements to trigger DNS requests, collected and analyzed DNS responses to realize the detection and classification of DNS servers. Marc [7] studied the classification of open DNS server. He analyzed the structure of servers based on one year of data, and classified open DNS server in accordance with device types and software versions. Shi [8] selected the Chinese IP address to identify whether they are DNS servers or not, and classified the recursive DNS servers in detail. The advantages of the active method lie in the fast recognition speed and freedom from collection of DNS traffic. However, there are recognition blind spots and recognition deviations.

Researches concerning passive classification also vary. Mahjoub D [9] advanced a technique for classifying domain names based on DNS traffic, which could identify malicious domain names or domain names associated with malicious activities. Sun [10] put forward a recursive DNS server online recognition method based on connectivity estimation, in which DNS traffic in a certain period of time was firstly collected and then analyzed, and finally the source IP connectivity and domain name connectivity were calculated, identified by Compared with the default threshold. The passive methods perform better in accuracy rate, however, fail to give accurate result sometimes in view of network fluctuations and the complexity of actual configuration of the DNS server.

Machine learning methods deserve further attention. Huang [11] rised a high-performance DNS server recognition model based on passive measurement. He extracted single-packet feature and multi-packet feature from DNS traffic, used estimation algorithms to calculate to meet online processing speed, and achieved classification through integrated learning classifiers and incremental learning classifiers. This method resulted in lower accuracy in order to satisfy the needs of online processing. Li [12] brought forward a recursive server identification method based on deep learning. This method analyzed the traffic direction feature, traffic statistical feature and protocol feature in DNS records, and employed multi-layer perceptrons to realize automatic learning of multi-dimensional features. This method outstands in effective identification of the recursive server, and in higher accuracy. Nonetheless it is inferior with regard to the complexity of features of machine learning. Zou [13] designed a DNS server software fingerprint recognition system based on machine learning. This system extracted DNS server software fingerprints and software version features from intercepted DNS traffic, and trained and classified them through decision trees. Aiming at the problem of DNS server identification, this paper proposes a DNS server classification method based on long-term behavioral features. This paper presents four classification features: protocol field feature of server response, traffic feature, dispersion feature of visiting user IP, dispersion feature of domain name, covering 19 attributes and realizing automatic learning and classification of multiple features.

## 3  DNS Server Classification Method Based on Long-Term Behavior Features

### 3.1  Features

The four types of features are as follows:

**DNS Protocol Field Feature:** In the header of DNS packets, there is a recursive available field (RA tag), in which a value of 1 indicates that the server can provide recursive resolution services, while a value of 0 suggests the opposite. In addition, authoritative response field (AA tag) setting to 1 indicates that the response is an authoritative response, while a value of 0 suggests the opposite. A combination of this field with the RA field helps determine whether the responder is an authoritative DNS server or a recursive one. Some other features are also needed to make a comprehensive judgment.

**DNS Server Traffic Feature:** The domain name resolution generally includes two major steps. The first step is that the client sends a DNS request message to the recursive DNS server, carring the domain name to be queried. The second step is that the recursive DNS server responds to the client with a DNS response message. In the latter step, the recursive server normally uses the iterative query method. For this reason, the recursive server has more traffic. Apart from that, the authoritative server with a higher domain level has relatively more visits than that of lower domain level.

Table 1. Feature table.

| Feature name | Description |
|---|---|
| ans_ttl_cnt (DNS protocol field feature) | Number of TTL types with different values |
| ans_class_cnt (DNS protocol field feature) | The number of different query types |
| AA0_RA0_rate (DNS protocol field feature) | Proportion of AA field set to 0 and RA field set to 0 |
| AA0_RA1_rate (DNS protocol field feature) | Proportion of AA field set to 0 and RA field set to 1 |
| AA1_RA0_rate (DNS protocol field feature) | Proportion of AA field set to 1 and RA field set to 0 |
| AA1_RA1_rate (DNS protocol field feature) | Proportion of AA field set to 1 and RA field set to 1 |
| OPCODE0_rate (DNS protocol field feature) | Proportion of OPCODE field set to 0 in response |
| OPCODE1_rate (DNS protocol field feature) | Proportion of OPCODE field set to 1 in response |
| OPCODE2_rate (DNS protocol field feature) | Proportion of OPCODE field set to 2 in response |
| RCODE0_rate (DNS protocol field feature) | Proportion of RCODE field set to 0 in response |
| RCODE2_rate (DNS protocol field feature) | Proportion of RCODE field set to 2 in response |
| RCODE3_rate (DNS protocol field feature) | Proportion of RCODE field set to 3 in response |
| ans_ip_cnt (DNS server traffic feature) | Number of destination IPs in response |
| ans_seg_cnt (Dispersion feature of access user IP) | Number of different network segments of destination IP in response |
| ans_domian_cnt (Dispersion feature of domain names) | The number of all domains in all domain name records |
| ans_domain1_cnt (Dispersion feature of domain names) | The number of top-level domains in all domain name records |
| ans_domain2_cnt (Dispersion feature of domain names) | The number of second-level domains in all domain name records |
| ans_domain3_cnt (Dispersion feature of do-main names) | The number of third-level domains in all domain name records |
| ans_cnt (Dispersion feature of domain names) | The total number of visits to the server |

**Dispersion Feature of Access User IP:** It can be perceived from the domain name resolution process that the recursive DNS server features large number of access, multiple user IP addresses, scattered user address segments, and high user IP fragmentation compared with the authoritative DNS server. Definition triples $(s, d, n)$ represent DNS response packets, where $s$ and $d$ represent the source IP and destination IP, and $n$ represents the domain name in the packet. Definition $E$ means all DNS data packet sets, thereby the user address set of DNS server A (IP is represented by $IP_A$) is $S = \{d|s = IP_A, (s, d, n) \in E\}$, and the number of addresses of access users of the DNS server is $|S|$. In this paper, IP dispersion features are represented by $|S|$.

**Dispersion Feature of Domain Names:** Different from authoritative DNS servers, recursive DNS servers are also larger in the number of domain names resolutions, scattered in sites, and highly dispersed in domain names [14]. The set of resolved domain names of server A (IP is represented by $IP_A$) is $S_n = \{n|s = IP_A, (s, d, n) \in E\}$, and the number of domain names resolved by the server is $|S_n|$. In this paper, the domain name dispersion is represented by three attributes including the number of first-level, second-level, third-level domain names $|S_{n1}|$, $|S_{n2}|$, $|S_{n3}|$, of which:

$$S_{n1} = \{n|n_1 = level1(n), s = IP_A, (s, d, n) \in E\},$$
$$S_{n2} = \{n|n_2 = level2(n), s = IP_A, (s, d, n) \in E\},$$
$$S_{n3} = \{n|n_3 = level3(n), s = IP_A, (s, d, n) \in E\}.$$

Specific characteristics are shown in Table 1.

### 3.2   Design of DNS Server Classification Algorithm Based on Long-Term Behavior Features

Based on the analysis of the features of DNS server identification and classification in the previous section, this paper proposes a DNS server classification method based on long-term behavioral features. Four supervised learning classifiers are used as tools for algorithm implementation, including support vector machine (SVM) [15], logistic regression (LR), random forest (RF) [16], and decision tree (DT) [17]. The algorithm maps a set of input vectors to a set of output vectors, compares the classification effects of different classifiers, and selects the classifier most suitable for the current situation. The classification process unfolds in the following steps as is shown in Fig. 1: collect the original domain name response data; preprocess the original data; extract the above four types of features, with a total of 19 attributes; use a classifier to train and classify the parameters.

Input layer: The input layer contains the four types of features described in Sect. 3.1, as shown in Table 1.

1) DNS protocol field feature: contains 12 attributes, which are the number of different TTL values in the response, the number of different query record

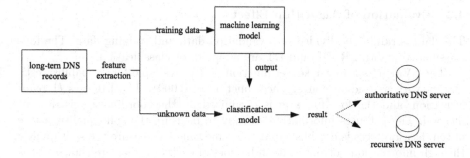

**Fig. 1.** Algorithm flow.

types in the response, proportion of the AA set to 0 and the RA set to 0 in all response, proportion of AA set to 0 and RA set to 1 in all response, proportion of AA set to 1 and RA set to 0 in all response, proportion of AA set to 1 and RA set to 1 in all response, proportion of OPCODE set to 0, 1, and 2 respectively in all response, proportion of RCODE set to 0, 2, and 3 respectively in all response.

2) Dispersion feature of access user IP: contains 2 attributes, which are the number of destination IPs of the reply message, and the number of different network segments where the user IP is located.

3) Dispersion feature of domain names: contains 4 attributes, which are the number of all domain names in the response packet, the numbers of top-level domains, second-level domain names, third-level domain names included in all domain names in the response packet.

4) DNS server traffic feature: the total access to the DNS server IP.

Output layer: The algorithm aims to divide IPs into authoritative DNS server IP and recursive DNS server IP, thereby the output layer categorized into two types.

## 4    Experimental Results and Analysis

### 4.1    Data Set

The experiment employs DNS records collected from an operator's export monitoring point, spanning 3 days with the number totaling approximately 450 million. The labeled 2506 source IPs are adopted as the data source for learning and evaluating the classifier. Among them, 1348 are marked as recursive DNS server IP, acquired from https://public-dns.info/ [18], and 1158 are labeled as authoritative DNS server IP, originating from NS records in DNS traffic.

In the training phase, 80% of the data is randomly selected as training data, and the remaining 20% is used for evaluation adopting 5-fold cross-validation method. The machine learning lib is employed and SCIKIT learning tool is used to assist processing with the code language of Python. Aiming at classification problem, this experiment selects the accuracy rate, recall rate and F1 value commonly used in classification algorithms as the standard.

## 4.2    Evaluation of Algorithm Effect

The data is randomly divided into validation data and training data. The four classifiers of SVM, LR, RF and DT are adopted for classification.

The SVM uses a linear kernel function and is available for probability estimation with a residual convergence function of 0.0001. The LR uses $l1$ regularization, and the loss function is optimized by the coordinate axis descent method. The DT uses information entropy to measure the split quality, a tree of ten decision trees is established at the same time to measure the split quality through information entropy. The indicators of each classifier are shown in the table below. In order to compare with the DNS server identification only using the protocol field information, the AA field and the RA field are used for the accuracy calculation of the marked DNS server, the results of which are also presented in the Table 2.

**Table 2.** Comparison of 4 classifier indicators.

|  | Precision | Recall | F1 | AUC | Train time |
|---|---|---|---|---|---|
| SVM | 0.9378 | 0.1737 | 0.2821 | 0.8290 | 1.3 |
| LR | 0.9325 | 0.2049 | 0.3218 | 0.8539 | 2.1 |
| DT | 0.9411 | 0.9451 | 0.9480 | 0.9341 | 1.4 |
| RF | 0.9426 | 0.9615 | 0.9570 | 0.9377 | 1.8 |
| Use only AA and RA fields | 0.7342 | | | | |

Experimental testing suggests that, the use of four features improves the accuracy rate by about 25% compared to using the protocol field. Among four classifiers, RF outstrips any other in the effect of classification making it more suitable as the classifier choice in the algorithm. After training, the accuracy rate reaches as high as 94%, and the recall rate and F1 value are above 95%.

## 4.3    Analysis of Results

After the completion of model training, SCIKIT is used to calculate the contribution of each feature vector attribute in the process, as shown in Fig. 2, where the feature attributes are numbered in the above order. It can be seen from the results that the TTL type number vector contributes the most to the classification algorithm, since the TTL values in the resolution record are generally stable concerning responses from authoritative DNS servers, while distinct in the range of variation in terms of response from the recursive DNS server. It can also be perceived from the results that the features of the AA and RA fields play a leading role, indicating that these fields can reflect the type of DNS server most of the time, but it does not mean they are reliable at all times. The role of the other fields, though smaller, still cannot be ignored. The bar graph of the contribution of Fig. 2 shows that in addition to the protocol field feature, the other three features also play a role, which verifies the experimental results of the previous section. Compared with sole selection of the field feature, choosing the four features increases the accuracy by 25%.

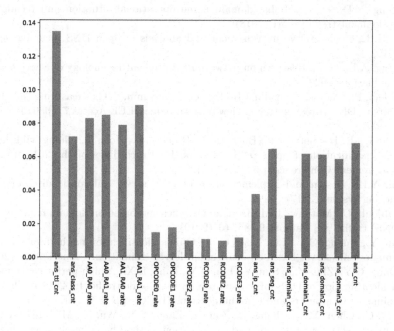

**Fig. 2.** Feature contribution.

## 5 Conclusion

DNS security constitutes an important research direction. Studying DNS server classification can help understand the information and attributes of DNS servers, furnish basic data for large network such as countries and operators, and provide support for the stable operation of the domain name system. This paper, with the help of mass data of long-term DNS records, proposes a DNS server classification method based on long-term behavior features, and reveals four features available for the identification of DNS server. Four types of classifiers including SVM, LR, RF, and DT are used to identify DNS servers, all of which can identify both authoritative and recursive DNS servers. Among them, the RF performs best with the accuracy and recall rate reaching as high as 94%.

# References

1. Mockapetris, P.V.: RFC1034: domain names-concepts and facilities (1987)
2. Mockapetris, P.V.: Rfc1035: domain names-implementation and specification (1987)
3. Zhang, Y.D., et al.: Global domain name operational situation and technology development trend report (2018)
4. Li, B.R., et al.: Active measurement and analysis of open DNS recursive server (2016)
5. Wang, A.L., et al.: Research on network measurement technology of active passive combination (2020)
6. Callejo, P., Cuevas, R., Vallina-Rodriguez, N., Rumin, A.C.: Measuring the global recursive DNS infrastructure: a view from the edge. IEEE Access **7**, 168020–168028 (2019)
7. Kührer, M., Hupperich, T., Bushart, J., Rossow, C., Holz, T.: Going wild: large-scale classification of open DNS resolvers. In: Proceedings of the 2015 Internet Measurement Conference, pp. 355–368 (2015)
8. Shi, Y.H.: Design and implementation of DNS server automatic discovery and tracking system (2017)
9. Mahjoub, D., Mathew, T.M.: Domain classification based on domain name system (DNS) traffic. US Patent 10,185,761 (2019)
10. Sun, Y., Huang, C.Y., Liu, X.M., et al.: Online identification method for recursive domain name server based on connection degree estimation. China Patent (2016)
11. Huang, C.Y., Zhu, Y.J., Sun, Y., Liu, Q.Y., Fang, B.X.: A high-performance model for identification of DNS infrastructures on large-scale traffic. In: 2019 IEEE Symposium on Computers and Communications (ISCC), pp. 1–7. IEEE (2019)
12. Li, G.C., Chen, X.X., Chang, P., Zou, X., Zang, T.N., Wu, X.: MFRdnsi: A DNS recursive server identification and classification method based on deep learning. In: Proceedings of the 2019 2nd International Conference on Information Science and Systems, pp. 27–32 (2019)
13. Zou, F.T., Zhou, J.L.P.B., et al.: A fingerprint recognition system and method of DNS server software based on machine learning (2017)
14. Gojmerac, I., Zwickl, P., Kovacs, G., Steindl, C.: Large-scale active measurements of DNS entries related to e-mail system security. In: 2015 IEEE International Conference on Communications (ICC) (2015)
15. Burges, C.J.C.: A tutorial on support vector machines for pattern recognition. Data Min. Knowl. Disc. **2**(2), 121–167 (1998). https://doi.org/10.1023/A:1009715923555
16. Liaw, A., Wiener, M., et al.: Classification and regression by randomforest. R News **2**(3), 18–22 (2002)
17. Ross Quinlan, J.: Induction of decision trees. Mach. Learn. **1**(1), 81–106 (1986)
18. Public DNS Server List. https://public-dns.info/. Accessed 14 Mar 2020

# Medical Image Segmentation Algorithm Based on Information Combined Level Set

Lin Ye, Liu Qingling[⊠], and Muhammad Ayoob Dars

Department of Information and Communication Engineering, Harbin Engineering University, Harbin 150001, China
lin_ye2019@163.com, liuqingling@hrbeu.edu.cn

**Abstract.** Medical images are difficult to observe due to blurred organ and tissue boundaries and uneven grayscale in the area, which interferes with the diagnosis of the disease. In this paper, a level set segmentation method combining information is proposed for the redundant effect caused by the segmentation of organ tissue in a complex environment. Simulation results show that, when the improved algorithm is used to segment medical images with blurred borders and uneven grayscale, although it requires a lot of calculation, it obviously sup- presses the redundancy effect and improves the accuracy and efficiency of the segmentation algorithm.

**Keywords:** Level set · Redundancy effect · Constraint relationship · Local minimum

## 1 Introduction

The level set theory [1–3] developed in recent years, using dynamic evolution curve to process its free topology to extract the target contour, can achieve a high recognition rate in the field of image segmentation. However, due to the blurring of organ tissue boundaries and uneven gray scale [4, 5], the redundancy effect is serious, resulting in that the organ tissue cannot be completely extracted from medical images, and the segmentation performance of the level set algorithm is seriously affected.

The stitching based on boundary and region information has been the research hotspot of the level set segmentation algorithm in recent years. Zhang et al. [6] proposed a selective binary Gaussian filter regularization level set method by combining the GAC model and the CV model (Selective Binary and Gaussian Filtering Regularised Level Set, SBGFRLS), to accurately and effectively segment gray-scale uniform images, but SBGFRLS model uses global statistical information, it is difficult to accurately locate the target boundary, especially for medical images with uneven gray-scale. Tian et al. [7] proposed an active contour model to segment targets in magnetic resonance images by integrating GAC model, LBF model and SBGFRLS model. However, when segmenting medical images with blurred borders and uneven grayscales, erroneous segmentation will occur. Wu [8] proposed a fast external force gradient vector convolution method for parametric active contour models and merged curve-based energy into it to segment the left ventricular myocardial boundary. Article [9] directly stitches together the energy

© Springer Nature Switzerland AG 2020
X. Chen et al. (Eds.): ML4CS 2020, LNCS 12488, pp. 413–418, 2020.
https://doi.org/10.1007/978-3-030-62463-7_38

terms of the two kinds of information without considering the constraint relationship between the two kinds of information. In [10], the stitched algorithm is applied to the MR image of the brain to deal with the problem of uneven grayscale, but the stitching algorithm is limited to the brain image, and the amount of parameters to be set when applying to other types of images is large.

In summary, this paper proposes a medical image segmentation method based on information combined level set, and adds an evolution method of momentum to the evolution process to achieve redundancy when segmenting medical images with uneven grayscale and blurred boundaries Effect, and the superiority of the algorithm is verified through simulation.

## 2 Information Combined Level Set Segmentation Algorithm

The edge information of the image refers to the gradient information of the current range of the reference image, which is helpful for the segmentation of the local position of the image, but the global segmentation can not be performed. Image region information refers to the statistical characteristics of the reference image, which is helpful for segmenting images with uniform gray levels, but segmenting the image boundary information or regions with uneven gray levels in the image is prone to errors. Therefore, many scholars try to merge the regional information and boundary information of the level set together to construct a mixed level set model algorithm.

Taking the CV model [11, 12] based on boundary information and the DRLSE model [13, 14] based on regional information as examples, the merged information directly constitutes a level set segmentation algorithm. The energy function of the algorithm is:

$$E_H(\phi) = \lambda_1 \int_\Omega g\delta(\phi)|\nabla\phi|dx + \lambda_2 \int_\Omega gH(-\phi)dx + \sum_{i=1}^{2} v_i \int_\Omega |I(x) - c_i|^2 M_i^x dx \quad (1)$$

The energy coefficient of the combined equation can only depend on the experiment or equation, and it is difficult to estimate the contribution of each energy term to the overall energy equation. When the four coefficients are set improperly, some energy terms will not work effectively. When applied to different kinds of pictures, it is difficult to apply different objects and methods.

To solve the above problems, the combined model is reduced to a simple equation containing fewer energy coefficients using Euler-Lagrange mathematical theory.

First, simplify the CV model

$$\frac{\partial\phi_{CV}}{\partial t} = \delta(\phi)[v_1(I - c_1)^2 - v_2(I - c_2)^2]$$

$$= \delta(\phi)(v_1 - v_2)[\frac{\sqrt{v_1}c_1 + \sqrt{v_2}c_2}{\sqrt{v_1} + \sqrt{v_2}} - I] \times [\frac{\sqrt{v_1}c_1 - \sqrt{v_2}c_2}{\sqrt{v_1} - \sqrt{v_2}} - I] \quad (2)$$

Part of the formula is a constant, and the meaningful part is reduced to

$$\frac{\partial\phi_{CV}}{\partial t} = v\delta(\phi)(\alpha c_1 + \beta c_2 - I) \quad (3)$$

Using Eq. (3) to replace the CV energy term in Eq. (1), the energy functional of the information-integrated level set algorithm is rewritten as:

$$E_H(\phi) = \lambda_1 \int_{\Omega} g\delta(\phi)|\nabla\phi|dx + \lambda_2 \int_{\Omega} gH(-\phi)dx$$

$$+ v \int_{\Omega} sgn(P_{CV})(ac_1 + \beta c_2 - I)H(-\phi)dx \qquad (4)$$

In the formula, $sgn(P_{CV})$ is defined as the direction controller to ensure that the evolution curve always moves to the target contour, and the energy decay. $P_{CV} = c_1 - c_2$

After the simplification of the CV model based on regional information, it is necessary to continue to simplify the DRLSE model based on boundary information. The DRLSE model contains the regional information energy term and the boundary information energy term that drive the evolution curve to the target boundary. The CV model directly covers the regional information energy term with the same effect in the level set model of the combined information, and finally the DRLSE model For the information energy level of the boundary information energy term and the CV model as the regional information energy term, the external energy functional simplification is:

$$E_H(\phi) = \lambda_1 \int_{\Omega} g\delta(\phi)|\nabla\phi|dx + v \int_{\Omega} f_{CV}H(-\phi)dx \qquad (5)$$

Although only the boundary indication function in the DRLSE model is replaced with the area indication function. This seems like a small improvement, but it is indeed an important replacement. The reason is that the area indicator function $f_{CV}$ not only uses one item to embed the area information into the model based on the boundary information, but also acts as a driving force for the level set evolution curve, thereby avoiding adding additional driving force and effectively reducing the energy coefficient in the equation The number. In addition, the area indicator function helps to quantitatively describe the relationship between the area information and the boundary information, and finally select the optimal coefficient suitable for the energy functional.

The above is just the external energy term of the mixed level set model. In order to prevent the initialization sensitivity of the level set's own evolution curve and the accumulation of errors after a period of time, resulting in rough or undesirable evolution curves, it needs to be similar to the traditional level set, adding length regular Term and distance regular term, the formula for the overall regular term is

$$E(\phi) = \lambda \int_{\Omega} g\delta_\varepsilon(\phi)|\nabla\phi|dx + v \int_{\Omega} fH_\varepsilon(-\phi)dx$$

$$+ \omega \int_{\Omega} \delta_\varepsilon(\phi)|\nabla\phi|dx + \mu \int_{\Omega} p(|\nabla\phi|)dx \qquad (6)$$

## 3   Simulation Experiment and Analysis

The resolution of the image is between 0.6 mm and 0.8 mm, the size is $512 \times 512$ pixels, and there are about 30 scan layers. The simulation experiment was performed under Windows 10 environment. The CPU uses Inter Core i5 processor, the memory is 16 GB, the graphics card uses Geforce GTX 1060, and is implemented using MATLAB R2016.

The experimental performance indicators in this paper are based on the level set simulation under different iterations, mainly based on the Jaccard similarity coefficient [15]—IOU

$$IOU\left(V_{gt}, V_{test}\right) = \frac{\left|V_{gt} \cap V_{test}\right|}{\left|V_{gt} \cup V_{test}\right|} = \frac{\left|V_{gt} \cap V_{test}\right|}{\left|V_{gt}\right| + \left|V_{test}\right| - \left|V_{gt} \cap V_{test}\right|} \tag{15}$$

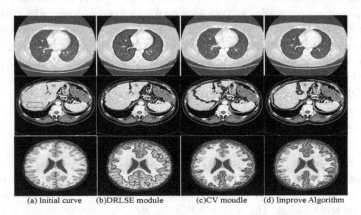

(a) Initial curve    (b)DRLSE module    (c)CV moudle    (d) Improve Algorithm

**Fig. 1.** Comparison and segmentation results of different algorithms

Figure 1 shows the results of segmentation of a given image using DRLSE model, CV model and multi-information combined level set algorithm model. Figure 1(a) is the initial position of the evolution curve, and Fig. 1(b) is the segmentation result of the DRLSE model. When the target boundary is a fuzzy weak boundary, the image gradient is small, resulting in the boundary indicator function unable to approach 0 When the curve moves, it will pass the target contour, continue to iterate and gradually away from the target contour, resulting in over-segmentation of the segmentation results. Figure 1(c) is the segmentation result of the CV model. Since the regional energy term drives the evolution curve to use the global statistical information of the image, the non-homogeneous image with uneven local grayscale is not considered, resulting in several additional small Area, under segmentation appears in the image segmentation result. Figure 1(d) is the segmentation result of the multi-information combined level set model. The algorithm combines the image boundary information, region information and feature point information, which plays a positive role in medical image segmentation and makes the evolution curve coincide with the target contour. Complete the segmentation task accurately.

Table 1 specifically shows the IOU and running time of four images under three models. By comparing the IOUs of several models, it can be found that the IOU of the multi-information combined level set algorithm on the four images is generally higher than that of the other two models. And its IOU value is relatively more stable, indicating that the segmentation result of the algorithm has the most overlapping area with the "gold standard", and the accuracy of segmentation is the highest in several models. Complex background images with weaker borders are more robust. In addition, it can be seen that the running time of the multi-information combined level set algorithm on each image is better than the DRLSE model. Although it is slightly inferior to the CV model, it can basically meet the actual needs. In summary, the multi-information combined level set algorithm model comprehensively utilizes the advantages of the composition model and makes up for the disadvantages of the composition model.

**Table 1.** Comparison of running time between improved model and traditional model

|  | DRLSE module | | CV module | | Improved method | |
|---|---|---|---|---|---|---|
|  | IOU | Time/s | IOU | Time/s | IOU | Time/s |
| I | 0.7922 | 3.012 | 0.9333 | 1.856 | 0.9862 | 2.754 |
| II | 0.9326 | 3.307 | 0.4859 | 1.698 | 0.9903 | 2.096 |
| III | 0.8593 | 3.055 | 0.9449 | 0.966 | 0.9869 | 1.319 |

## 4 Conclusion

This paper first introduces the shortcomings of the single information-driven evolution curve movement, and proposes a level set segmentation method that combines bounded boundary information and regional information. By segmenting a given image in different models, the evolution of the fuzzy interference of the boundary is realized. The correction of the curve improves the performance of image segmentation in complex environments. Then the KL divergence method based on the distribution information of feature points is combined to maximize the gap between regions. Finally, the idea of standardized stochastic optimization and momentum parameters is introduced to solve the problem of local minima that is prone to appear when the level set function is minimized. Experimental results show that the algorithm applied in a complex background not only improves the accuracy of image segmentation, but also improves the efficiency of image segmentation.

**Acknowledgements.** This paper is funded by the International Exchange Program of Harbin Engineering University for Innovation-oriented Talents Cultivation.

## References

1. Essadike, A., Ouabida, E., Bouzid, A.: Brain tumor segmentation with Vander Lugt correlator based active contour. Comput. Methods Programs Biomed. **160**, 103–117 (2018)

2. Soomro, S., Soomro, T.A., Choi, K.N.: An active contour model based on region based fitting terms driven by p-laplace length regularization. IEEE Access **6**, 58272–58283 (2018)
3. Khamechian, M.B., Saadatmand, T.M.: FoCA: a new framework of coupled geometric active contours for segmentation of 3D cardiac magnetic resonance images. Magn. Reson. Imaging **51**, 51–60 (2018)
4. Wu, J., Zhu, J.M., Zhang, H.: Segmentation method for gray uneven weak boundary vascular images. J. Comput. Appl. **36**(S1), 154–156 (2016)
5. Han, H.W., Feng, X.D.: Application of variational level set method to image segmentation with intensity inhomogenity. Comput. Eng. Appl. **53**(14), 203–208 (2017)
6. Zhang, K.H., Zhang, L., Song, H.H.: Active contours with selective local or global segmentation: a new formulation and level set method. Image Vis. Comput. **28**(4), 668–676 (2010)
7. Tian, Y., Duan, F.Q., Zhou, M.Q., Wu, Z.: Active contour model combining region and edge information. Mach. Vis. Appl. **24**(1), 47–61 (2013). https://doi.org/10.1007/s00138-011-0363-7
8. Wu, Y.W., Wang, Y.Q., Jia, Y.D.: Segmentation of the left ventricle in cardiac cine MRI using a shape-constrained snake model. Comput. Vis. Image Underst. **117**(9), 990–1003 (2013)
9. Wang, Z.F., Liu, Y.J.: Active contour model by combining edge and region information discrete dynamic systems. Adv. Mech. Eng. **9**(3) (2017). https://doi.org/10.1177/168781401 7692947
10. Yang, Y.Y., Jia, W.J., Shu, X.: Level set formulation on edge and region information with application to accurate lesion segmentation of brain magnetic resonance images. J. Optim. Theory Appl. **182**(2), 797–815 (2018). https://doi.org/10.1007/s10957-018-01451-1
11. Wang, B.T., Tian, X.D., Wang, Q.: Pulmonary nodule detection in CT images based on shape constraint CV model. Med. Phys. **42**(3), 1241–1254 (2015)
12. Song, Y., Wu, Y.Q., Dai, Y.M.: Automatic river target detection from remote sensing images based on image decomposition and distance regularized CV model. Comput. Electr. Eng. **54**, 285–295 (2016)
13. Wang, D.W.: Extremely optimized DRLSE method and its application to image segmentation. IEEE Access **7**, 119603–119619 (2019)
14. Ye, F., Li, W.R., Chen, J.Z.: Image fast segmentation algorithm based on saliency region detection and level set. J. Electron. Inf. Technol. **39**(11), 2661–2668 (2017)
15. Hwang, C.M., Yang, M.S., Hung, W.L.: New similarity measures of intuitionistic fuzzy sets based on the Jaccard index with its application to clustering. Int. J. Intell. Syst. **33**(8), 1672–1688 (2018)

# A Novel Two Dimensional Pythagorean Fuzzy Sets Model with Its Application in Multi-attribute Decision Making

Yi Fan and Fuyuan Xiao[✉]

School of Computer and Information Science, Southwest University, No. 2 Road,
Beibei District, Chongqing 400715, China
doctorxiaofy@hotmail.com, xiaofuyuan@swu.edu.cn

**Abstract.** Pythagorean fuzzy set (PFS) is a generalization of intuitionistic fuzzy set (IFS), which can express and handle uncertainty in uncertain environment more capable and is widely used in various fields. In the past, several methods of uncertainty measurement of PFSs have been proposed, however, some methods can not provide counterintuitive examples. Therefore, how to measure the uncertainty of PFSs is still an open question. Z-number is a new way to treat uncertainty and reliability of information, Z-number can greatly improve the accuracy and effectiveness of the information fusion process. However, one-dimensional index is not sufficient to express the expert's assessment. On account of this point, this paper proposes a new uncertainty measurement method between two-dimensional Pythagorean fuzzy sets by using Deng entropy. This method is based on the operation distribution of membership function, non-membership function and hesitation function of two PFSs. Numerical examples show that the proposed method can produce higher identification with more feasible, reasonable and superior result. In addition, by comparing the application of different methods in medical diagnosis, we find that the new algorithm is as effective as other methods. These results prove the practicability of this method in dealing with medical diagnosis problems.

**Keywords:** Uncertainty · Two dimensional Pythagorean fuzzy set · Deng entropy · Z-number · Classical discount method · Medical diagnosis · Multiple attribute decision making

## 1 Introduction

It is well known that the problems of dealing with uncertainty have attracted great attentions from researchers in decision theory [1–4] in recent years. Many

This research is supported by the National Natural Science Foundation of China (No. 62003280), Research Project of Education and Teaching Reform in Southwest University (No. 2019JY053), Fundamental Research Funds for the Central Universities (No. XDJK2019C085) and Chongqing Overseas Scholars Innovation Program (No. cx2018077).

© Springer Nature Switzerland AG 2020
X. Chen et al. (Eds.): ML4CS 2020, LNCS 12488, pp. 419–428, 2020.
https://doi.org/10.1007/978-3-030-62463-7_39

novel fuzzy techniques and systems have been presented for reasoning and managing with uncertainty [5–10]. As a generation Zadeh's fuzzy set (FS) theory, Atanassov initiated the concept of intuitionistic fuzzy sets (IFSs) [11]. Later, Yager extended the intuitionistic fuzzy sets into the Pythagorean fuzzy sets (PFSs) [12,13] in 2013. Since the membership space of PFS is larger than that of IFS, PFS becomes an effective mathematical tool to express uncertain information in decision-making. Z-number [14] is a new way to treat uncertainty and reliability of information, which is composed of two parts: one part is a restriction on values that can be assumed, and the other part is the reliability of the information. However, a simple reliability index is not enough to reasonably express the evaluation of experts. In information theory, Deng entropy [15] solves the problem of Shannon entropy's multi-subset limitation and always be used to measure the uncertainty between BPAs. It can measure the quality of multi-subset well. However, how to link the uncertainty between Deng entropy and PFSs is still an open question. With the development of fuzzy sets, medical diagnosis [16] is widely used to deal with uncertain information and more and more attention has been paid to it. However, how to improve the ability to deal with all kinds of uncertainty and inconsistency accurate decision-making [17,18] is still pending.

In order to solve this problem, this paper proposes a relatively simple uncertainty measurement model, which is called two-dimensional Pythagorean fuzzy set (TDPFS). Based on TDPFS, this paper proposed a new combination rule, which is defined by the operation of membership function, non-membership function and hesitation function of PFSs, rather than based on the distance measure or the relationship between membership function and non-membership function. Moreover, this paper designs a new decision making algorithm based on Deng entropy measure. The proposed algorithm mainly involves multiplication assignment between PFSs, and calculates the uncertainty between PFSs based on Deng entropy. Numerical examples show that the proposed algorithm has higher identification degree and more superiority than other representative methods. By comparing the application of different methods in medical diagnosis, we find that the proposed algorithm can solve the medical diagnosis problem as effectively as other methods. These results prove the practicability of this method in dealing with medical diagnosis.

## 2    Preliminaries

In this part, we will briefly introduce some existing methods to be used in this paper.

### 2.1    Pythagorean Fuzzy Set

**Definition 1.** *Let $X$ be a finite universe of discourse. A Pythagorean fuzzy set (PFS) [12,13] $A$ in $X$ is defined by the mathematical form:*

$$A = \{< x, A_Y(x), A_N(x) > | x \in X\}, \tag{1}$$

where
$A_Y(x) : X \to [0,1]$ and $A_N(x) : X \to [0,1]$
with the condition:

$$0 \leqslant A_Y^2(x) + A_N^2(x) \leqslant 1. \tag{2}$$

The $A_Y(x)$ is the degree of membership of the element $x \in X$; $A_N(x)$ is the degree of non-membership of the element $x \in X$.

Here, $C(X)$ is denoted as the commitment of the membership function, in which $C^2(x) = A_Y^2(x) + A_N^2(x)$.

For the PFS $A$ in $X$, a hesitancy function of $x \in X$ is defined by:

$$A_H(x) = \sqrt{1 - C^2(x)}, \tag{3}$$

which reflects the uncertainty in terms of the membership and non-membership grades of $x \in X$.

## 2.2 Z-Number

Z-number is a new tool to model uncertain information.

**Definition 2.** *A Z-number is an ordered pair of fuzzy numbers:*

$$Z = (A, B). \tag{4}$$

Z-number [14] was proposed by Zadeh (2011) as a new way to treat uncertainty and reliability of information. Z-numbers are composed of two parts: one part is a restriction on values that can be assumed, and the other part is the reliability of the information. The concept of Z-number as an ordered pair $Z = (A, B)$ of fuzzy numbers $A$ and $B$ is used, where $A$ is a fuzzy set used to describe the constraint, and $B$ is a fuzzy number to describe a constraint on a partial reliability of $A$. Z-number measures the reliability of sensor reports [19] based on different discount factors.

## 2.3 Dempster-Shafer Evidence Theory

Dempster-Shafer (D-S) evidence theory has many advantages to handle uncertain information, which has been widely used in various fields [20–22].

**Definition 3.** *A mass function is also called a basic probability assignment (BPA). Assume there are two BPAs indicated by $m_1$ and $m_2$. The propositions of $m_1$ and $m_2$ are denoted as $B_i$ and $C_j$, the Dempster's rule of combination is used to combine them as follows:*

$$m(A) = \begin{cases} \frac{\sum_{B_i \cap C_j = A} m_1(B_i) m_2(C_j)}{1-K}, & A \neq \emptyset; \\ 0, & A = \emptyset; \end{cases} \tag{5}$$

with

$$K = \sum_{B_i \cap C_j = \emptyset} m_1(B_i) m_2(C_j). \tag{6}$$

Dempster's combination rule [23] is used to obtain evidence of combination. The orthogonal sum of Eq. (9) can be extended to $n$ pieces of BPAs fusing $n - 1$ times at the same time. $K$ reflects the existence of conflict between evidences. Coefficient $1/1 - K$ is called normalization factor, its purpose is to avoid giving non-zero generality to empty set in synthesis. The larger the $K$ value is, the greater the conflict between evidences is.

### 2.4 Deng Entropy

**Definition 4.** *As a widely accepted measure of uncertainty measurement, Deng entropy is defined as follows:*

$$E_d(m) = - \sum_{A \subseteq X} m(A) \log_2 \frac{m(A)}{2^{|A|} - 1} \tag{7}$$

where $m$ is a mass function defined on the frame of discernment $X$, and $A$ is the focal element of $m$, $|A|$ is the cardinality of $A$. As shown in the above definition, Deng entropy, formally, is similar with the classical Shannon entropy, but the belief for each focal element $A$ is divided by a term $(2^{|A|} - 1)$ which represents the potential number of states in $A$ (of course, the empty set is not included). Through a simple transformation, it is found that Deng entropy is actually a type of composite measures, as follows:

$$E_d(m) = \sum_{A \subseteq X} m(A) \log_2(2^{|A|} - 1) - \sum_{A \subseteq X} m(A) \log_2 m(A) \tag{8}$$

where the term $\sum_{A \subseteq X} m(A) \log_2(2^{|A|} - 1)$ could be interpreted as a measure of total nonspecificity in the mass function $m$, and the term $-\sum_{A \subseteq X} m(A) \log_2 m(A)$ is the measure of discord of the mass function among various focal elements.

## 3    TDPFS: Two Dimensional Pythagorean Fuzzy Set

### 3.1    The Model of TDPFS

**Definition 5.** *A TDPFS, $R = (A, B)$, consists of two basic Pythagorean fuzzy sets. $A = \{< x_i, A_Y(x_i), A_N(x_i) >| x \in \chi\}$ is a classical Pythagorean fuzzy set, and $B = \{< x_i, B_Y(x_i), B_N(x_i) >| x \in \chi\}$ also is a classical Pythagorean fuzzy set, which is a measure of reliability of A.*

In Definition 5, $A$ is usually given by expert data or monitored by sensors. Similar to a single PFS [12,13], we can regard the quality of $A_Y(x_i)$ as membership of $x_i$, $A_N(x_i)$ as non-membership of $x_i$, and the rest part stand for the uncertainty. Part B, as a reliability measure of expert data. $B_Y(x_i)$ denotes support and express how reliable is; $B_N(x_i)$ denotes not-support express how unreliable is; $B_H(x_i) = \sqrt{1 - B_Y^2(x_i) - B_N^2(x_i)}$ express that there is no idea to measure the reliability. According to the definition of TDPFS, if a body of evidence is close to real value, its $B_Y(x_i)$ will be high and $B_N(x_i)$ will be low.

## 3.2   The Combination Rule of TDPFS

**Definition 6.** *The combination rule of TDPFS is defined as follows:*

$$A_Y(x_i) = A_Y(x_i) \times B_Y(x_i) + B_N(x_i) \times (1 - A_Y(x_i)) \tag{9}$$

$$A_N(x_i) = A_N(x_i) \times B_Y(x_i) + B_N(x_i) \times (1 - A_N(x_i)) \tag{10}$$

$$A_H(x_i) = A_H(x_i) \times B_Y(x_i) \tag{11}$$

The quality of $B_Y(x_i)$ is proportionally distributed to $A_Y(x_i)$ and $A_N(x_i)$, This value is the previous support of the subset. The mass of $B_N(x_i)$ is distributed in a single proportional $A_Y(x_i)$ and $A_N(x_i)$. This value is the edge support of this subset. In addition, the mass percentage of $B_Y(x_i)$ is redistributed to $A_H(x_i)$. The algorithm makes full use of expert data and the reliability of expert data, and assign membership and non-membership to different subsets. After normalization, more reasonable and higher-resolution expert data can be obtained. When all $B_N(x_i) = 0$, the combination of TDPFS degenerates the classical discount method. When all $B_Y(x_i) = 1$, the combination of TDPFS degenerates to Dempster combination rule.

# 4   Multi-attribute Decision Algorithm Based on TDPFS

The multi-attribute refers to the decision-making [24,25] of making a choice among the limited schemes which are made first, which has been successfully applied to many fields. In this paper, a new method of multi-attribute decision-making based on TDPFS model is designed. The new method can be effectively applied to complex multi-attribute decision-making and solve the problem of medical diagnosis reasonably. The detailed steps of this method are as follows:

Step1: Collect experts' data and reliability assessment of these data, use credibility $B_Y(x_i)$ and $B_N(x_i)$ to modify the initial evaluation results of expert by the proposed combination rule of TDPFS.

Step2: Assign the mass of $B_Y(x_i)$ to $A_Y(x_i)$ and $A_N(x_i)$ proportionally; assign the mass of $B_N(x_i)$ to the reverses of single subsets $A_Y(x_i)$ and $A_N(x_i)$ proportionally; reassign the percentage mass of $B_Y(x_i)$ to $A_H(x_i)$.

Step3: Through the combination of TDPFS, a set of standardized PFS is obtained, which contains experts' comprehensive evaluation of different goals and attributes.

Step4: Calculate the Deng entropy between the PFSs of each group of expert attributes and the decision results of each group of attributes one by one.

Step5: Averaged the Deng entropy under each target decision, and the smallest Deng entropy value is the result of multi-attribute decision-making.

Step6: The decision result of the multi-attribute problem is the attribute recognition corresponding to the minimum Deng entropy.

## 5   Numerical Examples

**Example Presentation:** $A$ and $B$ are two-dimensional Pythagorean fuzzy sets under this framework and $x_i$ (i = 1, 2, 3, 4, 5) represent five materials of building respectively, and $y$ is an unknown material. Their PFSs are given in Table 1, and we need to recognize the type of $y$ in A by finding the smallest Deng entropy $E_d(m)$ from them.

**Table 1.** The Pythagorean fuzzy sets of building materials.

| PFSs | $A$ | $B$ |
|------|-----|-----|
| $x_1$ | $< x_{1,A}, 0.173, 0.524 >$ | $< x_{1,B}, 0.818, 0.102 >$ |
| $x_2$ | $< x_{2,A}, 0.510, 0.365 >$ | $< x_{2,B}, 0.627, 0.125 >$ |
| $x_3$ | $< x_{3,A}, 0.495, 0.387 >$ | $< x_{3,B}, 0.603, 0.298 >$ |
| $x_4$ | $< x_{4,A}, 1.000, 0.000 >$ | $< x_{4,B}, 1.000, 0.000 >$ |
| $x_5$ | $< x_{5,A}, 0.732, 0.153 >$ | $< x_{5,B}, 0.556, 0.303 >$ |
| $y$ | $< y_A, 0.978, 0.003 >$ | $< y_B, 0.980, 0.012 >$ |

Modify the initial evaluation results of expert by the proposed combination rule of TDPFS. and calculate the Deng entropy of each group of materials, the results are in Table 2:

According to Table 2, we can see clearly that both the proposed combination rule of TDPFS and the classic discount method can effectively combine the opinions of experts with the reliability of expert data and can identify $y$

**Table 2.** Deng entropy by proposed and classical method.

| Material | Proposed method | Classical method |
|----------|-----------------|------------------|
| $x_1$ | 2.2520 | 2.6536 |
| $x_2$ | 1.8816 | 2.7981 |
| $x_3$ | 1.5977 | 2.8373 |
| $x_4$ | **0.0000** | **0.0000** |
| $x_5$ | 1.5955 | 2.7529 |
| $y$ | **0.2996** | **0.3734** |

type material as the fourth type of $x$ type material. The proposed method will identify the required decision more quickly and has a higher resolution in the multi-attribute decision-making problem [24, 25]. Therefore, the rationality of the proposed method is proved (Fig. 1).

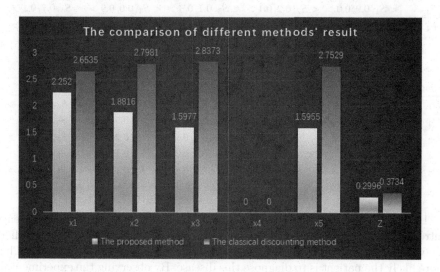

**Fig. 1.** The comparison of different method's result

# 6    Applications in Pattern Recognition

**Problem Statement:** Assume there are four patients: John, Jack, Cole, and Rain, denoted as P = $\{P_1, P_2, P_3, P_4\}$. Five symptoms: Temperature, Headache, Stomach pain, Cough, and Chest pain, are observed, denoted as S = $\{S_1, S_2, S_3, S_4, S_5\}$. Additionally, five diagnoses: Viral fever, Malaria, Typhoid, Stomach problems, and Chest problems, are defined, represented as D = $\{D_1, D_2, D_3, D_4, D_5\}$. Then, the Pythagorean fuzzy relations P $\longrightarrow$ S and D$\longrightarrow$ S are displayed in Tables 3, 4 and 5 below.

**Table 3.** Patients' symptoms given by the expert.

| Patient | $S_1$ | $S_2$ | $S_3$ | $S_4$ | $S_5$ |
|---|---|---|---|---|---|
| $P_1$ | $< S_1, 0.9, 0.1 >$ | $< S_2, 0.7, 0.2 >$ | $< S_3, 0.2, 0.8 >$ | $< S_4, 0.7, 0.2 >$ | $< S_5, 0.2, 0.7 >$ |
| $P_2$ | $< S_1, 0.0, 0.7 >$ | $< S_2, 0.4, 0.5 >$ | $< S_3, 0.6, 0.2 >$ | $< S_4, 0.2, 0.7 >$ | $< S_5, 0.1, 0.2 >$ |
| $P_3$ | $< S_1, 0.7, 0.1 >$ | $< S_2, 0.7, 0.1 >$ | $< S_3, 0.0, 0.5 >$ | $< S_4, 0.1, 0.7 >$ | $< S_5, 0.0, 0.6 >$ |
| $P_4$ | $< S_1, 0.5, 0.1 >$ | $< S_2, 0.4, 0.3 >$ | $< S_3, 0.4, 0.5 >$ | $< S_4, 0.8, 0.2 >$ | $< S_5, 0.3, 0.4 >$ |

**Table 4.** Reliability of patients' symptoms.

| Patient | $S_1$ | $S_2$ | $S_3$ | $S_4$ | $S_5$ |
|---|---|---|---|---|---|
| $P_1$ | $< S_1, 0.9, 0.1 >$ | $< S_2, 0.9, 0.0 >$ | $< S_3, 0.5, 0.3 >$ | $< S_4, 0.8, 0.2 >$ | $< S_5, 0.4, 0.3 >$ |
| $P_2$ | $< S_1, 0.9, 0.1 >$ | $< S_2, 0.2, 0.8 >$ | $< S_3, 0.3, 0.2 >$ | $< S_4, 0.2, 0.7 >$ | $< S_5, 0.6, 0.2 >$ |
| $P_3$ | $< S_1, 0.9, 0.0 >$ | $< S_2, 0.2, 0.1 >$ | $< S_3, 0.1, 0.8 >$ | $< S_4, 0.0, 0.9 >$ | $< S_5, 0.7, 0.1 >$ |
| $P_4$ | $< S_1, 0.6, 0.2 >$ | $< S_2, 0.5, 0.3 >$ | $< S_3, 0.7, 0.2 >$ | $< S_4, 0.0, 0.1 >$ | $< S_5, 0.5, 0.3 >$ |

**Table 5.** Patients' symptoms given by the expert.

| Diagnose | $S_1$ | $S_2$ | $S_3$ | $S_4$ | $S_5$ |
|---|---|---|---|---|---|
| $D_1$ | $< S_1, 0.3, 0.0 >$ | $< S_2, 0.3, 0.5 >$ | $< S_3, 0.2, 0.8 >$ | $< S_4, 0.7, 0.3 >$ | $< S_5, 0.2, 0.6 >$ |
| $D_2$ | $< S_1, 0.0, 0.6 >$ | $< S_2, 0.2, 0.6 >$ | $< S_3, 0.0, 0.8 >$ | $< S_4, 0.5, 0.0 >$ | $< S_5, 0.1, 0.8 >$ |
| $D_3$ | $< S_1, 0.2, 0.2 >$ | $< S_2, 0.5, 0.2 >$ | $< S_3, 0.1, 0.7 >$ | $< S_4, 0.2, 0.6 >$ | $< S_5, 0.2, 0.8 >$ |
| $D_4$ | $< S_1, 0.2, 0.8 >$ | $< S_2, 0.1, 0.5 >$ | $< S_3, 0.7, 0.0 >$ | $< S_4, 0.1, 0.7 >$ | $< S_5, 0.2, 0.7 >$ |
| $D_5$ | $< S_1, 0.2, 0.8 >$ | $< S_2, 0.0, 0.7 >$ | $< S_3, 0.2, 0.8 >$ | $< S_4, 0.1, 0.8 >$ | $< S_5, 0.0, 0.2 >$ |

In order to make a correct diagnosis for each patient, we calculate mean Deng entropy between symptom and symptom diagnosis for each patient. The smaller the value of Deng entropy is, the smaller the uncertainty of the diagnosis is. The more likely the patient is to diagnose this disease. By observing the experimental results of proposed method, it is obvious that the John has the least $E_d(m)$ 4.2204 for D1, Jack has the least $E_d(m)$ 5.9016 for D4, Cole has the least $E_d(m)$ 5.2751 for D3, and P4 has the least $E_d(m)$ 5.6103 for D1.

As shown in Table 6, obviously the proposed method generates the same results as Zhou et al.'s method [16] and Xiao's method [26], which can demonstrate that the proposed algorithm is capable of finishing medical diagnosis problems.

**Table 6.** The results generated by different methods.

| | $P_1$ | $P_2$ | $P_3$ | $P_4$ |
|---|---|---|---|---|
| Proposed method | Viral fever | Stomach problem | Typhoid | Viral fever |
| Xiao's method | Viral fever | Stomach problem | Typhoid | Viral fever |
| Zhou et al.'s | Viral fever | Stomach problem | Typhoid | Viral fever |

By comparing the results of the three methods in medical applications, we have determined that the prevalence of the four patients is the same as the other methods, proving the feasibility of the proposed algorithm. The new algorithm enlarges the uncertainty between PFSs by mapping the Deng entropy between attributes and decision-making. The Deng entropy increases the uncertainty ratio and plays a more important role in the diagnosis process. Make the recognition results of multi-attribute decision making more clear.

# 7   Conclusions

This paper propose a new model and a combination rule based on confidence measurement, which is called TDPFS and the TDPFS combination rule. Based on Deng entropy, the model is applied to multi-attribute decision-making.

TDPFS model extends the single discount index of Z-number, and two sets of Pythagorean fuzzy sets are used to express expert opinions. In the framework of two-dimensional expert data, the reliability of expert opinion is introduced. The combination rule in this paper is based on the direct operation of the membership function, non-membership function, and hesitation function of the two PFSs, rather than the distance measure or the relationship between the membership function and the non-membership function, it solves the irrationality of simple and reliable indicators, and obtains more reasonable and identification high degree of expert opinion. In numerical example, the TDPFS combination rule makes full use of the support of expert data to identify the target. Comparing with existing methods, the same recognition result is obtained and the recognition degree is higher. Moreover, the TDPFS model is applied to medical diagnosis, Deng entropy is used to replace the traditional distance, and Deng entropy of mapping attributes and decision is allocated through multiplication, which expands the uncertainty between PFSs and achieves the desired effect. The new algorithm has better resolution, which is helpful for identification and judgment in practical applications.

Overall, this study provides a new model called TDPFS. Based on it, a new combination rule is proposed, which can obtain more reasonable expert opinions. Moreover, a decision-making method based on Deng entropy to deal with uncertainty between PFSs is proposed, which has been successfully applied in medical diagnosis. The recognition result of this method is flexible and has broad application prospects.

# References

1. Yager, R.R.: On using the shapley value to approximate the Choquet integral in cases of uncertain arguments. IEEE Trans. Fuzzy Syst. **26**(3), 1303–1310 (2018)
2. Xiao, F.: EFMCDM: evidential fuzzy multicriteria decision making based on belief entropy. IEEE Trans. Fuzzy Syst. **28**, 1477–1491 (2020)
3. Gao, X., Deng, Y.: The negation of basic probability assignment. IEEE Access **7**, 107006–107014 (2019)
4. He, Z., Jiang, W.: An evidential Markov decision making model. Inf. Sci. **467**, 357–372 (2018)
5. Fei, L., Feng, Y., Liu, L.: Evidence combination using OWA-based soft likelihood functions. Int. J. Intell. Syst. **34**(9), 2269–2290 (2019)
6. Xiao, F.: Generalized belief function in complex evidence theory. J. Intell. Fuzzy Syst. **38**(4), 3665–3673 (2020)
7. Fu, C., Chang, W., Xue, M., Yang, S.: Multiple criteria group decision making with belief distributions and distributed preference relations. Eur. J. Oper. Res. **273**(2), 623–633 (2019)

8. Xiao, F.: A multiple-criteria decision-making method based on d numbers and belief entropy. Int. J. Fuzzy Syst. **21**(4), 1144–1153 (2019)
9. Cao, Z., et al.: Extraction of SSVEPs-based inherent fuzzy entropy using a wearable headband EEG in migraine patients. IEEE Trans. Fuzzy Syst. (2019). https://doi.org/10.1109/TFUZZ.2019.2905823
10. Han, Y., Deng, Y., Cao, Z., Lin, C.-T.: An interval-valued Pythagorean prioritized operator-based game theoretical framework with its applications in multicriteria group decision making. Neural Comput. Appl. **32**(12), 7641–7659 (2019). https://doi.org/10.1007/s00521-019-04014-1
11. Atanassov, K.T.: Intuitionistic Fuzzy Sets. STUDFUZZ, vol. 35, pp. 1–137. Springer, Heidelberg (1999). https://doi.org/10.1007/978-3-7908-1870-3
12. Yager, R.R.: Pythagorean membership grades in multicriteria decision making. IEEE Trans. Fuzzy Syst. **22**(4), 958–965 (2013)
13. Yager, R.R.: Properties and applications of Pythagorean fuzzy sets. In: Angelov, P., Sotirov, S. (eds.) Imprecision and Uncertainty in Information Representation and Processing. SFSC, vol. 332, pp. 119–136. Springer, Cham (2016). https://doi.org/10.1007/978-3-319-26302-1_9
14. Zadeh, L.A.: A note on Z-numbers. Inf. Sci. **181**(14), 2923–2932 (2011)
15. Deng, Y.: Deng entropy. Chaos Solitons Fractals **91**, 549–553 (2016)
16. Zhou, Q., Mo, H., Deng, Y.: A new divergence measure of Pythagorean fuzzy sets based on belief function and its application in medical diagnosis. Mathematics **8**(1) (2020). https://doi.org/10.3390/math8010142
17. Fei, L., Deng, Y.: Multi-criteria decision making in Pythagorean fuzzy environment. Appl. Intell. **50**(2), 537–561 (2019). https://doi.org/10.1007/s10489-019-01532-2
18. Fei, L., Feng, Y., Liu, L., Mao, W.: On intuitionistic fuzzy decision-making using soft likelihood functions. Int. J. Intell. Syst. **34**(9), 2225–2242 (2019)
19. Zhu, J., Wang, X., Song, Y.: Evaluating the reliability coeffcient of a sensor based on the training data within the framework of evidence theory. IEEE Access **6**, 30592–30601 (2018)
20. Fei, L., Xia, J., Feng, Y., Liu, L.: An ELECTRE-based multiple criteria decision making method for supplier selection using Dempster-Shafer theory. IEEE Access **7**, 84701–84716 (2019)
21. Zhang, H., Deng, Y.: Weighted belief function of sensor data fusion in engine fault diagnosis. Soft. Comput. **24**(3), 2329–2339 (2019). https://doi.org/10.1007/s00500-019-04063-7
22. Liu, Z.-G., Qiu, G., Mercier, G., Pan, Q.: A transfer classification method for heterogeneous data based on evidence theory. IEEE Trans. Syst. Man Cybern. Syst. (2019). https://doi.org/10.1109/TSMC.2019.2945808
23. Dempster, A.P.: Upper and lower probabilities induced by a multivalued mapping. In: Yager, R.R., Liu, L. (eds.) Classic Works of the Dempster-Shafer Theory of Belief Functions. STUDFUZZ, vol. 219, pp. 57–72. Springer, Heidelberg (2008). https://doi.org/10.1007/978-3-540-44792-4_3
24. Gou, X., Liao, H., Xu, Z., Min, R., Herrera, F.: Group decision making with double hierarchy hesitant fuzzy linguistic preference relations: consistency based measures, index and repairing algorithms and decision model. Inf. Sci. **489**, 93–112 (2019)
25. Liao, H., Mi, X., Xu, Z.: A survey of decision-making methods with probabilistic linguistic information: bibliometrics, preliminaries, methodologies, applications and future directions. Fuzzy Optim. Decis. Making **19**(1), 81–134 (2019). https://doi.org/10.1007/s10700-019-09309-5
26. Xiao, F., Ding, W.: Divergence measure of Pythagorean fuzzy sets and its application in medical diagnosis. Appl. Soft Comput. **79**, 254–267 (2019)

# Utilization of CNN-LSTM Model in Prediction of Multivariate Time Series for UCG

Yuteng Xiao[1] , Hongsheng Yin[1]([✉]) , Kaijian Xia[1], Yundong Zhang[2], and Honggang Qi[3]

[1] School of Information and Control Engineering, China University of Mining and Technology, Xuzhou 221116, China
xuzhouyhs@sina.com
[2] State Key Laboratory of Digital Multi-Media Chip Technology, Vimicro Corporation, 16/F Shining Tower, 35 Xueyuan Road Haidian District, Beijing 100191, China
[3] School of Computer Science and Technology, University of Chinese Academic of Sciences, Beijing 100049, China

**Abstract.** Underground Coal Gasification (UCG) is a serious attempt to clean and efficient use of coal. However, existing methods for predicting UCG are mostly single-variable and not very accurate, which hinders the development of UCG to some extent. Based on the spatio-temporal characteristics, we proposed the CNN-LSTM model. Initially, we reconstructed the UCG concentration data so that they represent spatial and temporal properties, respectively. Then, the spatial features are extracted from the spatial concentration matrices by the CNN at the bottom of the model. Thirdly, the LSTM captures the temporal dependencies among the spatial features. Finally, the final predictions are obtained from the last layer of the LSTM. Based on the data from China Huating coal UCG field test, the experiment shows that our model outperforms prior methods.

**Keywords:** Underground coal gasification · Prediction · CNN · LSTM

## 1 Introduction

Underground coal gasification (UCG), which converts coal into gas and subsequently removes sulfides and nitrogen oxides, bypassing the traditional coal-fired process, has the advantage of low pollution emissions and is one of the future directions of coal mining [1–3]. The principle of UCG is shown as Fig. 1. The compressor injects gasification agents (air, oxygen, different concentrations of enriched oxygen, carbon dioxide or water vapor) through the injection well, and the coal buried deep beneath the overburden is burned in situ directly and incomplete, and a thermochemical reaction occurs to gasify it into syngas. Subsequently, harmful gases such as sulphides, nitrogen oxides and micro-particles in the syngas are gas cleaned, and the solid waste is left deep underground. Finally, $H_2$, CO and $CH_4$ as the combustible gas are exported from the production wells and become clean feedstock gas for gas production, power generation and hydrogen production.

© Springer Nature Switzerland AG 2020
X. Chen et al. (Eds.): ML4CS 2020, LNCS 12488, pp. 429–440, 2020.
https://doi.org/10.1007/978-3-030-62463-7_40

However, since UCG is carried out in a high temperature and humidity environment underground, it is not intuitively monitored and controlled by humans, resulting in low and unsecured combustible gas yields [4]. Therefore, the prediction of combustible gas can improve the measurement and control of UCG [5]. Researchers have conducted a series of studies on the UCG prediction problem. Dufaux et al. [6] used the chemical equilibrium of the coal surface to predict the product gas composition. Laciak et al. [7] used a finite optimization method of Lagrangian multipliers to derive gasification product concentrations at different gasification temperatures based on known UCG temperatures and pressures and based on systematic Gibson energy minimization. Eftekhari et al. [8] extended the existing steady-state model to a transient model capable of describing alternating injections of air and steam in deep thin coal seams to predict UCG gas fractions. Subsequently, they validated the model with field test data to verify the accuracy of the model's gas fraction predictions [9]. In general, such prediction models can only predict the final UCG output, which is not very accurate and cannot be predicted at the next moment in the reaction process. Based on this, other scholars have used machine learning methods to predict the next moment of the UCG process. Kaur et al. [5] predicted the coal subsurface gasification calorific value and temperature data obtained under laboratory conditions for 7 h by a support vector machine method. Krzemien [10] predicted the temperature of the UCG process using a multivariate adaptive regression sample bars (MARS) method based on the data of different temperatures measured per hour during the experiment. Subsequently, he investigated the predictive performance of two different artificial neural network models (generalized regression model and multi-layer feedforward model) on the temperature data collected in the Wieczorek UCG test, using one-hour predictions to predict the syngas temperature at the generator outlet. The results showed that the generalized regression neural network (GRAN) has better predictive performance than the multi-layer feedforward network (MLFN) and MARS models [11]. However, the above prediction methods only predict UCG time series data in a single dimension, e.g. they predict the temperature or gas fraction at the next moment from the temperature or gas fraction at the previous period. However, the above methods can only predict univariate UCG time series data, such as temperature or gas composition. Literature on UCG multivariate time series prediction could not be found.

On the other hand, since deep learning methods, especially recurrent neural networks, can preserve the memory of recent events [12, 13], they have achieved good results in multivariate time-series prediction by processing before-and-after correlated data based on memory properties. Niu et al. [14] proposed a predictive framework based on a two-stage feature selection model, a deep learning model, and an error correction model is proposed that aims to efficiently capture the inherent nonlinearity of multi financial time series. Du et al. [15] proposed a new time attention coding decoding model for multivariate time series prediction problem. Inspired by the human attention mechanism (including the two-stage two-stage model) and the influence mechanism of target and non-target information, Liu et al. [16] proposed the DSTP-based RNN (DSTP-RNN) and DSTP-RNN-II long-term time series prediction methods, respectively.

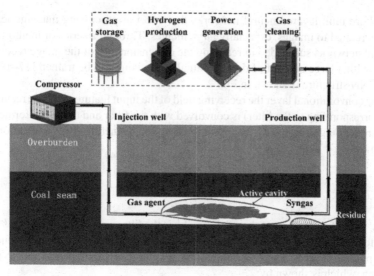

**Fig. 1.** The principle of UCG

In summary, it is feasible to apply deep learning methods to UCG multivariate prediction to address the problem of low accuracy and lack of multivariate prediction methods in prediction for UCG. Therefore, this paper applies CNN (Convolutional Neural Network) and the long short term memory (LSTM) model to the multivariate prediction of UCG in order to extract the spatio-temporal characteristics and improve the prediction accuracy.

## 2 Related Works

### 2.1 Convolutional Neural Networks

**Fig. 2.** Typical CNN structure

CNN are multilayered neural networks that excel at processing machine learning problems related to images, especially large images. Through a series of methods, convolutional networks succeeded in reducing the dimensionality of the image recognition problem with a large amount of data, eventually allowing it to be trained [17–19]. The typical CNN structure is shown in Fig. 2.

At the convolutional layer the receiving field of the input feature map (the rectangular region corresponding to the filter) is convolved with the filter and then transformed by a nonlinear mapping operated by the activation function. Generally, the convolution layer is shown by:

$$\mathbf{x}^l = g\left(\sum_{p \in \mathbf{M}_1} \mathbf{x}^{l-1} * \mathbf{k}_p^l + b^l\right) \tag{1}$$

Where $\mathbf{x}^l$ is the feature map of the $l$ th layer; $\mathbf{k}_p^l$ is the filter size of the $l$ th layer; $b^l$ is the bias vector; $*$ denotes the convolution operation and g denotes the activation function.

The pooling layer is used to compress the amount of data and parameters to reduce overfitting, which is shown by:

$$\mathbf{x}^l = down(\mathbf{x}^{l-1}) \tag{2}$$

Where $down(\bullet)$ represents the sub-sampling function, such as average-pooling and max-pooling.

Each node in the fully-connected layer is connected to all the nodes in the previous layer and is used to synthesize the features extracted from the previous side. It can be shown by:

$$\mathbf{x}^l = g(\mathbf{w}^l \mathbf{x}^{l-1} + \mathbf{b}^l) \tag{3}$$

Where $\mathbf{w}^l$ and $\mathbf{b}^l$ are the weight matrix and bias matrix respectively.

With the three special designs, including receptive fields, local connectivity, and shared weights, CNNs are capable of extracting spatial features effectively [20].

## 2.2  Long Short-Term Memory

LSTM is a variant of the Recurrent neural network (RNN) [21]. The neural network is essentially a nonlinear function approximator. RNN [22, 23] is a neural network with feedback connection, due to its implicit layer to join the feedback structure, that is, the output of the moment is not only related to the input of the moment, but also related to the output of the previous moment, equivalent to the depth of the neural network unfolded on the time series, making it more suitable for processing time series data. The structure of RNN is shown in Fig. 3.

Assuming that the input is $\mathbf{x}_t$, the hidden layer state is $\mathbf{h}_t$, $\mathbf{W}_x$ is the weight of the input layer to the hidden layer, $\mathbf{W}_h$ is the autoregressive weight of the hidden layer, and $\sigma$ is the activation function, then at the moment t:

$$\mathbf{h}_t = \sigma(\mathbf{W}_h \mathbf{h}_{t-1} + \mathbf{W}_x \mathbf{x}_t) \tag{4}$$

$h_t$ contains information on timing changes, i.e. the output of moment $t$ is determined jointly by the input of moment $t$ and the output of moment $t-1$.

However, RNN has the disadvantage of gradient disappearance and gradient explosion and cannot learn to rely on information in the long term, so LSTM was born.

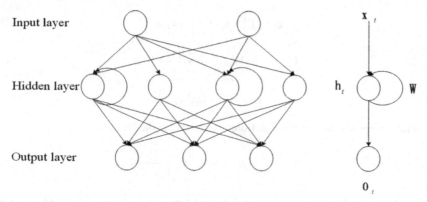

**Fig. 3.** RNN structure

LSTM draws on the long-time and short-time and forgetting characteristics of human neural memory and adds three controllers to the network, namely the input gate, the output gate and the forget gate. LSTM is often used for trajectory prediction [24], action classification [25] and time-series prediction [26], and its structure is shown in Fig. 4.

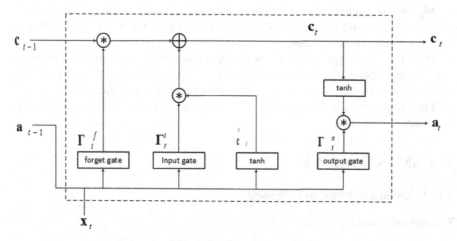

**Fig. 4.** LSTM structure

It can be seen from Fig. 4. There are three inputs to the LSTM at the $t$ moment: the input value $x_t$ of the network at the current moment, the output value $a_{t-1}$ of the LSTM at the previous moment, and the cell state $c_{t-1}$ at the previous moment; and two outputs: the output value $a_t$ of the current moment and the cell state $c_t$ at the current moment.

$$\Gamma_t^f = \sigma(W_f[a_{t-1}, x_t] + b_f) \tag{5}$$

$$\Gamma_t^i = \sigma(W_i[a_{t-1}, x_t] + b_i) \tag{6}$$

$$\hat{c}_t = \tanh(W_c[a_{t-1}, x_t] + b_c) \tag{7}$$

$$c_t = \Gamma_t^i * \hat{c}_t + \Gamma_t^f * c_{t-1} \tag{8}$$

$$\Gamma_t^o = \sigma(W_o[a_{t-1}, x_t] + b_o) \tag{9}$$

$$a_t = \Gamma_t^o * \tanh c_t \tag{10}$$

Where $\sigma$ represent sigmoid function, which can be shown as:

$$\sigma(x) = sigmoid(x) = \frac{1}{1 + e^{-x}} \tag{11}$$

$\hat{c}_t$ is used to describe the candidate value for cell state $c_t$, which is calculated based on the output state of the previous moment and the input state of the current. The tanh is a typical nonlinear activation function, expressed as:

$$\tanh(x) = \frac{e^x - e^{-x}}{e^x + e^{-x}} \tag{12}$$

$W_f$, $W_i$, $W_c$ and $W_o$ are the weighting matrix for each gate, respectively. $b_f$, $b_i$, $b_c$ and $b_o$ are bias term for each gate. For convenience, $[a_{t-1}, x_t]$ denotes connecting the two vectors into a longer vector.

From type (5) to type (10) can be seen: input gate control when to let the activation state into the deposit unit, that is, the writing of timing data; output gate control when to let the activation state out, that is, the output of timing data; forget gate control in whole or in part forgotten. The three gates work together to complete the learning of the rules in the data.

## 3   CNN-LSTM Model

### 3.1   Notation and Problem Statement

The entire datasets are expressed as:

$$X = \begin{bmatrix} x_1^1 & x_1^2 & \cdots & x_1^m \\ x_2^1 & x_2^2 & \cdots & x_2^m \\ \vdots & \vdots & & \vdots \\ x_n^1 & x_n^2 & \cdots & x_n^m \end{bmatrix} \tag{13}$$

Where $m$ represents the variables for input data, which include the concentration of $O_2$, $N_2$, $CO$, $CH_4$, $C_NH_M$ and $n$ represents total length of time.

We aim to use the data with time step $\omega$ to predict the concentration of $CO_2$ at $w + 1$. Therefore the input is expressed as:

$$
\mathbf{X}_{w-1} = \begin{bmatrix} \mathbf{x}_w^1 & \mathbf{x}_w^2 & \cdots & \mathbf{x}_w^m \\ \mathbf{x}_{w+1}^1 & \mathbf{x}_{w+1}^2 & \cdots & \mathbf{x}_{w+1}^m \\ \vdots & \vdots & & \vdots \\ \mathbf{x}_{w+t-1}^1 & \mathbf{x}_{w+t-1}^2 & \cdots & \mathbf{x}_{w+t-1}^m \\ \vdots & \vdots & & \vdots \\ \mathbf{x}_{n-2}^1 & \mathbf{x}_{n-2}^2 & \cdots & \mathbf{x}_{n-2}^m \\ \mathbf{x}_{n-1}^1 & \mathbf{x}_{n-1}^2 & \cdots & \mathbf{x}_{n-1}^m \end{bmatrix} = \begin{bmatrix} \mathbf{x}_w \\ \mathbf{x}_{w+1} \\ \vdots \\ \mathbf{x}_{w+t-1} \\ \vdots \\ \mathbf{x}_{n-2} \\ \mathbf{x}_{n-1} \end{bmatrix} (1 \le t \le n - w) \tag{14}
$$

where $\omega$ represents time step, $m$ represents the dimension of the input data and $n$ represents the total number of records per dimension. We use to represent the input data of each dimension respectively, which can be expressed as:

$$
\mathbf{x}_{w+t-1} = \begin{bmatrix} x_{t-1}^1 & x_{t-1}^2 & \cdots & x_{t-1}^m \\ x_t^1 & x_t^2 & \cdots & x_t^m \\ \vdots & \vdots & & \vdots \\ x_{w+t-1}^1 & x_{w+t-1}^2 & \cdots & x_{w+t-1}^m \end{bmatrix} \tag{15}
$$

The CNN-LSTM model aims to predict the current value of the target series $\hat{\mathbf{y}}_\omega = [\hat{\mathbf{y}}_w, \hat{\mathbf{y}}_{w+t-1}, \cdots, \hat{\mathbf{y}}_n]$ (the concentration of CO2) based on $\mathbf{X}_{\omega-1}$:

$$
\hat{\mathbf{y}}_\omega = F(\mathbf{x}_w, \mathbf{x}_{w+1}, \cdots, \mathbf{x}_{n-1}) \tag{16}
$$

where $F(\cdot)$ is a nonlinear mapping function we aim to learn.

### 3.2 The Proposed Network Architecture

In the UCG environment, the spatio-temporal function describing the concentration of the various gas components at any site is continuous in most cases. On the one hand, the concentration of the gas fraction of UCG is related to its historical value, called spatiotemporal correlation. On the other hand, the component concentrations are not statistically independent, and they can be expressed as:

$$
p_{O_2} + p_{N_2} + p_{CO} + p_{CH_4} + p_{C_N H_m} + p_{CO_2} = 1 \tag{17}
$$

where $p_{O_2}\, p_{N_2}\, p_{CO}\, p_{CH_4}\, p_{C_N H_m}$ and $p_{CO_2}$ represent the concentration of $O_2$, $N_2$, CO, $CH_4$, $C_N H_M$ and $CO_2$ respectively. Obviously, they are spatially correlated. Thus, knowledge of the predicted concentrations of UCG components is contained in both temporal and spatial correlations, both of which can be used to improve prediction accuracy. As shown in Sect. 2, CNN are capable of extracting spatial features effectively and LSTM is more suitable for processing time series data. We proposes CNN-LSTM model to predict multivariate time series for UCG.

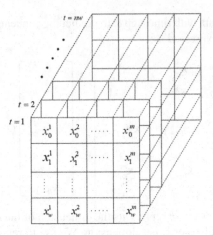

**Fig. 5.** The spatio-temporal sequence describing the concentration of UCG

In order to better represent the spatio-temporal characteristics of the concentrations of the UCG components, we reconstructed the data, which can be seen as a collection of images. It can be shown in Fig. 5.

The concentration data of the UCG components consisting of a time step $\omega$ is taken as a picture, corresponding to $\mathbf{x}_{w+t-1}$, which contains the spatial characteristics of the individual component concentrations. Obviously, there is also a temporal sequence between each image. Therefore, we first used the CNN model to deal with the spatiality of this data, followed by the LSTM model to deal with the temporal properties. The overall framework of the CNN-LSTM model is shown in Fig. 6. Because the spatial properties of the data in this paper are small, we use the basic CNN model of Fig. 2 and the LSTM model is shown in Fig. 4.

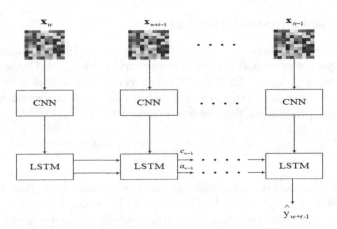

**Fig. 6.** Overall framework of the proposed CNN-LSTM model

In order to improve the stability of the calculation, the data are normalized and the range of input data is included in the [0, 1] interval, the formula is:

$$X = \frac{x - x_{min}}{x_{max} - x_{min}} \tag{18}$$

where $x$ is the actual value of the input variable; $x_{max}$ is the maximum value; $x_{min}$ is the minimum value and $X$ is the normalized value.

An inverse normalization of the predicted results using a trained prediction model leads to a comparative error analysis with actual values, with the formula:

$$Y = (x_{max} - x_{min})X + x_{min} \tag{19}$$

During the training process, we use mini-batch stochastic gradient descent (SGD) together with the Adam optimizer to minimize the mean absolute error (MAE) between the predicted value and real value:

$$MAE = \frac{1}{N} \sum_{t=1}^{N} |d_t - y_t| \tag{20}$$

## 4 Experiment

We implemented a CNN-LSTM model for UCG concentration prediction on the Ubuntu operating system, TensorFlow deep learning framework, Intel i7-8700 K 4* processor, 3.7 GHz CPU, 64 GB RAM and NVIDIA GTX 1080Ti and the programming language is python. Based on the concentration of various types of output gas collecting at the China Huating coal UCG field test, the paper studies and predicts concentration of $CO_2$ through the model.

The time-series data entered in this model is the data processed in Sect. 3, totaling 6770 min, taking previously 95% as training sets and others as test sets. To extract more features, the kernels of CNN are 12 and the kernel size is 3 * 3; the number of units of the first layer in LSTM is 128, and the number of units of the second layer in LSTM is 64. We get only the predicted value of the last cell from the second layer. The number of training iterations is 20,000; the initial learning rate is 0.006, and the batch size is 50.

This experiment is compared with BPNN and LSTM prediction methods respectively. Figure 7 is the comparison of real value and predicted value in three models. Compared with the BPNN and LSTM algorithms, the prediction of CO2 concentration based on CNN-LSTM model is in good agreement with the real value, which can better predict the CO2 concentration.

We use root mean square error (RMSE), mean absolute percent error (MAPE), symmetric mean absolute percentage error (SMAPE), Theil inequality coefficient (TIC), and correlation coefficient (CC), which are both measures of the error of deviation of the predicted value from the actual data and reflect the global error of the prediction. The corresponding standard definition is given by the following formula.

$$RMSE = \sqrt{\frac{1}{N} \sum_{t=1}^{N} (d_t - y_t)^2} \tag{21}$$

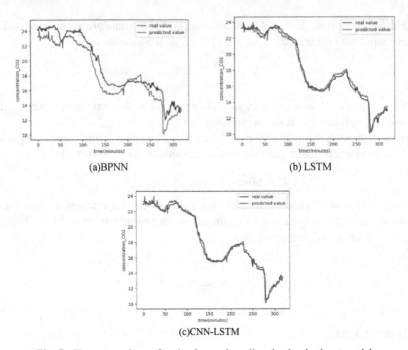

(a)BPNN                    (b) LSTM

(c)CNN-LSTM

**Fig. 7.** The comparison of real value and predicted value in three models

$$MAPE = 100 \times \frac{1}{N} \sum_{t=1}^{N} \left| \frac{d_t - y_t}{d_t} \right| \tag{22}$$

$$SMAPE = 100 \times \frac{2}{N} \sum_{t=1}^{N} \frac{|d_t - y_t|}{|d_t| + |y_t|} \tag{23}$$

$$TIC = \frac{\sqrt{\frac{1}{N} \sum_{t=1}^{N} (d_t - y_t)^2}}{\sqrt{\frac{1}{N} \sum_{t=1}^{N} d_t^2} + \sqrt{\frac{1}{N} \sum_{t=1}^{N} y_t^2}} \tag{24}$$

$$CC = \frac{\sum_{t=1}^{N} (y_t - \bar{y})(d_t - \bar{d})}{\sqrt{\sum_{t=1}^{N} (y_t - \bar{y})^2 \sum_{t=1}^{N} (d_t - \bar{d})^2}} \tag{25}$$

where $d_t$ is the concentration for CO2; $y_t$ is the predicted concentration for CO2; $N$ represents the number of data; $\bar{d}$ is average value of CO2 concentration; $\bar{y}$ is average value of predicted concentration for CO2. The smaller the values of MAE, RMSE, MAPE and SMAPE, the smaller the deviation of the predicted results from the actual values. A value of TIC closer to 0 indicates a higher prediction accuracy value; a value of CC closer to 1 indicates more errors and lower accuracy of the prediction.

**Table 1.** Performance of different methods in different evaluation

| Model | MAE | RMSE | MAPE | SMAPE | TIC | CC |
|---|---|---|---|---|---|---|
| BPNN | 1.1680 | 1.3741 | 6.9149 | 6.5660 | 0.0682 | 0.9757 |
| LSTM | 0.3339 | 0.4447 | 1.9625 | 1.9429 | 0.0232 | 0.9944 |
| CNN-LSTM | 0.2425 | 0.3444 | 1.4182 | 1.4073 | 0.0181 | 0.9962 |

The results are shown in Table 1. It can be illustrated that LSTM and our model are much better than BPNN. Because there are few spatial characteristics of the UCG concentration data, our method have a slight improvement compared with LSTM, which is the same as that represented in Fig. 7.

## 5 Conclusion

In this paper, we propose a novel CNN-LSTM model to predict the multivariate time series for UCG concentrated data. In general, they can better learn the spatial-temporal relationship and thus outperform other methods in UCG concentration prediction. Our reconstruction of the concentration data provides a developmental idea for further understanding of spatio-temporal relationships and for further exploring the application of CNN-LSTM and other methods in time series prediction. Numerous experiments have proven the following conclusions.

**Acknowledgements.** This work is supported "the Fundamental Research Funds for the Central Universities" under Grant 2019BSCX14; the National Key Research and Development Program of China (2016YFB0401904).

## References

1. Xiao, Y., et al.: Monitoring and control in underground coal gasification: current research status and future perspective. Sustainability **11**(1), 217 (2019)
2. Perkins, G.: Underground coal gasification - Part I: field demonstrations and process performance. Prog. Energy Combust. Sci. **67**, 158–187 (2018)
3. Perkins, G.: Underground coal gasification - Part II: fundamental phenomena and modeling. Prog. Energy Combust. Sci. **67**, 234–274 (2018)
4. Su, F.-Q., et al.: Monitoring and evaluation of simulated underground coal gasification in an ex-situ experimental artificial coal seam system. Appl. Energy **223**, 82–92 (2018)
5. Kacur, J., et al.: Utilization of machine learning method in prediction of UCG data (2017)
6. Pirard, J.P.: Modelling of the underground coal gasification process at Thulin on the basis of thermodynamical equilibria and isotopic measurements. Fuel **69**(5), 624–632 (1990)
7. Laciak, M., Kacur, J., Kostur, K.: The verification of thermodynamic model for UCG process (2016)
8. Eftekhari, A.A., et al.: Mathematical modeling of alternating injection of oxygen and steam in underground coal gasification. Int. J. Coal Geol. **150–151**, 154–165 (2015)

9. Eftekhari, A.A., et al.: Energy and exergy analysis of alternating injection of oxygen and steam in the low emission underground gasification of deep thin coal. Appl. Energy **208**, 62–71 (2017)

10. Krzemien, A.: Fire risk prevention in underground coal gasification (UCG) within active mines: temperature forecast by means of MARS models. Energy **170**, 777–790 (2019)

11. Krzemien, A.: Dynamic fire risk prevention strategy in underground coal gasification processes by means of artificial neural networks. Arch. Min. Sci. **64**(1), 3–19 (2019)

12. Lin, T., Horne, B.G., Giles, C.L.: How embedded memory in recurrent neural network architectures helps learning long-term temporal dependencies. Neural Netw. Off. J. Int. Neural Netw. Soc. **11**(5), 861–868 (1998)

13. Chen, P.-A., Chang, L.-C., Chang, F.-J.: Reinforced recurrent neural networks for multi-step-ahead flood forecasts. J. Hydrol. **497**, 71–79 (2013)

14. Niu, T., et al.: Developing a deep learning framework with two-stage feature selection for multivariate financial time series forecasting. Expert Syst. Appl. **148**, 113237 (2020)

15. Du, S.D., et al.: Multivariate time series forecasting via attention-based encoder-decoder framework. Neurocomputing **388**, 269–279 (2020)

16. Liu, Y.Q., et al.: DSTP-RNN: a dual-stage two-phase attention-based recurrent neural network for long-term and multivariate time series prediction. Expert Syst. Appl. **143**, 113082 (2020)

17. He, K., et al.: Deep residual learning for image recognition. In: 2016 IEEE Conference on Computer Vision and Pattern Recognition, pp. 770–778 (2016)

18. Szegedy, C., et al.: Going deeper with convolutions. In: 2015 IEEE Conference on Computer Vision and Pattern Recognition, pp. 1–9 (2015)

19. Krizhevsky, A., Sutskever, I., Hinton, G.E.: ImageNet classification with deep convolutional neural networks. Commun. ACM **60**(6), 84–90 (2017)

20. Zhu, Q., et al.: Learning temporal and spatial correlations jointly: a unified framework for wind speed prediction. IEEE Trans. Sustain. Energy **11**(1), 509–523 (2020)

21. Hochreiter, S., Schmidhuber, J.: Long short-term memory. Neural Comput. **9**(8), 1735–1780 (1997)

22. Sutskever, I., Vinyals, O., Le, Q.V.: Sequence to sequence learning with neural networks, vol. 4, pp. 3104–3112 (2014)

23. Graves, A.: Generating sequences with recurrent neural networks. Computer Science (2013)

24. Pei, Z., et al.: Human trajectory prediction in crowded scene using social-affinity long short-term memory. Pattern Recogn. **93**, 273–282 (2019)

25. Tufek, N., et al.: Human action recognition using deep learning methods on limited sensory data. IEEE Sens. J. **20**(6), 3101–3112 (2020)

26. Karim, F., et al.: Multivariate LSTM-FCNs for time series classification. Neural Netw. **116**, 237–245 (2019)

# An Imbalanced Learning Based Method for Esophageal Squamous Cell Carcinoma (ESCC) Distant Metastasis Predicting

Xuemei Li[1], Bo Yang[2(✉)], Lizhi Peng[2], Baosheng Li[3], and Jian Zhu[3]

[1] Educational Technology Center, Shandong College of Traditional Chinese Medicine, Yantai 264199, People's Republic of China
[2] Shandong Provincial Key Laboratory of Network Based Intelligent Computing, University of Jinan, Jinan 250022, People's Republic of China
yangbo@ujn.edu.cn
[3] Department of Radiation Oncology, Shandong Cancer Hospital affiliated to Shandong University, Jinan 250117, People's Republic of China

**Abstract.** Accurately predicting distant metastasis is very important for esophageal squamous cell carcinoma (ESCC) patients, as it is a key evidence to determine the degree of this disease. Traditional metastasis predictions are made by doctors manually, which is time and labor consuming. In recent years, some researchers try to predict ESCC distant metastasis using machine learning techniques. However, they ignored the factor that the problem of predicting ESCC distant metastasis is a highly imbalanced task. In this study, we try to build a new imbalanced learning based ESCC distant metastasis predicting scheme using blood cell analysis testing results. We first extract effective features from blood cell analysis testing data. And then we use over-sampling method to generate new positive instances (instances with ESCC distant metastasis). Finally, we apply k-nearest neighbor (k-NN) algorithms to predict unknown instances. Experimental results show that our method is effective in cases with few positive instances.

**Keywords:** Esophageal squamous cell carcinoma · Imbalanced data · Machine learning · Oversampling · Cost-sensitive learning

## 1 Introduction

According to statistics, the esophageal squamous cell carcinoma (ESCC) is one of the most common malignant tumors in the world, with the eighth highest incidence rate and the sixth highest mortality rate [1]. China is one of the countries with the highest incidence and mortality of ESCC in the world.

Determining cancer cells distant metastasis is very important for the diagnosis and treatment of ESCC patients. And it's also the basis for defining the

---

Supported by Shandong Provincial Key R&D Program under Grants No. 2017CXZC1206.

X. Chen et al. (Eds.): ML4CS 2020, LNCS 12488, pp. 441–453, 2020.
https://doi.org/10.1007/978-3-030-62463-7_41

stage of ESCC patients. At present, doctors mainly rely on imaging or pathological examination to determine the patients' cancer cells distant metastasize.
In this paper, we mainly take part in the study work, which use blood cell analysis of ESCC patients to predict the cancer cells with other organ metastasis
by imbalanced learning algorithms. In this way, patients can reduce curing time
and cost. Of course, doctors can also understand the patients' condition rapidly
and propose treatment plans. To solve this problem, it is necessary to propose
a model to predict distant metastasis in ESCC patients.

Due to the incidence rate, the data set of ESCC patients obtained from
Shandong Province Cancer Research Institute are imbalanced. Data imbalanced
distribution impose big challenges to standard classification algorithms. Because
the standard classifier algorithms are mostly based on three hypotheses:

- There's little difference in quantity of sample categories used for training
  model [2].
- For the dataset, the cost of misclassifying data is the same as that of the right
  one [2].
- The accuracy is an important criterion for evaluating algorithms [3].

For example, the standard classifiers focus on classifying all samples into a
majority class in a dataset with an imbalanced ratio of 99%. It is meaningless
that the accuracy of the classifier can reach to 99%. Most of the few samples are
misclassified but they are the focus of attention. In this paper, the proportion of
ESCC patients with other organ metastasis is relatively small, and the number
of ESCC patients without other organ metastasis is exactly the opposite. A large
number of predicted patients will be judged as without other organ metastasis if
we use the standard classification algorithm. The accuracy obtained in this way is
the highest, but this is obviously not in line with the reality. A patient with other
organ metastasis is diagnosed as without other organ metastasis, or a patient
without other organ metastasis is diagnosed as with other organ metastasis.
There is tremendous difference in the cost of these two misjudgments between a
patient with other organ metastasis and without. The latter will have a higher
expense because it may cause the patients are delayed in treatment.

The main contribution of this study is applied innovation. Blood cell analysis
is used to predict ESCC patients' cancer cells with other organ metastasis. We
address the imbalanced dataset at both the data level and algorithm level. In
the remainder of this paper, the following contents are as follows. Firstly, Sect. 2
introduces the related works on ESCC and medical problems solved by sampling
methods. Then, in Sect. 3 proposes a solution to the problem encountered. In
Sect. 4 explains the experimental process, including dataset, data preprocessing
and experimental design. Also experimental results and discussion are given. In
Sect. 5 summarizes the paper.

# 2    Related Work

In this section, firstly, we summarizes the medical literature related to ESCC. Secondly, this chapter describes the method used by machine learning to treat imbalanced data medically.

## 2.1    Current Status of Research on ESCC

It has been found that the distribution levels of SCC, CYFRA 21-1 and CEA in serum are positively correlated with the pathological stage of patients [4,5]. The levels of tumor markers have increased before cancer cells with distant metastasis. What's more, we found that the later the stage of ESCC patients, the higher both levels and positive rate of the serum D-dimer, CEA, CYFRA 21-1, CA72-4 and CA19-9. This result is of great significance for the diagnosis of ESCC patients.

The lymphatic metastasis rate of ESCC patients is as high as 20%–60% [6]. Postoperative recurrence and cancer cells metastasis are the main cause of death in patients. The preoperative cervical lymphatic metastasis directly affects the extent of lymphatic dissection [7]. Expanding the extent of lymphatic dissection helps to prolong patients' survival. There are many ways to predict lymphatic metastasis, but the sensitivity and specificity are extremely low. The paralaryngeal lymphatics are considered to be the sentinel lymphatics of esophageal carcinoma [8]. The statistical analysis software SPSS 22.0 is used to predict the cervical lymphatic metastasis, but the accuracy rate is no more than 50%. From this, it can be seen that predicting the cancer cells metastasis is a difficult problem in the medical field.

## 2.2    Sampling Methods Using for Medical Problems

Chawla proposed the Synthetic Minority Over-sampling Technique (SMOTE) algorithm in 2002 in order to solve the problem of imbalanced data [9,10]. The fundamental idea of the algorithm is to add new samples of artificial simulation to the dataset by analyzing and simulating a minority samples. Then, the raw data are balanced after adding new samples of artificial simulation. In addition, hybrid sampling has been proposed to improve the performance of the classifier. In other words, the hybrid sampling method combines with oversampling and undersampling methods [11]. The Synthetic Minority Over-sampling Technique + Edited Nearest Neighbor (SMOTE + ENN) and Synthetic Minority Over-sampling Technique + Tomek link (SMOTE + Tomeklink) hybrid sampling algorithms [12] are better than the simple sampling method, which are proposed by Gustavo et al. In this paper, the SMOTE + ENN algorithm achieves the best results through experimental comparison.

In the medical field, more and more diagnostic predictions are using sampling methods. Daisuke Ichikawa et al. used random undersampling to achieve balanced between positive and negative samples when predicting hyperuricemia [13]. Miriam Seoane Santos et al. introduce a new cluster-based oversampling method to predicting survival in patients with liver cancer [14]. The results show that this method is more effective than other predictive models currently applied to liver cancer. In the prediction of human breast and colon cancer, Abdul Majid et al. propose a way that increase the samples by the mega-trend diffusion (MTD) technique [15]. This model is very ideal and can be used as a tool to predict cancer.

## 3   Methods

A solution is proposed in this section for the problems encountered that prediction of distant metastasis by blood cell analysis of ESCC patients. Firstly, we have the dataset preprocessing. We used 75% of the samples as the training set and the remaining 25% as the testing set. Secondly, we select the features from all attributes. Thirdly, we make the positive and negative samples balance using oversampling algorithm because of imbalanced raw data. Then, we input the balanced data into the classifier for training predictive model. We also adopt two cost-sensitive learning algorithms that can directly focus on positive samples. At last, the testing set are used to evaluate the model using model evaluation indicators. The flow chart of proposed method is shown in Fig. 1.

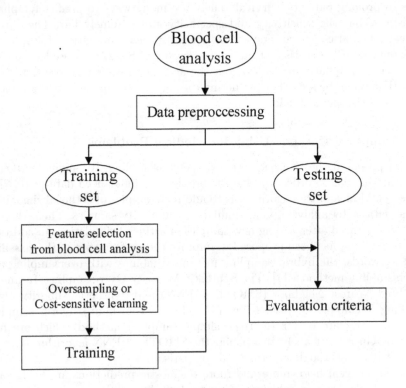

**Fig. 1.** Flow chart of proposed method

## 3.1    Data Preprocessing

The raw data come from Shandong Province Cancer Research Institute. The dataset are from 2013 to 2017 and include 2116 patients. Raw dataset include clinical information and others. Four tables are selected for the paper. It contains personal basic information, blood cell analysis, diagnosis table and surgical table. The basic personal information table contains basic information about the individual, but only the case number of each patient is used.

Raw data show that a one-to-many relation exists between ESCC patients and blood cell analysis. Patients check blood parameters at different times according to the doctor's request, in which the largest in the raw dataset. The patients who first tested the clinical M stage are selected from the diagnosis table according to the patients' existed diagnosis information. The clinical M stage is 0 so that the label is seted as 0. It means that the patients' cancer cells without organ metastasis. The clinical M stage was non-zero then seting the label to 1. It means that the patients' cancer cells with other organs metastasis. Selecting the blood cell analysis test data before the operation according to the operation time recorded by the patients in the operation table. Filtering the record on the day of the diagnosis or the time before the diagnosis if the patient no surgery is performed.

Invalid data are deleted in order to facilitate model analysis. On one hand, data samples with missing attribute values are deleted. On the other hand, we delete attributes with a large number of missing values. Some data are small number so that many of them become 0 after normalization. For these reasons, the original data are used in the experiment.

## 3.2    Feature Selection

The finished dataset are selected for feature selection. All data are not sampled and the feature weights are ordered using feature selection algorithm. We use weka 3.8 for the feature selection of the completed dataset. Two feature selection algorithms are used. An algorithm is ReliefFAttributeEval + Ranker. Another algorithm is CorrelationAttributeEval + Ranker. The attributes of the top eight of the feature sorting are taken as the characteristics of the dataset. Entering these data into the classifier and comparing the model evaluation metrics.

## 3.3    Oversampling Algorithm and Cost-Sensitive Learning

At present, there are two main solutions to the problem of dealing with unbalanced data classification. On one hand, we can make the data of the samples balance. At the data level, by using appropriate methods to reconstruct training samples in order to achieve data equalization. On the other hand, we can improve the standard algorithm or propose a new algorithm [16]. At the algorithm level, the existed classification algorithms are used to improve or propose a new classification algorithm [17]. So that a small number of samples get more attention and improve the accuracy of a few types of samples [18]. This article has worked on the data level and the algorithm level. At the algorithm level, we use the software of keel to implement cost-sensitive learning to solve data samples imbalanced problems.

There are three main strategies for processing raw dataset at the data level: oversampling, undersampling, and mixed sampling. Undersampling can easily lose important features in the dataset, and it is more suitable for larger sample sizes. There are less than 500 positive and negative samples in this paper. Therefore, it is not possible to equalize the original data ratio with an undersampling algorithm. In this study, several oversampling algorithms and hybrid sampling algorithms are tried. The best model evaluation indicators are selected through comparing the results.

### 3.4  Learning Model

We obtain a balanced dataset after using an oversampling algorithm. Then, it's inputed to the classification algorithm for model training.

### 3.5  Evaluation Model Indicator

As mentioned earlier, the accuracy rate is not a suitable evaluation criterion for the classifier when there is a imbalanced problem in the data samples and the misclassification cost is inconsistent. Traditional classification models don't reflect the classification performance of a few classes. We use the Area Under the Curve (AUC) and G-Mean as the evaluation model indicator. The AUC and G-Mean can be calculated using the classifier error rate to evaluate the performance of the classifier. The larger the AUC and G-Mean, the better the classification performance of the classifier.

## 4  Experiment

### 4.1  Dataset

There are 465 instances selected after cleaning up the dirty data. The result is shown in Fig. 2 below. There are 2116 patients including 613 patients with clinical M staging. Only 465 patients have complete blood cell analysis data after deleting the attribute *rdwcv* with severe data loss. Each patient is an instance that contains a total of 21 features and a label. The features are *pct, plt, mpv, pdw, gran1, ly0, mo1, gran0, ly1, wbc, eo0, mo0, mcv, hct, mchc, mch, hgb, ba1, eo1, rbc, ba0* and the lable of M stage. The meaning of each attribute is shown in Table 1. The situation of these 465 data samples is shown in Table 2. There are 349 patients' data as training set for training model. And the remaining 116 data are used as testing set to evaluate model performance. It can be clearly seen from Table 2 that the samples are imbalanced. The imbalanced data due to the incidence of patients treated in Shandong Province Cancer Research Institute.

Feature selection are performed using ReliefFAttributeEval + Ranker and CorrelationAttributeEval + Ranker. The results are shown in Fig. 3 and Fig. 4.

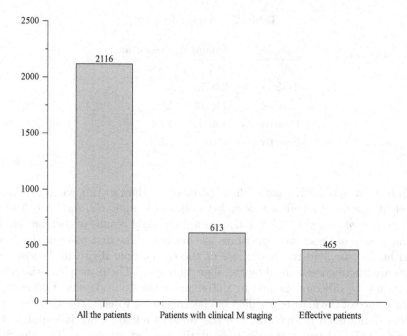

**Fig. 2.** Cleaning dirty data

**Table 1.** Blood cell analysis

| Property | Remark | Range | Unit |
|----------|--------|-------|------|
| pct | Plateletocrit | 1–2 | ml/L |
| plt | Platelet count | 125–350 | $*10^9$/L |
| mpv | Mean platelet volume | 7–11 | fL |
| pdw | Platelet distribution width | 15–17 | fL |
| gran1 | Neutrophil absolute count | 1.8–6.3 | $*10^9$/L |
| ly0 | Lymphoeyte percent | 20–50 | % |
| mo1 | Monocytes absolute count | 0.1–0.6 | $*10^9$/L |
| gran0 | Neutrophil percent | 40–75 | % |
| ly1 | Lymphocyte absolute count | 1.1–3.2 | $*10^9$/L |
| wbc | White blood cell | 3.5–9.5 | $*10^9$/L |
| eo0 | Eosinophils percent | 0.4–8 | % |
| mo0 | Monocytes percent | 3–10 | % |
| mcv | Mean corpuscular volume | 82–100 | fL |
| hct | Hematocrit | 35–45 | % |
| mchc | Mean corpuscular hemoglubin concentration | 316–354 | g/L |
| mch | Mean corpuscular hemoglubin | 27–34 | pg |
| hgb | Hemoglobin | 115–150 | g/L |
| ba1 | Baso absolute count | 0–0.06 | $*10^9$/L |
| eo1 | Eosinophil absolute count | 0.02–0.52 | $*10^9$/L |
| rbc | Red blood cell count | 3.8–5.1 | $*10^9$/L |
| ba0 | Basophils percent | 0–1 | % |
| M | Clinical M stage | 0,1 | null |

**Table 2.** Sample situation

| Samples | Quantity | Proportion |
|---|---|---|
| All | 465 | 100 |
| Training set | 348.75 | 75 |
| Test set | 116.25 | 25 |
| Positive set | 106 | 22.8 |
| Negative set | 359 | 77.2 |

It can be seen that sorting with reliefF is smoother than sorting with correlation. Top eight results of reliefF are *pdw, ly0, ly1, gran0, mpv, rbc, hgb, hct.* And the *mo0, mo1, hgb, eo1, gran0, hct, ba1, pdw* are top eight results of correlation. The sorting results of these two algorithms are different. The first place of the reliefF algorithm is *pdw*, and the first place of the correlation algorithm is *mo0*. The results are mainly determined by the algorithm itself. The principle of the reliefF algorithm is to subtract the feature difference of the same classification and plus the difference of the feature of different classifications. Finally, the feature sorting results are obtained by according to the value of reliefF. Correlation algorithm is based on the relevance of individual attributes and categories. The algorithm believes that effective features should be more relevant to classification, while invalid features are less relevant to classification. The top 8 attributes of the two feature selection algorithms are reserved for the latter experiments.

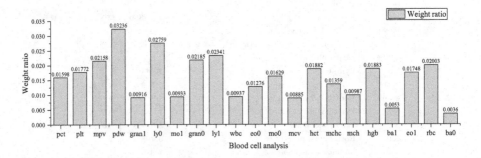

**Fig. 3.** ReliefFAttributeEval + Ranker

The results obtained after the feature selection and the original situation are shown in Table 3. The results received under the original data are slightly worse than those obtained after the feature selection. The results of reliefF feature selection are the best. These results are compared with the original situation, the AUC of the reliefF feature selection increases by 11.24%, and the G-Mean increases by 16.48%. The effectiveness of the feature selection algorithm are proved. These results can only be influenced by feature selection because the same model is used.

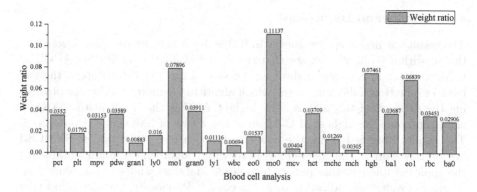

**Fig. 4.** CorrelationAttributeEval + Ranker

**Table 3.** Results of feature selection

| Algorithm | AUC | G-Mean |
| --- | --- | --- |
| Original | 0.4459 | 0.2031 |
| Correlation | 0.4727 | 0.3159 |
| ReliefF | 0.5583 | 0.3679 |

## 4.2   Experiment Design

The experiment is mainly divided into two steps. The first step is to solve the category imbalanced problem. In the second step, we select the most suitable ratio between positive and negative samples to achieve the best effect by increasing the proportion of positive samples.

For the first step, we make the ratio of the positive and negative samples reach 1:1 by using eight oversampling algorithms for the positive samples. Then we use the K-nearest neighbor (k-NN) algorithm (k = 5) to classify the balanced dataset. The eight oversampling algorithms are SMOTE + ENN [19], SMOTE + Tomeklink [20], SMOTE + Borderline1, SMOTE + Borderline2, ADASYN, SMOTE, RandomOverSampler and AWGSENN [21]. Finally, we use the k-NN classification calculates the AUC and G-Mean by using the scikit-learn API. Then, we use the two cost-sensitive learning algorithms of KEEL, namely, C_SVMCS-I and C45CS-I. We can calculated AUC and G-Mean by confusion matrix. The effect obtained after the feature selection is better than that without the feature selection. These results show that the training model is not as many features as possible, and they need to be compared with the experimental results.

In the second step, we adjust the positive sample oversampling ratio. The ratio of positive and negative samples increases from 0.8:1 to 1.5:1. Then we use the scikit-learn API to implement the k-NN classification. We can select the suitable oversampling ratio by comparing the obtained AUC and G-Mean.

## 4.3   Results and Discussion

The results of first step are shown in Table 4 whether at the data level or at the algorithm level. We can see that the model evaluation indicators AUC and G-Mean don't appear the highest at the same time. We can combine the two model evaluation indicators to see which algorithm is more suitable for processing the data set of this article. The highest AUC is the result of the SMOTE algorithm, and the highest of G-Mean is the result of SMOTE + Borderline1. The results obtained using the cost-sensitive learning algorithm C_SVMCS-I are slightly worse. The SMOTE + Borderline1 algorithm is more suitable for the topics studied in this paper. Insufficient data set and low incidence lead to model evaluation indicators in low accuracy. Previously, we also found that distant metastasis with ESCC patients were difficult to predict when examining medical literature. There are problems such as low sensitivity and low specificity.

**Table 4.** Results of oversampling and cost-sensitive learning

| Algorithm | AUC | G-Mean |
|---|---|---|
| SMOTE + ENN | 0.5236 | 0.4893 |
| SMOTE + Tomeklink | 0.5400 | 0.5279 |
| SMOTE + Borderline1 | 0.5397 | 0.5337 |
| SMOTE + Borderline2 | 0.5398 | 0.5318 |
| ADASYN | 0.5287 | 0.5155 |
| SMOTE | 0.5407 | 0.5254 |
| RandomOverSampler | 0.5373 | 0.5215 |
| AWGSENN | 0.4989 | 0.4989 |
| C_SVMCS-I | 0.4718 | 0.4698 |
| C45CS-I | 0.5218 | 0.5127 |

The results of adjusting the proportion of positive samples in Fig. 5 and Fig. 6 are based on the eight oversampling algorithms. We can see from Fig. 5 that all algorithms get the smallest AUC when ratio of positive and negative samples is 1.1:1. The changes in AUC seem to be disordered. However, the G-Mean values of SMOTE + Tomeklink, SMOTE + Borderline1, SMOTE + Borderline2, ADASYN, SMOTE and RandomOverSampler are very stable in Fig. 6. And most of them show good performance of the algorithm when the ratio at 1:1. SMOTE + ENN and AWGSENN are very sensitive to this data set because they drop very quickly after the ratio at 1:1, especially the AWGSENN algorithm. The AWGSENN algorithm is the latest algorithm, it may not be very mature. It can be seen the AUC and G-Mean obtained are ideal from the figure when the ratio of positive and negative samples from 0.9:1 to 1.1:1. In this interval, the AUC and G-Mean are the best.

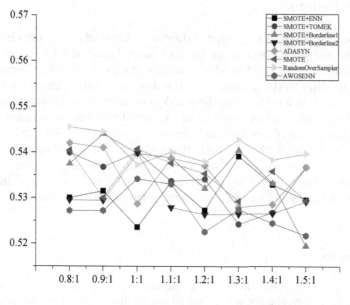

**Fig. 5.** AUC

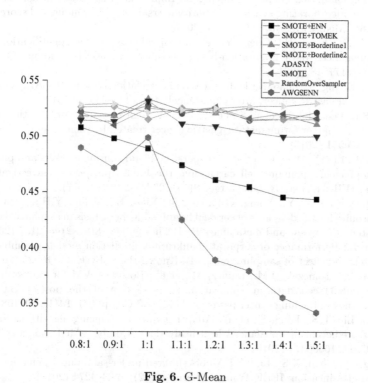

**Fig. 6.** G-Mean

# 5  Conclusions

Using blood cell analysis to predict distant metastasis of cancer cells in patients with esophageal squamous cell carcinoma is an imbalanced learning problem. A standard classifier will result in a large number of misclassifications because a few class samples due to the imbalance of the original dataset. This article mainly focuses on the data level to solve the problem of imbalanced classification. Eight oversampling algorithms are used in this paper. The result is the most suitable that the proportion of positive and negative samples between 0.9:1 and 1.1:1 no matter which algorithm is used. Regardless of the solution, this will help more ESCC patients to predict whether ESCC patients in late stage by blood cell analysis. We should try to use other methods to improve the specificity and sensitivity of the model in order to improve the predictive performance of the model.

# References

1. Chen, M.Q., Xu, B.H., Zhang, Y.Y.: Analysis of prognostic factors for esophageal squamous cell carcinoma with? Distant organ metastasis at initial diagnosis. J. Chin. Med. Assoc. **77**(11), 562–566 (2014)
2. Yan, S., Qian, W., Guan, Y., Zheng, B.: Improving lung cancer prognosis assessment by incorporating synthetic minority oversampling technique and score fusion method. Med. Phys. **43**, 2694–2703 (2017)
3. Peng, L., Zhang, H., Chen, Y., et al.: Imbalanced traffic identification using an imbalanced data gravitation-based classification model. Comput. Commun. **102**(C), 177–189 (2017)
4. Yu, J., Zhang, S., Zhao, B.: Differences and correlation of serum CEA, CA19-9 and CA72-4 in gastric cancer. Mol. Clin. Oncol. **4**(3), 441–449 (2016)
5. Go, S.I., Lee, M.J., Lee, W.S., et al.: D-Dimer can serve as a prognostic and predictive biomarker for metastatic gastric cancer treated by chemotherapy. Medicine **94**(30), e951 (2015)
6. Dan, H., Lin, X., Yan, C., et al.: Preoperative blood-routine markers and prognosis of esophageal squamous cell carcinoma: the Fujian prospective investigation of cancer (FIESTA) study. Oncotarget **8**(14), 23841–23850 (2017)
7. Leng, X.F., Zhu, Y., Wang, G.P., Jin, J., Xian, L., Zhang, Y.H.: Accuracy of ultrasound for the diagnosis of cervical lymph node metastasis in esophageal cancer: a systematic review and meta-analysis. J. Thorac. Dis. **8**(8), 2146–2157 (2016)
8. Zhou, L.: Performance of corporate bankruptcy prediction models on imbalanced dataset: the effect of sampling methods. Knowl.-Based Syst. **41**, 16–25 (2013)
9. Zieba, M., Tomczak, J.M., Lubicz, M., et al.: Boosted SVM for extracting rules from imbalanced data in application to prediction of the post-operative life expectancy in the lung cancer patients. Appl. Soft Comput. J. **14**(1), 99–108 (2014)
10. Li, J., Liu, L.S., Fong, S., et al.: Adaptive swarm balancing algorithms for rare-event prediction in imbalanced healthcare data. Comput. Med. Imaging Graph. S0895611116300374 (2016)
11. Kang, Q., Chen, X.S., Li, S.S.: A noise-filtered under-sampling scheme for imbalanced classification. IEEE Trans. Cybern. **47**(12), 4263–4274 (2017)

12. Richardson, A.M., Lidbury, B.A.: Enhancement of hepatitis virus immunoassay outcome predictions in imbalanced routine pathology data by data balancing and feature selection before the application of support vector machines. BMC Med. Inform. Decis. Mak. **17**(1), 121 (2017)
13. Ichikawa, D.: How can machine-learning methods assist in virtual screening for hyperuricemia? A healthcare machine-learning approach. J. Biomed. Inform. **64**, 20–24 (2016)
14. Santos, M.S., Abreu, P.H., Garca-Laencina, P.J., et al.: A new cluster-based over-sampling method for improving survival prediction of hepatocellular carcinoma patients. J. Biomed. Inform. **58**, 49–59 (2015)
15. Majid, A., Ali, S., Iqbal, M., et al.: Prediction of human breast and colon cancers from imbalanced data using nearest neighbor and support vector machines. Comput. Methods Programs Biomed. **113**(3), 792–808 (2014)
16. Lipitakis, A.D., Lipitakis, E.A.E.C.: On machine learning with imbalanced data and research quality evaluation methodologies. In: International Conference on Computational Science & Computational Intelligence. IEEE (2014)
17. Zhang, X., Hu, B.G.: A new strategy of cost-free learning in the class imbalance problem. IEEE Trans. Knowl. Data Eng. **26**(12), 2872–2885 (2014)
18. Cheng, F., Zhang, J., Wen, C., et al.: Large cost-sensitive margin distribution machine for imbalanced data classification. Neurocomputing **224**, 45–57 (2017)
19. Lee, W., Jun, C.H., Lee, J.S.: Instance categorization by support vector machines to adjust weights in AdaBoost for imbalanced data classification. Inf. Sci. **381**, 92–103 (2017)
20. Chan, T.M., Li, Y., Chiau, C.C., et al.: Imbalanced target prediction with pattern discovery on clinical data repositories. BMC Med. Inform. Decis. Mak. **17**(1), 47 (2017)
21. Pang, Y., Peng, L.Z., Chen, Z.X., Yang, B., et al.: Imbalanced learning based on adaptive weighting and Gaussian function synthesizing with an application on Android malware detection. Inf. Sci. **484**(1), 95–112 (2019)

# A Deep Learning Approach to Radar Extrapolation

Lei Yi[1](✉) ⓘ, Wei Tian[1] ⓘ, Xiang Wang[1,2], and Xiaolong Xu[1]

[1] Nanjing University of Information Science and Technology, Nanjing 210044, China
yilei.sunshine@gmail.com, {tw,wangxiang}@nuist.edu.cn, njuxlxu@gmail.com
[2] Institute for Climate and Application Research, Nanjing, China

**Abstract.** Nowcasting is really important for citizen and industry since it matters to everyone on social and economic activities and is significant for monitoring and forecasting disasters. As the foundation of the task, radar extrapolation is the most difficulty to be solved. However, the accuracy of the extrapolation is in a large bias. With the rapid development of computing power, recent researches show that artificial intelligence is a promising approach, especially deep learning approaches in learning accurate patterns and appear well suited for the task of extrapolation, given an ample account of radar echo maps. Since ConvLSTM has successfully introduced spatiotemporal sequence into the traditional LSTM, in this study, we modified a recurrent neural network (RNN) called bidirectional ConvLSTM (Bi-ConvLSTM) based on the ConvLSTM and used a reversed encoding-forecasting structure. The experiments show that the Bi-ConvLSTM achieved a better performance than the ConvLSTM.

**Keywords:** Radar extrapolation · Bi-ConvLSTM · ConvLSTM

## 1 Introduction

Nowcasting convective precipitation has long been an important problem in the field of weather forecasting [22]. As one of the most significant factors in water resource ecosystem, precipitation plays an important role in meteorological fields, which has a strong impact on human's daily lives as well as business such as agriculture and construction [7,10,11]. The variations in time and quantity of rainfall have the potential impact on the agriculture yield. Prior knowledge of rainfall behavior can help farmers and policy makers to minimize crop damage. Moreover, it plays an important role in disaster warning and relief. Therefore, the task of radar extrapolation, which is the predictive foundation of nowcasting, is in great significance.

At first, the meteorological observatory used cross correlation method to speculate future radar echo maps. This method utilizes the prior two frames maps to deduce the wind field which is a linear extrapolation of rigid body in essence. However, this technique has difficulty in solving the deformation

© Springer Nature Switzerland AG 2020
X. Chen et al. (Eds.): ML4CS 2020, LNCS 12488, pp. 454–463, 2020.
https://doi.org/10.1007/978-3-030-62463-7_42

problem and is unable to learn knowledge from historical data, which is quite helpless in the generation and extinction of the severe convection weather. Specifically, some computer vision techniques, especially optical flow based methods, have proven useful for making accurate extrapolation of radar maps [5,17]. The recent advanced system of the method is the Real-time Optical flow by Variational methods for Echoes of Radar (ROVER) algorithm proposed by the Hong Kong Observatory (HKO) for its Short-range Warning of Intense Rainstorms in Localized System (SWIRLS) [29]. This kind of method is precise in short time while the success of employing it is limited. During the extrapolation, the procedures of optical flow estimation and radar echo extrapolation are separate, which would lead to the accumulative error and make the set of hyper parameters more difficult to determine.

Recently, with the development of computer hardware, deep learning technologies have made a great breakthrough against the last century [12,25,30]. The huge success in deep learning attracts the meteorologist to explore the possibility of the application with such progressive technique in meteorological service field. From the perspective of computer science, the problem of radar extrapolation is essentially a spatiotemporal sequence prediction problem, which uses a sequence of several frame maps as input to deduce the future frame maps with a fixed number (usually greater than 1). However, the model would become more complicated with such learning problem since the sequence of time and space is in high dimension [1]. Moreover, due to the chaotic nature of the atmosphere, it is a great challenge to build an effective forecasting model of radar echo data.

The popular LSTM with encoder-decoder model provides a general framework for sequence-to-sequence learning problems by training temporally concatenated LSTMs since it has been proposed, one for the input sequence and another for the output sequence [2,4,9,16]. Shi et al. creatively extended the idea of FC-LSTM to Convolutional LSTM (ConvLSTM) which has convolutional structures in both the input-to-state and state-to-state transitions. They have proposed a novel end-to-end trainable radar extrapolation model by stacking multiple ConvLSTM layers with a structure of encoding-forecasting structure [18]. Based on the work of ConvLSTM, they later proposed another structure named TrajGRU [19,21]. They claimed that the deficiency of ConvGRU and other ConvRNNs is that the connection structure and weights are fixed for all the locations since the location-invariant filter in the convolution operation. They utilized the thought of optical flow to use a set of continuous optical flows to represent these "indices". They also improved the loss function that gives different value with different weights to reduce the extreme error. Experiments have shown that TrajGRU achieved a better result than the previous ConvLSTM. Patraucean et al. and Villegas et al. introduced optical flow into RNNs, which is lighted by the two-stream CNNs designed for action recognition. However, the optical flow images are difficult to use due to the high additional computational costs they brought in and the reduced prediction efficiency [15,20,24]. Wang et al. insisted that spatial appearances as well as temporal variations are both crucial structures. They proposed a new network called PredRNN with the core unit of Spatiotemporal

LSTM (ST-LSTM) unit, which extracts and memorizes spatial and temporal representations simultaneously [13,28]. However, the problem of gradients propagation would be prominent with the increasingly deep-in-time networks and the blur maps in predictions remain difficult to solve. The second year, Wang et al. continue to present PredRNN++, making their network deeper in time by leveraging a new recurrent structure named Causal LSTM with cascaded dual memories [26]. Meanwhile, they proposed a novel Gradient Highway Unit. The gradient highway units work seamlessly with the causal LSTMs, improving the performance of the model to adaptively capture the short-term and the long-term video dependencies. Besides, Wang et al. recently proposed another novel model called Eidetic 3D LSTM (E3D-LSTM) which integrates the 3D convolution into the RNN networks [27]. These can make the short term memory better than before and improve the long term memory with a self-attention module. Their experiments have shown the great coherence with the movement trend. Moreover, some researchers have been trying to employ generative antagonistic networks (GAN) to improve the spatiotemporal sequence forecasting to solve the blur problem in the latter prediction of the forecasting results [3].

In summary, the novel structures may be great in RGB image prediction while the prediction of each models yield different deficiencies when used in radar extrapolation. The most of all in problems is that: (1) the error of radar extrapolation is still big, especially after one hour, the definition of predictive radar map may reduce sharply; (2) the gradient disappearance problem is disturbing as well; (3) the extreme radar reflectivity factors always lead to the extreme error. In this paper, we solve the problem (1) by proposing a bidirectional structure based on the ConvLSTM with the encoding-decoding frame.

The remainder of this paper is organized as follows: Sect. 2 introduces the dataset of this task; Sect. 3 details the model of our Bi-ConvLSTM; Sect. 4 describes how the experiments we operation and Sect. 5 shows the result of our experiments. Finally, we conclude our work in Sect. 6.

## 2    Data

The data cover the whole year of 2018 come from a meteorological observatory located in the center area of Shijiazhuang in Hebei province. However, different from HKO-7 provided by HKO which consists of constant altitude plan position indicator (CAPPI) images, the data most meteorological observatories such as Shijiazhuang provided are radar reflectivity base data in storage of binary format and achieved every 6 minutes. The base data store 11 elevations scan information according to the polar coordinate system. Therefore, we have to transform these data into the Cartesian coordinate system through the cubic spline interpolation algorithm.

According to the meteorologist, the radar reflectivity factors below 15 are ground echoes, which is useless to the task. Therefore, we set the values below 15 to zero to reduce the noise. The ultimate data of each radar echo map is a $270 \times 340$ matrix of the lowest elevation with the resolution of $0.01° \times 0.01°$.

Then, we split the processed data grouped by date so we get 240 frames for a day. These frames continued to be divided into 6 groups, that is to say, 40 frames for a smaller group and there are 6 such groups for one day. We randomly select 4 groups as training, 1 as test and 1 as validation. In our work, we predicted 10 frames with prior 10 frames and we cropped continuous 20 frames in each group as sample, 10 for input and 10 for label. There are 997 samples in training set, 259 samples in test set and 259 samples in validation set.

# 3   Model

## 3.1   Long Short-Term Memory for Sequence Forecasting

Long Short-Term Memory (LSTM) as a modified RNN structure has been proven stable and effective for the tasks of prediction in various previous researches [6,8, 14,23]. This structure improves the deficiency of the Transitional RNN network which is a breakthrough in the history of RNN models. The main equation are shown in (1):

$$
\begin{aligned}
i_t &= \sigma(W_{xi}x_t + W_{hi}H_{t-1} + W_{ci} \odot C_{t-1} + b_i) \\
f_t &= \sigma(W_{xf}x_t + W_{hf}H_{t-1} + W_{cf} \odot C_{t-1} + b_f) \\
C_t &= f_t \odot C_{t-1} + i_t \odot \tanh(W_{xc}x_t + W_{hc}H_{t-1} + b_c) \\
o_t &= \sigma(W_{xo}x_t + W_{ho}H_{t-1} + W_{co} \odot C_t + b_o) \\
H_t &= o_t \odot \tanh(C_t)
\end{aligned}
\tag{1}
$$

where $\sigma$ is sigmoid activation function and the whole information flow direction inside the memory cell are determined by the input gate $i_t$, forget gate $f_t$, and output gate $o_t$.

## 3.2   Convolutional LSTM

Different from standard LSTMs, the ConvLSTM transforms the 2D image into 3D tensor, overcoming the limitation of traditional LSTMs which only use vector-variate representations as input (see Fig. 1) [18]. In the ConvLSTM, all the inputs $x_1, x_2, \cdots, x_t$, as well as the hidden states $H_1, H_2, \cdots, H_t$ and cell states $C_1, C_2, \cdots, C_t$ are belong to 3D tensors in $R^{C \times H \times W}$ where the first dimension C is either the number of channels or the number of feature maps and the last two dimensions are the height and width of the map size (H rows and W columns). The key equations of ConvLSTM are similar to the LSTM which are shown as follows:

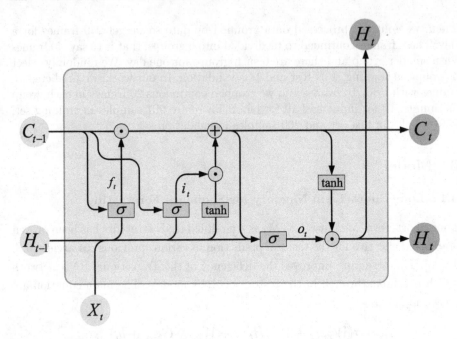

**Fig. 1.** This is the detail of the ConvLSTM. The arrows denote the information flow of the input.

$$g_t = \tanh(W_{xg} * x_t + W_{hg} * H_{t-1} + b_g)$$
$$i_t = \sigma(W_{xi} * x_t + W_{hi} * H_{t-1} + W_{ci} \odot C_{t-1} + b_i)$$
$$f_t = \sigma(W_{xf} * x_t + W_{hf} * H_{t-1} + W_{cf} \odot C_{t-1} + b_f)$$
$$C_t = f_t \odot C_{t-1} + i_t \odot g_t \qquad (2)$$
$$o_t = \sigma(W_{xo} * x_t + W_{ho} * H_{t-1} + W_{co} \odot C_t + b_o)$$
$$H_t = o_t \odot \tanh(C_t)$$

where $\sigma$ is sigmoid activation function, $*$ and $\odot$ respectively denote the convolutional operation and the element-wise multiplication. The whole information flow direction inside the memory cell are determined by the input gate $i_t$, forget gate $f_t$, input-modulation gate $g_t$ and output gate $o_t$. In this way, the vanishing gradient problem will be remitted when trapped in the memory.

### 3.3 Bi-ConvLSTM

As is shown in Fig. 2, we used a bidirectional structure based on the ConvLSTM for such consideration that the traditional method only captures the features from the past to the present for prediction while the features from the present to past may make sense in the prediction task. This structure may be more reasonable and accord with the actual situation. Here, the equations are same as the ConvLSTM while the difference is that the computation at each time sequence should be operated twice for one forward and the other backward, then the output should be operated a $1 \times 1$ convolution to reduce the dimension to the channel of hidden states on the concatenation of the hidden states from both forward and backward.

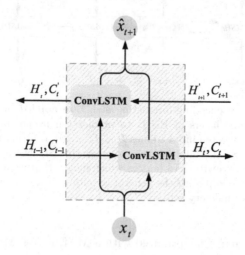

**Fig. 2.** This is the sketch of the Bi-ConvLSTM. The green round rectangles are the ConvLSTM mentioned above. The horizontal flows denote how the cell states and hidden states transmit. The output $\hat{x}_{t+1}$ is determined by $x_t$ based on both ordinal and reversed spatiotemporal sequences. (Color figure online)

### 3.4 Encoding-Forecasting Structure

The radar echo maps are similar to a spatiotemporal sequence $< S_1, S_2, \cdots >$. Our model can generate the most likely $k$-step forecasts $< \hat{I}_{t+1}, \hat{I}_{t+2}, \cdots, \hat{I}_{t+k} >$, according to the previous $J$ existing sequence including the current one: $< I_{t-J+1}, I_{t-j+2}, \cdots, I_t >$. Figure 3 illustrates the encoding-forecasting structure we used for our spatiotemporal sequence prediction problem with 10 observations to predict next 10 radar echo maps. We reversed the order of the forecasting network so that we could easily plug in additional RNN layers on the top of the structure. Between the layers, we add attention mechanism to enhance the ability of feature extraction.

## 4    Experiments

Our model is demonstrated to achieve a better performance on the radar extrapolation task which uses 10 frames to forecast 10 frames with same layers and

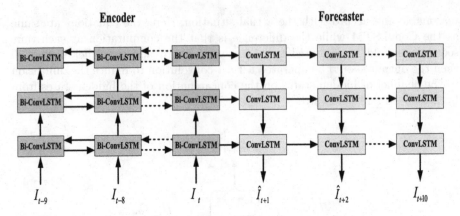

**Fig. 3.** Example of the encoding-forecasting structure applied in the paper. In the graph, we employ three layers of RNNs to predict ten frames of radar echo maps $\hat{I}_{t+1}, \hat{I}_{t+2}, \cdots, \hat{I}_{t+10}$ given ten input frames $I_{t-9}, I_{t-8}, \cdots, I_t$. The core cell in encoding structure is Bi-ConvLSTM while the one in Forecasting Structure is ConvLSTM. We initialized the parameters with zero tensors which means that we have no sense at the beginning. This is in accord with the general cognition of the public. The RNNs can be replaced with other memory cells.

hidden states. Our model is optimized with a MAE loss + MSE loss which are shown in (3).

$$loss = MAE + MSE$$

$$MAE = \frac{1}{n} \sum_{i=1}^{n} |y_i - \hat{y}_i|$$

$$MSE = \frac{1}{n} \sum_{i=1}^{n} (y_i - \hat{y}_i)^2$$

(3)

where $y_i$ is the true data, the $\hat{y}_i$ is the predicted value and $n$ represents the count of the dataset. All models are trained using Adam optimizer with a starting learning rate of 0.01 with some learning rate decay measures. The training process is stopped after 10000 iterations. The batch size of each iteration is set to 2 except that the last batch of one epoch may be less than it. All experiments are implemented in PyTorch and conducted on NVIDIA GeForce GTX 1660.

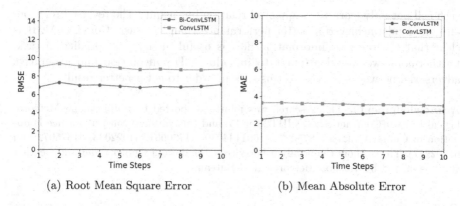

(a) Root Mean Square Error                    (b) Mean Absolute Error

**Fig. 4.** The root mean square error and mean absolute error of each forecasting time step

## 5    Results

We elevate the performance of the network using MAE, MSE and root mean square error (RMSE) and the equation of RMSE is shown as the follow:

$$RMSE = \sqrt{\frac{1}{n} \sum_{i=1}^{n} (y_i - \hat{y}_i)^2} \qquad (4)$$

where the parameters are the same meaning as the former formulation. The indexs of these two models are shown in Table 1. Specifically, the RMSE and

**Table 1.** errors performed by different models

| Model | MSE | RMSE | MAE |
|---|---|---|---|
| ConvLSTM | 79.404 | 8.909 | 3.447 |
| Bi-ConvLSTM | 48.463 | 6.960 | 2.672 |

MAE of each item step is shown in Fig. 4. It is shown that our proposed Bi-ConvLSTM model has achieved a greater performance than the basic ConvL-STM when used same hidden layers.

## 6    Conclusions

In this paper, we have proposed a novel recurrent network based on the Bi-ConvLSTM for radar extrapolation task. Bidirectional informance flow make each cell not only learn the sequential relation but acquire the reversed relation. and our model achieved a better performance than the basic ConvLSTM. It is clear that the reversed information flow is useful in sequence prediction task. Furthermore, we would attempt to introduce 3D convolution and generative adversarial networks (GAN) to this task in order to get a better result.

**Acknowledgments.** This research was jointly supported by Shijiazhuang Meteorological Bureau(Contract SJZQXJHT2019-45) and the National Natural Science Foundation of China (Grants 41875027, 41661144039, 41730961, 41922033, 41675072 and 61702277). The authors thankfully acknowledge the data of Dppler radar and weather stations offered by Taizhou Meteorological Bureau.

# References

1. Bhattacharjee, P., Das, S.: Temporal coherency based criteria for predicting video frames using deep multi-stage generative adversarial networks. In: Advances in Neural Information Processing Systems, pp. 4268–4277 (2017)
2. Cho, K., et al.: Learning phrase representations using RNN encoder-decoder for statistical machine translation. In: Proceedings of the 2014 Conference on Empirical Methods in Natural Language Processing (EMNLP), pp. 1724–1734 (2014)
3. Denton, E., Chintala, S., Szlam, A., Fergus, R.: Deep generative image models using a Laplacian pyramid of adversarial networks. In: NIPS 2015 Proceedings of the 28th International Conference on Neural Information Processing Systems - Volume 1, pp. 1486–1494 (2015)
4. Donahue, J., et al.: Long-term recurrent convolutional networks for visual recognition and description. IEEE Trans. Pattern Anal. Mach. Intell. **39**(4), 677–691 (2017)
5. Germann, U., Zawadzki, I.: Scale-dependence of the predictability of precipitation from continental radar images. Part I: description of the methodology. Mon. Weather Rev. **130**(12), 2859–2873 (2002)
6. Graves, A.: Generating sequences with recurrent neural networks. arXiv preprint arXiv:1308.0850 (2013)
7. He, J.S., Han, M., Ji, S., Du, T., Li, Z.: Spreading social influence with both positive and negative opinions in online networks. Big Data Min. Anal. **2**(2), 100–117 (2019)
8. Hochreiter, S., Schmidhuber, J.: Long short-term memory. Neural Comput. **9**(8), 1735–1780 (1997)
9. Karpathy, A., Fei-Fei, L.: Deep visual-semantic alignments for generating image descriptions. IEEE Trans. Pattern Anal. Mach. Intell. **39**(4), 664–676 (2017)
10. Kong, C., Luo, G., Tian, L., Cao, X.: Disseminating authorized content via data analysis in opportunistic social networks. Big Data Min. Anal. **2**(1), 12–24 (2018)
11. Kumar, S., Singh, M.: Big data analytics for healthcare industry: impact, applications, and tools. Big Data Min. Anal. **2**(1), 48–57 (2018)
12. Lotter, W., Kreiman, G., Cox, D.: Deep predictive coding networks for video prediction and unsupervised learning. In: ICLR 2017: International Conference on Learning Representations 2017 (2017)
13. Mathieu, M., Couprie, C., LeCun, Y.: Deep multi-scale video prediction beyond mean square error. In: ICLR 2016: International Conference on Learning Representations 2016 (2016)

14. Pascanu, R., Mikolov, T., Bengio, Y.: On the difficulty of training recurrent neural networks. In: Proceedings of the 30th International Conference on Machine Learning, pp. 1310–1318 (2013)
15. Patraucean, V., Handa, A., Cipolla, R.: Spatio-temporal video autoencoder with differentiable memory. arXiv preprint arXiv:1511.06309 (2015)
16. Ranzato, M., Szlam, A., Bruna, J., Mathieu, M., Collobert, R., Chopra, S.: Video (language) modeling: a baseline for generative models of natural videos. arXiv preprint arXiv:1412.6604 (2014)
17. Sakaino, H.: Spatio-temporal image pattern prediction method based on a physical model with time-varying optical flow. IEEE Trans. Geosci. Remote Sens. **51**(5), 3023–3036 (2013)
18. Shi, X., Chen, Z., Wang, H., Yeung, D.Y., Wong, W.K., Woo, W.C.: Convolutional LSTM network: a machine learning approach for precipitation nowcasting. In: NIPS 2015 Proceedings of the 28th International Conference on Neural Information Processing Systems - Volume 1, pp. 802–810 (2015)
19. Shi, X., et al.: Deep learning for precipitation nowcasting: a benchmark and a new model. In: Advances in Neural Information Processing Systems, pp. 5617–5627 (2017)
20. Simonyan, K., Zisserman, A.: Two-stream convolutional networks for action recognition in videos. Adv. Neural Inf. Process. Syst. **27**, 568–576 (2014)
21. Srivastava, N., Mansimov, E., Salakhudinov, R.: Unsupervised learning of video representations using LSTMs. In: Proceedings of the 32nd International Conference on Machine Learning, pp. 843–852 (2015)
22. Sun, J., et al.: Use of NWP for nowcasting convective precipitation: recent progress and challenges. Bull. Am. Meteorol. Soc. **95**(3), 409–426 (2014)
23. Sutskever, I., Vinyals, O., Le, Q.V.: Sequence to sequence learning with neural networks. In: Advances in Neural Information Processing Systems, 27, pp. 3104–3112 (2014)
24. Villegas, R., Yang, J., Hong, S., Lin, X., Lee, H.: Decomposing motion and content for natural video sequence prediction. In: ICLR 2017: International Conference on Learning Representations 2017 (2017)
25. Wang, S., Zhou, A., Yang, M., Sun, L., Hsu, C.H., et al.: Service composition in cyber-physical-social systems. IEEE Trans. Emerg. Top. Comput. **8**, 82–91 (2017)
26. Wang, Y., Gao, Z., Long, M., Wang, J., Yu, P.: PredRNN++: towards a resolution of the deep-in-time dilemma in spatiotemporal predictive learning. In: ICML 2018: Thirty-Fifth International Conference on Machine Learning, pp. 5110–5119 (2018)
27. Wang, Y., Jiang, L., Yang, M.H., Li, L.J., Long, M., Fei-Fei, L.: Eidetic 3D LSTM: a model for video prediction and beyond. In: ICLR 2019: 7th International Conference on Learning Representations (2019)
28. Wang, Y., Long, M., Wang, J., Gao, Z., Yu, P.S.: PredRNN: recurrent neural networks for predictive learning using spatiotemporal LSTMs. In: NIPS 2017 Proceedings of the 31st International Conference on Neural Information Processing Systems, pp. 879–888 (2017)
29. Woo, W.C.: Application of optical flow techniques to rainfall nowcasting. In: 27th Conference on Severe Local Storms (2014)
30. Zhou, J., et al.: Security-critical energy-aware task scheduling for heterogeneous real-time MPSoCs in IoT. IEEE Trans. Serv. Comput. **13**, 745–758 (2019)

# Peripheral Blood Leukocyte Image Segmentation Based on Deep Convolutional Neural Network

Yuanyuan Chen[1], Shenghua Teng[1(✉)], and Zuoyong Li[2(✉)]

[1] College of Electronics and Information Engineering, Shandong University of Science and Technology, Qingdao 266590, China
Chenyy242@163.com, tengshenghua@163.com
[2] Fujian Provincial Key Laboratory of Information Processing and Intelligent Control, College of Computer and Control Engineering, Minjiang University, Fuzhou 350121, China
fzulzytdq@126.com

**Abstract.** Counting and analysis of leukocyte on peripheral blood smears have great significance to assist doctors' pathological diagnosis. Thus, accurate and automatic segmentation of leukocyte from blood smear images is a fundamental task. To achieve this task, we proposed an end-to-end peripheral blood leukocyte segmentation method based on deep convolutional neural networks, where the encoder-decoder network is its basic structure. Specifically, the proposed method first utilizes the feature encoder to extract multi-scale leukocyte features. Then, the context-aware feature decoder is introduced to reconstruct the segmentation mask of leukocyte. Finally, each layer generates a prediction map in the decoder module, and all these prediction maps are averaged to be the final prediction map. To demonstrate the effectiveness of the proposed method, we performed qualitative and quantitative comparisons with several existing methods on two publicly available datasets and a private dataset. Experimental results validate that the proposed method is superior for peripheral blood leukocyte image segmentation in terms of accuracy and robustness.

**Keywords:** Leukocyte · Image segmentation · Deep neural network · Feature extraction

## 1 Introduction

Leukocyte (i.e., white blood cell), the basic component of peripheral blood, is usually used as an indicator for the diagnosis of certain blood diseases and inflammations such as lymphoma, leukemia, and dermatomyositis. When suffering from these diseases, the number and appearance of human peripheral blood leukocytes will change. In clinical practice, the diagnosis of those diseases usually is performed by medical professionals manually by visual examination of blood smears under the bright field microscopy. However, the manual diagnosis is a time-consuming and tedious task, and corresponding results are not accurate enough.

© Springer Nature Switzerland AG 2020
X. Chen et al. (Eds.): ML4CS 2020, LNCS 12488, pp. 464–473, 2020.
https://doi.org/10.1007/978-3-030-62463-7_43

The purpose of leukocyte segmentation is to extract leukocytes from a blood smear image, which consists of erythrocytes (i.e., red blood cells), leukocytes, platelets, and background. However, leukocyte image segmentation is still a challenging task due to some difficulties caused by cytoplasm's low contrast with image background, cell adhesion, and disturbance of staining impurities.

Recent years have witnessed a rapidly increasing interest in leukocyte segmentation tools based on computer-assisted methods, which are more efficient than manual analysis [1, 2]. Leukocyte image segmentation methods are roughly divided into two categories, i.e., traditional image processing-based methods and machine learning-based methods.

The traditional image processing-based methods include thresholding-based method [3], contour-based methods [4, 5], and saliency-based methods [6, 7]. The thresholding-based method [3] obtains the global threshold based on the Otsu (Maximum Between-Class Variance) method to achieve the segmentation of leukocyte nucleus. Zhou et al. [4] removed the background by combining color components and thresholds, then implemented contour detection based on Canny edge detection to complete the segmentation of leukocytes. Cao et al. [5] used a linear combination of color components to segment leukocytes by minimizing the fuzzy divergence. Visual attention mechanisms [6, 7] are introduced to solve the problem of leukocyte segmentation. The above mentioned methods are fast and intuitive, but when the background of the leukocyte image is more complicated, such as the presence of cell adhesion or a large number of staining impurities, the segmentation results of these methods are not ideal. Some methods based on machine learning [8–10] have optimized this situation. For example, Zheng et al. [8] used EM (expectation maximization) clustering and SVM (support vector machine) to achieve the segmentation of leukocyte, and it is more adaptable to changes in image lighting and color. Saidi et al. [9] used the random forest method based on region-growing to classify the pixels of interest for extracting leukocyte nucleus and cytoplasm. Recently, a self-supervised learning method [10] performs K-means clustering based unsupervised initial segmentation and SVM supervised segmentation refinement to achieve leukocyte segmentation. Although these methods work well on specific datasets, they are usually based on certain prior knowledge and may not be sufficient for all situations.

Recently, deep learning-based methods have brought higher performance in computer vision, such as semantic segmentation [11–15]. These successes have motivated our research of deep learning for leukocyte segmentation from blood smear images. In this paper, a novel end-to-end deep convolutional neural network is proposed, which can realize automatic leukocyte segmentation from complex backgrounds.

The remainder of this paper is organized as follows. Section 2 introduces the details of the proposed method. Experimental results are discussed in Sect. 3. The conclusion is drawn in Sect. 4.

## 2   The Proposed Method

Our research goal is to automatically extract leukocytes from peripheral blood images. Motivated by successful applications of encoder-decoder networks in semantic segmentation, we proposed a novel leukocyte segmentation network on the basis of U-Net [15]. Specifically, multi-scale leukocyte features are first extracted by the encoder module.

Then, the spatial information is reconstructed and fused gradually by the decoder module and the skip connections. Finally, each layer of the decoder module generates a prediction map, and all prediction maps are averaged as the final prediction map. The architecture of the proposed method is illustrated in Fig. 1.

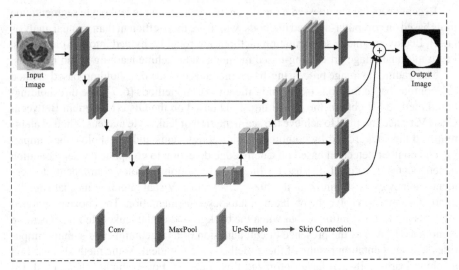

**Fig. 1.** Network architecture of the proposed method.

## 2.1    Context-Aware Feature Encoder

Given the outstanding performance of U-Net in terms of medical image segmentation, an improved encoder-decoder structure is proposed. The function of the proposed encoder is to realize feature extraction, which contains four layers, and each layer uses convolution and pooling operations. As the spatial dimension of feature maps gradually decreases, it is easier for deep encoders to capture longer range information. However, with the increase of the network's depth, the problem of gradient vanishing may be triggered. Motivated by ResNet [16], this problem can be solved well by fitting a residual map. As shown in Fig. 2, unlike traditional convolutional layers, the ResNet block is connected by a shortcut that can directly connect the input feature to the output.

In traditional directly connected convolutional layers (Fig. 2(a)), the mapping between input and output is defined as:

$$x_{i+1} = F(x_i), \tag{1}$$

where the variables $x_i$ and $x_{i+1}$ denote the input and output of the $i$-th layer, respectively. $F(\cdot)$ is a set of transformations consisting of convolutions, GN (group normalization) [17], and PReLU (Parametric Rectified Linear Unit) [18].

Unlike traditional directly connected convolutional neural networks, ResNet block (Fig. 2(b)) connects the input directly to the subsequent layer through a shortcut. This way protects the completeness of the information, and simplifies the learning goals and difficulty. The mapping between input and output is defined as:

$$x_{i+1} = PReLU(F(x_i) + I(x_i)), \qquad (2)$$

where the variables $x_i$ and $x_{i+1}$ denote the input and output of the $i$-th residual block, respectively. $I(\cdot)$ indicates identity function.

Considering the significant contribution of ResNet and its ideas to the research of convolutional neural networks, a modified ResNet architecture is utilized in the proposed encoder module. In the improved ResNet block, weight layers perform $3 \times 3$ convolution and $2 \times 2$ pooling operations, respectively.

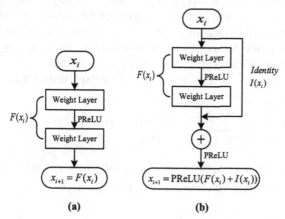

**Fig. 2.** Building blocks of difference networks: (a) traditional CNN, (b) ResNet [16].

## 2.2 Context-Aware Feature Decoder

After capturing the contextual information at multiple scales, the dimensions of these feature maps are reconstructed by using different resolutions. Four times of down-sampling is performed during encoding, and the decoder symmetrically performed convolution and up-sampling operations four times to reconstruct the advanced semantic feature maps with the resolution of the original images. In order to avoid a large amount of information loss, skip connections also are employed for information fusion. The skip connections connect the corresponding feature maps of the encoder to the feature maps of the decoder. This ensures that the finally reconstructed feature map merges more low-level features, and fuses the feature maps of different scales.

The novel structure is introduced, which generates corresponding local prediction maps for every decoder layer instead of a traditional decoder structure. Subsequently, all decoder layers' prediction maps are merged into a final prediction map.

The output loss of each layer can be propagated back to earlier layers in the decoder, which can also alleviate the gradient vanishing problem to some degree. The objective function of the merging process is calculated as:

$$\mathcal{L}_s = \sum_{m=1}^{M} \alpha_m L_s^{(m)}, \tag{3}$$

where $L_s^{(m)}$ indicates the loss function used in this paper. The variable $m$ is the number of encoder layers, and $\alpha_m$ is the fusion weight of the $m$-th encoder layer. We empirically set $\alpha_m = 0.25$ in our experiments.

### 2.3 Loss Function

Our research aims to separate leukocytes from complex backgrounds, such as staining impurities and erythrocytes. The idea of solving this problem is to convert the image segmentation into a binary classification problem. By performing pixel-level classification, the image is divided into leukocytes and non-leukocytes for achieving pixel-to-pixel mapping, and the task of leukocytes segmentation is completed.

A binary cross-entropy loss function is used in our experiments, which is defined as:

$$L_s = -\frac{1}{N} \sum_{i=1}^{N} \left( Y_{ij} \log \hat{Y}_{ij} + \left(1 - Y_{ij}\right) \log\left(1 - \hat{Y}_{ij}\right) \right), \tag{4}$$

where $Y_{ij}\left(Y_{ij} \in \{0, 1\}\right)$ and $\hat{Y}_{ij}\left(0 \le \hat{Y}_{ij} \le 1\right)$ are the binary ground truth label and the predicted value of the pixel with coordinate $(i, j)$, respectively.

## 3 Experimental Results

Comparative experiments on three datasets are used to validate the effectiveness of the proposed method in this section. We first introduce the datasets used in this paper and the evaluation criteria for the results of leukocyte segmentation. We then describe the data enhancement technology used to solve the sample shortage. Finally, we make qualitative and quantitative comparisons of the experimental results.

### 3.1 Dataset and Evaluation Metrics

Three datasets are adopted in this paper, i.e., Dataset1 [10], Dataset2 [7], and Dataset3. Dataset1 and Dataset2 are public datasets on the internet, where Dataset1 consists of 100 leukocyte images with sizes of 300 × 300 under standard staining conditions, and Dataset2 contains 300 leukocyte images with sizes of 120 × 120 under rapid staining conditions. With the help of the People's Hospital of Fujian Province, we constructed the Dataset3 consisting of 60 leukocyte images with sizes of 260 × 260 under standard staining conditions, and let the experts label their ideal segmentation results.

To evaluate performance, Precision [19], ME (Misclassification Error) [20], FPR (False Positive Rate) [21], FNR (False Negative Rate) [21], and Dice (Dice coefficient) [22] are used in this paper. These metrics can be defined as:

$$\text{Precision} = \frac{|G_l \cap P_l|}{|G_l|}, \tag{5}$$

$$\text{ME} = 1 - \frac{|G_n \cap P_n| + |G_l \cap P_l|}{|G_l| + |G_n|}, \tag{6}$$

$$\text{FPR} = \frac{|G_n \cap P_l|}{|G_n|}, \tag{7}$$

$$\text{FNR} = \frac{|G_l \cap P_n|}{|G_l|}, \tag{8}$$

$$\text{Dice} = \frac{2|G_l \cap P_l|}{|G_l| + |G_l|}, \tag{9}$$

where $G_l$ and $G_n$ are leukocyte and non-leukocyte regions of the ground truth, respectively; $P_l$ and $P_n$ are leukocyte and non-leukocyte regions of predicted results. Lower values of ME, FPR, and FNR mean better segmentation performance, while higher values of Precision and Dice means better segmentation performance.

## 3.2 Implementation Details

At present, there are fewer leukocyte images and less labeled images published on the internet. However, the segmentation method based on deep convolutional neural network requires a large number of training samples. In order to overcome this problem, we used data augmentation technology to enlarge the training samples. Operations such as horizontal flip, vertical flip, zoom, and rotation are used in this experiment to obtain sufficient training data.

The proposed method is implemented with Python based on Pytorch. Each dataset is randomly allocated to the training set and the testing set at a ratio of 7:3. During training, we employ adaptive moment estimation (Adam) for optimizing the deep convolution model.

## 3.3 Experimental Results and Analysis

To demonstrate the performance of the proposed method, it is qualitatively and quantitatively compared with three existing state-of-the-art methods including Zheng [8], Zhou [4], and U-Net [15] on three image datasets, i.e., Dataset1 [10], Dataset2 [7], and Dataset3.

**Qualitative Results.** Figures 3, 4 and 5 show visual leukocyte image segmentation results of applying different methods on the three image datasets, where columns from left to right are the results of Zheng [8], Zhou [4], U-Net [15], and the proposed method, respectively. For the Zheng and Zhou methods, in the results of peripheral blood leukocyte segmentation, due to the interference of complex backgrounds such as a large number of staining impurities and erythrocytes, it is easy to cause over-segmentation and reduce the accuracy of the segmentation results. Deep learning methods (U-Net and the proposed method) alleviate the misclassification of erythrocytes into leukocytes, but staining impurities still interfere with U-Net's prediction results. In contrast, the proposed method can segment leukocyte regions more accurately.

**Quantitative Results.** As can be seen from Table 1, the performances of several methods in terms of Precision [19], ME [20], FPR [21], FNR [21], and Dice [22] values are compared. The proposed method performs very well on Precision, ME, FPR, and Dice on Dataset1 [10], Dataset2 [7], and Dataset3, which is clearly superior to all other three methods. Of course, the proposed method obtains higher FNR values on Dataset1 and Dataset2 than Zheng's method, and obtains higher FNR values on Dataset3 than U-Net. The reason is that Zheng's method and U-Net usually suffer from more serious over-segmentation corresponding to lower FNR values. However, the proposed method generates segmentation results similar to ground truths, showing slighter over-segmentation with higher FNR values. In addition, it is not difficult to find that Zheng's method performs poorly on Dataset1 and Dataset3, and Zhou's method performs significantly lower on Dataset1 than in the other two datasets. In summary, compared with the other three methods, the proposed method greatly improves the segmentation performance of peripheral blood leukocyte, and has an excellent performance in terms of accuracy and stability.

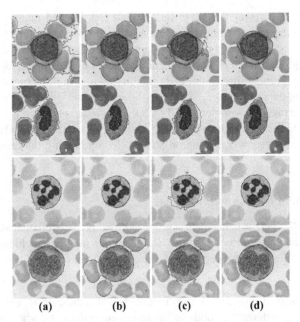

**Fig. 3.** Visual leukocyte image segmentation results on Dataset1: (a) Zheng [8], (b) Zhou [4], (c) U-Net [15], (d) Proposed method. (The blue line indicates the prediction result, and the red line indicates the ground truth) (Color figure online)

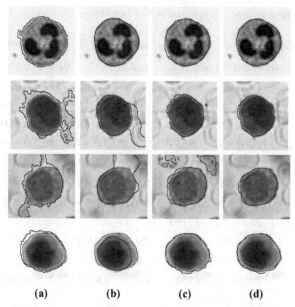

**Fig. 4.** Visual leukocyte image segmentation results on Dataset2: (a) Zheng [8], (b) Zhou [4], (c) U-Net [15], (d) Proposed method. (The blue line indicates the prediction result, and the red line indicates the ground truth) (Color figure online)

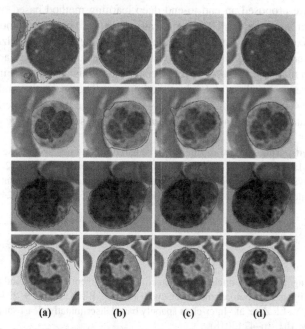

**Fig. 5.** Visual leukocyte image segmentation results on Dataset3: (a) Zheng [8], (b) Zhou [4], (c) U-Net [15], (d) Proposed method. (The blue line indicates the prediction result, and the red line indicates the ground truth) (Color figure online)

**Table 1.** Segmentation accuracy comparison of four methods on three leukocyte image datasets.

| Dataset | Method | Precision | ME | FPR | FNR | Dice |
|---|---|---|---|---|---|---|
| Dataset1 | Zheng [8] | 0.5361 | 0.2536 | 0.3261 | **0.0042** | 0.6693 |
| | Zhou [4] | 0.7394 | 0.1325 | 0.1365 | 0.1035 | 0.7605 |
| | U-Net [15] | 0.9276 | 0.0232 | 0.0215 | 0.0248 | 0.9498 |
| | Proposed | **0.9420** | **0.0117** | **0.0180** | 0.0163 | **0.9618** |
| Dataset2 | Zheng [8] | 0.8653 | 0.0522 | 0.0789 | **0.0003** | 0.9255 |
| | Zhou [4] | 0.8871 | 0.0632 | 0.0695 | 0.0542 | 0.9034 |
| | U-Net [15] | 0.9743 | 0.0159 | 0.0118 | 0.0270 | 0.9732 |
| | Proposed | **0.9803** | **0.0158** | **0.0096** | 0.0323 | **0.9737** |
| Dataset3 | Zheng [8] | 0.7150 | 0.1947 | 0.3072 | 0.0403 | 0.7969 |
| | Zhou [4] | 0.9077 | 0.0516 | 0.0742 | 0.0227 | 0.9389 |
| | U-Net [15] | 0.9294 | 0.0326 | 0.0465 | **0.0121** | 0.9571 |
| | Proposed | **0.9541** | **0.0276** | **0.0311** | 0.0227 | **0.9651** |

## 4  Conclusion

In this paper, we proposed an end-to-end deep learning method based on a modified encoder-decoder structure, which automatically segments peripheral blood leukocytes in a pixel-to-pixel manner. Specifically, the proposed method first utilizes a residual structure in the encoder module to better extract image features. Then, an accompanying local prediction map in each layer of the decoder module is generated. Finally, all prediction maps are averaged as the final prediction map. The qualitative and quantitative segmentation results on three image datasets show that the proposed method achieves more accurate and stable segmentation performance than several existing methods.

**Acknowledgments.** This work was supported in part by National Natural Science Foundation of China (61972187 and 61772254), Fujian Provincial Leading Project (2017H0030 and 2019H0025); Government GuidingRegional Science and Technology Development (2019L3009), and Natural Science Foundation of Fujian Province (2019J01756 and 2020J02024).

## References

1. Xing, F., Yang, L.: Robust nucleus/cell detection and segmentation in digital pathology and microscopy images: a comprehensive review. IEEE Rev. Biomed. Eng. **9**, 234–263 (2016)
2. Wen, J., Xu, Y., Li, Z., et al.: Inter-class sparsity based discriminative least square regression. Neural Netw. **102**, 36–47 (2018)
3. Madhloom, H.T., Kareem, S.A., et al.: An automated white blood cell nucleus localization and segmentation using image arithmetic and automatic threshold. J. Appl. Sci. **10**(11), 959–966 (2010)

4. Zhou, X., Li, Z., Feng, T., et al.: Leukocyte image segmentation based on adaptive histogram thresholding and contour detection. Curr. Bioinform. **15**(3), 187–195 (2020)
5. Cao, H., Liu, H., Song, E.: A novel algorithm for segmentation of leukocytes in peripheral blood. Biomed. Signal Process. Control **45**, 10–21 (2018)
6. Pan, C., Park, D.S., Yang, Y., Yoo, H.M.: Leukocyte image segmentation by visual attention and extreme learning machine. Neural Comput. Appl. **21**(6), 1217–1227 (2012)
7. Zheng, X., Wang, Y., Wang, G., Chen, Z.: A novel algorithm based on visual saliency attention for localization and segmentation in rapidly-stained leukocyte images. Micron **56**, 17–28 (2014)
8. Zheng, X., Wang, Y., Wang, G.: White blood cell segmentation using expectation-maximization and automatic support vector machine learning. J. Data Acquis. Process. **28**(5), 614–619 (2013)
9. Saidi, M., El Amine Bechar, M., et al.: Application of pixel selection in pixel-based classification for automatic white blood cell segmentation. In: Proceedings of the Mediterranean Conference on Pattern Recognition and Artificial Intelligence, pp. 31–38 (2016)
10. Zheng, X., Wang, Y., Wang, G., Liu, J.: Fast and robust segmentation of white blood cell images by self-supervised learning. Micron **107**, 55–71 (2018)
11. Everingham, M., Eslami, S.A., et al.: The pascal visual object classes challenge: a retrospective. Int. J. Comput. Vision **111**(1), 98–136 (2015)
12. Mottaghi, R., Chen, X., et al.: The role of context for object detection and semantic segmentation in the wild. In: Proceedings of the IEEE Conference on Computer Vision and Pattern Recognition, pp. 891–898 (2014)
13. Cordts, M., Omran, M., et al.: The cityscapes dataset for semantic urban scene understanding. In: Proceedings of the IEEE Conference on Computer Vision and Pattern Recognition, pp. 3213–3223 (2016)
14. Li, H., Zhao, X., Su, A., Zhang, H., et al.: Color space transformation and multi-class weighted loss for adhesive white blood cell segmentation. IEEE Access **8**, 24808–24818 (2020)
15. Ronneberger, O., Fischer, P., Brox, T.: U-Net: convolutional networks for biomedical image segmentation. In: Navab, N., Hornegger, J., Wells, William M., Frangi, Alejandro F. (eds.) MICCAI 2015. LNCS, vol. 9351, pp. 234–241. Springer, Cham (2015). https://doi.org/10.1007/978-3-319-24574-4_28
16. He, K., Zhang, X., Ren, S., Sun, J.: Deep residual learning for image recognition. In: Proceedings of the IEEE Conference on Computer Vision and Pattern Recognition, pp. 770–778 (2016)
17. Wu, Y., He, K.: Group normalization. In: Ferrari, V., Hebert, M., Sminchisescu, C., Weiss, Y. (eds.) ECCV 2018. LNCS, vol. 11217, pp. 3–19. Springer, Cham (2018). https://doi.org/10.1007/978-3-030-01261-8_1
18. He, K., Zhang, X., Ren, S., Sun, J.: Delving deep into rectifiers: surpassing human-level performance on ImageNet classification. In: Proceedings of the IEEE International Conference on Computer Vision, pp. 1026–1034 (2015)
19. Pont-Tuset, J., Marques, F.: Measures and meta-measures for the supervised evaluation of image segmentation. In: Proceedings of the IEEE Conference on Computer Vision and Pattern Recognition, pp. 2131–2138 (2013)
20. Yasnoff, W.A., Mui, J.K., Bacus, J.W.: Error measures for scene segmentation. Pattern Recogn. **9**(4), 217–231 (1977)
21. Fawcett, T.: An introduction to ROC analysis. Pattern Recogn. Lett. **27**(8), 861–874 (2006)
22. Singh, V.K., et al.: Conditional generative adversarial and convolutional networks for X-ray breast mass segmentation and shape classification. In: Frangi, A.F., Schnabel, J.A., Davatzikos, C., Alberola-López, C., Fichtinger, G. (eds.) MICCAI 2018. LNCS, vol. 11071, pp. 833–840. Springer, Cham (2018). https://doi.org/10.1007/978-3-030-00934-2_92

# Machine Learning Based Tongue Image Recognition for Diabetes Diagnosis

Xiaohui Lin[1], Zhaochai Yu[1(✉)], Zuoyong Li[1(✉)], and Weina Liu[2]

[1] Fujian Provincial Key Laboratory of Information Processing and Intelligent Control, College of Computer and Control Engineering, Minjiang University, Fuzhou 350121, China
31718657@qq.com, fzulzytdq@126.com
[2] Department of Computer Engineering, Fujian Polytechnic of Information Technology, Fuzhou 350003, China

**Abstract.** Tongue diagnosis of Traditional Chinese Medicine (TCM) is of great significance in the diagnosis of diabetes. To reduce the subjectivity of doctors in clinical diagnosis, this paper proposes a method for diabetic tongue image recognition based on machine learning. Specifically, the proposed method first transforms an image from RGB (red, green and blue) color space to HSV (hue, saturation, and value) color space, and the hue component is transformed into an image with higher difference degree. Secondly, image thresholding and morphological operation are performed on the image to obtain the initial tongue body region. Thirdly, the proposed method performs image thresholding and morphological operations on the value component to achieve image binarization, and retains the largest connected region as the final tongue body region. Fourthly, iterative image thresholding is used to further segment the tongue body region into tongue coating and tongue nature. Finally, the proposed method extracts tongue color features, and uses SVM (support vector machine) and KNN (k-Nearest Neighbor) to achieve image recognition based diabetes diagnosis. Experimental results show that SVM achieves higher diabetic image recognition accuracy than KNN.

**Keywords:** Diabetes · SVM · Tongue diagnosis

## 1 Introduction

At present, the number of diabetics has reached 120 million, of which about 9 million are suffering from complications of diabetes, which brings a heavy burden to the medical resources of our country. Tongue diagnosis of TCM is of great significance in the diagnosis of diabetes. Doctors make judgment and analysis on the condition of the tongue by observing various characteristics of the tongue, it has the characteristics of painless and convenient.

Tongue diagnosis of TCM depends on the clinical experience of doctors. Therefore, the results of doctors' diagnosis are subjective, which makes the results of tongue diagnosis different. Nowadays, with the improvement of science and technology, research on tongue diagnosis of TCM with the help of image processing technology has become possible.

© Springer Nature Switzerland AG 2020
X. Chen et al. (Eds.): ML4CS 2020, LNCS 12488, pp. 474–484, 2020.
https://doi.org/10.1007/978-3-030-62463-7_44

At present, image processing techniques are used by most researchers in the analysis and processing of tongue images [1–4]. Some researchers also pay attention to the research of automatic diagnosis of tongue diagnosis based on feature extraction [5–7]. The realization of automated diagnosis of tongue diagnosis in TCM, it includes feature extraction and feature-based disease classification [8]. In terms of disease classification, the SVM has its unique advantages in dealing with small samples and non-linear data problems [9]. Therefore, this paper uses image processing technology to extract tongue features, the SVM is used to establish a classifier, it compares with the traditional KNN algorithm [10]. Therefore, a diabetic tongue image recognition method based on machine learning is proposed.

## 2  The Proposed Method

The diabetic tongue image recognition method proposed in this paper, the flow chart of which is shown in Fig. 1. Following sub-sections will describe the theory and implementation of the proposed method.

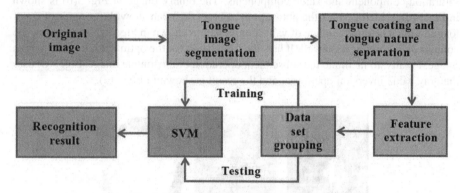

**Fig. 1.** Flow chart of the proposed method

### 2.1  Tongue Segmentation

Since the tongue image contains part of the upper and lower lips, the color of the lips is very close to the tongue body, these problems affect the accuracy of the tongue segmentation. Following sub-sections will describe the solution.

**Color Space Conversion.** In different color spaces, the distribution of pixel gray values of the tongue image is different. The original RGB image is converted to the HSV color space, and the three components of HSV are extracted as shown in Fig. 2. By observing the three components, the obvious differentiation can be found between tongue region and the face region of the hue component. Considering the difficulty of segmentation, it is obvious that the hue component is the easiest to process, however, there are large differences in the pixel values within the tongue region.

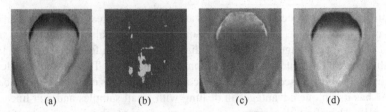

**Fig. 2.** HSV color space: (a) original image; (b) hue component; (c) saturation component; (d) value component

**Image Thresholding.** Because there is a large difference in the pixel value of the tongue region under the hue component, which is not conducive to tongue segmentation, two steps are proposed in this section. Firstly, binary thresholding is used to segment the hue component into binary image of the hue component, the binary image is shown in Fig. 3(a). Secondly, the binary image and the original hue component are subjected to matrix subtraction, the result in Fig. 3(b), which shows that there are lower and close pixel value in the tongue area. Otsu [11] thresholding is used to segment the Fig. 3(b), saturation component and value components. The binary image of Fig. 3(b) is shown in Fig. 3(d), which verifies the above point. Figure 3(d) can show the contour of tip of tongue well. The binary image of value component is shown in Fig. 3(c), it can show the contour of the tongue base well. It lays the foundation for the optimization of the tongue image. Finally, to facilitate the subsequent operation of the tongue image, Fig. 3(d) uses the gray value inversion operation, and the result is shown in Fig. 3(e).

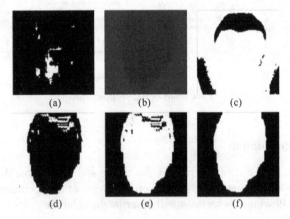

**Fig. 3.** Intermediate results from tongue segmentation: (a) binary thresholding; (b) Fusion result; (c) value component thresholding; (d) thresholding; (e) pixel inversion; (f) morphological operations

**Morphological Operations.** To get a more complete tongue image, this method uses morphological operations to fill these small holes in the tongue body area. According to the shape of the tongue image, this method selects circular structural elements and performs three morphological operations of "expansion", "closed operation" and "corrosion" to fill holes. Figure 3(f) shows the initial tongue image area extracted by this method.

**Tongue Optimization.** After the morphological operations, the initial tongue body image is obtained which contains the upper lip. As the value component of origin image in Fig. 2(d) shows the obvious dark area between the tongue and upper lip, so the following steps are proposed to extract exact tongue. Firstly, the value component of the initial tongue area is segmented using Otsu [11] thresholding, the result is shown in Fig. 4(a). Secondly, the corrosion is used to expand the gap, the result is shown in Fig. 4(b). Finally, retains the largest connected region, and performs the expansion, the result is shown in Fig. 4(c). Figure 4(d) shows the final tongue contour extracted by this method.

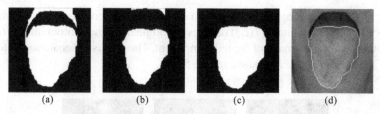

(a)                    (b)                    (c)                    (d)

**Fig. 4.** The process and results of tongue optimization: (a) hue component thresholding; (b) morphological operations; (c) the largest connected region; (d) final tongue contour

## 2.2 Tongue Nature and Tongue Coating Separation

The tongue is composed of tongue coating and tongue nature. The tongue is the muscle tissue of the tongue body, the layer of moss on the tongue is the tongue coating, and the disease can be diagnosed and predicted based on the characteristics of the tongue coating and tongue nature. In the process of clinical diagnosis, doctors analyze various characteristics of tongue coating and tongue nature to understand the organs of the human body and the operation of qi-blood. Therefore, the complete separation of tongue nature and tongue coating is an important step in the research of tongue diagnosis, and the following steps are proposed to achieve it. Firstly, we use the formula 1 to normalize the extracted red component of the RGB color space of the tongue image:

$$I_R = R \div (R + G + B) \tag{1}$$

where $I_R$ denotes the normalized image obtained. Secondly, the $I_R$ image uses the iterative method [12] to find the threshold, which is defined as:

$$T_0 = (Z_{max} + Z_{min}) \div 2 \tag{2}$$

$$T_{k+1} = (Z_a + Z_b) \div 2 \tag{3}$$

where $Z_{max}$ and $Z_{min}$ are denote the maximum gray value and minimum gray value of the $I_a$ image, the initial threshold is $T_0$. Thirdly, the image is divided into foreground and background according to the initial threshold $T_k$, find the average gray values $Z_a$ and $Z_b$ of the foreground and background, according to $Z_a$ and $Z_b$, find the new threshold $T_{k+1}$, if $T_k = T_{k+1}$, then $T_k$ is the obtained threshold, otherwise continue to calculate $T_{k+1}$.

Finally, according to the threshold obtained by the iterative method [12], the tongue nature and tongue coating are separated from the tongue body, which is defined as:

$$I_q(i,j) = \begin{cases} [0,0,0] & \text{if } I_a(i,j) \geq T_k \\ I(i,j) & \text{otherwise} \end{cases} \tag{4}$$

$$I_c(i,j) = \begin{cases} [0,0,0] & \text{if } I_a(i,j) \geq T_k \\ I(i,j) & \text{otherwise} \end{cases} \tag{5}$$

where $I_q(i,j)$ denotes tongue nature, $I_c(i,j)$ denotes tongue coating, $(i,j)$ denotes the pixel coordinates of the image, $I(i,j)$ denotes the original tongue contour image, $I_R(i,j)$ denotes the normalized image of the red component. The results of separation of tongue coating and tongue nature are shown in Fig. 5:

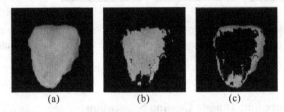

(a)               (b)               (c)

**Fig. 5.** Separation results of tongue nature and tongue coating: (a) tongue image; (b) tongue coating; (c) tongue nature

### 2.3   Feature Extraction

to the color characteristics of the tongue image can well reflect and predict the internal situation of the human body. Therefore, the color features of tongue are extracted to recognize diabetic tongue images in this paper. In the color space of an image, three color space models such as RGB, HSV and Lab (CIELAB) have their own advantages. After analysis the different color space characteristics of diabetic tongue images and healthy tongue images, 10 color features from tongue coating and tongue nature are chosen to diabetic tongue image recognition, where a denotes component a in Lab, b denotes component b in Lab. The color feature extraction results of 300 tongue images in this paper are shown in Table 1 and 2:

**Table 1.** Color characteristic values of tongue nature (mean ± standard deviation)

| Group | Red | Hue | Saturation | a | b |
|---|---|---|---|---|---|
| Diabetes | 141.8 ± 15.0 | 0.59 ± 0.23 | 0.38 ± 0.03 | 21.29 ± 2.34 | 7.94 ± 2.08 |
| Normal | 149.6 ± 13.9 | 0.53 ± 0.23 | 0.40 ± 0.03 | 24.07 ± 2.61 | 9.68 ± 2.12 |

**Table 2.** Color characteristic values of tongue coating (mean ± standard deviation)

| Group | Blue | Green | Hue | a | b |
|---|---|---|---|---|---|
| Diabetes | 110.4 ± 12.1 | 109.2 ± 13.0 | 0.55 ± 0.24 | 16.70 ± 2.20 | 5.91 ± 2.19 |
| Normal | 106.3 ± 11.6 | 106.1 ± 12.5 | 0.48 ± 0.21 | 17.12 ± 2.12 | 6.72 ± 2.03 |

## 2.4 Support Vector Machine (SVM)

Support Vector Machine (SVM) is a kind of supervised learning method [9], which is usually used to solve binary classification problems. Based on the principle of structural risk minimization. SVM looks for hyperplanes ($\omega \cdot X + b = 0$) that can correctly divide two types of samples and have the largest geometric separation. To maximize the hyperplane geometric spacing, the model represents the minimization of the neutralization in the solution formula 6:

$$\min_{\omega,b} \frac{1}{2}\|\omega\|^2 \ s.t. \ y_i(\omega \cdot x_i + b) - 1 \geq 0, \ i = 1, 2, \cdots, N \tag{6}$$

where $\omega$ denotes the weight, $b$ denotes the bias term, and $x_i$ denotes the sample feature.

By finding the optimal solutions $\omega'$ and $b'$, the hyperplane $(\omega' \cdot X + b' = 0)$ is obtained. When there are some sample points that cannot meet the constraint condition, they are outside the feasible solution area. Therefore, to correct the components of these classification errors, the model needs to be modified. The modified model is as follows:

$$\min_{\omega,b,\delta} \frac{1}{2}\|\omega\|^2 + C \sum_{i=1}^{N} \delta_i$$
$$s.t. \ y_i(\omega \cdot x_i + b) \geq 1 - \delta_i(\delta_i \geq 0) \tag{7}$$

where C denotes the penalty term, and $\delta_i$ denotes the slack variable.

The above is for the case where the samples to be classified are linear separable. This method uses a non-linear SVM. For non-linear problems, the kernel is used to replace the inner product to solve the problem.

Since the low-dimensional space vector is mapped to the high-dimensional space, the inner product operation $\varphi(x_i) \cdot \varphi(x_j)$ of the high-dimensional space will greatly increase the complexity of the algorithm, so by introducing the kernel $K(x_i, x_j)$, the kernel is expressed as follows:

$$K(x_i, x_j) = \varphi(x_i) \cdot \varphi(x_j) \tag{8}$$

# 3  Result

## 3.1  Experimental Data

In this experiment, a total of 100 diabetic tongue image samples and 200 normal human tongue image samples were collected. To objectively evaluate the performance of the SVM classifier, the collected experimental samples were grouped.

Step 1: From 200 normal human tongue samples, the training set R1 and test set T1 are randomly generated from the data set at a ratio of 7:3; From 100 tongue samples of diabetics, the training set R2 and the test set T2 are randomly generated from the data set at a ratio of 7:3.

Step 2: The training sets R1 and R2 are merged into the batch training sample data set Train1, and the training sets T1 and T2 are merged into the batch test sample data set Test1.

Step 3: Steps 1 and 2 are continuously repeated 4 times to obtain Date1 to Data5, a total of 5 sets of experimental data.

## 3.2  Parameter Selection

For the tongue coating and tongue nature feature values extracted from the experiment, to avoid "singular input samples", the Z-score (standard score) was used to convert the tongue coating tongue nature feature values into a sequence with mean 0 and variance 1. The Z-score is as follows:

$$y_i = \frac{x_i - \bar{x}}{s} \tag{9}$$

where $\bar{x}$ denotes the mean value of the original sample feature data, $s$ denotes the standard deviation of the original sample feature data, $x_i$ denotes the sample feature, then the mean value of the new sample feature data sequence $y_1, y_2, \cdots, y_n$ is 0, and the variance is 1.

In the experiment, radial basis function (RBF) with strong non-linear mapping capability is selected as the kernel of SVM, it is expressed as:

$$K(x_i, x_j) = exp\left(-g \left\| x_i - x_j \right\|^2\right) \tag{10}$$

where $g = \frac{1}{2}\sigma^{-2}$. When RBF is used as the kernel, the choice of penalty term C and parameter $g$ affects the classification ability of SVM.

This method uses the Scikit-learn machine learning package to build a prediction model [13], and uses the GridSearchCV function to return the optimized penalty term $C = 24.251$ and the parameter $g = 0.005524$.

The KNN classification algorithm of the control group. In this method, the K value is 1 to 10, and the classification accuracy under different K values is shown in Fig. 6:

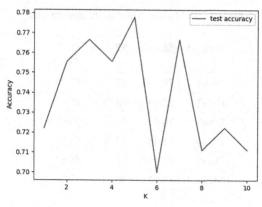

**Fig. 6.** Classification accuracy under different K values

It can be seen from Fig. 6 that the classification accuracy is the highest when the K value is 5, so the optimal value of K is selected in this experiment.

## 3.3  Experimental Results

**Performance Evaluation Standard.** To evaluate the performance of the classification algorithm, this paper uses sensitivity (SE), specificity (SP) and accuracy (ACC) to evaluate the recognition algorithm. The calculation formula is as follows:

$$SE = \frac{TP}{TP + FN} \times 100\% \tag{11}$$

$$SP = \frac{TN}{TN + FP} \times 100\% \tag{12}$$

$$ACC = \frac{TP + TN}{TP + FP + TN + FN} \times 100\% \tag{13}$$

where true positive (TP) denotes that the identification result of the diabetes sample is consistent with the true diagnosis result, and if it is inconsistent, it is denoted as false negative (FN). Similarly, the identification result of the normal tongue image sample is consistent with the true diagnosis result, it is expressed as a true negative (TN). Otherwise, it is expressed as a false positive (FP).

**Recognition Results and Analysis.** In this paper, we obtain 5 sets of data according to the classification criteria of the data set in the previous section. The SVM and KNN algorithms are used to train the five sets of data. Among them, the relevant parameters used in the classification and identification experiment, they come from the optimal values selected above. Experimental results obtained are shown in Table 3 and 4:

**Table 3.** SVM algorithm classification results

| Data set | SE (%) | SP (%) | ACC (%) |
|---|---|---|---|
| Data1 | 78.78 | 92.98 | 87.78 |
| Data2 | 82.14 | 88.71 | 86.67 |
| Data3 | 80.00 | 96.36 | 90.00 |
| Data4 | 81.48 | 87.30 | 85.56 |
| Data5 | 82.14 | 88.71 | 86.67 |
| Mean | 80.91 | 90.81 | 87.34 |

**Table 4.** KNN algorithm classification results

| Data set | SE (%) | SP (%) | ACC (%) |
|---|---|---|---|
| Data1 | 79.17 | 83.33 | 82.22 |
| Data2 | 77.27 | 80.88 | 80.00 |
| Data3 | 65.51 | 81.96 | 76.67 |
| Data4 | 69.56 | 79.10 | 76.67 |
| Data5 | 70.37 | 82.54 | 78.89 |
| Mean | 72.38 | 81.56 | 78.89 |

Based on the analysis of the results in Table 3 and Table 4, the average recognition ACC of the SVM classification results reached 87.34%, and the highest reached 90.00%, the average recognition ACC of the classification results of the KNN method is 78.89%, the average recognition ACC of SVM is significantly higher than that of KNN method, and the fluctuation of recognition ACC of the two algorithms is relatively stable, it also verified the validity of experimental data selection; The average SE of the SVM classification result is 80.91%, and the average SE of the KNN method classification result is 72.38%, it shows that SVM has a higher ability to correctly screen out diabetics; the average SP of SVM classification results is 90.81%, and the average SP of KNN classification results is 81.56%, it shows that the KNN method can easily divide normal samples into diabetes samples.

Based on the above analysis, SVM has better comprehensive performance than KNN method in classifying diabetic tongue image samples and normal tongue image samples.

## 4   Conclusion

In this paper, we proposed a method for diabetic tongue image recognition based on machine learning. In the stage of tongue image processing, the method based on the multi-color channel and threshold method, and combined with the characteristics of tongue image color distribution, the methods of tongue image segmentation and separation of tongue coating and tongue nature were proposed in this paper to solve the problem that the traditional threshold method is difficult to process the tongue image. In the stage of the classification and recognition, KNN was used as a control group to classify and evaluate 300 normal and diabetic tongue images in clinical data, and a conclusion was obtained that SVM has better overall performance. However, due to the limitations of the data, our identification method only can predict diabetics and non-diabetics. In the future, we will try to combine the dialectical laws of TCM and add the subjective data of patients to further subdivide the prediction results.

**Acknowledgment.** This work was supported in part by National Natural Science Foundation of China (61972187, 61772254), Fujian Provincial Leading Project (2017H0030, 2019H0025), Government Guiding Regional Science and Technology Development (2019L3009), and Natural Science Foundation of Fujian Province (2017J01768 and 2019J01756), and the Opening Foundation Projects in 2019 of Fujian Provincial Key Laboratory of Information Processing and Intelligent Control (Minjiang University) (MJUKF-IPIC201914), and Education and Research Project for Young and Middle-aged Teachers in Fujian Province (JT180406), President Fund Project of Minjiang University (103952019075).

# References

1. Ma, C.: Tongue Inspection Based on Image Processing Technology. Chongqing University, no. 5, p. 75 (2007)
2. Chen, H., Bu, J., Gong, Y., Lian, Y.: The algorithm of tongue body and tongue coating separation based on the dynamic threshold of multiple color channels. Beijing Biomed. Eng. (5), 466–469 (2006)
3. Zhang, L., Xu, J., He, T.: Progress in research and application of tongue image segmentation method. China J. Tradit. Chin. Med. Pharm. 25(4), 565–567 (2010)
4. Li, G., Shi, M., Li, F., Wang, Y.: An empirical study on tongue image detection. J. Shandong Univ. (Eng. Sci.) 40(5), 87–95 (2010)
5. Zhang, X., Wang, H., Gui, M., Hu, G.: Preliminary study of acne syndrom classification based on the characteristics of tongue in TCM. Beijing Biomed. Eng. 35(5), 464–468 (2016)
6. Zhang, Z., Liang, J., Chen, J., Yong, W., Wang, X., Peng, Y.: Study on the correlation between objective image recognition of tongue diagnosis and endoscopy classification of chronic gastritis. Acta Chin. Med. Pharmacol. 41(2), 119–121 (2013)
7. Shi, J., Shen, X., Zhang, D., Wu, W., Wu, J.: The study on the relationship between the glossoscopy objectivised image and differentiation type of patients with primary glomerulopathy. China J. Basic Med. Tradit. Chin. Med. 15(2), 120–122 (2009)
8. Yuan, L., Li, W., Yao, J., Song, H., Zheng, X.: Research progress of information processing technology on tongue diagnosis of traditional Chinese medicine. J. Shanghai Univ. Tradit. Chin. Med. 25(2), 80–86 (2011)
9. Ding, S., Qi, B., Tan, H.: An overview on theory and algorithm of support vector machines. J. Univ. Electron. Sci. Technol. China 40(1), 2–10 (2011)
10. Li, G., Fan, P., Yuan, L.: Continuous K -nearest neighbor processing based on speed and direction of moving objects in a road network. Telecommun. Syst. 55(3) (2014)
11. Otsu, N.: A threshold selection method from gray-level histograms. IEEE Trans. Syst. Man Cybern. 9(1), 62–66 (2007)

12. Perez, A., Gonzalez, R.C.: An iterative thresholding algorithm for image segmentation. IEEE Trans. Pattern Anal. Mach. Intell. **9**(6), 742–751 (1987)
13. Fisman, R., Iyengar, S.S., Simonson, K.I.: Gender differences in mate selection: evidence from a speed dating experiment. Q. J. Econ. **121**(2), 673–697 (2006)

# Remote Sensing Image Fusion Based on Multi-morphological Convolutional Neural Network

Meng Zhu, Jindong Xu(✉) (iD), and Zhaowei Liu

School of Computer and Control Engineering, Yantai University, Yantai 264000, China
jindong.xu@nlpr.ia.ac.cn

**Abstract.** In this paper, we propose a method for remote sensing image fusion based on multi-morphological convolutional neural network (MCNN). First, MCNN combines a local discrete cosine transform (LDCT) dictionary and a curvelet transform (CT) dictionary to form a decomposition dictionary. By adjusting the size of the threshold, the morphological component analysis (MCA) method is used to sparsely decompose the remote sensing image, the texture component and the cartoon component (segmented smooth component) are extracted from the remote sensing image, respectively. Secondly, combined with the convolutional neural network (CNN) that has achieved good results in the field of image processing. By inputting different forms of source images into the network, it outputs a fused image with end-to-end characteristics. The MCNN not only solves the problems that most existing models based on sparse algorithms are relatively complex and have high computational complexity, but also solves the problem that in the traditional deep learning fusion method, the original information in the source image is often lost when acquiring the image information. Compared with other commonly used remote sensing image fusion methods, experimental results show that MCNN has achieved good results in terms of objective evaluation indicators and subjective visual performance.

**Keywords:** Remote sensing image fusion · Sparse decomposition · Morphological component analysis · Convolutional neural network

## 1 Introduction

In recent years, with the continuous development of remote sensing and computer technology, remote sensing data has played a significant role in earth observation research. More and more remote sensing images are widely used in weather forecasting, land cover classification, environmental management, target detection and identification detection [1, 2]. However, due to the limitation of sensor technology, most earth observation satellites can only acquire panchromatic images (PAN) with high spatial resolution and low spectral resolution, and multispectral images (MS) with low spatial resolution and high spectral resolution. PAN images have rich spatial details and can accurately describe textures and shapes, but have less spectral information. MS images are rich in spectral

© Springer Nature Switzerland AG 2020
X. Chen et al. (Eds.): ML4CS 2020, LNCS 12488, pp. 485–495, 2020.
https://doi.org/10.1007/978-3-030-62463-7_45

details and can be used for accurate feature classification, but the resolution of imaging is low. The purpose of remote sensing image fusion is to make full use of the complementarity of spatial and spectral information, and fuse them to obtain remote sensing images with high spatial resolution and high spectral resolution. This process is also called pan-sharpening [3]. In the past few decades, researchers have proposed different fusion algorithms for this problem [4, 5]. It can be roughly divided into the following three branches, component substitution (CS) methods, multi-resolution analysis (MRA) methods, and model-based methods.

The CS method usually fuses the pixel gray space of the image, projects the MS image to another structural space, separate the spatial components of MS images and replace them with PAN image, and injects spatial information. Finally, the fusion image is obtained after inverse transformation. Commonly used component replacement methods include intensity-hue-saturation (IHS) [6, 7], principle component analysis (PCA) [8, 9], brovey transform [10], etc. This method runs fast and effectively improves the spatial resolution of the fused image, but the mismatch of the spectral range of the PAN image and the MS image will cause different degrees of spectral distortion [11].

MRA method decomposes PAN image at multiple scales, to obtain low-frequency and high-frequency parts, respectively. And the high-frequency parts are injected into the MS image, obtains a fused image after inverse transformation. The difference between MRA method and the CS method is to filter the PAN image in order to find more suitable high-frequency information. Commonly used MRA methods mainly include discrete wavelet transformation (DWT) [12, 13], non-subsampled contourlet transform (NSCT) [14, 15], high pass filter (HPF) [16] and so on. Most fusion methods based on multi-resolution analysis will strengthen the fusion rules according to different fusion targets, so as to obtain better fusion results [17, 18]. Compared with the CS-based method, the MRA method improves the spectral resolution and retains the spectral information effectively of the MS image. However, due to the number of image decompositions and the non-uniqueness of the filtering parameters, the fusion image may have problems such as spatial distortion and ringing effects.

Model-based methods achieve image fusion through some optimization problems. In recent years, with the rapid development and widespread attention of compressed sensing technology [19], sparse expression as an effective signal processing tool in the signal world, has been rapidly and widely applied to the field of remote sensing image processing [20], and has become a hot topic in research. Sparse expression is derived from the sparse coding idea of mammalian primary visual cortex [21], and its purpose is to express most or all signals with as few atoms as possible in a given over-complete dictionary. Li and Yang [22] proposed a fusion algorithm based on sparse representation, which established PAN image and MS image as a degradation model of high-resolution MS image, and used a basis pursuit algorithm to reconstruct high-resolution MS image, realize the fusion of PAN and MS. This method can preserve the spatial and spectral information well, but the application of the data will be limited and the model will be more complicated. Zhu et al. [23] proposed a sparse reconstruction (SparseFI) method based on the sparse coding of MS on high-spatial and low-spatial resolution dictionaries constructed from PAN image. This method avoids constructing a dictionary from MS images with high spatial resolution. However, due to the large spectral difference between the PAN and the MS, it is easy to cause spectral distortion of the fused image. Yang et al. [24] considered that the spectral information of MS image can be fully utilized, and proposed a fusion method that uses high-frequency and low-frequency dictionaries of MS image to obtain high-resolution image. Starck et al. [25, 26] proposed a signal-based sparse representation method that is MCA. It has achieved good results in image decomposition, image repair and other directions [27–29]. It combines the advantages of many classic transformation dictionaries basis to decompose the image into multiple morphological components. When describing and decomposing images, it can be made sparse to obtain better image fusion effect.

This paper uses the MCA method, through the LDCT dictionary and CT dictionary to sparsely decompose PAN image and MS image into texture component and cartoon component respectively, in order to retain the features of the source image to the maximum. Blend the texture component and the cartoon component separately, and the fused texture and cartoon components as the next input. So that in the subsequent fusion process, the complexity of image processing is reduced, and the characteristics of the processing target become simpler and clearer. In recent years, with the rapid development and widespread application of deep learning technology, more and more researchers have introduced deep learning into image processing, such as target detection [30], multi-focus image fusion [31], remote sensing image classification [32, 33], remote sensing image fusion [34–36] and so on. CNN is one of the most convenient and effective methods in deep learning models [37–39]. In this paper, based on the deep CNN fusion framework proposed by Ye et al. [40], by combining with MCA and through further improvement processing, to get a more optimized fusion image, a better fusion effect is achieved than existing methods.

## 2 Remote Sensing Image Fusion Based on MCNN

### 2.1 Fusion Process

The purpose of remote sensing image fusion is to obtain MS images with high spatial resolution by fusing PAN images with high spatial resolution and MS images with high spectral resolution. However, due to the limitations of satellite technology and the particularity of remote sensing images, there is often some noise and interference when acquiring remote sensing images. Because the noise has high frequency, it is more obvious in the PAN image with high spatial resolution. Most fusion algorithms treat the source image as a single component and directly process it, without considering that the characteristics of images with different attributes are very different, so some details in the source image are often ignored. When we use MCA method for remote sensing image fusion, the source image is sparsely decomposed into fine-scale texture components and smooth cartoon components. Considering that image fusion is to extract and fuse the most effective information in the source image, a fusion scheme was formulated as follows:

1. Combine the LDCT base $D_1$ and CT base $D_2$ as MCA decomposition dictionary $D = [D_1, D_2]$, perform MCA decomposition on the input source images PAN and MS respectively, and extract the texture and cartoon components of the two images. The

texture and cartoon components of the PAN image are represented as $x_{PAN}^T$ and $x_{PAN}^C$ ($T$ represents texture and $C$ represents cartoon), the texture and cartoon components of the MS image are represented as $x_{MS}^T$ and $x_{MS}^C$, respectively.

2. According to the rules of image fusion, use the fusion framework of the CNN to fuse $x_{PAN}^T$ and $x_{MS}^T$ as inputs, and output the fused texture component $x_{Fusion}^T$. In the same way, $x_{PAN}^C$ and $x_{MS}^C$ are fused as inputs to obtain the fused cartoon component $x_{Fusion}^C$.

3. The texture component $x_{Fusion}^T$ and cartoon component $x_{Fusion}^C$ output in step2 are used as the input image again. After fusion processing of CNN, the final output fusion image $x_{Fusion}$ is obtained.

The flowchart of MCNN based remote sensing image fusion method is shown in Fig. 1.

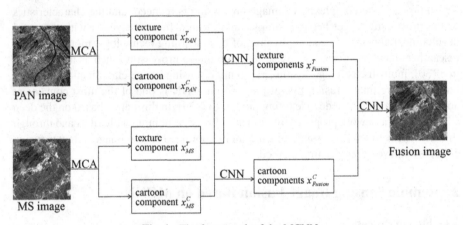

**Fig. 1.** The framework of the MCNN.

## 2.2 Fusion Algorithm

**MCA Processing.** Formulate the sparse decomposition of the PAN image and the MS image, expressed by Eq. (1) and Eq. (2) respectively.

$$x_{PAN} = x_{PAN}^T + x_{PAN}^C = D_1 \alpha_{PAN}^T + D_2 \alpha_{PAN}^C = \left( \alpha_{PAN}^T + \alpha_{PAN}^C \right) D \tag{1}$$

$$x_{MS} = x_{MS}^T + x_{MS}^C = D_1 \alpha_{MS}^T + D_2 \alpha_{MS}^C = \left( \alpha_{MS}^T + \alpha_{MS}^C \right) D \tag{2}$$

where, $\alpha_{PAN}^T$, $\alpha_{PAN}^C$, $\alpha_{MS}^T$, $\alpha_{MS}^C$ represents the corresponding decomposition coefficient.

**FusionCNN.** During the fusion process, it is assumed that *Fusion(a, b)* is the pixel at the position $(a, b)$ of the fusion image, the pixels corresponding to the PAN and MS images are PAN$(a, b)$ and MS$(a, b)$. $T(a, b)$ and $C(a, b)$ are the corresponding pixels of the texture component and the cartoon component, respectively. Since the convolution operation is a linear operation, the output of each layer is a linear function of the input of the previous layer. In order to improve the quality of the fused image, alleviate the problem of over-fitting, and reduce the interdependence between parameters, a non-linear operation is performed after the convolution. The convolution operation is defined as Eq. (3).

$$Fusion = \text{ReLU}\,(x \circ w) \tag{3}$$

where $x$ is the input to the convolution, $w$ is the convolution kernel, and ReLU is the non-linear activation function.

$$\text{ReLU}(x) = max\{0, x\} \tag{4}$$

Fusion network includes 10 convolutional layers, of which 6 convolutional layers is $n \times n$ $(n > 1)$, and the others are $(1 \times 1)$. This function of fusion $Fusion$ $(PAN, MS) \rightarrow Fusion$ is represented by this network. Let $T^k$ $(k = 1, 2, 3)$ and $C^k$ $(k = 1, 2, 3)$ denote the output of the $k$th convolutional layer of the texture component and the cartoon component, respectively.

$$T^k(a, b) = \text{ReLU}\left(T^{k-1}(a, b) \circ w_T^k\right)$$

$$C^k(a, b) = \text{ReLU}\left(C^{k-1}(a, b) \circ w_C^k\right) \tag{5}$$

The final fused image $Fusion$ can be expressed by the Eq. (6).

$$Fusion(a, b) = \left[T^k(a, b), C^k(a, b)\right] \circ w_{Fusion} \tag{6}$$

In the entire process of convolution operation, each layer refers to the fusion result of the previous layer during fusion, this fusion method can also be called multi-scale image fusion. The final fusion image $Fusion$ obtained by the convolution operation is the result of image fusion of not the same scales.

## 3  Experimental Results and Analysis

In order to prove the effectiveness and robustness of MCNN, the experiment selected two sets of remote sensing images with different terrain areas for fusion, the spatial resolution and spectral wavelength of these remote sensing images have their own characteristics. The first group (Fig. 1(a)–(j)) is located in the Jilin Province, covering various types of vegetation, building land, traffic roads, etc. Its image size is $256 \times 256$ pixels. The second experimental data (Fig. 2(a)–(j)) were taken from the MS image with a resolution of 19.5 m in the CBERS-01 image and the PAN image with a resolution of 15 m in the Landsat ETM + image. The experimental area is located in Zhuhai City, Guangdong Province, and includes forest land, river, agricultural land, etc. The image size is $512 \times 512$ pixels.

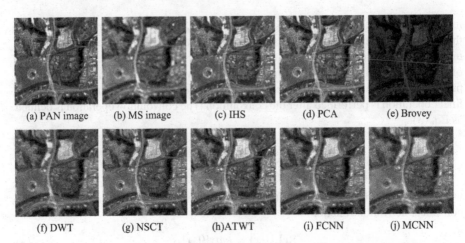

(a) PAN image    (b) MS image    (c) IHS    (d) PCA    (e) Brovey

(f) DWT    (g) NSCT    (h)ATWT    (i) FCNN    (j) MCNN

**Fig. 2.** Fusion results of different methods in the first set of remote sensing data.

## 3.1 Evaluation Index

In order to verify the performance objectively of different methods, we selected the following five commonly used objective evaluation indicators to evaluate the experimental result images: relative average spectral error index (RASE), root mean square error (RMSE), erreur relative globale adimensionnelle de synthese (ERGAS), spectral information divergence (SID), peak signal-to-noise ratio (PSNR).

(1)  Relative average spectral error index (RASE)
     Reflects the average spectral performance of all channels. The higher the spectral quality of the fused image, the lower the RASE, and the ideal value is 0.
(2)  Root Mean Square Error (RMSE)
     Calculate the difference in pixel values between the fused image and the reference image. The smaller the difference, the better the spectrum is maintained, the ideal value is 0.
(3)  Erreur Relative Globale Adimensionnelle de Synthese (ERGAS)
     Evaluate the spatial and spectral quality of the fused image. The smaller the value, the better the fused image quality, the ideal value is 0.
(4)  Spectral information divergence (SID)
     A spectral classification method that measures the difference between two spectra based on information theory. The smaller the value, the smaller the difference, that is, the better the spectral retention, the ideal value is 0.
(5)  Peak signal-to-noise ratio (PSNR)
     It reflects the noise and distortion level of the image. The larger the PSNR value, the closer the image is to the reference image, the better the quality.

## 3.2 Experimental Results and Analysis

Set the parameters of the morphosis sparse decomposition algorithm on the source image in the experiment, let $N_{iter} = 300$, $\lambda_{min} = 10^{-6}$, $\gamma = 1$. In order to more accurately extract multiple forms of the source image and reduce unnecessary interference information, the PAN image and the MS image are sparsely decomposed under different thresholds. So as to ensure that the texture information and spectral information of the image are fully obtained, the threshold setting range is [16/512, 64/512] when extracting the texture components. When extracting cartoon components, the threshold setting range is [128/512, 256/512].

Eight fusion methods including IHS, PCA, Brovey, DWT, NSCT, ATWT, SparseFI, and FusionCNN (FCNN) were compared with the methods in this paper. The results of different fusion methods in the first set of remote sensing data are shown in Fig. 2. From the comparison experiments, it can be seen that compared with the source image, the 8 methods all improve the quality of the fused image. However, the Fig. 2(c)–(i) show different degrees of distortion. Figure 2(c) and (d) show the situation of fusion oversaturation. Figure 2(e)–(h) have different levels of spectral distortion. The texture and spectral information of Fig. 2(i) are well preserved, but MCNN is superior to Fig. 2(i) in the retention of spectral information.

Table 1 lists the objective evaluation index of the results produced by different fusion of the first set of data. The bolds numbers in Table 1 indicate the optimal values of each evaluation index. It can be seen that MCNN is not the best in SID, and is the best result in other indicators. In terms of comprehensive visual effects, MCNN is superior to other algorithms.

**Table 1.** Numerical evaluation of experimental data fused results in Jilin

|  | RASE close to 0 | RMSE close to 0 | ERGAS close to 0 | SID close to 0 | PSNR The bigger the better |
|---|---|---|---|---|---|
| IHS | 38.6201 | 37.0125 | 9.9329 | 0.0600 | 38.6002 |
| PCA | 37.8751 | 36.2985 | 9.4701 | 0.0425 | 38.9898 |
| Brovey | 81.9278 | 78.5174 | 69.0353 | **0.0048** | 23.5589 |
| DWT | 22.9932 | 22.0361 | 6.0844 | 0.0681 | 48.9716 |
| NSCT | 26.0420 | 24.9580 | 6.8998 | 0.0680 | 46.4814 |
| ATWT | 25.0864 | 24.0422 | 6.4494 | 0.0588 | 47.2291 |
| FCNN | 19.2577 | 18.4561 | 5.0788 | 0.0326 | 52.5174 |
| MCNN | **18.2311** | **17.4722** | **4.6541** | 0.0533 | **53.6131** |

Figure 3 show the fusion results of different methods in the second set of remote sensing data. Figure 3(c)–(i) all have different degrees of spatial structure blurring and spectral distortion, especially in Fig. 3(e)–(g), the spectral distortion is severe. From the subjective visual effect, MCNN is better than other algorithms.

From the indicators in Table 2, The bolds numbers indicate the optimal values of each evaluation index. Compared with the other eight methods, MCNN has better evaluation index results. Combined with subjective visual evaluation, MCNN is superior to other methods.

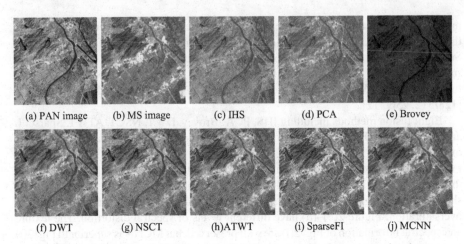

(a) PAN image     (b) MS image     (c) IHS     (d) PCA     (e) Brovey

(f) DWT     (g) NSCT     (h)ATWT     (i) SparseFI     (j) MCNN

**Fig. 3.** Fusion results of different methods in the second set of remote sensing data.

**Table 2.** Numerical evaluation of experimental data fused results in Guangdong

|  | RASE close to 0 | RMSE close to 0 | ERGAS close to 0 | SID close to 0 | PSNR The bigger the better |
|---|---|---|---|---|---|
| IHS | 32.2736 | 40.3886 | 8.0462 | 0.0617 | 36.8543 |
| PCA | 32.0348 | 40.0897 | 8.0022 | **0.0387** | 37.0029 |
| Brovey | 74.3512 | 93.0463 | 55.2522 | 0.0433 | 20.1633 |
| DWT | 33.0915 | 41.4122 | 8.2235 | 0.1058 | 36.3538 |
| NSCT | 35.3656 | 44.2581 | 8.7908 | 0.1305 | 35.0245 |
| ATWT | 33.8932 | 42.4154 | 8.4950 | 0.2298 | 35.8750 |
| SparseFI | 30.2172 | 37.8150 | 7.5885 | 0.2164 | 38.1711 |
| MCNN | **28.2787** | **35.3892** | **7.1689** | 0.1559 | **39.4971** |

Through the experimental evaluation of different satellite data, and the tested remote sensing images contain different types of ground features, it proves the effectiveness and robustness of MCNN.

## 4   Conclusion

In this paper, we propose a remote sensing image fusion based on multi-morphological convolutional neural network. MCNN combines the morphological components of the image with the CNN, which avoids the loss of source image information and enhances the extraction of input image details. In the method, the texture and cartoon components of the remote sensing image are captured through MCA. By extracting and fusing different morphologies to obtain richer original image information, the CNN is used to fuse these features to obtain the final fusion image. Compared with the classical fusion methods, MCNN can better maintain the spatial and spectral information of the source image, which greatly improves the fusion effect and obtains a more satisfactory fusion image.

**Acknowledgment.** This research is funded by the Natural Science Foundation of Shandong (ZR2019MF060, ZR2017MF008), A Project of Shandong Province Higher Educational Science and Technology Key Program (J18KZ016), and the Yantai Science and Technology Plan (2018YT06000271).

# References

1. Zhao, W., Jiao, L.C., et al.: Superpixel-based multiple local CNN for panchromatic and multispectral image classification. IEEE Trans. Geosci. Remote Sens. **55**(7), 4141–4156 (2017)
2. Han, J.W., Zhang, D.W., Cheng, G., Guo, L., Ren, J.C.: Object detection in optical remote sensing images based on weakly supervised learning and high-level feature learning. IEEE Trans. Geosci. Remote Sens. **53**(6), 3325–3337 (2015)
3. Vivone, G., Alparone, L., et al.: A critical comparison among pansharpening algorithms. IEEE Trans. Geosci. Remote Sens. **53**(5), 2565–2586 (2015)
4. Garzelli, A.: A review of image fusion algorithms based on the super resolution paradigm. IEEE Trans. Geosci. Remote Sens. **8**(10), 797 (2016)
5. Ghassemian, H.: A review of remote sensing image fusion methods. Inf. Fusion **32**(PA), 75–89 (2016)
6. Carper, W.J., Lillesand, T.M., Kiefer, P.W.: The use of intensity-hue-saturation transformations for merging SPOT panchromatic and multispectral image data. Photogram. Eng. Remote Sens. **56**(4), 459–467 (1990)
7. Yang, Y., Wan, W.G., et al.: Remote sensing image fusion based on adaptive IHS and multiscale guided filter. IEEE Access **4**(1), 4573–4582 (2016)
8. Pohl, C., Van Genderen, J.L.: Review article multisensor image fusion in remote sensing: concepts, methods and applications. Int. J. Remote Sens. **19**(5), 823–854 (1998)
9. Shahdoosti, H.R., Ghassemian, H.: Combining the spectral PCA and spatial PCA fusion methods by an optimal fifilter. Inf. Fusion **27**(C), 150–160 (2016)
10. Zhou, H.Z., Wu, S., Mao, D.F., et al.: Improved Brovey method for multi-sensor image fusion. J. Remote Sens. **16**(2), 343–360 (2012)
11. Thomas, C., Ranchin, T., Wald, L., Chanussot, J.: Synthesis of multispectral images to high spatial resolution: a critical review of fusion methods based on remote sensing physics. IEEE Trans. Geosci. Remote Sens. **46**(5), 1301–1312 (2008)
12. Pajares, G., Cruz, J.M.D.L.: A wavelet-based image fusion tutorial. Pattern Recogn. **37**(9), 1855–1872 (2004)
13. Pradhan, P.S., King, R.L., Younan, N.H., Holcomb, D.W.: Estimation of the number of decomposition levels for a wavelet-based multiresolution multisensor image fusion. IEEE Trans. Geosci. Remote Sens. **44**(12), 3674–3686 (2006)
14. Cunha, A.L.D., Zhou, J., Do, M.N.: The nonsubsampled contourlet transform: theory, design, and applications. IEEE Trans. Image Process. **15**(10), 3089–3101 (2006)
15. Yang, Y.T., Zhu, M., He, B.G., et al.: Fusion algorithm based on improved projected gradient NMF and NSCT. Optics Precis. Eng. **19**(5), 1143–1150 (2011)
16. Chavez, P.S., Sides, S.C., Anderson, J.A.: Comparison of three different methods to merge multiresolution and multispectral data: Landsat TM and SPOT panchromatic. Photogramm. Eng. Remote Sens. **57**(3), 265–303 (1991)

17. Shao, Z.F., Liu, J., Cheng, Q.M.: Fusion of infrared and visible images based on focus measure operators in the curvelet domain. Appl. Opt. **51**(12), 1910–1921 (2012)
18. Lin, H., Tian, Y.F., Pu, R.L., Liang, L.: Remotely sensing image fusion based on wavelet transform and human vision system. Signal Process. Image Process. Pattern Recog. **8**, 291–298 (2015)
19. Donoho, D.L.: Compressed sensing. IEEE Trans. Inf. Theory **52**(4), 1289–1306 (2006)
20. Li, S.T., Yin, H.T., Fang, L.Y.: Remote sensing image fusion via sparse representations over learned dictionaries. IEEE Trans. Geosci. Remote Sens. **51**(9), 4779–4789 (2013)
21. Olshausen, B.A., Field, D.J.: Emergence of simple-cell receptive field properties by learning a sparse code for natural images. Nature **381**, 607–609 (1996)
22. Li, S., Yang, B.: A new pan-sharpening method using a compressed sensing technique. IEEE Trans. Geosci. Remote Sens. **49**(2), 738–746 (2011)
23. Zhu, X.X., Bamler, R.: A sparse image fusion algorithm with application to pan-sharpening. IEEE Trans. Geosci. Remote Sens. **51**(5), 2827–2836 (2013)
24. Yang, X.M., Jian, L.H., Yan, B.Y., et al.: A sparse representation based pansharpening method. Future Gener. Comput. Syst. **88**, 385–399 (2018)
25. Starck, J.L., Elad, M., Donoho, D.L.: Redundant multiscale transforms and their application for morphological component analysis. Adv. Imaging Electron. Phys. **132**, 287–348 (2004)
26. Starck, J.L., Elad, M., Donoho, D.L.: Image decomposition via the combination of sparse representation and a variational approach. IEEE Trans. Image Process. **14**(10), 1570–1582 (2005)
27. Elad, M., Starck, J.L., Querre, P., Donoho, D.L.: Simultaneous cartoon and texture image inpainting using morphological component analysis (MCA). Appl. Comput. Harmonic Anal. **19**(3), 340–358 (2005)
28. Yong, X.Y., Ward, R.K., Birch, G.E.: Artifact removal in EEG using morphological component analysis. In: IEEE International Conference on Acoustics, Speech and Signal Processing, ICASSP 2009, pp. 345–348 (2009)
29. Fadili, M.J., Starck, J.L.: Em algorithm for sparse representation-based image inpainting. IEEE International Conference on Image Processing. ICIP 2005, pp. 61–64. Genoa, Italia (2005)
30. Redmon, J., Divvala, S., Girshick, R., Farhadi, A.: You only look once: unified, real-time object detection. In: IEEE Conference on Computer Vision and Pattern Recognition, CVPR 2016, pp. 779–788 (2016)
31. Liu, Y., Chen, X., Peng, H., Wang, Z.F.: Multi-focus image fusion with a deep convolutional neural network. Inf. Fusion **36**, 191–207 (2017)
32. Romero, A., Gatta, C., Camps-Valls, G.: Unsupervised deep feature extraction for remote sensing image classification. IEEE Trans. Geosci. Remote Sens. **54**(3), 1349–1362 (2016)
33. Liang, H.M., Li, Q.: Hyperspectral imagery classification using sparse representations of convolutional neural network features. Remote Sens. **8**(2), 99 (2016)
34. Masi, G., Cozzolino, D., Verdoliva, L., et al.: Pansharpening by convolutional neural networks. Remote Sens. **8**(7), 594 (2016)
35. Rao, Y.Z., He, L., Zhu, J.W.: A residual convolutional neural network for pan-shaprening. In: International Workshop on Remote Sensing with Intelligent Processing., RSIP 2017, pp. 1–4 (2017)
36. Wei, Y.C., Yuan, Q.Q., Shen, H.F., Zhang, L.P.: Boosting the accuracy of multispectral image pansharpening by learning a deep residual network. IEEE Geosci. Remote Sens. Lett. **14**(10), 1795–1799 (2017)
37. Yuan, Q.Q., Wei, Y.C., Meng, X.C., et al.: A multiscale and multidepth convolutional neural network for remote sensing imagery pan-sharpening. IEEE J. Sel. Top. Appl. Earth Observ. Remote Sens. **11**(3), 978–989 (2018)

38. Zhong, J.Y., Yang, B., et al.: Remote sensing image fusion with convolutional neural network. Sens. Imag. **17**(1), (2016)
39. Dong, C., Loy, C.C., He, K., Tang, X.: Image super-resolution using deep convolutional networks. IEEE Trans. Pattern Anal. Mach. Intell. **38**(2), 295–307 (2016)
40. Ye, F., Li, X., Zhang, X.: FusionCNN: a remote sensing image fusion algorithm based on deep convolutional neural networks. Multimedia Tools and Applications **78**(11), 14683–14703 (2018). https://doi.org/10.1007/s11042-018-6850-3

# From Sentiment to Well-Being: Internet Users in an Uncertain and Risky Environment

Ge Zhan[1(✉)], Shuang Geng[2], and Ruowen Wang[3]

[1] Shenzhen Pianpian Technology, Shenzhen, Guangdong, China
gs@szu.edu.cn
[2] College of Management, and Greater Bay Area International Institute for Innovation,
Shenzhen University, Shenzhen, Guangdong, China
[3] Beijing Normal University-Hong Kong Baptist University United International College,
Zhuhai, Guangdong, China

**Abstract.** This research is to investigate emerging perspectives on online user well-being and provide a new metric to capture this construct by using machine learning based sentiment analysis. In study 1, a corpus was compiled consisting of over 10,000 online reviews (in Chinese) collected from an online forum with Huawei's Smart Watch users. In study 2, a sentiment analysis was conducted and the online reviews were rated by college students in southern China. The factor analysis and text analytics indicate that the new metric captures the concept of community member well-being, and the corpus can by used to predict well-being in the virtual community context.

**Keywords:** Sentiment analysis · Well-being · Factor analysis

## 1 Introduction

Sentiment analysis (SA) has been widely used for analyzing text data for about twenty years since Pang, Lee and Vaithyanathan's (2002) seminal work. Recently, there is a surge in studies adopting SA to examine users' attitude expressed in texts toward an entity [1, 2]. Machine learning based approach for SA is a young and active research area from computer science and computational linguistics. Pang, Lee & Vaithyanathan (2002) made an early effort in apply machine learning, including Naïve Bayes, Maximum Entropy, Support Vector Machine, for binary sentiment classification of movie reviews. Following that, a wide array of efforts has been made on text feature selection and algorithm improvement [3–7]. Evidently, SA is an effective approach to capture users' attitudes and emotional states.

User well-being is a key concept in social psychological studies, while the detection of user well-being in various contexts of human-computer interaction is still in its infancy. Online platforms such as Twitter and Weibo, where users communicate their emotions and contribute ideas and suggestions, have been explored in recent studies on user well-being [8–10].

X. Chen et al. (Eds.): ML4CS 2020, LNCS 12488, pp. 496–502, 2020.
https://doi.org/10.1007/978-3-030-62463-7_46

The aim of this research is to investigate emerging perspectives on user well-being and provide a new metric to capture this construct by using machine learning based sentiment analysis. The contributions of this study include i) compiling of a corpus that captures both social psychology and positive psychology perspectives of well-being; ii) validating the new measurement with secondary and first-hand data.

## 2 Related Work

### 2.1 Sentiment Analysis for Users' Discourse

Sentiment analysis (SA) is a relatively young research field emerged in recent years that uses machine learning and natural language processing techniques. There have been many contributions that adopted SA to classify users' discourses into sub-categories, i.e. positive, neutral, and negative. By associating the users' sentiment attitude with specific product or service features, brand and product managers obtain clue for product feature evaluation, improvement, and design [11].

Despite that SA has become a popular method for digging deep into customers' perceptions, few studies have related SA with users' well-being in virtual communities. There are several plausible reasons. First, a majority of research on SA was carried out within the scope of computer science and natural language processing area, and mainly focuses on the proposition of SA algorithms. Second, researchers interested about users' well-being in virtual communities preferred to adopt empirical approach which has become a routine and non-doubtable method for typical social science studies. There is a lack of interdisciplinary studies which may help to facilitate the effective integration between machine learning approach and user self-reported approach for users' well-being study. We seek to address this research gap in this study.

### 2.2 Social Psychology Perspective

From the middle and late 20$^{th}$ century, the study of quality of life began to be combined with the movement of social indicators. In the process of trying to construct a subjective indicator system of quality of life, sociologists and psychologists have developed the meaning of quality of life for subjective well-being, which is a combination of "life satisfaction" and net affection [12]. Special attention has been paid to the study of sociological factors such as family background, education level, and income that affect happiness. This concept emphasizes the significance of intangible spiritual living standards for human survival and social development.

Social contribution, as a key indicator of subjective well-being, refers to the degree to which individuals make satisfactory judgments about their contribution to society in general. Diener subdivided life satisfaction, thinking that life satisfaction includes overall life satisfaction and satisfaction in specific areas [12]. At the same time, the Diener study also found that the weight of the importance of different individuals' perceptions of different areas is different, and happy individuals are more likely to give the best parts of life the greatest weight, and unhappy individuals may give the worst parts of life the greatest weight.

## 2.3  Positive Psychology Perspective

The origins of positive psychology research can be traced back to the 20th century Terman's research on genius, and Jung's research on the meaning of life. In 1950s to 1960s, Maslow, Rogers and other Psychologists began to study the positive side of human nature. The purpose of studying positive psychology is to help people live happier, more active and healthy lives. The involvement of positive psychology has expanded the scope of subjective well-being research [13, 14].

Personality environment interaction model suggests that the variation of the feelings is the sum of the individual, context, and interaction variation, and subjective well-being is the result of the interaction of personality and situation [12]. In other words, some people may have a stronger tendency to respond to positive emotions, and the occurrence of positive events can bring a stronger sense of happiness [15]. Others are more sensitive to negative situational stimuli, or negative situations. The occurrence of the event will have a greater negative impact on their subjective well-being [16].

# 3  Study 1

The goal of study 1 is to investigate the extent to which supervised models can provide insights into the member well-being in online communities. The corpus consists of over 10,000 online reviews (in Chinese) which were collected from an online forum with Huawei's Smart Watch users. These users are exposed to an uncertain and risky online environment, as they keep providing heart rate, sleep pattern, and other private data to the mobile device. This challenging environment provides an ideal setting for collecting both negative and positive sentiment data in that users are actively expressing their feeling and emotions. Table 1 presents representative terms used by online users for each component of VLS. Table 2 shows sample online reviews given by the users.

**Table 1.** Online community well-being in community corpus.

|  | Components | Representative terms |
|---|---|---|
| Virtual life satisfaction | Social contribution | Harmonious, growth, positive, appreciate, love, friend, work hard, support, learn, knowledge, patient, truth |
|  | Hedonic experience | Fun, pleasure, funny, interesting |
|  | Self acceptance | Satisfied, relax, optimistic, lucky, feeling, good, self, mind, ability, experience, like |

# 4  Study 2

First, we subsampled a set of online reviews and employed machine learning method to tag the sentiment direction (positive and negative) and scores for each review [17, 18].

**Table 2.** Sample reviews in the online community.

| Components | Sample reviews |
|---|---|
| Social contribution | It is your positive energy that truly changes your destiny |
| | This is my first post. My pleasure to become a Huawei fans |
| | The person who helps you most, in fact, is yourself |
| | Your greatest contribution to mankind is to make yourself happy. Your greatest contribution to yourself is to be mentally strong |
| Hedonic experience | Very interesting |
| | That's so funny. Hahaha… |
| Self acceptance | Luck extraordinary |
| | Bloom in the cold wind |
| | My mistakes<br>True beauty is being yourself |
| | I am a vulgar person |

We used Senta, an open-source sentiment classification system established by Baidu employing Bidirectional Long Short-Term Memory (BiLSTM) model for our sentiment analysis task. LSTM model is a type of RNN model which fixes the vanishing gradient problem. The model learns to store the relevant content and drop the irrelevant ones based on training. BiLSTM is a type of recurrent neural network model that accepts input of review sentence and output three results (0 represents negative, 1 represents positive), probabilities for positive sentiment (range from 0 to 1), and probabilities for positive sentiment (range from 0 to 1). The performance metrics of BiLSTM are competitive with recent studies which reported over 85% accuracy on SST2 and IMDb datasets [19].

Second, eight college students in southern China were recruited to rate "virtual life satisfaction"(VLS) for a sub-sample of online reviews (10% of the full sample). They were also asked to judge the sentiment of each online review. Rating scores range from 1 (low VLS) to 5 (high VLS). Table 3 indicates the distribution of the rated scores with sentiment polarity.

**Table 3.** Descriptive statistics of sentiment polarity with number of rated reviews.

| | No. of rated reviews | Sentiment | No. of rated reviews | Sentiment |
|---|---|---|---|---|
| 1 | 193 | Negative | 64 | Positive |
| 2 | 399 | Negative | 203 | Positive |
| 3 | 102 | Negative | 71 | Positive |
| 4 | 130 | Negative | 113 | Positive |
| 5 | 10 | Negative | 21 | Positive |
| Total | 834 | | 472 | |

Table 4 presents the correlation coefficients between variables. Post VLS (VLS for the post title) and review VLS are strongly correlated. Review VLS has a much stronger correlation with sentiment scores than title VLS. Regression results (Table 5) indicate that sentiment positively and significantly relates to review VLS.

**Table 4.** Correlation matrix.

|   |   | 1 | 2 | 3 | 4 |
|---|---|---|---|---|---|
| 1 | Post VLS | 1.000 | | | |
| 2 | Review VLS | 0.492 | 1.000 | | |
| 3 | Sentiment | 0.093 | 0.173 | 1.000 | |
| 4 | Positive | 0.123 | 0.191 | 0.884 | 1.000 |
| 5 | Negative | −0.123 | −0.191 | −0.884 | −1.000 |

**Table 5.** Regression model: virtual life satisfaction (VLS) for online review.

|   | Model 1 | | Model 2 | |
|---|---|---|---|---|
| Constant | 1.04*** | 0.07 | 0.96*** | 0.07 |
| Post VLS | 0.58*** | 0.03 | 0.57*** | 0.03 |
| Sentiment | | | 0.28*** | 0.05 |
| N | 1314 | | 1304 | |
| F | 418.95*** | | 226.64*** | |
| $R^2$ | 0.24 | | 0.26 | |

Our metric of online community well-being includes affection and satisfaction dimensions, in line with the definition of well-being. Posi_neg (positive minus negative sentiment) is the net affection. Factor analysis results in Fig. 1 and Table 6 indicate that the measurement metric is valid.

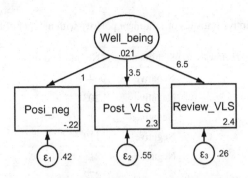

**Fig. 1.** Two-dimension model of online community well-being: net affection and satisfaction.

**Table 6.** Cronbach's alpha values.

| Dimensions of well-being | | Alpha |
|---|---|---|
| Satisfaction | Post VLS | 0.65 |
| | Review VLS | |
| Affection | Sentiment | 0.91 |
| | Posi-neg | |

## 5 Discussion

In this paper, a new measurement of online user well-being has been proposed on the basis of sentiment analysis and psychology literature. The factor analysis and text analytics conducted in the context of Smart Watch online reviews indicate that the new metric captures the concept of community member well-being, and the corpus can by used to predict well-being in the virtual community context. This study enriches the ongoing works on online user sentiment and happiness [20].

**Acknowledgement.** This work was supported by the National Natural Science Foundation of China [71901150] and China Postdoctoral Science Foundation Grant [2016M600677, 2019M663083], Guangdong Province Postgraduate Education Innovation Plan (2019SFKC46).

## References

1. Pang, W., Lee, L., Vaithyanathan, S.: Thumbs up?: Sentiment classification using machine learning techniques. In: Proceedings of the ACL-02 Conference on Empirical Methods in Natural Language Processing (EMNLP 2002), p. 10. Association for Computational Linguistics (2002)
2. Medhat, W., Hassan, A., Korashy, H.: Sentiment analysis algorithms and applications: a survey. Ain Shams Eng. J. **5**(4), 1093–1113 (2014)
3. Bai, X.: Predicting consumer sentiments from online text. Decis. Support Syst. **50**(4), 732–742 (2011)
4. Basari, A.S.H., Hussin, B., Ananta, I.G.P., Zeniarja, J.: Opinion mining of movie review using hybrid method of support vector machine and particle swarm optimization. Procedia Eng. **53**, 453–462 (2013)
5. Wang, G., Sun, J., Ma, J., Xu, K., Gu, J.: Sentiment classification: the contribution of ensemble learning. Decis. Support Syst. **57**, 77–93 (2014)
6. Zhai, Z., Liu, B., Wang, J., Xu, H., Jia, P.: Product feature grouping for opinion mining. IEEE Intell. Syst. **27**(4), 37–44 (2011)
7. Zhang, Z., Ye, Q., Zhang, Z., Li, Y.: Sentiment classification of Internet restaurant reviews written in Cantonese. Expert Syst. Appl. **38**(6), 7674–7682 (2011)
8. Quercia, D., Ellis, J., Capra, L., Crowcroft, J.: Tracking "gross community happiness" from tweets. In: Proceedings of the ACM 2012 Conference on Computer Supported Cooperative Work (CSCW 2012), pp. 965–968 (2012)

9. Schwartz, H.A., et al.: Predicting individual well-being through the language of social media. In: Biocomputing 2016: Proceedings of the Pacific Symposium (BIOC 2016), pp. 516–527 (2016)
10. Wang, J., Feng, Y., Naghizade, E., Rashidi, L., Kwan, H., Lee, K.: Happiness is a choice: sentiment and activity-aware location recommendation. In: WWW 2018 Companion: The 2018 Web Conference Companion (WWW 2018). ACM, New York (2018)
11. Ortigosa, A., Martin, J.M., Carro, R.M.: Sentiment analysis in Facebook and its application to e-learning. Comput. Hum. Behav. **31**, 527–554 (2014)
12. Diener, E., Suh, E.M., Lucas, R.E., Smith, H.L.: Subjective well-being: three decades of progress. Psychol. Bull. **125**(2), 276–302 (1999)
13. Ryff, C.D.: Happiness is everything, or is it? explorations on the meaning of psychological well-being. J. Pers. Soc. Psychol. **57**(6), 1069 (1989)
14. Springer, K.W., Hauser, R.M.: An assessment of the construct validity of Ryff's scales of psychological well-being: Method, mode, and measurement effects. Soc. Sci. Res. **35**(4), 1080–1102 (2006)
15. Keyes, C.L.M.: Social well-being. Soc. Psychol. Q. **61**(2), 121–140 (1998)
16. Shapiro, A., Keyes, C.L.M.: Marital status and social well-being: Are the married always better off? Soc. Indic. Res. **88**(2), 329–346 (2008)
17. Li, Z., Zhao, H., Liu, Q., Huang, Z., Mei, T., Chen, E: Learning from history and present: next-item recommendation via discriminatively exploiting user behaviors. In: Proceedings of the 24th ACM SIGKDD International Conference on Knowledge Discovery & Data Mining (SIGKDD 2018), pp. 1734–1743. ACM, New York (2018)
18. Lindner, A., Hall, M., Niemeyer, C., Caton, S.: BeWell: a sentiment aggregator for proactive community management. In: Proceedings of the 33rd Annual ACM Conference Extended Abstracts on Human Factors in Computing Systems (HFCS 2015), pp. 1055–1060. ACM, New York (2015)
19. Hameed, Z., Garcia-Zapirain, B.: Sentiment classification using a single-layered BiLSTM model. IEEE Access **8**, 73992–74001 (2020)
20. Syed, B., Indurthi, V., Shah, K., et al.: Ingredients for happiness: Modeling constructs via semi-supervised content driven inductive transfer learning. In: Proceedings of the 2nd Workshop on Affective Content Analysis (AffCon 2019). AAAI, Honolulu, Hawaii (2019)

# Channel Parameters Extraction Based on Back Propagation Neural Network

Huiting Li[1]([✉]), Liu Liu[1]([✉]), Jiachi Zhang[1], Kai Wang[1], Yiqian Li[1], Tao Zhou[1], and Jiahui Qiu[2]

[1] School of Electronic and Information Engineering,
Beijing Jiaotong University, Beijing 100044, China
lhtlucky1997@163.com, {liuliu,zhangjiachi,18120133,18125038,
taozhou}@bjtu.edu.cn
[2] Network Technology Research Institute of China United Network Communications
Group Co., Ltd., Beijing 100048, China
qiujh21@chinaunicom.cn

**Abstract.** Accurately mastering the information of wireless channel characteristics is of great significance for improving the spectrum utilization rate and channel capacity. This paper studies the application of back propagation neural network (BPNN) in channel parameters extraction based on QuaDriGa platform. In this paper, the QuaDriGa platform is used to generate the Channel Impulse Response (CIR) in urban scenes, and SAGE algorithm is used to extract channel parameters such as delay spread, azimuth angle (AOA, AOD) in horizontal dimension and elevation angle (EOA, EOD) in vertical dimension. Then BPNN is trained with sample data to extract different channel parameters. The results show that there is little difference between the prediction results of BPNN model and SAGE algorithm, so BPNN model can replace SAGE algorithm to extract channel parameters for MIMO channel simulation. In addition, the time complexity of the two methods is also compared. The results show that BPNN has higher time complexity than SAGE algorithm. Besides, the simulation results of three common error back propagation algorithms are compared. The results show that the L-M algorithm has the lowest mean square error and the best effect in training BPNN model.

**Keywords:** MIMO channel · Back propagation neural network · SAGE · Error back propagation algorithm

## 1  Introduction

The demand of high-data-rate application is expected to grow explosively in the fifth generation (5G) wireless communication systems. The explosion of data and the demand for fast and efficient data exchange put forward higher requirements for the spectrum utilization and capacity of wireless communication systems. In order to improve the spectrum utilization rate and realize ultra-large capacity transmission, the premise is to accurately grasp the wireless channel characteristic information and establish an accurate channel model [1].

© Springer Nature Switzerland AG 2020
X. Chen et al. (Eds.): ML4CS 2020, LNCS 12488, pp. 503–515, 2020.
https://doi.org/10.1007/978-3-030-62463-7_47

There exist two traditional modeling methods for wireless channels: one is geometry-based stochastic model (GBSM) and another is deterministic model. The method of GBSM predicts the fading of the received signal according to the probability distribution experience, and then reconstructs the channel impulse response (CIR) by the stochastic method. The disadvantage of this method is that it cannot accurately predict the CIR of a specific geographical location. The deterministic model method uses electromagnetic field theory to determine CIR for a certain geographic location. The main problem of this method is the large amount of computation and poor universality. Once the environment changes, the original model will no longer be applicable [2]. Therefore, the traditional channel modeling method has various shortcomings.

With the development of artificial intelligence, the combination of neural network and wireless communication has attracted the attention of many scholars. In [3], channel characteristics such as path loss are analyzed based on the radial basis function (RBF) artificial neural network model. The results show that the neural network model can accurately replay measured data, and the parameters extracted from the model can be used for time-varying channel simulation. In [4], a depth Multiple-Input Multiple-Output (MIMO) detector is built based on depth learning. The iterative process of gradient projection method is expanded into a depth neural network, which compromises the performance and complexity. In [5], time-varying channel modeling method is studied based on multilayer neural network. Simulation results show that neural networks can track the channel characteristics well. However, the focus of the above research mainly focuses on the application of machine learning method to solve the problem of wireless channel modeling. For the description of statistical characteristics of wireless channel parameters, only the azimuth component of horizontal dimension is considered, but the pitch component of vertical dimension is not considered, which is not completely match with the actual wireless channel.

In this paper, the application of back propagation neural network (BPNN) in channel parameters extraction is studied. The QuaDriGa simulation platform is used to generate the CIR in urban scenes, and SAGE algorithm is used to extract channel parameters such as delay spread, azimuth angle in horizontal dimension and elevation angle in vertical dimension. Then BPNN is trained with sample data to extract channel parameters, and the results are compared with SAGE algorithm. In addition, the time complexity of BPNN and SAGE algorithms is also discussed, and the prediction effects of three common error back propagation algorithms are compared.

## 2  Traditional MIMO Channel Parameter Estimation Algorithm

In order to complete the integrity description of 3D MIMO channel, channel basic parameters such as delay spread and Azimuth of Arrival (AOA) need to be obtained through channel parameter extraction technology. Traditional channel parameter estimation algorithms can be mainly classified into three categories: spatial spectral estimation algorithm, parameter subspace-based estimation algorithm and deterministic parameter estimation algorithm [6]. The performance of spectral estimation algorithms is excellent, but depending on the quality of incoming signals, its estimation ability will be greatly reduced or even completely invalidated under low SNR conditions. The parameter subspace estimation algorithm has fast computation speed, but it needs the cooperation of spatial smoothing technology and its performance is poor. Determination parameter estimation algorithms, such as Expectation-Maximization (EM) algorithm and space-alternating generalized expectation maximization (SAGE) algorithm, are mainly used to estimate multi-dimensional parameters and still have better performance under low SNR [7]. Because the research work in this paper needs to obtain three-dimensional channel parameters, SAGE algorithm is mainly used. The principle is as follows [8]:

Assuming that signal $X_l(t), \ldots, X_L(t)$ are independent of each other, the signal on the $l$-$th$ path is the superposition of the transmission signal $S(t; \theta_l)$ and the noise signal $N_l(t)$.

$$X_l(t) = S(t; \theta_l) + N_l(t) \tag{1}$$

Where $\theta_l = \left[\theta_{AOA,l}, \theta_{AOD,l}, \phi_{EOA,l}, \phi_{EOD,l}, \tau_l, \alpha_l\right]$ is the estimation parameter matrix, $\theta_{AOA,l}, \theta_{AOD,l}, \phi_{EOA,l}, \phi_{EOD,l}$ are the horizontal arrival azimuth, horizontal departure azimuth, vertical arrival pitch angle and vertical departure pitch angle on the $l$-$th$ path respectively, $\tau_l$ and $\alpha_l$ are the delay spread and complex amplitude.

According to the principle of maximum likelihood estimation (ML), in order to find the required parameters, it is necessary to find $\theta_l$ so that the likelihood function $\Lambda(\theta_l; x_l)$ obtains the maximum value:

$$(\theta_l)_{ML}(x_l) \in \arg\max\nolimits_{\theta_l} [\Lambda(\theta_l; x_l)] \tag{2}$$

The conditional expectation can be calculated from $Y(t) = y(t)$ and the last estimated values $\theta'$ of $\theta$:

$$x_l(t; \theta') = E_{\theta'}\left[X_l(t)|Y(t) = y(t)\right]$$

$$= y(t) - \sum_{l'=1, l' \neq l}^{L} s(t; \theta'_{l'}) \tag{3}$$

$x_l(t; \theta')$ obtained by conditional expectation is used to reevaluate the propagation wave parameter vector $\theta_l$ through maximum likelihood estimation to obtain:

$$\theta''_l = (\theta_l)_{ML}\left[x_l(t; \theta'_l)\right] \tag{4}$$

SAGE algorithm divides the parameter matrix space set into parameter subsets with smaller dimensions. Each iteration only updates the parameter subsets. It divides the parameter matrix $\theta_l = \left[\theta_{AOA,l}, \theta_{AOD,l}, \phi_{EOA,l}, \phi_{EOD,l}, \tau_l, \alpha_l\right]$ into subsets such as $(\theta_{AOA,l}, \tau_l), (\theta_{AOA,l}, \alpha_l)$. The values of the parameter matrix obtained by one iteration are such as follows:

$$\tau''_l = \arg\max\nolimits_{\tau_l}\left[\theta'_{AOA,l}, \theta'_{AOD,l}, \phi'_{EOA,l}, \phi'_{EOD,l}, \tau_l, \alpha'_l\right]$$

$$\theta''_{AOA,l} = \arg\max\nolimits_{\theta_{AOA,l}}\left[\theta_{AOA,l}, \theta'_{AOD,l}, \phi'_{EOA,l}, \phi'_{EOD,l}, \tau'_l, \alpha'_l\right] \tag{5}$$

The parameter matrix $\theta_l$ of the $l$-$th$ path required finally can be obtained through multiple iterations of the parameter matrix that can make the objective function obtain the maximum value.

# 3 Neural Network Modeling MIMO Channel

## 3.1 The Principle of Simulating MIMO Channel by Neural Network

The communication channel can be regarded as a nonlinear complex system. When a nonlinear system is modeled by neural network, it is assumed that $x$ represents the $N_x$-dimensional input signal and $y$ represents the $N_y$-dimensional output signal. The mapping relationship between $y$ and $x$ may be highly nonlinear and multidimensional. In general, the relationship can be defined by a formula: $y = f(x, \omega, b)$, where $\omega, b$ representing the weight and the threshold of the neural network model respectively. The neural network is used to predict the channel model, which can learn the input and output data of the nonlinear system and then approach the input and output relationship.

According to Kolmogorov theorem [9], the three-layer network can approach almost any function after sufficient learning, so the three-layer structure neural network shown in Fig. 1 can be established for learning and training. Using neural network to learn and predict channel parameters can reduce the amount of data calculation and has high precision and strong flexibility.

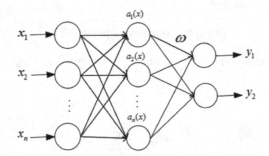

**Fig. 1.** Structure diagram of three-layer neural network

Figure 2 shows a neural network-based channel regression model. This model combines the advantages of neural network and regression analysis, and can better reveal the quantitative relationship between variables in complex channel environment. The model divides the actually measured data into two parts: training set and test set. The training set of the input layer can be channel parameters or variables related to the environment. The training parameters of the output layer come from actual measurement or classical model. Then, the NN algorithms such as back propagation algorithm or Newton algorithm are used to train the training set in the hidden layer, and fitting values of corresponding parameters are obtained in the output layer, thereby estimating channel parameters or establishing a channel model; Finally, the fitting value is compared with the test set, and some evaluation indexes such as residual error and root mean square error are used to obtain the accuracy of applying NN algorithm to wireless channel simulation. In addition, cross-validation can be used to verify the generalization ability of the model.

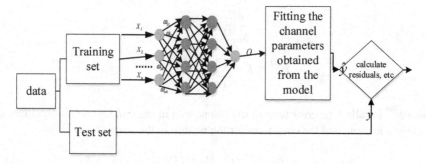

**Fig. 2.** General model of neural network in wireless channel simulation

## 3.2 The Principle of Back Propagation Neural Network

Back propagation neural network (BPNN) is used to extract the channel parameters. BPNN is a multilayer feedforward neural network trained by the error back propagation algorithm. Error reverse transmission is used to find the source of the error, and then $\omega$ and $b$ are updated according to the principle of gradient descent.

Before training by BPNN, the input data need to be preprocessed and the value range of the original data can be mapped to the [0, 1] interval. Then the normalized sample data is transferred in the direction of input layer → hidden layer → output layer, and the output signal is generated through nonlinear transformation. When the activation function of neurons in each layer adopts sigmod function, it is expressed as:

$$f(x) = 1/\left(1 + e^{-\left(W^T x + b\right)}\right) \tag{6}$$

Where, $W = \left[w_{ji}^{(p)}\right]$ represents the network weight matrix, $x$ represents the input vector, and $b = \left[b_j^{(p)}\right]$ is the network threshold matrix, $b_j^{(p)}$ represents the threshold of the $j$-th neuron in the $p$-th layer. The update formula is as follows:

$$W(t + 1) = W(t) - \eta \frac{\partial E(t)}{\partial W(t)} \tag{7}$$

$$b(t + 1) = b(t) - \eta \frac{\partial E(t)}{\partial b(t)} \tag{8}$$

Where $E(t)$ represents the total error, $t$ represents the number of training times, and $\eta \in (0, 1)$ represents the step factor. Comparing the actual output with the expected output, if not equal, calculating the error $E(t)$ and transferring to the reverse transmission process of the error. The output error of the output layer is transmitted in the reverse direction, and the error is distributed to all units to calculate the residual error of each node:

$$\frac{\partial E(t)}{\partial w_{ji}^{(p)}} = x_{ji}^{(p)} \delta_j^{(p)} \tag{9}$$

$$\frac{\partial E}{\partial b_j^{(p)}} = \delta_j^{(p)} \tag{10}$$

$$\delta_j^{(p)} = f' \cdot \sum_k w_{jk}^{(p+1)} \delta_k^{(p+1)} \tag{11}$$

Where $\delta_j^{(p)}$ is called the error term of the $j$-th neuron in the $p$-th layer, $f$ is the output of the current neuron, and the error term for the neuron in the output layer is:

$$\delta_k = f'(x_k) \cdot \left[ t_k - f(x_k) \right] \tag{12}$$

The error signal obtained from each layer is used to update $\omega$ and $b$ so that the error decreases in the gradient direction. The above process is repeated until the obtained error is less than the set value or reaches the alternative training requirements such as training times [10].

Therefore, the training steps of BPNN can be summarized as follows:

1) Normalized preprocessed channel parameters;
2) Forward transmission of signals, and calculation of network error $E$;
3) Reverse transmission of errors, allocating the errors to all units of each layer, and calculating the residual error of each node;
4) Updating and iterating the $\omega$ and $b$ according to the error signals obtained by each layer, so that the error decreases along the gradient direction;
5) Repeat the training process of 1) to 4) until the error is less than the set value or the iterative training requirements such as training times are met.

## 4 Simulation and Verification

### 4.1 Using BPNN to Predict and Simulate

In this paper, the QuaDriGa simulation platform is used to generate wireless channels in urban scenes. The QuaDRiGa is a random channel modeling method based on three-dimensional geometry. It can simulate 25 kinds of scenes such as urban and suburban [11]. The platform is used to generate CIR in urban scenes, and SAGE algorithm is used to extract channel parameters such as time delay spread, arrival/departure azimuth and elevation angle. Then BPNN is trained with sample data to extract channel parameters, and the results are compared with SAGE algorithm. The simulation parameter is shown in Table 1. BPNN training is carried out by MATLAB, and the simulation parameters are set as shown in Table 2.

**Table 1.** Parameters of simulation

| Parameters | Values |
| --- | --- |
| Simulation model | 3GPP_38.901 |
| Propagation scenario | Urban |
| Center frequency | 28 GHz |
| LOS/NLOS | LOS&NLOS |
| Transmitter antenna | Height 10 m |
| Receiver antenna | Height 1.5 m |
| Antenna scale | 16*32 MIMO |
| Antenna type | Omnidirectional Antenna |

**Table 2.** Parameters of Bpnn model

| Parameters | Values |
| --- | --- |
| Number of nodes per layer | 80 |
| Number of hidden layer | 1 |
| Iteration times | 1000 |
| Setting error | 0.1 |
| Training set | 70% |
| Test set/verification set | 15% |

Figure 3 shows the comparison results of angle domain parameters predicted by BPNN model and SAGE algorithm, which are respectively represented by solid lines and dashed lines. The horizontal axis indicates that the receiver moves 20 m continuously from the starting point, and the vertical axis indicates the angle results. The antenna scale of the simulation is 16*32 MIMO. In order to facilitate the observation and comparison of the effect, only the data on four propagation paths are extracted from the graph for comparison. It can be seen that the difference between the prediction effect of the model and SAGE is very small. The maximum root mean square values (RMSEs) of AOA and AOD between the predicted value of BPNN model and the actual value are 1.47° and 1.38° respectively, while the maximum angel extension of horizontal dimension is between 33°–41°, and the error value is very small relative to the maximum angle extension. Similarly, the maximum RMSEs of EOA and EOD are 0.74° and 0.89° respectively, while the maximum angle extension in vertical dimension is between 7.6°–24°. The results show that the parameters extracted from the BPNN model fit quite well with actual data. Figure 4 shows the comparison effect between the delay spread predicted by BPNN model and SAGE algorithm. The maximum RMSEs of the delay spread between

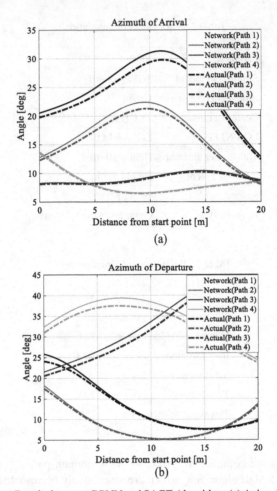

**Fig. 3.** Comparison Results between BPNN and SAGE Algorithm. (a) Azimuth of Arrival (AOA), (b) Azimuth of Departure (AOD), (c) Elevation of Arrival (EOA), (d) Elevation of Departure (EOD)

BPNN model and the actual value is 0.04 ms, with little difference, indicating that there is little difference between the prediction results of BPNN model and SAGE algorithm, so BPNN model can replace SAGE algorithm to extract channel parameters for MIMO channel simulation.

## 4.2 Time Complexity Comparison of Different Methods

SAGE algorithm needs many iterations and its calculation complexity is $O(n \log n)$, where $n$ is the number of sequence points. BPNN is trained by algorithm and its time complexity is $O(n^3)$, where $n$ is the number of network nodes. Complexity $O(n^3) > O(n \log n)$, therefore BPNN has higher time complexity. The time required

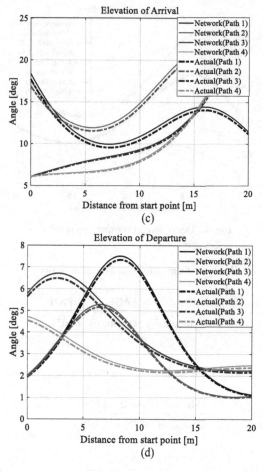

**Fig. 3.** (*continued*)

for the two methods to run with MATLAB is shown in Table 4. It can be seen that BPNN takes longer than SAGE algorithm. This is because BPNN needs to learn the training process repeatedly. Besides, the time required for neural network training is related to the algorithm and other factors (Table 3).

## 4.3  Comparison of Different Algorithms

Common propagation algorithms include Levenberg-Marquardt algorithm, Bayesian regularization algorithm and scaled conjugate gradient algorithm. These three algorithms are used to train BPNN respectively, and the mean square errors (MSEs) obtained are shown in Fig. 5 and Table 5. In Fig. 5, the horizontal axis represents BPNN training times, the vertical axis represents MSEs, and the circle represents the minimum MSEs

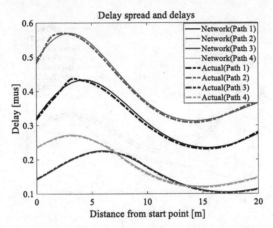

**Fig. 4.** Delay spread comparison results

**Table 3.** Comparision of time complexity

| Method | SAGE | BPNN |
|---|---|---|
| Time complexity | $O(n \log n)$ | $O(n^3)$ |

**Table 4.** Comparision of run time

| Method | SAGE | BPNN |
|---|---|---|
| Run time | 15.76 s | 27.30 s |

obtained after the verification data set passes through the trained network model. It can be seen that the MSEs obtained by Bayesian regularization algorithm and scaled conjugate gradient algorithm are $6.45 \times 10^{-3}$ and $9.17 \times 10^{-3}$, which is higher than the L-M algorithm. Therefore, the accuracy of L-M algorithm is higher.

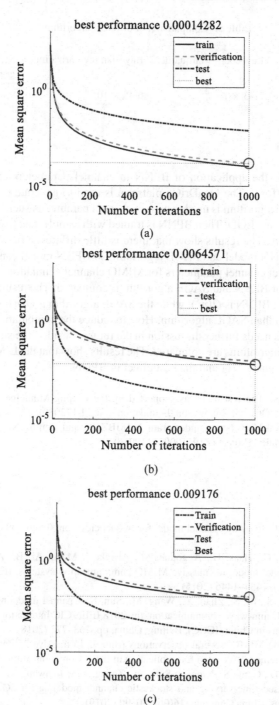

**Fig. 5.** Mean Square Error Comparision. (a) L-M algorithm, (b) Bayesian regularization algorithm, (c) scaled conjugate gradient algorithm

**Table 5.** Comparision of different algorithms

| Algorithms | Levenberg-Marquardt | Bayesian regularization | Scaled conjugate gradient |
|---|---|---|---|
| Mean square error | $1.43 \times 10^{-4}$ | $6.45 \times 10^{-3}$ | $9.17 \times 10^{-3}$ |

## 5 Conclusion

This paper studies the application of BPNN in channel parameters extraction based on QuaDriGa platform. The QuaDriGa platform is used to generate the CIR in urban scenes, and SAGE algorithm is used to extract channel parameters such as delay spread, AOA, AOD, EOA and EOD. Then BPNN is trained with sample data to extract different channel parameters. The results show that there is little difference between the prediction results of BPNN model and SAGE algorithm, so BPNN model can replace SAGE algorithm to extract channel parameters for MIMO channel simulation. In addition, the time complexity of BPNN and SAGE algorithm is compared. The results show that the time complexity of BPNN is $O(n^3)$, while the SAGE algorithm is $O(n \log n)$, BPNN has more computation than SAGE algorithm. How to reduce the computational complexity of neural networks needs further discussion in the future. Besides, the simulation results of three common algorithms are compared. The results show that the L-M algorithm has the lowest mean square error and the best effect in BPNN.

**Acknowledgements.** The research was supported by the Beijing Municipal Natural Science Foundation-Haidian Original Innovation Foundation (No. L172030), Fundamental Research Funds for the Central Universities under grant 2018JBZ102 and Beijing Nova Program Interdisciplinary Cooperation Project (Z191100001119016).

## References

1. GPP TR 38.901: Study on channel model for frequencies from 0.5 to 100 GHz, Rel. 14.1.1, July 2017
2. Wu, S., Wang, C., Haas, H., Aggoune, M., Alwakeel, M.M., Ai, B.: A Non-stationary wideband channel model for massive MIMO communication systems. IEEE Trans. Wirel. Commun. **14**(3), 1434–1446 (2015)
3. Sun, N., Geng, S., Li, S., Zhao, X., Wang, M., Sun, S.: Channel modeling by RBF neural networks for 5G mm-wave communication. In: 2018 IEEE/CIC International Conference on Communications in China (ICCC), Beijing, China, pp. 768–772 (2018)
4. Zhang et al.: 3D MIMO: several observations from 32 to massive 256 antennas based on channel measurement. Accepted for publication in IEEE Communications Magazine (2017)
5. Zhao, X., Fei, D., Geng, S., Ningyao Sun, Yu., Zhang, Z.F., Jianwang, G.: Neutral network and GBSM-based time-varying and stockastic channel modeling for 5G millimeter wave communications. China Commun. **16**(06), 80–90 (2019)
6. Ye, H., Li, G.Y., Juang, B.H.F.: Power of deep learning for channel estimation and signal detection in OFDM systems. IEEE Wirel. Commun. Lett. **7**(1), 114–117 (2017)

7. Ma, W.-D.K., Lewis, J.P., Bastiaan Kleijn, W.: The HSIC bottleneck: deep learning without back-propagation. Knowl. Discov. Data Mining **8**, 5 (2019)
8. Richter, A., Thoma, R.S.: Joint maximum likelihood estimation of specular paths and distributed diffuse scattering. In: Proceedings of the IEEE 61st Vehicular Technology Conference (VTC-Spring), vol. 1. Stockholm, Sweden, May/June 2005, pp. 11–15 (2005)
9. Silvio, F.B., Pinto, R.C.L.: Multi-step knowledge-aided iterative conjugate gradient algorithms for DOA estimation. Circuits, Syst. Signal Process. **38**(8) (2019)
10. Popescu, I., Nafornita, I., Constantinou, P.: Comparison of neural network models for path loss prediction. In: (WiMob 2005), IEEE International Conference on Wireless and Mobile Computing, Networking And Communications, vol. 1, pp. 44–49. IEEE (2005)
11. Jaeckel, S., et al.: QuaDRiGa: a 3-D multi-cell channel model with time evolution for enabling virtual field trials. IEEE Trans. Antennas Propag. **62**(6), 3242–3256 (2014)

# Automatic Vibration Waveform Recognition Using Convolutional Neural Network

Wei Ding[1], Heng Liu[2], Dijing Wang[1], and Yong Jiang[1(✉)]

[1] Key Laboratory of Earthquake Geodesy, Institute of Seismology,
China Earthquake Administration, 430071 Wuhan, China
tingwhere@163.com, wangdijin@126.com, 653122705@qq.com
[2] School of Information and Management, Guangxi Medical University, 530021 Nanning, China
35354682@163.com

**Abstract.** A rapid verification model of intrusion signals in the vibration monitoring system can be used to improve the speed and accuracy of vibration identification and microseism research. Existing waveform recognition methods rely on preprocessing the waveform signal such as clipping and aligning. The requirements of high signal-to-noise ratio for waveform signal make those methods inaccurate in term of accuracy and speed. By using the improved VGG16 model for seismic waveform recognition to identify two types of vibration signals, a model for intelligent recognition of vibration signals based on convolutional neural networks is employed to identify 2 types of vibration waveform recognition. The test accuracy rate of the model was 96.8%, and the single frame detection time was 0.069 s. The experiments show that this algorithm can deal with the phase alignment problem and classify small sample size waveform data effectively.

**Keywords:** Vibration waveform recognition · VGG16 · Convolutional neural network · Deep learning

## 1 Introduction

Waveform recognition technology is widely used in the fields of intelligent fault diagnosis, clinical electrocardiographic monitoring, and digital instruments [1]. Identifying and classifying the waveform recorded by related instruments can help stakeholders to understand related research knowledge. Different sensors can be employed to observe various physical quantities continuously. The collected data can be expanded in the time domain to form various observation waveforms. With the development of embedded technology and artificial intelligence, design the algorithm and related model with fast and effective waveform recognition will improve the development of intelligent instruments, e.g., seismic data acquisition, which recorded non-periodic low-frequency signal. The analysis and classification of waveform characteristics is the prerequisite for vibration monitoring and early warning, which can be distinguished from natural vibration or perimeter intrusion signals.

© Springer Nature Switzerland AG 2020
X. Chen et al. (Eds.): ML4CS 2020, LNCS 12488, pp. 516–525, 2020.
https://doi.org/10.1007/978-3-030-62463-7_48

Waveform signal identification methods include frequency domain analysis, three-dimensional spectral array, and wavelet analysis. The preprocessing processes of above-mentioned methods are inefficient and tedious. For example, the initial signals need to be processed adapting twice harmonic and aligned with each other. The time-domain waveform of the vibration signal can directly reflect the collected data. If the signal can be directly analyzed, the classification task can be completed without secondary analysis. Machine learning techniques have been developed to perform efficient automatic classification systems [2]. This paper use a model based on VGG16 architecture to perform transfer learning and explore classification for waveform of vibration signals. The data used in this paper is a subset of two types of waveform image data, which is collected using vibration warning instrument and composed of 2 classes, 600 training images, and 300 test images. The test accuracy rate of the model was 98.3%, and the single frame detection time was 0.069 s, which can satisfy the requirements of vibration alarm.

## 2  Related Work

The methods of waveform recognition are usually divided into two categories: 1, Matching waveform template in time domain to recognize, i.e., aligning, comparing and matching the time domain characteristic parameters of a waveform to the standard template waveforms. For example, Maximum correntropy criterion (MCC) provides a criterion for non-Gaussian signal processing to calculate the maximum correlation coefficient between the waveform under test and the template waveform [3]. But this method is unable to distinguish subtle differences in the waveform. In order to improve the stability presents an algorithm based on maximum close distance (MCD) correlation matching, which combines multi sub-templates matching and target movement prediction [4]. If the amplitude or width of the detected signal exceeds twice the signal in MCD template, the correlation coefficient will be greater than 1 and unable to recognize signals. Hausdorff distance can be used as a metric to identify image similarity, which has strict requirements for the waveform division [5]. The abovementioned methods need to align the waveform to be measured with various waveform templates, which cannot identify waveform rapidly and efficiently in automatic detection system. 2, Analyzing and extract waveform characteristics in frequency domain, i.e., use Fourier transform, wavelet transform and other methods to extract the frequency domain features of the signal waveform, and compare with the existing template of frequency domain features. For example, Wang et al. extracted features to establish a two-dimensional time-frequency domain space, and applied subspace projection method for waveform recognition [6]. But this method is depended on complicated calculation to choose two-dimensional window function parameters optimally.

The techniques of deep learning to realize waveform recognition mainly include self-organizing mapping network [7], BP neural network [8], and support vector machine [9]. The requirements of abovementioned methods have the following characteristics: high discrimination for the sample dataset, high signal-to-noise ratio for the waveform, time-consuming to train, and complex neural network structure, which increases power consumption and reduces the calculation rate of hardware system.

## 3  Design Principle

The waveform recognition algorithm consists of three steps: 1, waveform data preprocessing; 2, improving model based on the deep convolutional neural network VGGNet; 3, classifying images using model.

### 3.1  Data Precession

The phases of waveform data preprocessing are elaborated as follows:

1) Removing error and missing points;
2) Truncating some waveform data without vibration;
3) Adapting window smooth filtering to increase the accuracy of the waveform image.

### 3.2  VGGNet Model

Neural network can simulate the structure and function of human brain, which has adaptive learning ability and has important applications in the research of pattern recognition, e.g., classification recognition based on memory association and optimized calculation. Convolutional Neural Network (CNN) is a feed-forward neural network. It is consist of cross-stacking convolutional layers, convergence layers, and fully connected layers. It is trained to handle image information based on the backpropagation algorithm. When using fully connected feedforward neural networks to train images, there will be a problem that there are too many parameters and it is difficult to extract local invariant features. The convolutional neural network has three structural characteristics: local connections, weight sharing and equal variation represent, which make the convolutional neural network to maintain translation, scaling and rotation invariance. VGG16 is a type of convolutional neural network [10]. It explores the relationship between the depth of the convolutional neural network and its performance. The max pooling layers are used to reduce the size of the picture by repeatedly stacking. As shown in Fig. 1, by repeatedly stacking $3 \times 3$ small convolution kernels and $2 \times 2$ maximum pooling layers, VGG16 constructs a 16-layer deep convolutional neural network, i.e., 16 layers using convolutional layer (conv) in addition to the max pooling layer represented by solid frame and dotted frame separately. Its main characteristics are: 1) The concept of raising the convolutional layer to the convolutional block: the convolutional block is composed of 2–3 convolutional layers, which makes the network have a larger receptive field while reducing the network parameters, and multiple use of ReLu activation function has more linear transformation and stronger learning ability; 2) Multi-scale is used for data enhancement during training and prediction: the same picture is scaled to different sizes during training, and randomly cropped to $224 \times 224$, which can increase the amount of data. In the hidden layer of the VGG-16 convolutional neural network, an activation function ReLU is used to make the neural network have a nonlinear mapping ability. The ReLU function is an unsaturated function whose operation result is relatively sparse, which can reduce the back propagation error and accelerate the convergence speed of the network.

Keras provides the TensorFlow backend engine and APIs of machine learning library that allow users to focus more on model design and conduct model experiments faster. We implemented our model in Keras using ImageNet weights provided by the library.

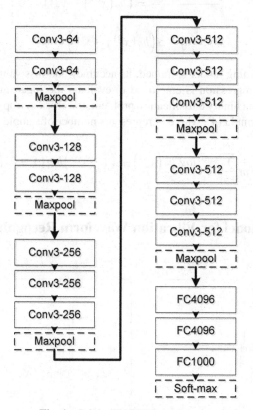

**Fig. 1.** Original VGG16 architecture

## 3.3  Model Optimization Algorithm

As one of the most commonly used gradient descent optimization algorithm for convolutional neural networks, stochastic gradient descent (SGD) is employed in our work. The advantages of SGD are described as follows: gradient is updated iteratively at each sampling round; compared with non-random algorithms, SGD can use the redundant information more efficiently, which can be represented as formula (1)–(3), where $h_\theta$ represents cross-entropy and $i$ represents the category number:

The objective function $J$ for a sample is shown as follows:

$$J^{(i)}(\theta_0, \theta_1) = \frac{1}{2}\Big(h_\theta\big(x^{(i)}\big) - y^{(i)}\Big)^2 \tag{1}$$

The partial derivative of the objective function can be calculated in formula (2) and (3):

$$\frac{\Delta J^{(i)}(\theta_0, \theta_1)}{\theta_j} = \left(h_\theta\left(x^{(i)}\right) - y^{(i)}\right) \tag{2}$$

$$\theta_j := \theta_j - \alpha\left(h_\theta\left(x^{(i)}\right) - y^{(i)}\right)x_j^{(i)}) \tag{3}$$

After a deep learning model is trained, its accuracy and loss value need to be evaluated. The logistic regression is employed to evaluate the logarithmic loss function. Because this work is a binary classification problem, the cross-entropy loss function can be adapted using formula (4), where m represents number of sample.

$$J(\theta) = -\frac{1}{m}\sum_{i=1}^{m}[y^{(i)}\log h_\theta\left(x^{(i)}\right) + \left(1 - y^{(i)}\right)\log\left(1 - h_\theta\left(x^{(i)}\right)\right)] \tag{4}$$

## 4   Improved Model for Vibration Waveform Recognition

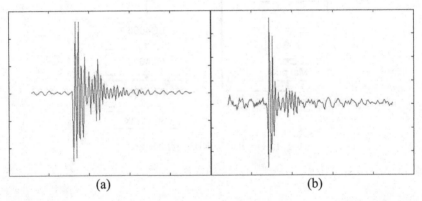

|           |           |
|:---------:|:---------:|
|    (a)    |    (b)    |

**Fig. 2.** Vibrations signals collected by warning system from various source, (a) from footstep (b) from metal knock.

Figure 2 shows the footstep signal and metal knock signal images collected by the early warning monitoring terminal separately. The duration and amplitude of the signals are different for each image. Note that, in order to make the trained model have better generalization, we did not normalize the amplitude and delete the time and amplitude axis marks in the figure in the data set preprocessing. Design model to detect and classify alarm from footsteps and metal knocking, and implement alarm detection only for metal tapping intrusions. The model needs to be trained on the vibration data set to achieve the optimal detection effect. Since the images in vibration dataset are simple and solid-color, it is possible to reduce excessive convolutional layers.

VGGNET has the following deficiencies for wave recognition:

1) The number of network layers is large, the amount of calculation during training is large, and the convergence is slow, which require a large amount of training sample data;
2) There are many network parameters, more memory is required to store the network, and the training time is relatively longer;
3) VGGNet uses SoftMax classifier to classify images. SoftMax is an extension of logistic regression to solve multi-classification problems. There is no special optimization for waveform recognition. The characteristics of waveforms are large intra-class differences and small differences between classes. Therefore, the following improvements can be made: 1. the number of fully connected layer can be reduced to lessen network parameters from more connections; 2. to maintain image feature extraction performance, reduce the number of convolutional layers and the amount of calculation.

An improved VGG16 model for vibration image is employed for image dataset. The number of weight parameters of VGG-16 is 138 million, of which 3 fully connected layer parameters are highly concentrated. The parameters of VGG16 are designed for 1000 classification categories, and this study only focuses on 2 classification categories. Therefore, it is proposed to replace the original three fully connected layers with two fully connected layers. In order to improve the recognition accuracy and efficiency of the model, the first fully connected layer is 4 096, and the second fully connected layer is 2. As shown in Fig. 3, the improved model use a small convolution kernel (3 × 3), two convolutional layers with a max pooling layer to make the first three convolutional pooling structures of VGG16: increasing the number of convolution kernels (64, 128, 256) using powers of 2. To avoid excessive calculation, the fully connected layer does not use the huge first three-layer structure of VGG16. The model can be divided into two phases, i.e., training phase and classifying phase.

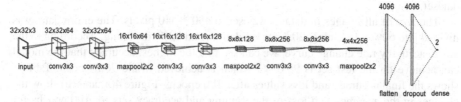

**Fig. 3.** Architecture of improved VGG16 model for vibration waveform recognition

In training phase: the first convolutional network consists of 2 convolutional layers and 1 max pooling layer. The size of the convolution kernels of the two convolutional layers is 3 × 3, the number of convolution kernels is 64, and the step size is 1 × 1. The input size of the first convolutional layer is 32 × 32 × 3, and the output size is 32 × 32 × 64. The input and output size of the second convolutional layer are adjusted to 32 × 32 × 64 to receive the output of a convolutional layer. The layer near two convolutional layers is a 2 × 2 max pooling layer. Since the step size is 2, after the maximum pooling, the output size becomes 16 × 16 × 64. The structures of the second and third convolutional networks are similar to that of the first one. The size of the convolution kernel of the two convolutional layers is also 3 × 3, but the number of output channels after these two convolutional layers has changed to 8 × 8 × 128 and 4 × 4 × 256 separately.

In classifying phase, the results are input to the Flatten layer, which is used to flatten the input, i.e., convert the multi-dimensional input to one-dimensional data. The weight is initialized by He Normal instead of Xavier, and a dropout layer is added to prevent the model from overfitting and improve the generalization of the model. Since it is a binary classification problem, the last layer has only one neuron with activation function sigmoid. At this time, the number of weight parameters is reduced to 2.83 million, which reduces the amount of calculation. Note that, the ReLu layer has been omitted for simplicity in Fig. 3. All the convolutional layers adopt the SAME architecture, i.e., the input of each layer of convolution will fill a circle of pixels with a value of 0 on the outermost layer, so the length and width of the final convolution result will be consistent with the input.

## 5  Result and Discussion

### 5.1  Performance of Improved Model

The ADXL345 digital acceleration sensor is used to collect vibration data. The Keras 2.2 with Tensorflow backend engine is implemented in Python 3.6 as the algorithm running platform. In order to match the sampling rate with the information processing rate, the sensor sampling rate is set to 50 points per second. The training data in the experiment includes two categories: footstep interference signals and metal knock signals collected by the perimeter early warning monitoring system. We collected 600 mixed samples of two types of waveforms as dataset for classifying and testifying the waveforms. In addition to collect waveforms images directly from early warning monitoring system, no noise suppression method is added during the pre-processing stage of generating dataset.

The size of all images in dataset was set to 450 * 340 pixels. The entire dataset is divided into 60% for training, 20% for validation, and 20% for testing. According to the above accuracy rate change curve, we can see that the highest accuracy that the model can achieve on the test set is 99.03%, which is achieved in the 96th epoch. Table 1 shows the final accuracy and loss values after 100 epochs. Figure 4(a) and (b) show the changes in the accuracy and loss of the training and accuracy sets for 100 epochs for original VGG16 model and improved VGG16 model separately. Figure 4 shows that the accuracy of the improved model is significantly improved, and the loss is significantly reduced. Applying the trained model stored with h5 style, the detection time of a single frame in test dataset was 0.069 s, which is faster than our recent work using STA/LTA [11].

**Table 1.** Accuracy and loss for 100 epochs for improved model

| Dataset | Accuracy | Loss |
|---|---|---|
| Training dataset | 0.9903 | 0.0230 |
| Test dataset | 0.9680 | 0.0046 |

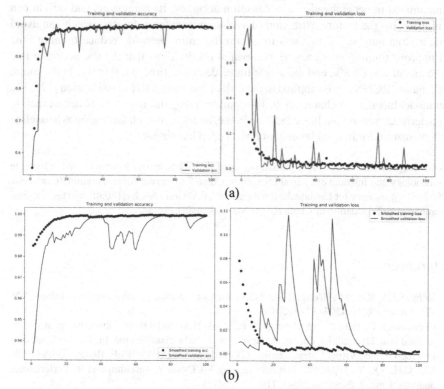

**Fig. 4.** The accuracy and loss of the training and accuracy sets for 100 epochs for (a) original VGG16 model and (b) improved VGG16 model

## 5.2  Discussion

The improved VGG16 model avoids the operation of waveform alignment between the waveform and the template which commonly used in other waveform recognition algorithms. It is testified that the proposed improved model offers significant improvement, such as robustness to noise and recognition accuracy. The number of training samples for the neural network in this paper is small, which can affect the classification accuracy. However, under the influence of these unfavorable factors, the model can automatically detect wave with various characteristic in a certain accuracy rate, which reflects the feasibility of the algorithm.

# 6 Conclusion

Using a dataset with a small sample size to classify the vibration signals, this paper applies artificial intelligence to vibration waveform recognition and proposes an improved model of vibration waveform classification based on VGG16 convolutional neural network. In this model, the fully connected layers and convolutional layers from the VGG-16 model are improved. By adjusting the depth of the fixed layer during training phase, the model is maximized in generalization and detection accuracy. Its parameters and weight can be migrated to the feature extraction of vibration waveform defects, which can avoid using a large number of data sets to retrain the entire network, reduce training time and improve training efficiency. Experimental results show that the test accuracy rate of the model was 96.8%, and the single frame detection time was 0.069 s. In the future work, Faster-RCNN can be applied to the task of multi objective classification [12]. The new model has several characteristics for pattern recognition such as better extracting characteristics, recommending candidate frame, and detecting classification as a baseline implementation for image classification with deep learning.

**Acknowledgements.** This work is partially sponsored by the Scientific Research Fund of Institute of Seismology and Institute of Crustal Dynamics, China Earthquake Administration, (Grand No. IS201726156), Science for Earthquake Resilience of CEA (Grant No. XH18030), and the National Natural Science Foundation of China (Grant No. 61903089).

# References

1. Kong, S.-H., Kim, M., Hoang, L.M.: Automatic LPI radar waveform recognition using CNN. IEEE Access **6**(1), 4207–4219 (2018)
2. Orenstein, E.C., Beijbom, O., Peacock, E.E., Sosik, H.M.: WHOH-Plankton - a large scale fine grained visual recognition enchmark dataset for plankton classification. In: The 3rd Workshop on Fine-Grained Visual Categorization (CVPR), bs/1510.00745. IEEE, Boston, USA (2015)
3. Liu, C.H., Qi, Y., Ding, W.: The data-reusing MCC-based algorithm and its Performance analysis. Chin. J. Electron. **25**(4), 719–725 (2016)
4. Liu, P., Wang, C.P., Fu, Q.: An anti-occlusion tracking algorithm based on MCD correlation matching. Adv. Mater. Res. **718–720**(1), 2005–2009 (2013)
5. Wei, G., Lin, R., Wang, H.: Distance and similarity measures for hesitant interval-valued fuzzy sets. J. Intell. Fuzzy Syst. **27**(1), 19–36 (2014)
6. Wang, J., Zhang, S.H.: Detection of transient signals using matched subspace projection. Syst. Eng. Electron. **2**(1), 22–24 (2000)
7. Chiu, C.C., Hsu, C.L., Liau, B.Y., Lan, C.Y.: Real-time electrocardiogram waveform classification using self-organization neural network. In: Lim, C.T., Goh, J.C.H. (eds.) 13th International Conference on Biomedical Engineering (ICBME), pp. 578–581. Springer, Heidelberg (2008). 10.1007/978-3-540-92841-6_142
8. Wu, Q.M., Wei, M.: A mathematical expression for air ESD current waveform using BP neural network. J. Electrostat. **71**(2), 125–129 (2012)
9. Jia, D., Zhang, D., Naimin, L.: Pulse waveform classification using support vector machine with gaussian time warp edit distance kernel. Comput. Math. Methods Med. **15**(2), 947254 (2014)

10. Simonyan, K., Zisserman, A.: Very deep convolutional networks for large-scale image recognition. In: The 3rd International Conference on Learning Representations (ICLR), San Diego, CA, USA (2015). arXiv:1409.1556
11. Ding, W., Wu, T., Liao, C., Wang, J., Jiang, Y.: A high resolution ADC model combined pipeline and sigma -delta oversampling architecture. In: The 18th IEEE International Conference on Cognitive Informatics & Cognitive Computing (ICCICC), pp. 350–355. IEEE, Milan, Italy (2019)
12. Lokanath, M., Kumar, K.S., Keerthi, E.S.: Accurate object classification and detection by faster-RCNN. IOP Conf. Ser. Mater. Sci. Eng. 263(5), 052028 (2017)

# A Prediction Method for Soil Moisture Time Series

XiaoXia Yang[1], Song Jia[2], and Chengming Zhang[1](✉)

[1] School of Information Science and Engineering, Shandong Agricultural University,
Tai'an 271018, China
chming@sdau.edu.cn
[2] Tai'an Branch of China Unicom, Tai'an 271000, China

**Abstract.** Soil moisture prediction with high quality has important guiding effect on agricultural production. Aiming at the problem that the time series of soil moisture is non-stationary, a prediction method which combine wavelet transform and improved Back Propagation Neural Network (BPNN) is proposed in this paper. First obtained several relatively stable data sequence with different scales by decompose the original time series using the discrete wavelet transform, then use BPNN to predict each sequence respectively, finally reconstruct to obtain final prediction result. Aiming at the problem that BPNN has slow converging speed and easy to fall into the local optimal solution, proposed an optimization method by adjusting the momentum factor and learning rate adaptively in the learning process. The experiment has been carried out in KenLi town, the research region of "BoHai Barn" in ShanDong province, using data from 10 observation stations. We use 3 prediction methods respectively to predict the original time series, experimental results explicit that the proposed method has higher prediction accuracy (compared with the other two methods, increased by 9.5% and 31% respectively) and fewer iterations (compared with the other two methods, increased by 70% and 77% respectively).

**Keywords:** BP neural network · Soil moisture · Wavelet transform

## 1 Introduction

Soil moisture is an important factor in the process of energy exchange between land and atmosphere, and plays a key role in meteorology, hydrology and agriculture. As an indispensable nutrient for plant growth, soil moisture is one of the most important factors leading to the growth and yield of crops. Therefore, high quality prediction of soil moisture is the basic requirement for guiding agricultural production. With the development of the Internet of things, the ground observation network is more and more used in the field of agriculture, which makes it possible to obtain long time series soil moisture observation data in the observation area. These time series data contain the development rule information of the data.

© Springer Nature Switzerland AG 2020
X. Chen et al. (Eds.): ML4CS 2020, LNCS 12488, pp. 526–537, 2020.
https://doi.org/10.1007/978-3-030-62463-7_49

In the aspect of time series data prediction, scholars have done a lot of research work and put forward many different prediction methods. These methods can be divided into physical methods, statistical models and artificial intelligence methods. Different methods have their own application occasions. Physical methods use correlation to predict time series, such as precipitation, runoff, temperature and other physical quantities to predict soil moisture, Li et al. (2013) use the correlation between soil moisture and meteorological factors, circulation factors to select the main influencing factors, use these factors to establish regression equation to predict soil moisture, and the prediction accuracy of the model reaches 80%; Based on the soil moisture observation data of nearly 30 years in Northeast China, Sun et al. (2014) used cast clustering method to divide the region according to the difference of the soil moisture, analyzed the temporal and spatial variation characteristics of soil moisture in each region, screened the key factors that have significant impact on the soil moisture, and used them to establish multiple linear feedback equation, so as to establish the prediction model of soil moisture in each region. The prediction error of this model is lower than 13%. The accuracy prediction of these physical method depends on the data related to soil moisture, which are often of many kinds and some data are difficult to obtain. Romano et al. (2014) proposed that the output results of physical models can often be used as auxiliary input data of statistical models. The purpose of statistical method is to describe the correlation between historical data (Tascikaraoglu et al. 2014). AR, ARMA, ARIAM, SARIMA and other models are used for time series prediction (Tascikaraoglu et al. 2014; Huang and Chalabi 1995; Erdem and Shi 2011; Su et al. 2014; Lydia et al. 2016). These time series prediction methods based on statistics often assume that there is a linear correlation between time series data. However, the time series data of soil moisture is nonlinear, so it is difficult to get high prediction accuracy by using statistical model.

In order to overcome these problems, artificial intelligence method is more and more used in time series prediction. Ho et al. (2002) used neural network model and ARIMA model for time series model prediction at the same time. The results show that the neural network model has better prediction accuracy than ARIMA model in the case of random and nonlinear time series distribution. Li et al. (2010) used feed forward ANN model to predict time series, and compared with ARIMA prediction results, the results showed that ANN prediction results were more accurate.

As a typical artificial neural network, BPNN is widely used in time series prediction because of its good nonlinear mapping ability. Yang (2012), Wang et al. (2011) used BPNN to predict the time series data, and obtained ideal prediction results.

In this paper, make full use of the existing soil moisture time series data to predict the soil moisture. According to the characteristics of non-stationary soil moisture, use the combination of discrete wavelet transform and BPNN to predict the soil moisture. At the same time, aiming at the problem of slow iterative speed and easy to fall into local optimum of BPNN, an improved method of BPNN is proposed.

Finally, the experiment was carried out based on data from 20 observation sites in Kenli County, Dongying City, Shandong experimental area of "Bohai granary" from 2013 to 2014, and different prediction methods were used in the same group of data for comparative experiment to verify the effectiveness of the proposed method.

## 2  Experimental Area and Data Preprocessing

### 2.1  Experimental Area

In 2013, China launched the "Bohai granary" agricultural science and technology project. Through the transformation of more than 40 million mu of medium and low yield fields and more than 10 million mu of salt alkali wasteland in the Bohai Rim region, achieve the goal of increasing grain by 6 billion kilogram in 2017 and 10 billion kilogram in 2020. In this paper, Kenli County, Dongying City, Shandong project area of "Bohai granary" is chosen as the experimental area. The experimental area is located at the mouth of the lower reaches of the Yellow River in the Yellow River Delta area in the northeast of Shandong Province. Its location is 37°24'–38°10' N and 118°15'–119°19'E. The experimental area is located in the temperate monsoon climate area. It is dry and cold in winter, humid and hot in summer, with four distinct seasons. The annual average precipitation is 547.2 mm, of which 67.8% is in summer and only 3.6% in winter. There are two types of soil in this area: tidal soil and salinized soil, which are light in texture, poor in nutrients and heavy in salinization.

### 2.2  Data Acquisition

The time series data is generated by the monitoring stations arranged in the experimental area. There are 10 monitoring stations arranged in the experimental area. Figure 1 shows the arrangement of the experimental area and the monitoring stations. The soil types of the 10 monitoring points are saline soil. The observation frequency of each monitoring stations is 1 h. This paper selects the data of the monitoring stations from January 1, 2013 to December 31, 2014, with a total of 175200 data records.

**Fig. 1.** Experimental area and monitoring stations

## 2.3 Data Preprocessing

Due to sensor failure or transmission, there is a small part of missing data, whose value is 0 (data default value), which will affect the prediction accuracy. For these missing data, the Euclidean distance is used to select the closest six groups of data on the same monitoring station for weighted average.

$$D_k = \omega_{k+3}D_{k+3} + \omega_{k+2}D_{k+2} + \omega_{k+1}D_{k+1} + \omega_{k-1}D_{k-1} + \omega_{k-2}D_{k-2} + \omega_{k-3}D_{k-3}$$

$$(1)$$

Among them, Dk is the missing data, Dk + i is the data that lags Dk in time, Dk−i is the data that leads Dk in time, and ωk-3 is the corresponding weight. In this paper, we take all ω equal (1/6).

In this way, we get 10 groups of time series data, and use statistical methods to analyze the distribution characteristics of the data of 10 monitoring points. Figure 2 shows the distribution of the observation data of 4 monitoring stations. It can be seen from the figure that the data distribution of the 4 monitoring stations is approximately the same. Further, the data of 10 observation stations at the same time are calculated and averaged, and 17520 time series data are obtained. These data are used as experimental data to study the prediction method.

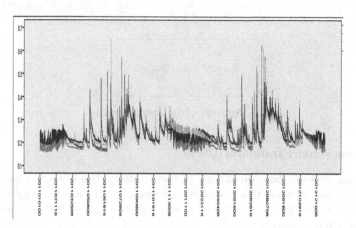

**Fig. 2.** Data distribution of some monitoring stations

## 3 Prediction Model

Figure 3 illustrate the structure of the prediction model. Soil moisture time series are decomposed into several subsequences by DWT. Then train the subsequences separately use different structure of BPNN, and use the trained BPNN to predict each subsequence. Finally, use wavelet reconstruction to obtain the prediction result.

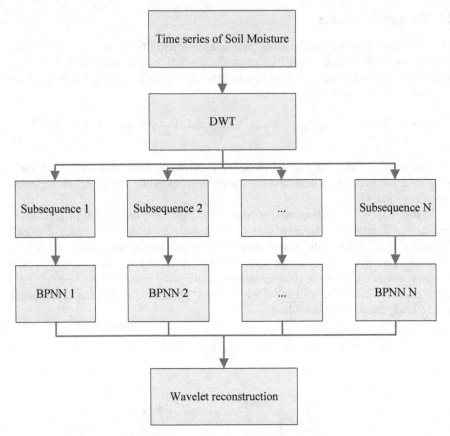

**Fig. 3.** Structure of the prediction model

## 3.1 Discrete Wavelet Transform (DWT)

Soil moisture data has the characteristic of nonstationary. Use DWT to transform the nonstationary soil moisture time series into several relatively stationary subsequences with different scales (Meng et al. 2016; Hu et al. 2015). Then use BPNN to predict the subsequences respectively.

Daubechines (db) wavelet has good characteristics for non-stationary time series. According to the value of n, different db series wavelets from db1 to db10 have different processing effects. The larger the n-band division effect, the better the calculation time, the longer the calculation time. According to the selection method of wavelet function proposed by Sang et al. (2008), choose db3 wavelet to perform DWT on soil moisture time series data:

$$W_f(j, k) = \alpha_0^{-\frac{j}{2}} \int_{-\infty}^{+\infty} f(t)\phi^*(\alpha_0^{-j}t - k\tau_0)dt \tag{2}$$

Among this, $f(t)$ is the soil moisture time series, $W_f(j,k)$ is the coefficients of DWT, $\phi^*$ (x) is the complex conjugate function of mother wavelet, j is the time scale, k is the location parameters. Figure 4 is the DWT result of db3, in which the decomposition level is selected as 8 levels through repeated experiments, d1–d8 and a8 are the subsequences.

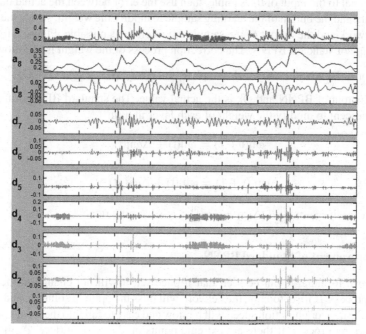

**Fig. 4.** DWT result

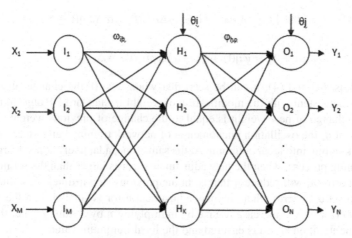

**Fig. 5.** Structure of standard BPNN

## 3.2 Standard BPNN

The main idea of BP neural network is: for a given learning sample, make the input of the network equal to the input of the sample, then use the error between the actual output of the network and the output of the learning sample to modify the weight, so that the output of the network and the output of the sample are as close as possible. Figure 5 illustrate the structure of standard BPNN. The correction process of the weight is as follows: if E is the error of BPNN, then for each sample P, the error $Ep = \frac{1}{2} \sum_{k=1}^{N} (T_k - O_k)^2$, Where $T_k$ is the output of BPNN and Ok is the actual value. The standard BP algorithm uses gradient descent to modify the weight of each layer in turn.

$$\omega ij(t + 1) = \omega ij(t) + \Delta \omega ij(t) \tag{3}$$

$$\varphi ij (t + 1) = \varphi ij (t) + \Delta \varphi ij (t) \tag{4}$$

Where $\Delta \omega ij = -\eta \frac{\partial E}{\partial \omega_{ij}}$ , $\Delta \varphi ij = -\eta \frac{\partial E}{\partial \omega_{ij}}$, $\eta$ is learning rate.

## 3.3 Improvement of Standard BPNN

In the above formula, it can be seen that $\eta$•once set, it will not change in the whole learning process. If $\eta$•set too big, the network will miss the optimal solution, and if $\eta$•set too little, the network iteration speed will be slow and easy to fall into the local optimum. In the learning process, if there are extreme points on the error surface, $\Delta \omega ij = 0$ or $\Delta \varphi i = 0$, then $\omega ij(t + 1) = \omega ij(t)$or $\varphi ij(t + 1) = \varphi ij(t)$, at this time, the BPNN is easy to fall into the local optimal value and stop learning. Based on these problems, formula (5) and (6) are modified by momentum factor as follows:

$$\omega ij(t + 1) = \omega ij(t) + (1 - \alpha)\Delta \omega ij(t) + \alpha \Delta \omega ij(t - 1) \tag{5}$$

$$\varphi ij(t + 1) = \varphi ij(t) + (1 - \alpha)\Delta \varphi ij(t) + \alpha \Delta \varphi ij(t - 1) \tag{6}$$

From Eqs. (3) and (4), it can be seen that when $\alpha = 0$, the change of weight is completely consistent with the modification of weight in standard BP algorithm; when $\alpha = 1$, the change of new weight is equal to the change of previous weight. Through adjustment of $\alpha$, the oscillation phenomenon of network training can reduce and avoid the network falling into local optimum. At the same time, adaptively adjust learning rate in the learning process, when the new adjustment error is larger than the set maximum adjustment error, it will reduce $\eta$•that is, in the previous step $\eta$•times a coefficient less than 1, and set $\alpha$ to zero; when the new adjustment error is smaller than the previous adjustment error, it will increase its size by multiplying $\eta$ by a value greater than 1. In this way, the iteration speed is ensured and the local minimum value of the network is avoided effectively.

### 3.4   Designment of BPNN Structure

**(1) Network hierarchy**
According to kolrnogorov, a 3 layers BPNN can approach any non-linear function, so this paper selects 3 layer BP neural network to predict soil moisture, that is, the network contains one hidden layer.

**(2) Node number of input layer, output layer and hidden layer**
Here the number of nodes in the network output layer is set as one. In the process of parameter estimation of ARIMA model proposed by Hui Liu (2012), autocorrelation function (ACG), partial autocorrelation function (PACF) are used to determine the number of nodes in the input layer of neural network. According to the method, ACF and PACF of each subsequence are analyzed firstly, and the number of input layer nodes is obtained. The selection of hidden layer nodes is based on Hecht-Nelson method: if the number of input layer nodes is n, the number of hidden layer nodes is $2n + 1$. Table 1 shows the structure of BP neural network of each subsequence.

**Table 1.** Structure BPNN for each subsequence

|       | A8 | D1 | D2 | D3 | D4 | D5 | D6 | D7 | D8 |
|-------|----|----|----|----|----|----|----|----|----|
| 输入层 | 5  | 4  | 8  | 6  | 4  | 3  | 2  | 3  | 4  |
| 隐含层 | 11 | 9  | 17 | 13 | 9  | 7  | 5  | 7  | 9  |
| 输出层 | 1  | 1  | 1  | 1  | 1  | 1  | 1  | 1  | 1  |

**(3) The other parameter**
Maximum number of iterations: 500, initial learning rate: 0.01, expected error 1e−4. Figure 5 and Fig. 6 are the neural network simulation results of A8 and D8 respectively. Table 2 is the comparison table of neural network prediction results of each subsequence.

## 4   Experimental Results and Discussion

In order to verify the effectiveness of the method proposed in this paper, we have carried out a comparative experiment of the other three methods. Method 1 is the method proposed in this paper, and is called DB-IBP. In method 2, the improved BPNN is used to predict the original time series data directly, which is called IBP. Method 3. First, we use Daubechines wavelet to transform the original time series into discrete wavelet, then use standard BPNN to predict, at last use wavelet reconstruction to obtain the prediction result, which is called DB-BP. Table 3 is the comparison table of three methods.

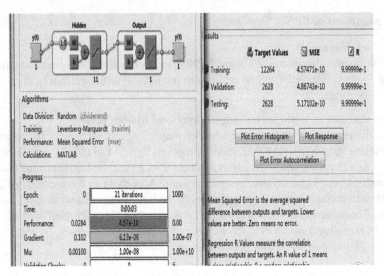

**Fig. 6.** Predict result of subsequence A8

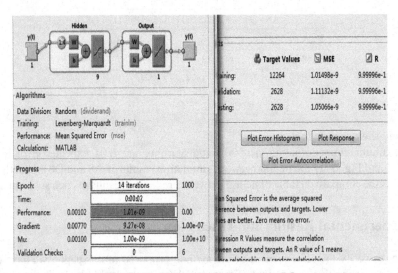

**Fig. 7.** Predict result of subsequence D8

Figure 8 shows the prediction results using IBP method. In this method, we use the best neural network structure through repeated experiments. One hidden layer is used in the prediction network, with 2 nodes in the input layer, 10 nodes in the hidden layer and 1 node in the output layer. Table 4 shows the prediction results of DB-BP method, and the selection scheme of neural network structure of each layer is the same as that of DB-IBP.

**Table 2.** Comparison of prediction results for each subsequence

|  | A8 | D1 | D2 | D3 | D4 | D5 | D6 | D7 | D8 |
|---|---|---|---|---|---|---|---|---|---|
| Number of iterations | 21 | 32 | 15 | 17 | 28 | 39 | 36 | 26 | 14 |
| MSE | 4.9e−10 | 1.5e−10 | 5.0e−10 | 1.6e−10 | 1.3e−10 | 1.1e−9 | 2.4e−10 | 9.1e−10 | 1.1e−9 |

**Table 3.** Comparison table of the 3 methods

| Method number | Abbreviation | Method description |
|---|---|---|
| 1 | DB-IBP | First use Daubechines to decomposed the soil moisture time series into several subsequences. Then train the subsequences separately use different structure of improved BPNN, and use the trained BPNN to predict each subsequence. Finally, use wavelet reconstruction to obtain the prediction result |
| 2 | IBP | Use the improved BPNN to predict soil moisture time series |
| 3 | DB-BP | First use Daubechines to decomposed the soil moisture time series into several subsequences. Then train the subsequences separately use different structure of standard BPNN, and use the trained BPNN to predict each subsequence. Finally, use wavelet reconstruction to obtain the prediction result |

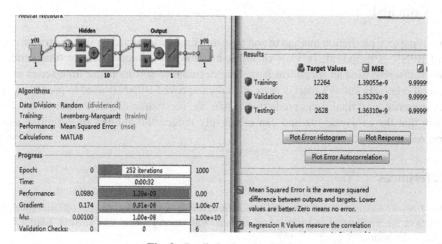

**Fig. 8.** Prediction result of IBP

**Table 4.** Prediction result of each subsequence for DB-BP

|  | A8 | D1 | D2 | D3 | D4 | D5 | D6 | D7 | D8 |
|---|---|---|---|---|---|---|---|---|---|
| Number of iterations | 35 | 37 | 71 | 20 | 26 | 43 | 36 | 42 | 21 |
| MSE | 9.2e−10 | 1.3e−9 | 6.4e−10 | 1.4e−10 | 4.2e−9 | 6.3e−9 | 7.2e−10 | 1.3e−9 | 9.8e−10 |

**Table 5.** Comparison results of the 3 methods

|  | Average number of iterations | Total number of iterations | Average MSE |
|---|---|---|---|
| DB-IBP | 25 | 228 | 4.2e−10 |
| IBP | 252 | 252 | 1.4e−9 |
| DB-BP | 37 | 331 | 1.8e−9 |

Table 5 shows the comparison results of the 3 methods. From Table 5, it can be seen that the DB-IBP method proposed in this paper is 9.5% less than the IBP method in terms of the number of iterations, and 31% less than the DB-BP method, which can effectively reduce the iteration time. In terms of prediction accuracy, the method proposed in this paper is 70% higher than IBP method, 77% higher than DB-IBP method, and has higher prediction accuracy.

## 5  Conclusion

In this paper, a prediction method of soil moisture time series based on wavelet transform and BPNN is proposed. In view of the nonstationarity of soil moisture time series, firstly, discrete wavelet transform is used to decompose soil moisture time series into several relatively stable subsequences, then the improved BPNN is used to predict each subsequence, and finally the prediction results are reconstructed. Experimental results show that the proposed method can effectively reduce the number of iterations and improve the prediction accuracy.

## References

Qiu, J., et al.: Comparison of temporal trends from multiple soil moisture data sets and precipitation: the implication of irrigation on regional soil moisture trend. Int. J. Appl. Earth Obs. Geoinformation **48**, 17–27 (2016)

Romano, N.: Soil moisture at local scale: measurements and simulations. J. Hydrol. **516**, 6–20 (2014)

Sun, Q., Liu, J., Liang, H.: Analysis of regional soil moisture forecasting model in northeast China. J. Natural Res. **29**(6), 1065–1075 (2014)

Li, Y., et al.: Research on forecasting method for soil moisture during spring sowing period in Northeast Area of China. Agric. Res. Arid Areas **33**(6), 178–183 (2013)

Hui, L., et al.: Comparison of two new ARIMA-ANN and ARIMA-Kalman hybrid methods for wind speed prediction. Appl. Energy **98**, 415–424 (2012)

Hui, L., et al.: A hybrid model for wind speed prediction using empirical mode decomposition and artificial neural networks. Renew. Energy, **48**, 545–556 (2012)

Bouzgou, H., et al.: Multiple architecture system for wind speed prediction. Appl. Energy **88**, 2463–2471 (2011)

Cadenas, E., et al.: Wind speed forecasting in three different regions of Mexico, using a hybrid ARIMA-ANN model. Renew. Energy **35**, 2732–2738 (2010)

Shi, H., Yang, J., Ding, M., Wang, J.: A short-term wind power prediction method base on wavelet decomposition and BP neural network. Autom. Electric Power Syst. **35**(16), 44–48 (2011)

Liu, L., Ye, W.: Precipitation prediction of time series model based on BP artificial neural network. J. Water Res. Water Eng. **21**(5), 156–159 (2010)

Hu, S., Zhang, Z.: Fault prediction for nonlinear time series based on neural network. ACTA Automatica Sinica **33**(7), 744–748 (2007)

Hecht-Nielsen, R.: Kolmogorov's mapping neural network existence theorem. In: Proceedings of the International Conference on Neural Networks, Publishing (1987)

Na, Y.: Prediction of soil moisture from characteristic meteorological elements by BP neural network. Chin. J. Soil Sci. **42**(6), 1324–1329 (2011)

Pandhian, S.M., et al.: A comparative analysis and time series forecasting of monthly stream flow data using hybrid model. Jurnal Teknologi **76**(13), 67–74 (2015)

Jin, J., et al.: Forecasting natural gas prices using wavelets, time series, and artificial neural networks. Plos One **10**(11) (2015)

Qi, L., Wang, X., Xiaolong, X., Dou, W., Li, S.: Privacy-aware cross-platform service recommendation based on enhanced locality-sensitive hashing. IEEE Trans. Network Sci. Eng. (2020a). https://doi.org/10.1109/TNSE.2020.2969489

Wang, X., Yang, L.T., Wang, Y., Ren, L., Deen, M.J.: ADTT: a highly-efficient distributed tensor-train decomposition method for IIoT big data. IEEE Trans. Ind. Inf. (2020). https://doi.org/10.1109/tii.2020.2967768

Qi, L., He, Q., Chen, F., Zhang, X., Dou, W., Ni, Q.: Data-driven web APIs recommendation for building web applications. IEEE Trans. Big Data (2020b). https://doi.org/10.1109/TBDATA.2020.2975587

Zhong, W., Yin, X., Zhang, X., Li, S., Dou, W., Wang, R., Qi, L.: Multi-dimensional quality-driven service recommendation with privacy-preservation in mobile edge environment. Comput. Commun. (2020). https://doi.org/10.1016/j.comcom.2020.04.018

Chi, X., Yan, C., Wang, H., Rafique, W., Qi, L.: Amplified LSH-based recommender systems with privacy protection. Concurrency Comput. Pract. Exp. (2020). https://doi.org/10.1002/CPE.5681

Wang, X., Yang, L.T. Kuang, L., Liu, X., Zhang, Q., Jamal Deen, M.: A tensor-based big data-driven routing recommendation approach for heterogeneous networks. IEEE Network Mag. **33**(1), 64–69 (2019)

# Research on Automatic Target Detection and Recognition System Based on Deep Learning Algorithm

Qinghui Zhang$^{(\boxtimes)}$, Hongbin Xu, Zhengyu Li, Xiaobin Liu, Yuxi Li, and Yingjie Jiao

Xi'an Modern Control Technology Research Institute, Xi'an 710065, China
zqh820@zju.edu.cn

**Abstract.** Automatic target detection and recognition is the cornerstone of the intelligent unmanned systems to realize higher-level tasks. In this paper, the deep learning algorithm of Faster R-CNN was studied in depth, and the target detection model is designed combining the RPN network and the fast R-CNN. The target detection and recognition device with the ability of image acquisition and intelligent processing was also designed. Combining the device with the Faster R-CNN model, the automatic target detection and recognition system was developed. At last, the VGG-16 model was adopted for training the detection model, and the system was used for target detection experiments. The results show that the recognition accuracies of the system for the visible light images of trucks and tanks are 89.7% and 90.3%, respectively, and that for infrared images of tanks is 63.7%. Therefore, a good recognition effect has been achieved. This work provides a reference for the application of deep learning algorithms in the field of automatic target detection and recognition.

**Keywords:** Automatic target detection and recognition · Deep learning algorithm · Faster R-CNN

## 1 Introduction

Target recognition and detection is one of the basic capabilities of intelligent un-manned systems. Research on automatic and efficient target detection frameworks and methods in complex environments is the cornerstone of intelligent systems to achieve more complex and higher-level tasks such as situation cognition, intention recognition, and decision planning [1].

Traditional target detection and recognition methods [2–5] include statistical pattern recognition, syntactic pattern recognition, knowledge-based target recognition, template matching recognition, neural network, etc. For image detection, knowledge-based method and template matching method are often used. The knowledge-based method uses the feature information of the target, including background, contour, symmetry and edge texture, to detect and identify the target. In order to improve the detection accuracy, multiple features are usually selected for fusion to detect the target, rather than a single feature. The recognition method based on template matching establishes the target image

© Springer Nature Switzerland AG 2020
X. Chen et al. (Eds.): ML4CS 2020, LNCS 12488, pp. 538–548, 2020.
https://doi.org/10.1007/978-3-030-62463-7_50

template library, and compares and matches the image to be detected with the template library. The regions with high similarity are considered as the identified target. In more advanced target detection and recognition, the two methods are usually combined. That is, the knowledge-based target recognition method is used to detect the approximate area of the target based on the target characteristics, and then the template matching method is used to accurately locate and recognize the target. Compared with the two methods used alone, the results of target detection and recognition is improved. However, the traditional target recognition methods have limitations. The knowledge-based method relies too much on pre-set target features, and the limited features are difficult to adapt to all detection environments and are less versatile. The template matching method is very dependent on the pre-established template library. Even the same type of target has its diversity and shooting angle changes, and the template library obviously cannot adapt to such changes. Therefore, it is difficult for traditional methods to achieve good target detection and recognition effect under complex background.

At present, with the development of neural network [6–10], deep learning and other technologies and the progress of key devices, image recognition based on high-performance special processing equipment is being applied more and more widely in the fields of national defense, security and people's livelihood. Among them, image recognition based on deep learning technology has been popularized in transportation, security, logistics and other fields. Furthermore, by combining such technologies with high-performance low-power processors and neural network engines, edge computing for image recognition and processing is realized, which not only reduces communication bandwidth and storage capacity requirements, but also improves the performance of the system [11].

The target detection and recognition method based on deep convolutional neural network has obvious advantages over traditional target recognition methods in terms of both recognition accuracy and versatility. Target detection and recognition algorithms based on convolutional neural networks mainly include two types of one-stage target detection and two-stage target detection. Classic single-stage target detection algorithms include YOLO series [12], SSD [13], FPN [14], etc. Its advantage lies in its fast operation speed, but there are problems such as the unbalanced proportion of positive and negative samples and the lower detection accuracy than the two-stage target detection algorithm. The two-stage target detection algorithm is represented by R-CNN series and adopts the candidate box extraction operation, which has a high detection accuracy but a slow recognition operation speed. Therefore, algorithms with superior performance, such as Fast R-CNN [15] and Faster R-CNN [16], have also been developed. Among them, the Faster R-CNN model integrates the generation of candidate regions, feature extraction, classification and location refinement into a deep neural network, which is 10 times Faster than the Fast R-CNN model and greatly improves the detection efficiency while ensuring the identification accuracy. Therefore, Faster R-CNN has a very wide application in the fields of traffic, ocean target detection, scene identification, medical field, gap detection and other target detections [17–23]. However, these applications are carried out in the laboratory and the target detection tests are based on the desktop computer, the actual use of target detection platform for outdoor testing is rare.

In this paper, a target detection model is designed based on the Faster R-CNN deep learning algorithm with both accuracy and efficiency, and an automatic target detection and recognition system is developed by combining it with the advanced image acquisition and intelligent processing platform. At last, the system is used to carry out target detection experiments and good recognition results are obtained.

## 2    Deep Learning Algorithm Based on Faster R-CNN

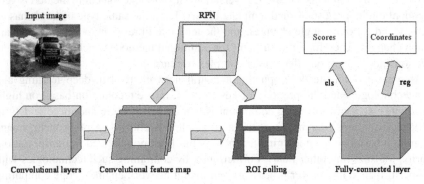

**Fig. 1.** Faster R-CNN network framework

The R-CNN series model is a deep learning method based on regional suggestion, which is widely used in the target detection field [17]. The R-CNN model fundamentally solves the feature selection problem of traditional detection methods. Through the strong learning ability of CNN, features of each layer from shallow to deep are extracted, and then combined with the generated candidate boxes to complete the detection task. R-CNN uses a selective search method to generate thousands of candidate regions for an image, and then carries out convolution operation on the image in each candidate region to extract features. The classifier is trained to get object categories, and the size of the candidate target box is fine-adjusted through bounding box regression (BBR). Fast R-CNN still uses the selective search method to search for the candidate boxes, but it is to perform convolution operation on the full image to extract features, normalizes the feature map through ROI pooling, and finally trains the classification and regression jointly to obtain the final candidate boxes [15]. Meanwhile, Faster R-CNN innovatively improves the selective search method, assigns the task of searching candidate boxes to the neural network, and adds a region proposal network (RPN) to share the previous convolution calculation [16]. Compared with R-CNN and Fast R-CNN, Faster R-CNN overcomes the redundant operations generated when extracting convolutional features. At the same time, because RPN is integrated into GPU operations, the detection and recognition speed of the algorithm is greatly improved without reducing the accuracy. The network framework of Faster R-CNN is shown in Fig. 1.

## 2.1   Region Proposal Network (RPN)

The RPN is placed after the last convolutional layer and shares complete image convolution features with the detection network, so that candidate regions can be extracted without any cost [24]. The RPN schematic diagram is shown in Fig. 2. RPN feature modeling uses a fully convolutional network, which can predict the target score and target boundary at each location. After end-to-end training [20], RPN generates high-quality area candidate boxes, and then inputs them to the detection network to complete the test. RPN and Fast R-CNN share the convolutional layers to synthesize a network.

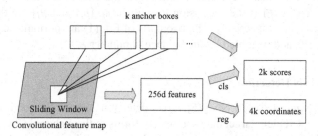

**Fig. 2.**  RPN structure diagram

Slide a small window on the convolutional feature map output by the last convolutional layer to generate the suggested area box [18]. The features in each sliding window are mapped to corresponding low-dimensional features, which are passed to two parallel full connection layers, namely the box regression layer and the box classification layer respectively. Each sliding window position simultaneously predicts the recommended area, and the maximum recommended number of sliding windows at each position is k. In this way, the reg layer outputs 4k coordinates of k boxes, and the cls layer outputs 2k probability scores of k boxes, which predict it is the object or a non-object. The k suggestions are parameterized into k reference boxes, which are called anchor points. In order to cope with objects of different sizes, three different types of sliding Windows are adopted, with the length-width ratio of 1:1, 1:2 and 2:1. Two scales, $128^2$ and $256^2$, are used to zoom sliding windows, and a total of six types of sliding windows are adopted. Then, the candidate windows of 6 sizes and the 256-dimensional feature vector are input to the regression layer and the classification layer of the window, respectively, and the position correction of the 6 candidate windows extracted for the center point of the sliding window and the score of whether it belongs to the foreground target are obtained respectively. Finally, all the candidate windows extracted by the sliding window are subjected to the maximum suppression process to remove the windows with higher repetition, and the 200 candidate windows with the highest score are extracted as candidate regions and suggested to be entered into the subsequent Fast R-CNN.

To train the RPN network, the anchor point of each candidate box is given a binary label to indicate whether the candidate box is the target. The anchor point with the highest overlap between a candidate box and the real target box is assigned as a positive sample label. When the overlap rate between a suggestion box and all the real boxes is less than

0.3, the anchor point will be used as a negative sample label. According to the above definition, the loss function [23] of the image is

$$L(\{p_i\}, \{t_i\}) = \frac{1}{N_{cls}} \sum_i L_{cls}(p_i, p_i^*) + \lambda \frac{1}{N_{reg}} \sum_i p_i^* L_{reg}(t_i, t_i^*) \tag{1}$$

where, $i$ is the number of the anchor point and $p_i$ is the target prediction probability of anchor point $i$. If the anchor point is a positive sample label, $p_i = 1$, otherwise, $p_i = 0.t_i$ is the coordinate vector of the predicted bounding box and $t_i^*$ is that of the ground-truth box associated with a positive anchor point. $L_{cls}$ is the classification loss function and $L_{reg}(t_i, t_i^*) = R(t_i - t_i^*)$ is the regression loss function. $\{p_i\}$ and $\{t_i\}$ are the output of the cls and reg layer, respectively. The two parts of Eq. (1) are normalized by $N_{cls}$ and $N_{reg}$, respectively, and weighted by the balance parameter $\lambda$.

## 2.2 Fast R-CNN Detection

Integrating RPN into Fast R-CNN [25], target detection and recognition is divided into five steps. First, a vehicle target data set that conforms to the standard (PASCAL VOC) is constructed, and the targets in the training data set and the test data set are labeled respectively. Then the VGG-16 network is trained on the ImageNet dataset to initialize the model. And then, the calibrated training data set is put in the pre-trained network for training, and the method of stochastic gradient descent and reverse regression is adopted to continuously learn, so as to adjust the RPN region to extraction network parameters. In the fourth step, the training data set is input to the target feature parameters obtained by the network training for the softmax classifier parameter target common convolution layer, and the test data set is input to the network for detecting the target classification. Finally, the average accuracy MAP is calculated, and the training is performed repeatedly if the conditions are not met until the accuracy of the target detection reaches the expected effect.

# 3 Design of Automatic Target Recognition Device

The automatic target recognition device is mainly composed of visible light camera, infrared thermal camera, laser rangefinder, servo module, electrical control module, intelligent image processing system, etc., which can realize the automatic detection, identification and tracking of targets in complex background during day and night, as well as the ranging. The electrical structure of the automatic target recognition device is shown in Fig. 3.

The electrical control module realizes the electrical connection and function realization of all components of the target automatic identification device through logic circuit control, which is mainly composed of interfaces, servo sensors, rotating components, circuit boards, etc.

- The interfaces mainly consist of power supply interface, communication inter-face, video interface and test interface to realize the connection between the device and the external equipment.

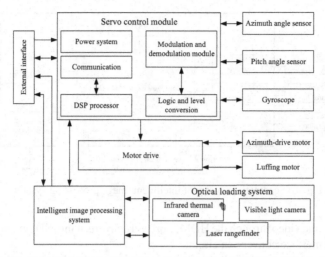

**Fig. 3.** Electrical structure diagram of automatic target recognition device

- Servo sensors include gyroscopes, azimuth, pitch angle sensors, etc., to achieve inertial space stability and position angle monitoring of the photoelectric platform.
- The rotating components include the torque motors of the two axis and a con-fluence ring, which respectively provide power for the movement of the platform and realize signal transmission when the platform rotates.
- The circuit boards include a servo control board, a power board and an intelligent image processing board. The servo control board achieves the functions of stability control, external communication, signal acquisition, drive, etc. The power supply of each component in the system is realized by the power board. The image processing board can realize the functions of the video acquisition and switching output, the algorithm operation of the automatic target recognition and tracking, etc.

The core processor of the intelligent image processing system selects the artificial intelligence image processing platform with excellent performance, which has strong image processing capability and can run the target automatic recognition algorithm based on deep learning efficiently. Image processing circuit is mainly composed of image processing platform, PAL video capture card, SDI video capture card and output video interface conversion module. The structure of the intelligent image processing system is shown in Fig. 4.

The image processing circuit collects the high-definition video output by the visible light camera and the standard-definition video output by the infrared thermal camera using the SDI and the PAL video capture card respectively and sends them to the image processing platform through the PCIe port. The GPU and the CPU in the image processing platform are utilized to perform the target recognition algorithm based on deep learning and the tracking algorithm, which can generate the target missed distance data and send it to the servo control circuit to realize the real-time tracking processing. The processor performs corresponding character superposition according to the processed

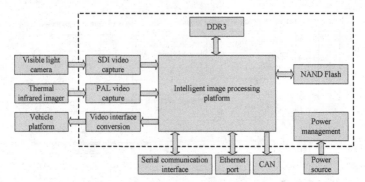

**Fig. 4.** Schematic diagram of intelligent image processing system

result, outputs the video through the HDMI port and converts it into SDI video for display by the video interface conversion circuit.

The visible light camera is used to obtain environmental images under sufficient daylight conditions. It uses a full HD camera with a resolution of 1920 (H) × 1080 (V) and a 1/2.8-in. back-illuminated CMOS sensor with a pixel size of 3.45 μm. The pixel number of the target in the image is

$$N = \frac{f}{p} \times \frac{H}{L} \tag{2}$$

where, $f$ is the focal length of the camera, $p$ is the pixel size of the sensor, $H$ is the target size and $L$ is the distance between the target and the camera.

In this paper, the longest focal length of visible light camera is 110.25 mm. for $L = 1.5$ km, the target with a size of 2.3 m can present 49 pixels in the image.

The infrared thermal imager is an uncooled long-wave system, which is mainly used to obtain environmental images when the lighting conditions are insufficient. The pixel size of the infrared sensor is 17 μm, with a resolution of 640 (H) × 512 (V). The maximum focal length of the camera is set to 150 mm and a 2x electronic zoom function is applied. Then, for $L = 1.5$ km, a target with a size of 2.3 m can present 27 pixels in the image.

The laser rangefinder is responsible for accurate distance measurement of the target within the range of 100–5000 m, and the measurement error $\sigma$ is less than 5 meters.

The servo module provides power for the azimuth and pitch motion during the search process, and enables the target recognition device to output images stably. It adopts the direct torque motor and the direct-drive transmission mode, which has the advantages of stall-rotating operation, high response speed, low no-load speed, large low-speed torque, good characteristic linearity and small torque fluctuation. It is suit-able for low-speed and large torque load and can cooperate with the gyroscope to provide higher stability accuracy.

## 4   Results and Analysis of Target Detection

To improve the training efficiency, the VGG-16 model pre-trained on the ImageNet dataset was used to pre-train the RPN and Fast R-CNN networks, and the parameters were adjusted repeatedly so that the two networks could share the convolutional layer. The training was carried out on a desktop workstation. The visible light image data set of vehicles was 10,000 and that of tank was 6,000. The infrared image data set of tanks was 1,000. Among them, 80% of each data set was taken as the training data set and 20% was the test data set.

The pre-photographed images were detected using the trained Faster R-CNN mod-el on the desktop workstation and the results were shown in Fig. 5. It can be seen from the figure that the Faster R-CNN algorithm can achieve a good detection effect for different vehicles and tanks, and can detect the target even when the colors of the target and the environment are close.

**Fig. 5.** Target recognition results on desktop workstation using Faster R-CNN

Loaded the trained Faster R-CNN detection model into the automatic target recog-nition device to form an automatic target detection and recognition system, and used it to perform target search and recognition in the wild environment. During the search process, the photoelectric turntable of the system swung back and forth within a certain range, and the truck and tank targets continued to move. Figure 6, Fig. 7 and Table 1 are the detection and recognition results of the vehicle and tank targets through the visible light camera and the infrared camera channel, respectively.

**Fig. 6.** Recognition results of visible light image

**Fig. 7.** Recognition results of infrared image

**Table 1.** results of target detection

| Target type | Image type | Total number | Missed number | False number | Accuracy |
|---|---|---|---|---|---|
| Truck | Visible Light Image | 350 | 22 | 14 | 89.7% |
| Tank | | 279 | 23 | 4 | 90.3% |
| Tank | Infrared Image | 526 | 191 | 0 | 63.7% |

It can be seen from the figure that the system has a good recognition effect on trucks and tanks, even when the truck is blocked by trees. From Table 1, it can be seen that using faster R-CNN to detect trucks and tanks, the accuracy of visible light image detection is relatively high, reaching 89.7% and 90.3% respectively. However, the infrared image recognition accuracy of tanks is lower than them, which is 63.7%. This may be caused by the small number of data set.

## 5   Conclusion

In this paper the Faster R-CNN deep learning algorithm that combined the RPN network with the Fast R-CNN was deeply studied and the target detection model was designed based on this. Combining the model with the image acquisition and intelligent processing platform, the target automatic detection and identification system was developed. At last, target detection experiments were carried out using the sys-tem. The results show that the recognition accuracies of truck and tank visible light images are 89.7% and 90.3%, respectively, and the recognition accuracy of infrared image of tanks is 63.7%. Good results are achieved in the experiments. This work provides a reference for the application of deep learning algorithms in the field of automatic target detection and recognition.

**Acknowledgments.** The authors are highly thankful for National Key Research Program (2019YFB1706001), Industrial Internet Innovation Development Project (TC190H468), National Natural Science Foundation of China (61773001).

# References

1. Yang, F., Rao, Y.: Vision-based intelligent vehicle road recognition and obstacle detection method. Int. J. Pattern Recognit Artif Intell. **34**(07), 1–15 (2020)
2. Entezami, A., Sarmadi, H., Behkamal, B., et al.: Big data analytics and structural health monitoring: a statistical pattern recognition-based approach. Sensors (Basel) **20**(8), 1–17 (2020)
3. Siemiatkowski, M.: Application of syntactic pattern recognition approach in design and optimisation of group machining systems. Solid State Phenom. **957**, 342–347 (2010)
4. Xue, J., Shuwen, X., Shui, P.: Knowledge-based target detection in compound Gaussian clutter with inverse Gaussian texture. Digit. Signal Proc. **95**, 1–9 (2019)
5. Kannan, S.: Intelligent object recognition in underwater images using evolutionary-based Gaussian mixture model and shape matching. SIViP **14**(5), 877–885 (2020). https://doi.org/10.1007/s11760-019-01619-w
6. Hong, S., Lv, C., Zhao, T., et al.: Cascading failure analysis and restoration strategy in an interdependent network. J. Phys. A: Math. Theor. **19**(49), 195101 (2016)
7. Hong, S., Wang, B., Ma, X., et al.: Failure cascade in interdependent network with traffic loads. J. Phys. A: Math. Theor. **48**(48), 485101 (2015)
8. Hong, S., Zhu, J., Braunstein, L.A., et al.: Cascading failure and recovery of spatially interdependent networks. J. Stat. Mech. Theory Exp. **10**, 103208 (2017)
9. Hong, S., Yang, H., Zhao, T., et al.: Epidemic spreading model of complex dynamical network with the heterogeneity of nodes. Int. J. Syst. Sci. **11**(47), 2745–2752 (2016)
10. Hong, S., Zhang, X., Zhu, J., et al.: Suppressing failure cascades in interconnected networks: Considering capacity allocation pattern and load redistribution. Mod. Phys. Lett. B **5**(30), 1650049 (2016)
11. Wang, J., Liu, C., Fu, T., et al.: Research on automatic target detection and recognition based on deep learning. J. Vis. Commun. Image Represent. **60**, 44–50 (2019)
12. Redmon, J., Divvala, S., Girshick, R., et al.: You only look once: unified, real-time object detection. In: 2016 IEEE Conference on Computer Vision and Pattern Recognition (CVPR), pp. 779–788, Las Vegas (2016)
13. Liu, W., et al.: SSD: Single Shot MultiBox Detector. In: Leibe, B., Matas, J., Sebe, N., Welling, M. (eds.) ECCV 2016. LNCS, vol. 9905, pp. 21–37. Springer, Cham (2016). https://doi.org/10.1007/978-3-319-46448-0_2
14. Lin, T., Dollár, P., Girshick, R., et al.: Feature pyramid networks for object detection. In: 2017 IEEE Conference on Computer Vision and Pattern Recognition (CVPR), pp. 936–944. IEEE: Honolulu (2017)
15. Girshick, R.: Fast R-CNN. In: 2015 IEEE International Conference on Computer Vision (ICCV), pp. 1440–1448. IEEE: Santiago (2015)
16. Ren, S., He, K., Girshick, R., et al.: Faster R-CNN: towards real-time object detection with region proposal networks. IEEE Trans. Pattern Anal. Mach. Intell. **39**(6), 1137–1149 (2017)
17. Zuo, Z., Yu, K., Zhou, Q., et al.: Traffic signs detection based on faster R-CNN. In: 2017 IEEE 37th International Conference on Distributed Computing Systems Workshops (ICDCSW), pp. 286–288. IEEE, Atlanta (2017)
18. Tian, Q., Wang, M., Zhang, Y., et al.: A research for automatic pedestrian detection with ACE enhancement on fasters R-CNN. In: 2018 11th International Congress on Image and Signal Processing. BioMedical Engineering and Informatics (CISP-BMEI), pp. 1–9. IEEE, Beijing (2018)
19. Mou, X., Chen, X., Guan, J., et al.: Marine target detection based on improved faster R-CNN for navigation radar PPI images. In: 2019 International Conference on Control. Automation and Information Sciences (ICCAIS), pp. 1–5. IEEE, Chengdu (2019)

20. Xinyu, L., Xiaochun, L., Rongfeng, C., et al.: Application of the faster R-CNN algorithm in scene recognition function design. In: 2019 15th International Conference on Computational Intelligence and Security (CIS), pp. 16–19. IEEE, Macao (2019)
21. Tobias, R.R., Jesus, L.C.D., Mital, M.E., et al.: Faster R-CNN model with momentum optimizer for RBC and WBC variants classification. In: 2020 IEEE 2nd Global Conference on Life Sciences and Technologies (LifeTech), pp. 235–239. IEEE, Kyoto (2020)
22. Tang, J., Mao, Y., Wang, J., et al.: Multi-task enhanced dam crack image detection based on faster R-CNN. In: 2019 IEEE 4th International Conference on Image. Vision and Computing (ICIVC), pp. 336–340. IEEE, Xiamen (2019)
23. Liu, Y.: An improved faster R-CNN for object detection. In: 2018 11th International Symposium on Computational Intelligence and Design (ISCID), pp. 119–123. IEEE, Hangzhou (2018)
24. Shi, J., Chang, Y., Changhang, X., et al.: Real-time leak detection using an infrared camera and Faster R-CNN technique. Comput. Chem. Eng. **135**, 106780 (2020)
25. Liu, B., Zhao, W., Sun, Q.: Study of object detection based on faster R-CNN. In: 2017 Chinese Automation Congress (CAC), pp. 6233–6236. IEEE, Jinan (2017)

# Correction to: Machine Learning Assisted OSP Approach for Improved QoS Performance on 3D Charge-Trap Based SSDs

Zongwei Zhu(iD), Chao Wu(iD), Cheng Ji(iD), and Xianmin Wang(iD)

## Correction to:
Chapter "Machine Learning Assisted OSP Approach for Improved QoS Performance on 3D Charge-Trap Based SSDs" in: X. Chen et al. (Eds.): *Machine Learning for Cyber Security*, LNCS 12488, https://doi.org/10.1007/978-3-030-62463-7_9

In the published version the subfigures (g) and (h) in Fig. 7 have been removed.

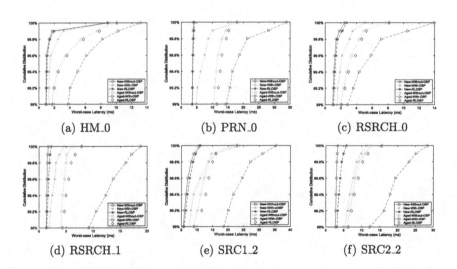

(a) HM_0     (b) PRN_0     (c) RSRCH_0

(d) RSRCH_1     (e) SRC1_2     (f) SRC2_2

The updated version of this chapter can be found at
https://doi.org/10.1007/978-3-030-62463-7_9

# Author Index

Printed in the United States
By Bookmasters